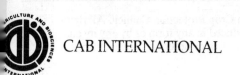

CAB INTERNATIONAL

BCPC BRITISH CROP PROTECTION COUNCIL

The UK Pesticide Guide

GW00630799

Revised Edition

see page iv

Editor – R. Whitehead B.A., M.Sc.

CAB INTERNATIONAL (CABI), is an international, intergovernmental organization established in 1928. CABI is dedicated to improving human welfare worldwide through the dissemination, application and generation of scientific knowledge in support of sustainable development, with emphasis on agriculture, forestry, human health and the management of natural resources, and with particular attention to the needs of developing countries. For further information please contact CAB INTERNATIONAL, Wallingford, Oxon OX10 8DE, UK.

The British Crop Protection Council (BCPC), is a self-supporting limited company with charity status which was formed in 1968 to promote the knowledge and understanding of the science and practice of crop protection. The corporate members include government departments and research councils; advisory services; associations concerned with the farming industry, agrochemical manufacturers, agricultural engineering, agricultural contracting and distribution services; universities; scientific societies; organizations concerned with the environment; and some experienced independent members. Further details available from BCPC, 49 Downing Street, Farnham GU9 7PH.

ISBN 0 85199 113 0

Cover design by David Pratt
Printed and bound in Great Britain at the University Press, Cambridge

CONTENTS

Editor's Note   v

Disclaimer   vi

Changes since 1995 edition   vii

SECTION 1 INTRODUCTORY INFORMATION

About the UK Pesticide Guide   3
How to use the Guide   5
Pesticide Legislation   8
    The Food and Environment Protection Act 1985 (FEPA) and Control of Pesticides
        Regulations 1986 (COPR)   8
    The 'Authorisation' Directive   8
    The Control of Substances Hazardous to Health Regulations 1988 (COSHH)   9
    Certificates of Competence – the roles of BASIS and NPTC   9
    Maximum Residue Levels   10
Approval (on-label and off-label)   13
Guidance on the use of pesticides   20
Chemicals subject to the Poisons Law   22
Occupational exposure limits   23
Agricultural pesticides and water   24
Use of herbicides in or near water   25
Use of pesticides in forestry   26
Pesticides used as seed treatments   29
Aerial application of pesticides   32
ULV application   40
Adjuvants   43
Poisoning by pesticides – first aid measures   61

SECTION 2 CROP/PEST GUIDE

Crop Index   67
Crop/Pest Guide   71

SECTION 3 PESTICIDE PROFILES   157

SECTION 4 APPENDICES

Appendix 1   Suppliers of pesticides and adjuvants   593
Appendix 2   Useful contacts   602
Appendix 3   Keys to crop and weed growth stages   604
Appendix 4   A. Key to label requirements for protective clothing   611
              B. Key to numbered label precautions   612
Appendix 5   Key to abbreviations and acronyms   615

Index of Proprietary Names of Pesticides   618

UK PESTICIDE GUIDE 1996

Revised Edition

Some important information was omitted from the Pesticide Profiles section of the 1996 edition of *The UK Pesticide Guide*, caused by a computer typesetting error. This **Revised Edition** contains the missing information, and the opportunity has also been taken to update some other details.

Although not incorrect, the information on some pages of the original version was incomplete and potentially misleading. To avoid confusion, you are strongly advised to destroy the original version.

EDITOR'S NOTE

This is the ninth annual edition of *The UK Pesticide Guide*. It is the most comprehensive catalogue of available pesticides and adjuvants in the UK for use in agriculture, horticulture, forestry and amenity situations. Most people concerned with the supply, storage and use of pesticides in these sectors, or those involved in teaching about pesticides, now use this *Guide* as a standard reference.

The *Guide* is fully revised for every edition so that the information on product availability and new legislation is as up to date as possible. For this edition, extra efforts have been made to remove obsolete entries, but changes in the commercial world can cause confusion. When a manufacturer or supplier changes company name, product approvals have to be amended, and new approval numbers are allocated. This means that, for a season or two, identical products can be on the market, differing only in approval number and the name of the supplier. So long as they remain available and legally usable, all identities will be listed in this *Guide*, provided the supplier agrees. Where an approval for a listed product will finally expire during 1996, the expiry date is shown and (new for this edition) information is given about any replacement.

Over 130 new products are listed, while about 220 have been deleted, usually because they are no longer available. For the first time these additions and deletions have been listed in a special section. Recent approval changes affecting the labels of sulfonylurea herbicides, isoproturon products and organophosphorus insecticides used on carrots are reflected in amendments to the relevant profiles. Active ingredients new to this edition include:

- anilazine – a triazine fungicide
- diphacinone – an anticoagulant rodenticide
- fenbuconazole – a triazole fungicide
- fenpyroximate – an oxime acaricide
- fomesafen – a diphenyl ether herbicide
- methoprene – an insect growth regulator
- propaquizafop – a broad spectrum grass weed herbicide
- tebufenpyrad – a carboxamide acaricide

Other changes in this edition are the inclusion of:

- the revised Long-term Arrangements for Extension of Use (off-label)
- additional Maximum Residue Levels (MRLs)
- guidance on safeguarding bees
- a brief summary of the functions of BASIS and NPTC
- a new appendix containing an enlarged list of abbreviations and acronyms

The editor is grateful to all the agrochemical manufacturers and suppliers without whose help and cooperation the *Guide* could not be published. Many people also offer helpful comment and criticism, as well as pointing out errors. Such comments are gratefully received and suggestions for improvement will always be welcomed.

R. Whitehead
Editor

DISCLAIMER

Every effort has been made to ensure that the information in this book is complete and correct at the time of going to press but the Editor and the publishers do not accept liability for any error or omission in the content, or for any loss, damage or other accident arising from the use of products listed herein. Omission of a product does not necessarily mean that it is not approved and available for use.

The information has been selected from official sources and from suppliers' labels and product manuals of pesticides approved under the Control of Pesticides Regulations 1986 and available for pest, disease and weed control applications in the United Kingdom.

It is essential to follow the instructions on the approved label before handling, storing or using any crop-protection product. Approved 'off-label' uses are made entirely at the risk of the user.

The contents of this publication are based on information received up to October 1995.

CHANGES SINCE 1995 EDITION

Pesticides and adjuvants added or deleted since the 1995 edition are listed below. Every effort has been made to ensure the accuracy of this list, but late notifications to the Editor are not included.

Products Added

New products are shown here. In addition, products that were listed in the previous edition whose MAFF Approval number *and* supplier have changed are also included. Where only the MAFF number has changed, the product is not listed.

PRODUCT	MAFF NO.	SUPPLIER	PRODUCT	MAFF NO.	SUPPLIER
50/50 Liquid Mosskiller	07191	Vitax	Dicotox Extra	05330	R-P Env.
Actellic 40 WP	06932	Zeneca	Duk 110	06266	DuPont
Admire	07481	Bayer	Duk 51	06764	DuPont
Allicur	06468	Bayer	Dyrene	06982	Bayer
Ally WSB	06588	DuPont	Endeavour	07385	Bayer
Antec Tomcat	07171	Antec	Exit	07632	Bayer
Apache	06748	Zeneca	Falcon	07025	Cyanamid
Apex 5 E	05730	Sandoz Spec.	FS Thiram 15% Dust	07428	Ford Smith
Arma	ADJ0200	Interagro	Greenmaster Extra	07594	Levington
Avadex Excel 15G	07117	Monsanto	Headland Judo	07387	Headland
Barbarian	07625	Barclay	Headland Quilt	ADJ0188	Headland
Barclay Canter	07350	Barclay	Headland Zebra Flo	07442	Headland
Barclay Clinger	ADJ0198	Barclay	Headland Zebra WP	07441	Headland
Barclay Cluster	07467	Barclay	Hinge	04929	Quadrangle
Barclay Dingo	07642	Barclay	Hydraguard	03278	Agrichem
Barclay Dryfast XL	ADJ0121	Barclay	Isoproturon 500	06718	AgrEvo
Barclay Gallup	05161	Barclay	ISK 375	07455	ISK Biosciences
Barclay Hat-Trick	07579	Barclay			
Barclay Haybob	07597	Barclay	I.T. Bentazone	07415	I.T.Agro
Barclay Meadowman	07639	Barclay	I.T. Glyphosate	07212	I.T.Agro
Barclay Mecrop	07578	Barclay	Indar 5EW	07580	AgrEvo
Barclay Metaza	07354	Barclay	Konker	03988	BASF
Beetup	05299	MTM Agrochem	Landgold Cyproconazole	06463	Landgold
Bellmac Plus	07521	United Phosphorus	Landgold Deltaland	07480	Landgold
			Landgold Lambda-C	06097	Landgold
Bellmac Straight	07522	United Phosphorus	Lentagran WP	07556	Sandoz
			Lightning	ADJ0177	Batson
Betosip Combi	07235	Sipcam	Luxan 9363	07359	Luxan
Calidan	06536	R-P Agric.	Luxan Isoproturon 500	07437	Luxan
Camppex	06809	United Phosphorus	Flowable		
			Luxan MCPA 500	07470	Luxan
Comulin	ADJ0164	Batson	Marshal suSCon	06978	Fargro
Contact	ADJ0191	Interagro	Masai	07452	Cyanamid
Crackshot	07542	Cyanamid	Match	07185	Cyanamid
Crusader S	05174	DowElanco	MC Flowable	06295	United Phosphorus
Danadim Dimethoate 40	07351	Cheminova	Microthiol Special	06268	PBI
DB Straight	07523	United Phosphorus	Mitac HF	07358	AgrEvo
			Monceren IM Flowable	06731	Bayer
Devcol Morgan 6% Metaldehyde Slug Killer	07422	Nehra	Monicle	07375	Bayer
			Moot	06990	Sandoz

(vii)

PRODUCT	MAFF NO.	SUPPLIER	PRODUCT	MAFF NO.	SUPPLIER
MSS Flotin 480	07616	Mirfield	Sponsor	07136	AgrEvo
MSS Mirlin	07636	Mirfield	Stamina	ADJ0202	Interagro
Multispray	H5165	AgrEvo Environ	Stampede	06327	Zeneca
Myriad	07584	R-P Agric.	Standon Bentazone	06895	Standon
Nico Soap	07517	United Phosphorus	Standon Chlorpyrifos	07633	Standon
			Standon Cyproconazole	07751	Standon
Octolan	06256	Sandoz	Standon Flusilazole Plus	07403	Standon
Output	ADJ0163	Zeneca	Standon Tebuconazole	07408	Standon
Palette	06691	Zeneca	Stefes Banlene Plus	07688	Stefes
Pastor	07440	DowElanco	Stefes CAT 80	ADJ0182	Stefes
Patafol	07397	Mirfield	Stefes CCC 640	06993	Stefes
PP 375	06681	Zeneca	Stefes Cover	ADJ0187	Stefes
Prospect	06541	DuPont	Stefes Derosal WDG	07658	Stefes
Prosper	07532	United Phosphorus	Stefes Medimat 2	07249	Stefes
			Stefes Phenoxylene 50	07612	Stefes
Quickstep	07389	Sandoz	Stefes Poraz	07528	Stefes
Raxil S	06974	Bayer	Sugar Beet Herbicide	07525	United Phosphorus
Raxil S	07484	Zeneca			
Redipon Extra	07518	United Phosphorus	Sulfosate	06750	Zeneca
			Sulphur Flowable	07526	United Phosphorus
Redlegor	07519	United Phosphorus	Systhane 6 Flo	07334	AgrEvo
			Tattoo C	07623	AgrEvo
Reflex T	07423	Zeneca	Thiovit	05572	Sandoz
Rizolex Nova	07434	AgrEvo	Toppel 10	06632	United Phosphorus
Ryda	ADJ0168	Interagro			
Salute	07660	United Phosphorus	Tripart Orbis	ADJ0204	Tripart
			Tritox	07764	Levington
Salvo	07092	Zeneca	Unicrop Zineb	02279	Unicrop
Seedox SC	07591	Uniroyal	UPL Diuron 80°	07619	United Phosphorus
Sequel	07624	Promark			
Sibutol	07487	Zeneca	UPL Linuron 45% Flowable	07435	United Phosphorus
Slippa	ADJ0206	Interagro	Uplift	07527	United Phosphorus
Sorex Fly Spray RTU	H5718	Sorex			
Sorex Warfarin 250 ppm Rat Bait	07371	Sorex	Vassgro Non-Ionic	ADJ0190	Vass
			Vitaflo	06379	Uniroyal
Sorex Warfarin 500 ppm Rat Bait	07372	Sorex	Vitaflo Extra	07048	Uniroyal
Sorex Warfarin Sewer Bait	07373	Sorex	X-Spor SC	06310	United Phosphorus
Source	07427	Chiltern			

Products Deleted

The appearance of a product name in the following list may not necessarily mean that it is no longer available. In cases of doubt refer to the supplier.

PRODUCT	MAFF NO.	SUPPLIER	PRODUCT	MAFF NO.	SUPPLIER
Ace Sodium Chlorate (Fire Suppressed) Weedkiller	06413	Ace	Agritox 50	06161	R-P Agric.
			Ambush C	00087	ICI
Actellic Smoke Generator No 20	00018	ICI	Aphox	00106	ICI
			Apollo 50 SC	07242	AgrEvo
Agrichem 2,4-D	04098	Agrichem	Apollo 50 SC	03996	Schering
Agrichem MCPA-50	04097	Agrichem	Astix	06174	R-P Agric.
Agricultural Hoppit	05477	Nilco	Banlene Plus	07245	AgrEvo

PRODUCT	MAFF NO.	SUPPLIER
Banlene Plus	03851	Schering
Barclay Dryfast	ADJ0123	Barclay
Barclay Gallup	05020	Barclay
Bavistin	00217	BASF
Bayleton 5	00222	Bayer
Baytan	00225	Bayer
Baytan Flowable	03845	ICI
Baythroid	04273	Bayer
BB Chlorothalonil	03320	Brown Butlin
Beetomax	03129	Fine
Betanal Progress	07111	AgrEvo
Betanal Tandem	07254	AgrEvo
Betanal Tandem	03857	Schering
Betaren 120	03822	PBI
BH 2,4-D Ester 50	05390	R-P Env.
BH Prefix D	03688	R-P Env.
Blex	00284	ICI
Bonzi	02155	ICI
Bos MH 180	06502	Uniroyal
Brestan 60	07304	AgrEvo
Brestan 60	00325	Hoechst
Bushwacker Spray	05249	R-P Env.
Casoron G	00448	ICI
Casoron G4	02406	Nomix-Chipman
Casoron G4	05734	Vitax
Catoff	04400	Nilco
CDA Castaway Plus	04758	R-P Env.
CDA Rovral	04679	R-P Env.
CDA Supertox 30	04664	R-P Env.
Ceratotect	03554	MSD Agvet
Cerevax	06641	Zeneca
Cerevax	02500	ICI
Cerevax Extra	02501	ICI
Cerevax Extra	06642	Zeneca
Cheetah R	07308	AgrEvo
Cheetah R	04932	Hoechst
Chemtech Cypermethrin	04827	Chemtech
Chemtech Nuarimol	04517	Chemtech
CIA Dimethoate	05937	CIA
CIA Diquat 200	05551	CIA
CIA Fentin 475-WP	05550	CIA
CIA MCPA 50	05874	CIA
CIA Paraquat 200	05520	CIA
CIA Propyzamide	05483	CIA
Competitor	07310	AgrEvo
Competitor	06605	Hoechst
Conquest	07311	AgrEvo
Conquest	05679	Hoechst
Crossfire	05329	R-P Env.
Cudgel	04349	ICI
Cultar	05055	ICI
Curb Garden Powder	03983	Sphere
Cymag	00623	ICI
Danadim 40	07040	Cheminova
Deloxil 400	07314	AgrEvo
Deloxil 400	06613	Schering
Digermin	05823	Isagro
Dimilin WP	03810	ICI
Disyston FE-10	00714	Bayer
Dithane 945	00719	Rohm & Haas
Dithane Dry Flowable	04251	Rohm & Haas
Docklene	07257	AgrEvo
Docklene	03863	Schering
Dogoff	05490	Nilco
Dorado	05700	ICI
DowElanco Adjuvant		DowElanco
Du-Ter 50	07317	AgrEvo
Du-Ter 50	06416	Hoechst
Endspray	02799	PBI
Ethywet	ADJ 0102	TopFarm
Farmatin 560	07320	AgrEvo
Farmatin 560	06417	Hoechst
Fazor	05461	Uniroyal
Ferrax	02827	ICI
Fire Suppressed Sodium Chlorate Weedkiller	05128	Strathclyde
Fonofos Seed Treatment	04051	ICI
Force ST	05109	ICI
Forestry Bee	ADJ 0098	TopFarm
Fumispore	04745	LCB
Fusarex Granules	00955	ICI
Gamma-Col	00964	ICI
Gammasan 30	00969	ICI
Garlon 2	03767	ICI
Glyfosate-360	05319	TopFarm
Glyphosate-360	06316	Clayton
Glytex	04230	Bayer
Gramonol Five	00995	ICI
Gramoxone 100	07260	AgrEvo
Gramoxone 100	03867	Schering
Gramoxone 100	00997	ICI
Gunner	04547	Quadrangle
Hallmark	04466	ICI
Halloween	01268	Mandops
Hele Stone	01269	Mandops
Ibis	06816	R-P Agric.
Javelin	06192	R-P Agric.
Javelin Gold	06200	R-P Agric.
Kerb 50 W	02986	Rohm & Haas
Kerb Flo	02759	Rohm & Haas
Kerb Granules	01136	Rohm & Haas
Killgerm Ratak Cut Wheat Rat Bait	04448	Killgerm
Killgerm Ratak Rat Pellets	04446	Killgerm
Killgerm Ratak Whole Wheat Rat Bait	04447	Killgerm
Legumex Extra	07262	AgrEvo
Legumex Extra	03869	Schering
Liquid Linuron	01556	PBI
Luxan Chloridazon	06279	Luxan
Luxan Dimethoate 40	06558	Luxan
Mascot Contact Turf Fungicide	02711	Rigby Taylor
Meld	04914	BASF

PRODUCT	MAFF NO.	SUPPLIER	PRODUCT	MAFF NO.	SUPPLIER
Mitac 20	07265	AgrEvo	Spreadite Liquid	ADJ0015	DowElanco
Mitac 20	03870	Schering	Spudweed	04965	PBI
Mowchem	05333	R-P Env.	Stantion	06025	R-P Agric.
Narsty	06003	Mandops	Sumicidin	02568	Shell
Neminfest	05822	Isagro	Systhane 6 Flo	03921	Rohm & Haas
Nemispor	05506	Isagro			
New Kotol	06212	R-P Agric.	Systhane 6 W	04570	Rohm & Haas
Nimrod	01498	ICI			
Nomix 2,4-D Herbicide	06394	Nomix-Chipman	Thinsec	02463	ICI
			Thiovit	02125	PBI
Non-Ionic 90	ADJ0100	TopFarm	Throttle	05204	Quadrangle
Parable	03805	ICI	Tiptor	07295	AgrEvo
Patrol	03317	ICI	Tiptor	06971	Schering
Penncozeb WDG	07095	Cyanamid	Tolkan Liquid	06172	R-P Agric.
Penncozeb WP	07089	Cyanamid	Tombel	04124	Hortichem
Phenoxylene 50	07163	AgrEvo	Top Farm Carbendazim 435	05307	TopFarm
Phenoxylene 50	03854	Schering			
Pirimor	01594	ICI	Top Farm Chlormequat 640	05323	TopFarm
Portman Betalion	04677	Portman			
Portman Chlormequat 400	01523	Portman	Top Farm Chlorothalonil 500	05926	TopFarm
Portman Chlormequat 460	02549	Portman			
Portman Chlormequat 700	03465	Portman	Top Farm CMPP	05468	TopFarm
Portman Chlortoluron	03068	Portman	Top Farm Cypermethrin 10	05624	TopFarm
Portman Dimethoate 40	01527	Portman			
Portman Glider	04695	Portman	Top Farm Dimethoate	05936	TopFarm
Portman Glyphosate 360	04699	Portman	Top Farm Diquat 200	05552	TopFarm
Portman Isotop	03434	Portman	Top Farm Ethofumesate 200	05987	TopFarm
Portman Propachlor 50 FL	02784	Portman			
Portman Supaquat	03466	Portman	Top Farm Fentin 475-WP	05504	TopFarm
Portman Trifluralin	05751	Portman	Top Farm IPU 500	05978	TopFarm
Portman Weedmaster	06018	Portman	Top Farm MCPA 500	05930	TopFarm
PP Captan 80	05826	ICI	Top Farm Paraquat	05075	TopFarm
Puma X	07332	AgrEvo	Top Farm Paraquat 200	05519	TopFarm
Puma X	06768	Hoechst	Top Farm Propiconazole 250	05938	TopFarm
Ramrod Granular	05806	Hortichem			
Rappor	04900	DowElanco	Top Farm Propyzamide 500	05484	TopFarm
Rappor Plus	04899	DowElanco			
Regulex	03010	ICI	Top Farm Toluron 500	05986	TopFarm
Rentokil Methyl Bromide	05646	Rentokil	Top Farm Trifluralin 470	05471	TopFarm
Reply	07084	Cyanamid	Toro	05927	Sipcam
Reply	07084	Cyanamid	Triplen	05897	Sipcam
Ridento Ready-to-Use Rat Bait	03804	Ace	Triplen Combi	05939	Sipcam
			Tropotox	06179	R-P Agric.
Roundup	03947	AgrEvo	Tropotox Plus	06156	R-P Agric.
Safers Insecticidal Soap	06064	Koppert	Unicrop Dimethoate 40	06549	UCP
Sakarat Special	02844	Killgerm	Unicrop Flowable Atrazine	02268	UCP
Scattercarb	01886	C-Vet			
Sedanox	05033	Bayer	Unicrop Flowable Simazine	05447	UCP
Seedox SC	05935	DowElanco			
Septal WDG	05985	Schering	Weedex S2 FG	04223	Hortichem
Seradix 1	06191	R-P Agric.	Wireworm FS Seed Treatment	05640	DowElanco
Seradix 2	06193	R-P Agric.			
Seradix 3	06194	R-P Agric.	Wireworm Liquid Seed Treatment	05641	DowElanco
Sorexa Plus Rat Bait	01986	Sorex			
Spinnaker	07023	Cyanamid	Zolone Liquid	06206	R-P Agric.
Sportak 45 HF	05724	Schering			

SECTION 1
INTRODUCTORY INFORMATION

ABOUT THE UK PESTICIDE GUIDE

Purpose

The primary aim of this book is to provide a practical guide to what pesticides, plant growth regulators and adjuvants the farmer or grower can realistically and legally obtain in the UK, and to indicate the purposes for which they may be used. It is designed to help in the identification of products appropriate for a particular problem, and a Crop/Pest guide is included to facilitate this. In addition to uses recommended on product labels, details are provided of those uses which do not appear on labels but which have been granted off-label approval. Such uses are not endorsed by the manufacturer, and are undertaken entirely at the risk of the user.

As well as identifying the products available, the book provides guidance on how to use them safely and effectively, but without giving details of doses, volumes, spray schedules or approved tank mixtures. Other sections provide essential background information on a whole range of pesticide-related issues including legislation, codes of practice, poisons and treatment of poisoning, products for use in special situations, and weed and crop growth stage keys.

While we have tried to cover all other important factors, this book does not provide a full statement of product recommendations. **Before using any pesticide product it is essential that the user should read the label carefully and comply strictly with the instructions it contains.**

Scope

Individual profiles are provided of over 480 distinct active ingredients or mixtures thereof. Each profile has a list of available approved products for use in agriculture, horticulture (including amenity horticulture), forestry and areas near water, and the supplier from whom each may be obtained. Within these fields all types of pesticide covered by the Control of Pesticides Regulations 1986 are included. This embraces acaricides, algicides, fungicides, herbicides, insecticides, lumbricides, molluscicides, nematicides and rodenticides, together with plant growth regulators used as straw shorteners, sprout inhibitors and for various horticultural purposes. The total number of pesticide products covered in the Guide is about 1300.

In addition, the introductory section gives information on more than 70 authorized adjuvants which, although not themselves pesticides, may be added to pesticides to improve their effectiveness in use.

The Guide does **not** cover home garden, domestic, food storage, public health or animal health uses.

Sources of Information

The information for this edition has been drawn from these authoritative sources:

- approved labels and product manuals received from the suppliers of pesticides up to September 1995
- the MAFF/HSE publication *Pesticides 1995*
- entries in *The Pesticides Register*, published monthly by MAFF and HSE, listing new UK approvals (including off-label approvals), up to and including the issue for July 1995
- MAFF lists of approval expiries

Criteria for Inclusion

To be included in the Guide, a product must meet the following conditions:
- it must have MAFF/HSE Approval under the 1986 Regulations
- information on the approved uses must have been provided by the supplier
- the product must be expected to be on the UK market in 1996

Active ingredients which have been banned in the period since 1987 are listed, but no further details are given.

When a company changes its name, whether by merger, take-over or joint venture, it is obliged to re-register its products in the name of the new company, and new MAFF numbers are assigned. Where stocks of the previously registered products remain legally available, both old and new numbers are included in the Guide and will remain until approval for the former lapses or stocks are notified as exhausted.

Products that have been withdrawn from the market and whose approval will finally lapse during 1996 are identified in each profile. After the indicated date, sale, storage or use of the product *bearing that approval number* becomes illegal. Where there is a direct replacement product, it is indicated.

HOW TO USE THE GUIDE

The book consists of four main sections:
- Introductory Information
- Crop/Pest Guide
- Pesticide Profiles
- Appendices

Introductory Information

This section summarises legislation covering approval, storage, sale and use of pesticides in the UK. It also provides lists of products approved for use in or near water, in forestry, and as seed treatments. Products specifically approved for aerial application or for use with ULV applicators are tabulated. This section also gives detailed information on more than 70 authorised adjuvants which, although not themselves pesticides, may be added to pesticides to improve their effectiveness. Finally, there is a summary of first aid measures if pesticide poisoning should be suspected.

Crop/Pest Guide

This section (p.65) enables the user to identify which active ingredients are approved for a particular crop/pest combination. The crops are grouped as shown in the Crop Index (p.67), but indications from the Crop/Pest Guide must always be followed up by reference to the specific entry in the pesticide profiles section. This is because chemicals indicated as having uses in cereals, for example, may only be approved for use on winter wheat and barley, not for other cereals, and the profile entry must be consulted to find out which particular products are approved for which particular crops. Because of differences in the wording of product labels it may sometimes be necessary to refer to broader categories of organism in addition to specific organisms (e.g. annual grasses and annual weeds in addition to annual meadow grass).

Pesticide Profiles

Each active ingredient has a separate numbered profile entry, as does each mixture of active ingredients. The entries are arranged in alphabetical order using the common names approved by the British Standards Institution. Where an active ingredient is available only in mixtures, this is stated. The ingredients of the mixtures are themselves ordered alphabetically and the entries appear at the correct point for the first named ingredient.

Within each profile entry a table lists the products approved and available on the market, in the following style:

Product name	Main supplier	Active ingredient content	Formulation type	Registration No. *
3 Dursban 4	DowElanco	480 g/l	EC	05735
4 Lorsban T	Rigby-Taylor	480 g/l	EC	05970
5 Spannit	PBI	480 g/l	EC	01992

* Normally refers to registration with MAFF. In a few cases where products registered with the Health and Safety Executive are used in agriculture, HSE numbers are quoted, e.g. H0462

Many of the product names may be registered trade-marks but no special indication of this is given. Individual products are indexed under their **entry number** in the Index of Proprietary Names at the back of the book. The main supplier indicates the marketing outlet through which the product may be purchased. Full addresses and telephone/fax numbers of suppliers are listed in Appendix 1. For mixtures, the active ingredient contents are given in the same order as they appear in the profile heading. The formulation types are listed in full in the key to abbreviations and acronyms (Appendix 5).

The **Uses** section lists all approved uses (on-label and off-label) notified by October 1995, giving both the principal target organisms and the recommended crops or situations. Where there is an important condition of use, or the approval is off-label, this is shown in parentheses. Numbers in square brackets refer to the numbered products in the table above. Thus a typical use approved for Dursban 4 and Lorsban T (products 3 and 4) but not for Spannit (product 5) appears as:

Cutworms in LETTUCE (outdoor only) [3, 4]

Below the **Uses** paragraph, **Notes** are listed under the following headings. Unless otherwise stated any reference to dose is made in terms of product rather than active ingredient. Where notes refer to particular products, rather than to the entry generally, this is indicated by numbers in square brackets as described above

Efficacy	Factors important in making the most effective use of the treatment. Certain factors, such as the need to apply chemicals uniformly, the need to spray at appropriate volume rates and the need to prevent settling out of active ingredient in spray tanks, are assumed to be important for all treatments and are not emphasized in individual profiles.
Crop safety/ Restrictions	Factors important in minimizing the risk of crop damage including statutory conditions relating to the maximum permitted number of applications. Recommended timing of treatment in relation to crop growth stage is included, with reference to standardized growth stage keys where possible (GS numbers — see Appendix 3). Where a statutory 'latest time of application' has been established this is given in the section 'Latest application/Harvest interval (HI)'.
Special precautions/ Environmental safety	Notes are included in this section where products are subject to the Poisons Law (see p. 22), where Maximum Exposure Limits apply (see p. 23) and where the label warns about organophosphorus and/or anticholinesterase compounds. Where the label specifies a Hazard Class this is noted, together with the associated risk phrases. Any other special operator precautions are specified.
	Environmental hazards are also noted here, including potential dangers to livestock, game, wildlife, bees and fish. The need to avoid drift onto neighbouring crops and to wash out equipment thoroughly after use are important with all pesticide treatments, but may receive special mention here if of particular significance.

Protective clothing/ Label precautions	The COSHH regulations require that wherever there is a label recommendation for use of protective clothing it should be preceded by the phrase '*Engineering control of operator exposure must be used where reasonably practicable in addition to the following personal protective equipment:*' and followed by '*However, engineering controls may replace personal protective equipment if a COSHH assessment shows that they provide an equal or higher standard of protection.*'

These phrases are not repeated in the profiles but an indication is given of the items of protective clothing which may be required in using the various products. Without a greater degree of uniformity in the wording of labels, it is not possible to show the items required for different operations e.g. handling concentrate, spraying with hand lance, cleaning equipment etc. However, the items which may be needed for one or other of the operations involved in using a product are given as a series of letters on the first line of this section and refer to the items of protective clothing listed in Appendix 4.

Other label precautions not concerned with choice of protective clothing are given on the second line as a list of numbers referring to the key to standard phrases in Appendix 4. **The letters referring to protective clothing and the numbered precautions are given for information only and should not be used for the purpose of making a COSHH assessment without reference to the current product label.**

Withholding period	Any requirements relating to pesticides used on grazing land with potentially harmful effects on livestock are given here.
Latest application/ Harvest interval (HI)	The 'Latest time of application' given in this section is laid down as a statutory condition of use for individual crops. Where quoted in terms of the period which must elapse between the last application and harvesting for human or animal consumption this 'Harvest Interval' is also given.
Approval	Includes notes on approval for aerial or ULV application and 'Off-label' approvals, giving references to the official approval document numbers (OLA numbers), copies of which can be obtained from ADAS or NFU and must be consulted before the treatment is used.

Where a product has been withdrawn by its manufacturer, and its approval will finally expire in 1996, the expiry date is shown here, together with the MAFF number of its direct replacement, if any.

MRLs	If MRLs have been set for any/all of the active ingredients in the profile, they are listed here, or else a cross-reference is given to the "parent" profile where they may be found.

PESTICIDE LEGISLATION

The Food and Environment Protection Act 1985 (FEPA) and Control of Pesticides Regulations 1986 (COPR)

FEPA introduced statutory powers to control pesticides with the aims of protecting human beings, creatures and plants, safeguarding the environment, ensuring safe, effective and humane methods of controlling pests and making pesticide information available to the public. Control of pesticides is achieved by COPR, which lay down the Approvals required before any pesticide may be sold, stored, supplied, advertised or used, and allow for the general requirements set out in various Consents, which specify the conditions subject to which approval is given. Consent A relates to advertisement, Consent B to sale, supply and storage, Consent C (i) to use and Consent C (ii) to aerial application of pesticides. The conditions of the Consents may be changed from time to time. Details are given in the MAFF/HSE reference book *Pesticides 1995* (revised annually) and are updated in the monthly publication *The Pesticides Register*.

The controls currently in force include the following :

- Only approved products may be sold, supplied, stored, advertised or used
- Only products specifically approved for the purpose may be applied from the air
- A recognised Storeman's Certificate of Competence is required by anyone who stores for sale or supply pesticides approved for agricultural use
- A recognised Certificate of Competence is required by anyone who gives advice when selling or supplying pesticides approved for agricultural use
- Users of pesticides must comply with the Conditions of Approval relating to use
- A recognised Certificate of Competence is required by all contractors and persons born after 31 December 1964 applying pesticides approved for agricultural use (unless working under direct supervision of a certificate holder)
- Only those adjuvants authorised by MAFF may be used (see Adjuvants, pp. 43–60)
- Regarding tank-mixes, 'no person shall combine or mix for use two or more pesticides which are anti-cholinesterase compounds unless the approved label of at least one of the pesticide products states that the mixture may be made; and no person shall combine or mix for use two or more pesticides if all the conditions of the approval relating to this use cannot be complied with'

The 'Authorisation' Directive

European Council Directive 91/414/EEC, known as the 'Authorisation' Directive, is intended to harmonise national arrangements for the authorisation of plant protection products within the European Union. It became effective on 25 July 1993. Under the provisions of the Directive, individual Member States will be responsible for authorisation within their own territory of products containing active substances that appear in a list agreed at Community level. This list, to be known as Annex I, will be created over a period of time by review of existing active ingredients and authorisation of new ones.

Individual Member States are amending their national arrangements and legislation in order to meet the requirements of Directive 91/414/EEC. In the UK this has been achieved by the Plant Protection Products Regulations 1995 (PPPR) which came into force on 17 April 1995.

Meanwhile existing product approvals will be maintained under COPR, and new ones may be granted for products containing active ingredients already on the market by 25 July 1993. Products containing new active substances not on the market at this date may be granted provisional approval in advance of Annex I listing of their active ingredients.

The Directive also provides for a system of mutual recognition of products registered in other Member States, subject to a number of constraints.

The Control of Substances Hazardous to Health Regulations 1988 (COSHH)

The COSHH regulations, which came into force on 1 October 1989, were made under the Health and Safety at Work Act 1974 and are also important as a means of regulating the use of pesticides. The regulations cover virtually all substances hazardous to health, including those pesticides classed as Very toxic, Toxic, Harmful, Irritant or Corrosive, other chemicals used in farming or industry and substances with occupational exposure limits. They also cover harmful micro-organisms, dusts and any other material, mixture or compound used at work which can harm people's health.

The original Regulations, together with all subsequent amendments, have been consolidated into a single set of regulations: The Control of Substances Hazardous to Health Regulations 1994 (COSHH 1994).

The basic principle underlying the COSHH regulations is that the risks associated with the use of any substance hazardous to health must be assessed before it is used and the appropriate measures taken to control the risk. The emphasis is changed from that pertaining under the Poisonous Substances in Agriculture Regulations 1984 (now repealed), whereby the principal method of ensuring safety was the use of protective clothing, to the prevention or control of exposure to hazardous substances by a combination of measures. In order of preference the measures should be :

a. substitution with a less hazardous chemical or product
b technical or engineering controls (e.g. the use of closed handling systems etc.)
c. operational controls (e.g. operators located in cabs fitted with air-filtration systems etc.)
d. use of personal protective equipment (PPE), which includes protective clothing

Consideration must be given as to whether it is necessary to use a pesticide at all in a given situation and, if so, the product posing the least risk to humans, animals and the environment must be selected. Where other measures do not provide adequate control of exposure and the use of PPE is necessary, the items stipulated on the product label must be used as a minimum. It is essential that equipment is properly maintained and the correct procedures adopted. Where necessary, the exposure of workers must be monitored, health checks carried out and employees must be instructed and trained in precautionary techniques. Adequate records of all operations involving pesticide application must be made and retained for at least 3 years.

Certificates of Competence – the roles of BASIS and NPTC

COPR, COSHH and other legislation places certain obligations on those who handle and use pesticides. Minimum standards are laid down for the transport, storage and use of pesticides and the law requires those who act as storekeepers, sellers and advisors to hold recognised Certificates of Competence (see above).

BASIS is an independent Registration Scheme for the pesticide industry, recognised under COPR. It is responsible for organising training courses and examinations to enable such staff to obtain a Certificate of Competence.

In addition, BASIS undertakes annual assessment of pesticide supply stores, enabling distributors, contractors and seedsmen to meet their obligations under the Code of Practice for Suppliers of Pesticides. Further information can be obtained from BASIS (see Appendix 2).

Certain spray operators also require Certificates of Competence (see above). These are issued by the National Proficiency Tests Council (see Appendix 2).

Maximum Residue Levels

Even if they are used correctly, a relatively small number of pesticides are liable to leave residues in foodstuffs. Where residues can occur, statutory limits, known as Maximum Residue Levels (MRLs), have been established. MRLs are not safety limits. They are intended primarily as a check that good agricultural practice is being followed and to allow international trade to take place. Product approval is granted only if the likely residues present no risk to health. Although it is an offence to put into circulation any produce where the MRL is exceeded, eating food containing residues above these levels does not automatically imply a risk to health.

The UK has set statutory MRLs since 1988. The European Union intends eventually to introduce MRLs for all pesticide/commodity combinations. These are being introduced initially by a series of priority lists but will subsequently be covered by the review programme under Directive 91/414 (see above).

MRLs apply to imported as well as to home-produced foodstuffs. Details of those set for the first list of 87 active ingredients are contained in *The Pesticides (Maximum Residue Levels in Crops, Food and Feeding Stuffs) Regulations 1994*, effective from 26 July 1994. A further 14 active ingredients (10 of which are included in this Guide) are covered by an amendment to these Regulations which came into force on 30 June 1995. Both these Statutory Instruments are available from HMSO.

The MRLs applicable to the active ingredients included in this Guide are shown in the relevant pesticide profiles (Section 3). However, it is essential to consult the Regulations for full details and definitions, and for information on chemicals not currently marketed in Britain.

MRLs are shown by crop in the pesticide profiles (Section 3). To make the information more manageable, crops have been grouped according to the structure used in the Regulations and reproduced below. Where the same MRL applies to all crops in a group, only the group name is given. Thus *pome fruits 0.5* indicates that a MRL of 0.5 mg/kg of food has been set for all crops in the pome fruit sub-group (apples, pears, quinces). *Apples, pears 0.5* indicates that either a MRL has not been set for quinces, or that a different level has been set and appears elsewhere in the list. *Fruits 0.5* indicates the level set for the entire fruits group.

Group	Sub-group	Crops/products
FRUITS	Citrus fruits	Grapefruit, lemons, limes, mandarins (including clementines & similar hybrids), oranges, pomelos, others
	Tree nuts	Almonds, brazil nuts, cashew nuts, chestnuts, coconuts, hazelnuts, macadamia nuts, pecans, pine nuts, pistachios, walnuts, others
	Pome fruits	Apples, pears, quinces, others
	Stone fruits	Apricots, cherries, peaches, nectarines, plums, others
	Berries & small fruit	Grapes, strawberrries, cane fruits (blackberries, loganberries, raspberries, others), bilberries, cranberries, currants, gooseberries, wild berries, others
	Miscellaneous fruit	Avocados, bananas, dates, figs, kiwi fruit, kumquats, litchis, mangoes, olives, passion fruit, pineapples, pomegranates, others
VEGETABLES	Root & tuber vegetables	Beetroot, carrots, celeriac, horseradish, Jerusalem artichokes, parsnips, parsley root, radishes, salsify, sweet potatoes, swedes, turnips, yams, others
	Bulb vegetables	Garlic, onions, shallots, spring onions, others
	Fruiting vegetables	Tomatoes, peppers, aubergines, cucumbers, gherkins, courgettes, melons, squashes, watermelons, sweet corn, others
	Brassica vegetables	Broccoli, cauliflowers, Brussels sprouts, head cabbages, Chinese cabbage, kale, kohlrabi, others
	Leafy vegetables/herbs	Lettuces (cress, lamb's lettuce, lettuce, scarole), spinach, beet leaves, watercress, witloof, herbs (chervil, chives, parsley, celery leaves), others
	Legume vegetables	Beans (with pods), beans (without pods), peas (with pods), peas (without pods), others
	Stem vegetables	Asparagus, cardoons, celery, fennel,

	globe artichokes, leeks, rhubarb, others
Fungi	Wild and cultivated mushrooms
PULSES	Beans, lentils, peas, others
OILSEEDS	Linseed, peanuts, poppy seed, sesame seed, sunflower seed, rape seed, soya bean, mustard seed, cotton seed, others
POTATOES	Early potatoes, ware potatoes
TEA	
HOPS	
CEREALS	Wheat, rye, barley, oats, triticale, maize, rice, others
ANIMAL PRODUCTS	Meat, fat and preparations of meat, milk, dairy produce, eggs

APPROVAL (ON-LABEL AND OFF-LABEL)

Approvals are normally granted only in relation to individual products and for specified uses. It is an offence to use non-approved products or to use approved products in a manner which does not comply with the specific conditions of approval.

Statutory Conditions of Use

Statutory conditions have been laid down for the use of individual products and may include:

- field of use
- crop or situations for which treatment is permitted
- maximum individual dose
- maximum number of treatments
- maximum area or quantity which may be treated
- latest time of application or harvest interval
- operator protection or training requirements
- environmental protection requirements
- any other specific restrictions relating to particular pesticides.

All products must now display these statutory conditions of use in a 'statutory box' on the label.

Types of Approval

There are three types of approval:

- *Full* (granted for an unspecified period)
- *Provisional* (granted for a specified period)
- *Experimental Permit* (granted for the purposes of testing and developing new products, formulations or uses). Products with only an Experimental Permit do not appear in this Guide

The official list of approved products, including both full and provisional approvals, is the MAFF/HSE Reference Book 500, *Pesticides 1995*. Details of new full and provisional approvals, and of amendments to existing approvals, as well as off-label approvals are published in *The Pesticides Register*.

Withdrawal of Approval

Approvals may be reviewed, amended, suspended or revoked at any time. Revocation may occur for various reasons, such as commercial withdrawal or failure to meet data requirements. Where there are no safety concerns, a new approval will normally be issued for up to 2 years to allow the using up of stocks by persons other than the approval holder. Where safety considerations make it necessary, however, immediate revocation may take effect.

Approval of Commodity Substances

Some chemicals have minor uses as pesticides but are predominantly used for non-pesticidal purposes. Approval is granted to such commodity substances for use only. They may not be sold, supplied, stored or advertised as pesticides unless specific approval has been granted. 17 such substances have been approved for certain specified uses as laid down in Annex D of *Pesticides 1995*. Of these, carbon dioxide, formaldehyde, methyl bromide, strychnine hydrochloride and sulfuric acid are approved for agricultural use and are included in this Guide.

Off-label Approval

Products may legally be used in a manner not covered by the printed label in several ways:
- in accordance with the "Off-label Arrangements" (see below)
- in accordance with a specific off-label approval (SOLA). SOLAs are uses for which approval has been sought by individuals or organisations other than the manufacturers. The Notices of Approval are published by MAFF and are widely available from ADAS or NFU offices. Users of SOLAs must first obtain a copy of the relevant Notice of Approval and comply strictly with the conditions laid down therein
- in tank mixture with other approved pesticides in accordance with Consent C(i) made under FEPA. Full details of Consent C(i) are given in Annex A of *Pesticides 1995*, but there are two essential requirements for tank mixes. Firstly, all the conditions of approval of all the components of a mixture must be complied with. Secondly, no person may mix or combine pesticides which are cholinesterase compounds unless allowed by the label of at least one of the pesticides in the mixture
- in conjunction with authorised adjuvants
- in reduced spray volume under certain conditions
- the use of certain herbicides on specified set-aside areas (subject to review by end March 1997)
- by mutual recognition of a use fully approved in another Member State of the European Union and authorised by PSD

Although approved, off-label uses are not endorsed by manufacturers and such treatments are made entirely at the risk of the user.

The Off-label Arrangements

Since 1 January 1990, arrangements have been in place allowing many approved products to be used for additional specific minor uses. These were revised to current standards in December 1994 to become *The Revised Long Term Arrangements for Extension of Use 1995*, and are valid until 31 December 1999. This review resulted in the loss of certain extrapolations (including those that were granted an extension until December 1995) and the inclusion of several new areas of use. *For this reason it is important that users read the new arrangements thoroughly before proceeding with pesticide uses that were allowed under the previous Long-Term Off-Label Arrangements.*

The arrangements are set out in full at Annex C of *Pesticides 1995*. The following is a summary of the main points and uses.

Specific restrictions for extension of use

Certain restrictions are necessary to ensure that the extension of use does not increase the risk to the operator, the consumer or the environment:

Extensions of use may be made only from label recommendations. Extrapolations may not be made from specific off-label approvals.

Safety precautions and statutory conditions relating to use as specified on the label must be observed. The application method must be as stated on the product label and in accordance with relevant Codes of Practice and COSHH requirements. Where application of a pesticide under these arrangements is made by hand-held equipment or broadcast air-assisted sprayers particular restrictions apply (consult Annex C of *Pesticides 1995*).

Pesticides must only be used in the same situation as that on the product label ie outdoor or protected. Approval for use on tomatoes, cucumbers, lettuces, chrysanthemums and mushrooms include protected crops unless otherwise stated. Use on other protected crops must be specifically permitted on the label.

Off-label use in or near water, or by aerial application, is not permitted under the arrangements.

Rodenticides are not included, nor is use on land not intended for cropping.

Pesticides classed as harmful, dangerous or extremely dangerous to bees must not be used off-label on any flowering crop. Where use is recommended on-label on flowering crops of peas, cereals or oilseed rape this relates only to those crops.

The user of a pesticide under these arrangements must ensure that such use does not result in a breach of any statutory MRL.

The arrangements apply as follows:

Non-edible crops and plants

Subject to the specific restrictions set out above, pesticides approved for use on any growing crop may be used on commercial holdings and forest nurseries on: (i) ornamental crops including hardy nursery stock, plants, bulbs, flowers and seed crops where neither the seed nor any part of the plant is to be consumed by humans or animals; (ii) forest nursery crops prior to final planting out.

In addition, pesticides approved for use on any growing *edible* crop (except seed treatments) may be used on non-ornamental crops grown for seed subject to the same consumption restrictions as above (but not including seed crops of potatoes, cereals, oilseeds, peas and beans).

Pesticides (except seed treatments) approved for use on oilseed rape may be used on hemp grown for fibre.

Farm forestry and rotational cropping

Subject to the specific restrictions set out above, *herbicides* approved for use on cereals may be used in the first five years of establishment in farm forestry on land previously under arable cultivation or improved grassland.

In addition, herbicides approved for use on cereals, oilseed rape, sugar beet, potatoes, peas and beans may be used in the first year of regrowth after cutting in coppices established on land previously under arable cultivation or improved grassland.

Nursery fruit crops

Subject to the specific restrictions set out above, pesticides approved for use on any crop for human or animal consumption may be used on commercial holdings on nursery fruit trees, nursery vines prior to final planting out, bushes, canes and non-fruiting strawberries

15

provided any fruit harvested within one year is destroyed. Applications must not be made if fruit is present.

Crops used partly or wholly for consumption by humans or livestock

Subject to the specific restrictions set out above, pesticides may be used on commercial holdings on crops in the first column below if they have been approved for use on the crop(s) opposite them in the second column. These extrapolations do not include uses in store, which are subject to separate arrangements (see below).

MINOR USE		CROPS ON WHICH USE IS APPROVED	ADDITIONAL SPECIAL CONDITIONS
Arable crops			
Poppy (grown for oilseed)		Sunflower	
Sesame		Sunflower	
Mustard		Oilseed rape	
Linseed		Oilseed rape	
Evening primrose		Oilseed rape	
Honesty		Oilseed rape	
Linola		Oilseed rape, linseed	
Flax (oilseed and fibre)		Oilseed rape, linseed	
Borage (grown for oilseed)		Oilseed rape	Seed treatments are not permitted
Grass seed crop		Wheat, barley, oats, rye, triticale	Treated crops must not be grazed or cut for fodder. Seed treatments are not permitted
Grass seed crop		Grass for grazing or fodder	
Rye		Wheat, barley	Treatments applied before second node detectable stage only
Triticale		Wheat, barley	
Durum wheat		Wheat	
Lupins		Combining peas, field beans	
Fruit crops			
Almond	Application to the orchard floor ONLY	Apple, cherry, plum	For herbicides used on the orchard floor ONLY
Chestnut		Apple, cherry, plum	
Hazelnut		Apple, cherry, plum	
Walnut		Apple, cherry, plum	
Quince		Apple, pear	
Crab apple		Apple, pear	

MINOR USE	CROPS ON WHICH USE IS APPROVED	ADDITIONAL SPECIAL CONDITIONS
Almond Chestnut Hazelnut Walnut	Products approved for use on two of the following: almond, chestnut, hazelnut, walnut	
Nectarine Apricot	Peach Peach	
Blackcurrant Blackberry Rubus species (e.g. tayberry, loganberry) Dewberry	Redcurrant Raspberry Raspberry Raspberry	
Redcurrant Whitecurrant Bilberry Cranberry	Blackcurrant Blackcurrant, redcurrant Blackcurrant, redcurrant Blackcurrant, redcurrant	
Vegetable crops Parsley root Fodder beet Mangel Horseradish Parsnip Salsify Swede Turnip	Carrot, radish Sugar beet Sugar beet Carrot, radish Carrot Carrot, celeriac Turnip Swede	
Garlic Shallot	Bulb onion Bulb onion	
Aubergine	Tomato	
Squash Pumpkin Marrow Watermelon	Melon Melon Melon Melon	
Broccoli Calabrese Roscoff cauliflower Collards	Calabrese Broccoli Cauliflower Kale	
Lamb's lettuce, frise, radicchio Beet leaves Cress	Lettuce Spinach Lettuce	

17

MINOR USE	CROPS ON WHICH USE IS APPROVED	ADDITIONAL SPECIAL CONDITIONS
Scarole	Lettuce	
Leaf herbs and edible flowers*	Lettuce, spinach, parsley, sage, mint, tarragon	
Edible podded peas (e.g. mange-tout, sugar snap)	Edible podded beans	
Runner beans	Dwarf French beans	
Rhubarb	Celery	
Cardoon	Celery	
Edible fungi other than mushroom (e.g. oyster mushroom)	Mushroom	

* This extension of use applies to the following leaf herbs and edible flowers: angelica, balm, basil, bay, borage, burnet (salad), caraway, chamomile, chervil, chives, clary, coriander, dill, fennel, fenugreek, feverfew, hyssop, land cress, lovage, marjoram, marigold, mint, nasturtium, nettle, oregano, parsley, rocket, rosemary, rue, sage, savory, sorrel, tarragon, thyme, verbena (lemon), woodruff.

Application in store on crops used partly or wholly for human or animal consumption

Subject to the specific restrictions set out above, pesticides may be used on commercial holdings on the crops listed in the first column below if they have been approved for use in store on the crops listed opposite them in the second column. *Seed treatments are not covered by this arrangement.*

MINOR USE	CROPS ON WHICH USE IS APPROVED
Rye	Wheat
Barley	Wheat
Oats	Wheat
Buckwheat	Wheat
Millet	Wheat
Sorghum	Wheat
Triticale	Wheat
Dried peas	Dried beans
Dried beans	Dried peas
Mustard	Oilseed rape
Sunflower	Oilseed rape
Linola	Oilseed rape
Flax	Oilseed rape
Honesty	Oilseed rape

MINOR USE	CROPS ON WHICH USE IS APPROVED
Poppy (grown for oilseed)	Oilseed rape
Borage (grown for oilseed)	Oilseed rape
Evening primrose	Oilseed rape
Sesame	Oilseed rape
Linseed	Oilseed rape

GUIDANCE ON THE USE OF PESTICIDES

The information given in the *UK Pesticide Guide* provides some of the answers needed to assess health risks, including the hazard classification and the level of operator protection required. However, the Guide cannot provide all the details needed for a complete hazard assessment, which must be based on the product label itself and, where necessary, the Health and Safety Data Sheet and other official literature.

Detailed guidance on how to comply with the Regulations is available from several sources:

1. *Pesticides: Code of practice for the safe use of pesticides on farms and holdings*, 1990. (ISBN 0 11 242892 4)

The Code of Practice is available from HMSO in a package, which includes additional up-to-date guidance on the consents to pesticide use, user certification, protection of honey bees, management of field margins and avoiding problems with pesticide wastes, together with an agricultural safety chart.

The Code of Practice gives comprehensive guidance on how to comply with COPR and the COSHH regulations under the headings User training and certification, Planning and preparation, Working with pesticides, Disposal of pesticide waste and Keeping records. In particular it details what is involved in making a COSHH Assessment. The principal source of information for making such an assessment is the approved product label. In most cases the label provides all the necessary information but in certain circumstances other sources must be consulted and these are listed in the Code of Practice.

2. Other codes of practice:

- *Code of Practice for the use of Approved Pesticides in Amenity & Industrial Areas* (ISBN 1 871140 12 9)
- *The Safe Use of Pesticides for Non-Agricultural Purposes – Approved Code of Practice* (ISBN 0 11 885673 1)
- *Code of Practice for Suppliers of Pesticides to Agriculture, Horticulture and Forestry* (MAFF Booklet PB 0091)
- *Code of Good Agricultural Practice for the Protection of Soil* (MAFF Booklet PB 0617)
- *Code of Good Agricultural Practice for the Protection of Water* (MAFF Booklet PB 0587)

3. Other guidance and practical advice:

HSE (from HSE Books – see Appendix 2)
- *A step by step guide to COSHH assessment*, HS(G) 97 (ISBN 0 11 886 379 7)
- *Food and Environment Protection Act 1985/Control of Pesticides Regulations 1986. An open learning course* (ISBN 0 11 8857 43 6)
- *COSHH in agriculture*, AS 28
- *COSHH in forestry*, AS 30
- *Occupational Exposure Limits 1995*, Guidance Note EH 40/95 (ISBN 0 7176 0876 X)
- *Biological monitoring of workers exposed to organophosphorus pesticides*, Guidance Note MS 17 (ISBN 0 11 883951 9)

MAFF (from HMSO – see Appendix 2)

- *Pesticides 1995,* Reference Book 500 (ISBN 0 11 242964 5)
- *Pesticides Register* (monthly) (ISSN 0955 7458)

British Agrochemicals Association (see Appendix 2)

- *The plain man's guide to pesticides and the COSHH assessment*
- *Amenity Handbook: A guide to the selection and use of amenity pesticides*
- *Think Water – Keep it Clean: Practical advice for pesticide users to protect the water environment*
- *Pesticides and the Law: A guide to the user*
- *Container Rinsing*
- *Pesticide Disposal*

British Crop Protection Council (see Appendix 2)

- *The Pesticide Manual* (ISBN 0 948404 79 5)
- *Hand-operated Sprayers Handbook* (ISBN 0 948404 30 2)
- *Boom Sprayers Handbook* (ISBN 0 948404 50 7)
- *Fruit Sprayers Handbook* (ISBN 0 948404 60 4)

National Rivers Authority (see Appendix 2)
- *The Prevention of Pollution of Controlled Waters by Pesticides* (Leaflet PPG9)

Protection of Honey Bees

Honey bees are a source of income for their owners and important to farmers and growers as pollinators of their crops. It is irresponsible and unnecessary to use pesticides in such a way that may endanger them. Pesticides vary in their toxicity to bees, but those that present a special hazard are classed as "harmful", "dangerous" or "extremely dangerous" to bees. Products so classified carry a specific warning in the precautions section of the label. These are indicated in this *Guide* in the protective clothing/label precautions section of the pesticide profile by the numbers 48, 48a, 49 or 50, depending on which phrase applies. The phrases are stated in full in Appendix 4.

Where use of a product that is hazardous to bees is contemplated, there are simple guidelines to follow, all of which will be stated on the product label. Most important of these are:

- give local beekeepers as much warning of your intention as possible
- avoid spraying crops in flower or where bees are actively foraging
- keep down flowering weeds

The British Beekeepers' Association (see Appendix 2) can give further advice on protecting honey bees from pesticides.

CHEMICALS SUBJECT TO THE POISONS LAW

Certain products in this book are subject to the provisions of the Poisons Act 1972, the Poisons List order 1982 and the Poisons Rules 1982 (copies of all these are obtainable from HMSO). These Rules include general and specific provisions for the storage and sale and supply of listed non-medicine poisons.

The chemicals approved for use in the UK and included in this book are specified under Parts I and II of the Poisons List as follows :

Part I Poisons (sale restricted to registered retail pharmacists)

aluminium phosphide	sodium cyanide
chloropicrin	strychnine
methyl bromide	

Part II Poisons (sale restricted to registered retail pharmacists and listed sellers registered with local authority)

aldicarb	
carbofuran (a)	methomyl (f)
chlorfenvinphos (a,b)	nicotine (c)
demeton-S-methyl	oxamyl (a)
dichlorvos (a, e)	paraquat (d)
disulfoton (a)	phorate (a)
fentin acetate	quinalphos
fentin hydroxide	sulfuric acid
fonofos (a)	thiometon
formaldehyde	triazophos (a)
mephosfolan	zinc phosphide

(a) Granular formulations containing up to 12% w/w of this, or a combination of similarly flagged poisons, are exempt
(b) Treatments on seeds are exempt
(c) Formulations containing not more than 7.5% of nicotine are exempt
(d) Pellets containing not more than 5% paraquat ion are exempt
(e) Preparations in aerosol dispensers containing not more than 1% a.i. are exempt
    Materials impregnated with dichlorvos for slow release are exempt
(f) Solid substances containing not more than 1% w/w a.i. are exempt

OCCUPATIONAL EXPOSURE LIMITS

A fundamental requirement of the COSHH Regulations is that exposure of employees to substances hazardous to health should be prevented or adequately controlled. Exposure by inhalation is usually the main hazard, and in order to measure the adequacy of control of exposure by this route various substances have been assigned occupational exposure limits.

There are two types of occupational exposure limits defined under COSHH: Occupational Exposure Standards (OES) and Maximum Exposure Limits (MEL). The key difference is that an OES is set at a level at which there is no indication of risk to health; for a MEL a residual risk may exist and the level takes socio-economic factors into account. In practice, MELs have been most often allocated to carcinogens and to other substances for which no threshold of effect can be identified and for which there is no doubt about the seriousness of the effects of exposure.

OESs and MELs are set on the recommendations of the Advisory Committee on Toxic Substances (ACTS). Full details are published by HSE in *EH 40/95 Occupational Exposure Limits 1995* ISBN 0 7176 0876 X.

As far as pesticides are concerned, OESs and MELs have been set for relatively few active ingredients. This is because pesticide products usually contain other substances in their formulation, including solvents, which may have their own OES/MEL. In practice inhalation of solvent may be at least as important (if not more so) than that of the active ingredient. These factors are taken into account in the approval process of the pesticide product under the Control of Pesticide Regulations. This indicates one of the reasons why a change of pesticide formulation usually necessitates a new approval assessment under COPR.

AGRICULTURAL PESTICIDES AND WATER

Even when diluted, some pesticides are potentially dangerous to fish and other aquatic life. Not only this, but it requires only a small amount of contamination to breach the stringent European Union water quality standards which have set a maximum admissible level in drinking water for any pesticide, regardless of its toxicity, at 1 part in 10,000 million. As little as 250 g could be enough to cause the daily supply to a city the size of London to exceed the permitted levels.

The Food and Environment Protection Act 1985 (FEPA) places a special obligation on users of pesticides to "safeguard the environment and in particular avoid the pollution of water". Under the Water Resources Act 1991 it is an offence to pollute any controlled waters (watercourses or groundwater), either deliberately or accidentally. Protection of controlled waters from pollution is the responsibility of the National Rivers Authority (NRA).

Users of pesticides therefore have a duty to adopt responsible working practices and, unless they are applying herbicides in or near water (see next section), to prevent them getting into water. Guidance on how to achieve this is given in the MAFF Code of Good Agricultural Practice for the Protection of Water (see p. 20).

The duty of care covers not only the way in which a pesticide is sprayed, but also its storage, preparation and disposal of surplus, sprayer washings and the container. Products in this *Guide* that are a major hazard to fish, other aquatic life or aquatic higher plants carry label precautions 51, 51a, 52 or 53 (see Appendix 4) in their profile, depending on which safety phrase applies. Those with a specific label restriction to prevent direct spray contamination of surface waters or ditches have precaution 54a.

Advice on any water pollution problems and practical guidance for users is available from NRA and BAA (see Appendix 2)

USE OF HERBICIDES IN OR NEAR WATER

Products in this Guide approved for use in or near water, as at 31 October 1994, are listed below. Before use of any product in or near water the appropriate water regulatory body (National Rivers Authority/Local Rivers Purification Authority or, in Scotland, the local River Purification Board) should be consulted. Always read the label before use.

CHEMICAL	PRODUCT	WEEDS CONTROLLED	SAFETY INTERVAL BEFORE IRRIGATION
asulam	Asulox	docks, bracken	nil
2,4-D	Atlas 2,4-D Dormone	water weeds, dicotyledon weeds on banks	3 wk
dalapon + dichlobenil	Fydulan G	weeds in or near water	–
dichlobenil	Casoron G Casoron G-SR	aquatic weeds	2 wk
diquat	Midstream Reglone	aquatic weeds	10 d
fosamine-ammonium	Krenite	woody weeds	nil
glyphosate	Barclay Gallup Amenity Glyphogan Helosate Roundup Roundup Biactive Roundup Biactive Dry Roundup Pro Roundup Pro Biactive Spasor Stirrup	Reeds, rushes, sedges, waterlilies, grass weeds on banks	nil
maleic hydrazide	Bos MH 180 Regulox K	suppression of grass growth on banks	3 wk
terbutryn	Clarosan 1FG	aquatic weeds	7 d

USE OF PESTICIDES IN FORESTRY

The table below lists those products, the labels of which refer to approval for use in forestry.

CHEMICAL	PRODUCT	USE
aluminium phosphide	Amos Talunex Phostoxin Phostek	Mole and rabbit control
aluminium ammonium sulfate	Guardsman	Reduction of damage by deer, hares, rabbits
ammonium sulfamate	Amcide Root-Out	Rhododendron and other woody weed control
asulam	Asulox	Bracken control
atrazine	Atlas Atrazine MSS Atrazine 50FL	Weed control in conifers
chlorpyrifos	Dursban 4 Lorsban T	Weevil and beetle control in transplant lines and cut logs
clopyralid	Dow Shield	Weed control in conifers and hardwoods (off-label)
2,4-D	Dicotox Extra	Perennial and woody weed control
2,4-D + dicamba + triclopyr	Broadshot	Perennial and woody weed control
dalapon + dichlobenil	Fydulan G	Grass and perennial weed control
dicamba	Tracker	Bracken control
dichlobenil	Casoron G Casoron G4 Prefix D	Annual and perennial weed control in plantations of certain tree species established for at least 2 years
diflubenzuron	Dimilin WP	Caterpillar control on forest trees
diquat + paraquat	PDQ	Pre-planting weed control,

CHEMICAL	PRODUCT	USE
	Parable	pre-emergence control in nurseries, directed sprays or ring weeding in plantations. Firebreak maintenance
fosamine-ammonium	Krenite	Woody weed control in forestry land (not conifer plantations)
gamma-HCH	Lindane Flowable	Various insect pests in nursery beds, transplant lines, plantations and cut timber
glufosinate-ammonium	Challenge	Control of annual and perennial weeds (directed sprays)
glyphosate	Barbarian Barclay Gallup Amenity Glyphogan Helosate Hilite Roundup Roundup Biactive Roundup Biactive Dry Roundup Pro Roundup Pro Biactive Stefes Glyphosate Stefes Kickdown Stefes Kickdown 2 Stirrup	Control of annual, perennial and woody weeds (directed sprays and wiper application). Chemical thinning
isoxaben	Flexidor 125 Gallery 125	Pre-emergence weed control in forestry transplants, second year undercuts, forest and woodland plantings
paraquat	Barclay Total Dextrone X Gramoxone 100	Pre-planting weed control, pre-emergence control in nurseries, directed sprays or ring-weeding in plantations. Firebreak maintenance
permethrin	Permasect 25 EC	Pine weevil in conifers (off-label)
propaquizafop	Falcon	Control of annual and perennial grass weeds

CHEMICAL	PRODUCT	USE
propyzamide	Headland Judo Kerb Flo Kerb Granules Kerb 50W	Grass weed control in forest trees
resmethrin	Turbair Resmethrin Extra	Pine looper moth control in pines
simazine	Ashlade Simazine 50FL Gesatop 500SC Gesatop 50WP MSS Simazine 50FL	Annual weed control in nursery beds and transplant lines
triclopyr	Garlon 4 Timbrel	Woody weed control
ziram	Aaprotect	Bird and animal repellent

PESTICIDES USED AS SEED TREATMENTS
(including treatments on seed potatoes)

Information on the target pests for these products can be found in the relevant pesticide profile in Section 3.

CHEMICAL	PRODUCT	FORMULATION	CROP(S)
aluminium ammonium sulphate	Guardsman STP Seed Dressing Powder	DP	Field crops, vegetables, fruit crops, ornamentals
2-aminobutane	Hortichem 2-aminobutane	VP	Potatoes
bendiocarb	Seedox SC	FS	Maize, sweetcorn
benomyl	Benlate	WP	Field beans, peas
bitertanol + fuberidazole	Sibutol	FS	Wheat
carbendazim + tecnazene	Hickstor 6 + MBC	DS	Potatoes
	Hortag Tecnacarb	DS	Potatoes
	New Arena Plus	DS	Potatoes
	Tripart Arena Plus	DS	Potatoes
carboxin + gamma-HCH + thiram	Vitavax RS	FS	Oilseed rape
carboxin + imazalil + thiabendazole	Vitaflo Extra	FS	Barley, oats
carboxin + thiabendazole	Vitaflo	FS	Rye, wheat
chlorfenvinphos	Birlane LST	LS	Wheat
ethirimol + flutriafol + thiabendazole	Ferrax	FS	Barley
fenpropimorph + gamma-HCH + thiram	Lindex-Plus FS	FS	Oilseed rape
fonofos	Fonofos Seed Treatment	FS	Barley, wheat, ryegrass

CHEMICAL	PRODUCT	FORMULATION	CROP(S)
fuberidazole + triadimenol	Baytan Flowable	FS	Cereals
gamma-HCH	Gammasan 30	FS	Cereals
	Kotol FS	FS	Cereals
gamma-HCH + thiabendazole + thiram	Hysede FL	FS	Broccoli, Brussels sprouts, cabbages, oilseed rape, swedes, turnips
gamma-HCH + thiram	Hydraguard	LS	Cabbage, Brussels sprouts, kale, swedes, turnips, oilseed rape
guazatine	Panoctine	LS	Wheat, barley, oats
guazatine + imazalil	Panoctine Plus	LS	Barley, oats
hymexazol	Tachigaren 70WP	WP	Sugar beet
imazalil	Fungazil 100 SL	SL	Potatoes
imazalil + pencycuron	Monceren IM Flowable	FS	Potatoes
imazalil + thiabendazole	Extratect Flowable	FS	Potatoes
imidacloprid	Gaucho	WS	Sugar beet
iprodione	Rovral WP	WP	Brassicas, oilseed rape, potatoes, linseed, flowers
	Rovral Liquid FS	FS	Oilseed rape, linseed, potatoes
mancozeb + zinc oxide	Mazin	WP	Potatoes
metalaxyl	Apron 350FS	FS	Oilseed rape
	Polycote Universal		Oilseed rape, various horticultural crops
metalaxyl + thiabendazole	Apron T69WS	WS	Grass seed
	Polycote Select	WS	Carrots, grass seed
metalaxyl + thiabendazole + thiram	Apron Combi 453FS	FS	Beans, peas

CHEMICAL	PRODUCT	FORMULATION	CROP(S)
pencycuron	Monceren DS	DS	Potatoes
	Monceren Flowable	FS	Potatoes
prochloraz + tolclofos-methyl	Rizolex Nova	KK	Potatoes
tebuconazole + triazoxide	Raxil S	LS	Barley
tecnazene	Various products, see tecnazene entry	Various	Potatoes
tecnazene + thiabendazole	Hytec Super	DP	Potatoes
	Storite SS	SS	Potatoes
tefluthrin	Force ST	CS	Sugar and fodder beet
thiabendazole	Storite Clear Liquid	LS	Potatoes
	Storite Flowable	FS	Potatoes
	Hykeep	DP	Potatoes
thiabendazole + thiram	Hy-TL	FS	Peas
	Hy-Vic	FS	Various field crops and vegetables
thiram	Agrichem Flowable Thiram	FS	Beans, peas, maize, vegetables, flowers
tolclofos-methyl	Rizolex	DS	Potatoes
	Rizolex Flowable	FS	Potatoes

AERIAL APPLICATION OF PESTICIDES

Only those products specifically approved for aerial application may be so applied and they may only be applied to specific crops or for specified uses. A list of products approved for application from the air was published as Annex B in *Pesticides 1995*. The list given below is taken mainly from this source with updating from issues of the *Pesticides Register* and product labels.

It is emphasized that the list is for guidance only — reference must be made to the product labels for detailed conditions of use which must be complied with. The list does not include those products which have been granted restricted aerial approval limiting the area which may be treated.

Detailed rules are imposed on aerial application regarding prior notification of the Nature Conservancy, water authorities, bee keepers, Environmental Health Officers, neighbours, hospitals, schools etc. and the conditions under which application may be made. The full conditions are available from MAFF and must be consulted before any aerial application is made.

CHEMICAL	PRODUCT	CROPS/USES
asulam	Asulox	Grassland, forestry, non-crop land (bracken control)
benalaxyl + mancozeb	Barclay Bezant	Potatoes
benomyl	Benlate Fungicide	Wheat, barley, oilseed rape
carbendazim	Ashlade Carbendazim FL	Winter wheat, winter and spring barley, oilseed rape
	Bavistin DF	Winter wheat, winter and spring barley, winter rye, field and dwarf beans, oilseed rape, onions
	Carbate Flowable	Wheat, barley, field beans, oilseed rape
	Delsene 50DF	Winter wheat, winter and spring barley, field beans, oilseed rape
	Hinge	Winter wheat, barley
	Tripart Defensor FL	Winter wheat, winter and spring barley, oilseed rape

CHEMICAL	PRODUCT	CROPS/USES
carbendazim + chlorothalonil	Bravocarb	Winter wheat, winter barley, spring barley
carbendazim + maneb	Ashlade Mancarb FL	Cereals
	Campbell's MC Flowable	Winter and spring wheat, winter and spring barley
	Headland Dual	Wheat, barley
	Tripart Legion	Winter wheat, winter barley
carbendazim + maneb + tridemorph	Cosmic FL	Winter wheat, winter and spring barley
chlormequat	Ashlade 460 CCC	Wheat, winter barley, oats, rye
	Ashlade 700 CCC	Wheat, winter barley, oats, rye
	Atlas 3C: 645	Wheat, oats
	Atlas Chlormequat 46	Wheat, oats
	Atlas Chlormequat 700	Wheat, oats
	Atlas Terbine	Wheat, oats
	Atlas Tricol	Wheat, oats
	Barleyquat B	Barley
	Bettaquat B	Wheat
	CCC 700	Wheat, oats, winter barley
	Hyquat 70	Wheat, oats, winter barley
	Mandops Chlormequat 700	Wheat, oats, winter barley
	Quadrangle Chlormequat 700	Wheat, oats
	Tripart Brevis	
	Uplift	Wheat, oats, winter barley

CHEMICAL	PRODUCT	CROPS/USES
chlormequat + choline chloride	Ashlade 5C Atlas 5C Chlormequat	Wheat, oats Wheat, oats
	Ashlade 500 7C Atlas Chlormequat 460:46 Atlas Quintacel MSS Mircell	Wheat, oats Wheat, oats Wheat, oats Wheat, oats
	New 5C Cycocel	Wheat, oats, triticale, barley, rye
2-chloroethyl phosphonic acid	Cerone	Winter barley
	Stefes Stance	Winter barley
chlorothalonil	Barclay Corrib 500	Wheat, barley, potatoes, field beans
	Bombardier	Wheat, potatoes
	Bravo 500	Winter wheat, potatoes, field beans, peas
	Jupital	Winter wheat, potatoes, field beans, peas
	Mainstay	Winter wheat, potatoes, field beans, peas
	Sipcam UK Rover 500	Wheat, potatoes
	Tripart Faber	Wheat, potatoes
chlorothalonil + cymoxanil	Guardian	Potatoes
chlorotoluron	Ashlade Tol-7	Winter wheat, winter barley, durum wheat, triticale
	Dicurane 700 SC	Winter wheat, winter barley, durum wheat, triticale
	Tripart Ludorum	Winter wheat, winter barley, durum wheat, triticale
chlorpropham + linuron	Profalon	Daffodils, narcissi, tulips

CHEMICAL	PRODUCT	CROPS/USES
copper oxychloride	Cuprokylt	Potatoes
cymoxanil + mancozeb	Ashlade Solace	Potatoes
	Curzate M	Potatoes
	Fytospore	Potatoes
	Standon Cymoxanil Extra	Potatoes
	Stefes Blight Spray	Potatoes
	Systol M	Potatoes
cymoxanil + mancozeb + oxadixyl	Ripost Pepite	Potatoes
	Trustan	Potatoes
	Trustan WDG	Potatoes
demeton-S-methyl	Campbell's DSM	Cereals, peas, potatoes, sugar beet, carrots, brassicas, field beans
dimethoate	Atlas Dimethoate 40	Cereals, peas, root crops, brassicas, top fruit, bush fruit, soft fruit
	BASF Dimethoate 40	Cereals, peas, ware potatoes, sugar beet
	Danadim dimethoate 40	Cereals, peas, potatoes, beet crops
	MTM Dimethoate 40	Wheat, barley, oats, peas, potatoes, fodder beet, sugar beet, mangels
disulfoton	Disyston P 10	Brassicas, beans, sugar beet, carrots
fenitrothion	Dicofen	Cereals, peas
	Unicrop Fenitrothion 50	Cereals, peas
fenpropimorph	Corbel	Wheat, barley, oats, rye, triticale, field beans

CHEMICAL	PRODUCT	CROPS/USES
fentin hydroxide	Ashlade Flotin	Potatoes
flamprop-M-isopropyl	Commando	Wheat, barley, rye, triticale, durum wheat
	Stefes Flamprop	Wheat, barley, durum wheat, rye, triticale
iprodione	Rovral Flo	Oilseed rape, peas
linuron	Afalon	Potatoes, carrots, parsnips, celery, parsley
mancozeb	Agrichem Mancozeb 80	Potatoes
	Ashlade Mancozeb FL	Potatoes, wheat, barley, oats, rye, triticale
	Barclay Manzeb 80	Potatoes
	Dithane 945	Potatoes
	Dithane Dry Flowable	Potatoes
	Headland Zebra WP	Potatoes
	Luxan Mancozeb Flowable	Potatoes
	Manzate 200	Potatoes
	Unicrop Flowable Mancozeb	Potatoes, wheat, barley, oats, rye, triticale
	Unicrop Mancozeb	Potatoes
mancozeb + metalaxyl	Fubol 75WP	Potatoes
	Osprey 58WP	Potatoes
mancozeb + oxadixyl	Recoil	Potatoes
maneb	Agrichem Maneb 80	Potatoes
	Ashlade Maneb Flowable	Potatoes
	Campbell's X-Spor SC	Potatoes

CHEMICAL	PRODUCT	CROPS/USES
	Headland Spirit	Barley, winter wheat, potatoes
	Maneb 80	Potatoes
	Stefes Maneb DF	Potatoes
	Unicrop Maneb 80	Barley, potatoes
maneb + zinc oxide	Mazin	Potatoes
metaldehyde	Doff Agricultural Slug Killer with Animal Repellent	All crops
	Doff Horticultural Slug Killer Blue Mini Pellets	All crops
	Doff Metaldehyde Slug Killer Mini Pellets	All crops
	Hardy	All crops
	Mifaslug	All crops
	Optimol	All crops
	PBI Slug Pellets	All crops
	Quad Mini Slug Pellets	All crops
	Superflor 6% Metaldehyde Slug Killer	All crops
	Tripart Mini Slug Pellets	All crops
	Unicrop 6% Mini Slug Pellets	All crops
methabenzthiazuron	Tribunil	Barley, winter wheat, autumn sown spring wheat, winter oats, winter rye, triticale, perennial grass, ryegrass
methiocarb	Decoy	All crops
	Draza	All crops

CHEMICAL	PRODUCT	CROPS/USES
	Exit	All crops
monolinuron	Arresin	Potatoes, dwarf beans, leeks
phorate (granules only)	MTM Phorate	Field and broad beans
phosalone	Zolone Liquid	Brassica seed crops
pirimicarb	Aphox	Cereals, peas, potatoes, sugar beet, beans, leaf brassicas, maize, swedes, oilseed rape, turnips, carrots
	Barclay Pirimisect	Wheat, barley, triticale, oats, rye
	Phantom	Cereals, peas, potatoes, sugar beet, beans, leaf brassicas, maize, sweetcorn, swedes, oilseed rape, turnips, carrots
	Stefes Pirimicarb	Cereals, peas, potatoes, sugar beet, beans, leaf brassicas, maize, sweetcorn, swedes, oilseed rape, turnips, carrots
propiconazole	Mantis 250EC	Wheat, barley, rye, oats
	Radar	Wheat, barley, rye, oats, triticale
	Standon Propiconazole	Wheat, barley, rye, oats
	Stefes Restore	Wheat, barley, rye, oats
	Tilt 250EC	Wheat, barley, rye, oats, triticale
propiconazole + tridemorph	Tilt Turbo 475EC	Wheat, barley
sulfur	Thiovit	Sugar beet
terbutryn	Prebane 500SC	Cereals, rye, triticale
triadimefon	Bayleton	Sugar beet, cereals, brassicas (including turnips and swedes)

CHEMICAL	PRODUCT	CROPS/USES
	Standon Triadimefon 25	Wheat, barley, rye, oats
tri-allate	Avadex BW Granular	Wheat, barley, winter beans, peas
	Avadex Excel 15G	Winter wheat, barley, winter beans, peas
trichlorphon	Dipterex 80	Beet crops, brassicas, spinach
tridemorph	Calixin	Barley, oats, winter wheat, swedes, turnips
	Standon Tridemorph 750	Barley, oats, winter wheat, swedes, turnips
triforine	Saprol	Barley
zineb	Unicrop Zineb	Potatoes

ULV APPLICATION

Certain pesticide products are specifically formulated for use with ULV applicators of various types, such as mistblowers, aerosol projectors, CDA sprayers and fogging machines. Others may be used without dilution as ULV treatments or may be diluted for application at higher volume rates. Products of both types are included in the table below.

CHEMICAL	PRODUCT	CROP
aluminium phosphide	Phostek	grain stores
asulam	Asulox	bracken in hill land, forestry and non-crop land
chlorotoluron	Dicurane 700 SC	cereals
	Tripart Ludorum	cereals
		cereals
chlorpropham	BL 500	potatoes
	Warefog 25	potatoes
chlorpropham + propham	Atlas Indigo	potatoes
	Luxan Gro-Stop	potatoes
	Pommetrol M	potatoes
	Nomix Turf Selective Herbicide	turf
dichlorophen	Halophen RE 49	potatoes
dichlorvos	Darlingtons Dichlorvos	mushroom houses
	Nuvan 500 EC	poultry houses
diuron + glyphosate	Touché	ornamental trees and shrubs
fenitrothion + permethrin + resmethrin	Turbair Grain Store Insecticide	grain stores
glyphosate	Apache	non-crop
	Barclay Gallup Amenity	non-crop
	Clarion	non-crop
	CDA Spasor	non-crop
	Glyphogan	non-crop
	Helosate	non-crop
	Hilite	non-crop
	Roundup	non-crop
	Roundup Biactive	non-crop

CHEMICAL	PRODUCT	CROP
	Roundup Biactive Dry	non-crop
	Roundup Four 80	non-crop
	Roundup Pro	non-crop
	Roundup Pro Biactive	non-crop
	Spasor	non-crop
	Stampede	non-crop
	Stefes Complete	non-crop
	Sting CT	non-crop
	Stirrup	non-crop
	Sulfosate	non-crop
	Touchdown	non-crop
	Touchdown LA	non-crop
imazalil + thiabendazole	Extratect Flowable	potatoes
imazapyr	Arsenal	non-crop
	Arsenal 50	non-crop
	Turbair Rovral	lettuce, tomatoes, chrysanthemums
isoproturon	Auger	cereals
	Stefes IPU	cereals
methyl bromide	Methyl Bromide 100	grain stores
	Sobrom BM 100	grain stores
methyl bromide + amyl acetate	Fumyl-O-Gas	field crops, protected crops, mushroom compost
methyl bromide + chloropicrin	Methyl Bromide 98	field crops, protected crops, mushroom compost
	Sobrom BM 98	field crops, protected crops, mushroom compost
oxycarboxin	Plantvax 75	glasshouse ornamentals
pencycuron	Monceren Flowable	potatoes
permethrin	Turbair Permethrin	vegetables
phenothrin	Sumithrin 10 Sec	livestock houses
pirimiphos-methyl	Actellifog	glasshouse crops
propyzamide	Kerb Flo	forestry
	Kerb 50W	forestry

CHEMICAL	PRODUCT	CROP
pyrethrins	Alfadex	livestock houses, farm buildings
	Dairy Fly Spray	livestock houses, farm buildings
	Killgerm ULV 400	farm buildings
	Pybuthrin 33	refuse tips
pyrethrins + resmethrin	Pynosect 30 Fogging Solution	glasshouse crops
resmethrin	Turbair Resmethrin Extra	vegetables, soft fruit, ornamentals, mushrooms
simazine + trietazine	Remtal SC	peas, beans, lupins
tecnazene	Nebulin	potatoes
tecnazene + thiabendazole	Storite SS	potatoes
terbuthylazine + terbutryne	Opogard 500 SC	peas, beans, lupins
terbutryn	Prebane 500 SC	cereals
tetramethrin + phenothrin	Killgerm ULV 500 Insecticide	farm buildings, grain stores, refuse tips
thiabendazole	Storite Flowable	potatoes
	Storite Clear Liquid	potatoes
tolclofos-methyl	Rizolex Flowable	potatoes

ADJUVANTS

Pesticide adjuvants are not themselves classed as pesticides but under the Control of Pesticides Regulations 'No person shall use a pesticide in conjunction with an adjuvant except in accordance with the conditions of the approval given originally in relation to that pesticide, or as varied subsequently by lists of authorised adjuvants published by the Ministers'.

Lists of authorised adjuvants are published in *The Pesticides Register* at intervals, the latest appearing in Issue 8 1994 with amendments in subsequent issues. Products notified by suppliers as available in 1995 are listed below.

Adjuvant product labels must be consulted for details of compatible chemicals or products, rates etc. but the table below provides a summary of the label information to indicate the area of use of the adjuvant. Protective clothing requirements and label precautions refer to the keys given in Appendix 4 and may include warnings about products harmful or dangerous to fish.

	PRODUCT	SUPPLIER	ADJ. NO.
	Actipron	Batson	0013
	Actipron	Bayer	0013
Type	Adjuvant oil containing 97% highly refined mineral oil		
Use with	Asulox		Dow Shield + Goltix WG
	Basagran		Goltix WG
	Benlate		Laser
	Betanal E		Pilot
	Betanal E + Goltix WG		Roundup
	Checkmate		
	Checkmate + Goltix WG		
Label precautions	21, 28, 29, 54, 63, 65		

	Activator 90	Newman	0062
Type	Non-ionic wetter/spreader containing 750 g/l alkylphenyl hydroxypolyoxyethylene		
Use with	Any spray for which additional wetter is approved and recommended. Contact supplier for further details		
Label precautions	A, C/6a, 6b, 18, 21, 28, 29, 36, 37, 53, 63, 65		

	PRODUCT	SUPPLIER	ADJ. NO.
	Adder	R P Agric.	0019

Type	Adjuvant oil containing 97% refined mineral oil

Use with	Ambush C + Checkmate	Checkmate + Goltix WG
	Asulox*	Checkmate + Thiovit
	Basagran	Checkmate + Toppel 10
	Butisan S + Checkmate	Dow Shield + Goltix WG
	Checkmate	Gesaprim 500 FW
	Checkmate + Decis	Goltix WG
	Checkmate + Dow Shield	phenmedipham
	Checkmate + Fubol 58 WP	Pilot
	Checkmate + gamma-HCH	Roundup
	(SC formulations)	

*Not in forestry

Label precautions	21, 28, 29, 54, 63, 65, 70

	Agral	ICI	0033
	Agral	Zeneca	0154

Type	Non-ionic wetter/spreader containing 948 g/l alkyl phenol ethylene oxide

Use with	Any spray for which additional wetter is approved and recommended

Label precautions	A, C/6a, 6b, 12c, 18, 21, 30, 36, 37, 54, 63, 65, 70

	Agropen	Ideal	0143

Type	Adjuvant oil containing 95% emulsifiable vegetable oil

Use with	Compatible with most pesticide formulations. See label for details

Label precautions	A, C/6a, 6b, 16, 21, 29, 36, 37, 54, 63, 66, 70, 78

	Agstock Addwett Agstock		0056

Type	Wetter/spreader containing 80% polyethanoxy (15) tallow amine

Use with	glyphosate

Label precautions	A, C/5c, 6a, 6b, 18, 21, 25, 28, 29, 36, 37, 53, 63, 65

	PRODUCT	SUPPLIER	ADJ. NO.
	Arma	Interagro (UK) Ltd	0200
Type	A penetrating adjuvant containing 500 g/litre alkoxylated fatty amine + 500 g/litre polyoxyethylene monolaurate		
Use with	Cereal growth regulators, cereal herbicides and a range of other pesticides		
Label precautions	A, C/5c, 6b, 14, 18, 29, 36, 37, 52, 54a, 63, 70, 78		

	PRODUCT	SUPPLIER	ADJ. NO.
	Ashlade Adjuvant Oil	Ashlade	0135
Type	Adjuvant oil containing 99% highly refined mineral oil		
Use with	Goltix WG		phenmedipham
Label precautions	6a, 6b, 6c, 18, 21, 28, 29, 36, 37, 53, 63, 66		

	PRODUCT	SUPPLIER	ADJ. NO.
	Atlas Adherbe	Atlas	0023
Type	Adjuvant oil containing 83% highly refined mineral oil		
Use with	Atlas Atrazine Atlas Brown Checkmate Goltix WG Goltix WG + Protrum K		Grasp Laser Pilot Protrum K
Label precautions	A, C/6a, 6b, 6c, 18, 21, 28, 29, 36, 37, 53, 60, 63, 66		

	PRODUCT	SUPPLIER	ADJ. NO.
	Atlas Adjuvant Oil	Atlas	0021
Type	Adjuvant oil containing 95% highly refined mineral oil		
Use with	Goltix WG Goltix WG + Protrum K		phenmedipham Protrum K
Label precautions	A, C/6a, 6b, 6c, 18, 21, 28, 29, 36, 37, 53, 60, 63, 66		

	PRODUCT	SUPPLIER	ADJ. NO.
	Atlas Bandrift	Atlas	–
Type	A spray drift retardant containing a non-ionic polyamide dispersed in oil		

	PRODUCT	SUPPLIER	ADJ. NO.

Use with	Many water soluble, emulsifiable or wettable powder herbicides

	Atlas Courier Atlas	0185
Type	Sticker/extender containing 12.5% w/w polyacrylamide	
Use with	Any approved pesticide in horticulture, forestry, industrial and non-crop situations, and with herbicides on cereals.	
Label precautions	A, C, H, M/10a, 30, 37, 54, 63, 66	

	Axiom Batson	0136
Type	A self-emulsifying adjuvant oil containing 97% highly refined mineral oil	
Use with	A range of herbicides and fungicides. See label for details	
Label precautions	22, 28, 29, 54, 63, 65	

	Barclay Actol Barclay	0126
Type	Adjuvant oil containing 99% highly refined paraffinic oil	
Use with	Any approved pesticide for which the addition of a spraying oil is recommended	
Label precautions	28, 30, 53, 63, 66	

	Barclay Clinger Barclay	0198
Type	Spray deposit protector containing 96% w/v poly-1-p-menthene	
Use with	Barclay Gallup, Barclay Gallup Amenity and a wide range of approved fungicides and insecticides. See label for details	
Label precautions	28, 30, 53, 63, 65	

	Barclay Dryfast Barclay **XL**	0121
Type	Wetter, sticker, extender and anti-transpirant containing di-l-p-menthene and nonyl phenol ethylene oxide condensate	
Use with	A wide range of approved pesticides and nutrients. See label for details	

PRODUCT	SUPPLIER	ADJ. NO.

Label precautions	30, 35(4 wk), 53, 63, 65

Bond	Newman	0037

Type	Sticker/extender containing 450 g/l synthetic latex
Use with	A wide range of contact herbicides, fungicides and insecticides. Contact supplier for further details
Label precautions	A, C/6a, 6b, 18, 21, 28, 29, 36, 37, 53, 63, 65

Citowett	BASF	0047

Type	Non-ionic wetter/spreader containing 99–100% alkyarylpolyglycol ether
Use with	Any spray for which additional wetter is approved and recommended
Label precautions	A, C/6a, 6b, 18, 21, 29, 36, 37, 53, 63, 66

Clifton Alkyl 90	Clifton	0118

No details available

Clifton Glyphosate Additive	Clifton	0024

Type	Wetter/spreader containing 850 g/l tallow amine ethoxylate
Use with	glyphosate
Label precautions	A, C/5c, 6a, 6b, 12a, 18, 21, 25, 29, 36, 37, 53, 63, 66, 70, 78

Clifton Wetter	Clifton	0028

Type	Non-ionic wetter/spreader containing 250 g/l alkyl phenol ethoxylate
Use with	Numerous fungicides, herbicides, insecticides and nutrients. See label for details
Label precautions	A, C/6a, 6b, 18, 21, 36, 37, 53, 63

PRODUCT	SUPPLIER	ADJ. NO.

Codacide Oil	Microcide	0011

Type — Adjuvant oil containing 95% emulsifiable vegetable oil

Use with — All approved pesticides and tank mixes. See label for details

Label precautions — 28, 29, 54, 63, 66

Compatibility Agent 2	Ciba Agric.	MAFF 03530

Type — Compatibility agent containing 770 g/l phosphate ester complex

Use with — Dicurane 700 SC — Prebane 500 SC when being sprayed in liquid fertilizer

Label precautions — A, C/7, 12c, 18, 21, 30, 36, 37, 53, 63, 66, 70, 78

Comulin	Batson	0164

Type — Self-emulsifying adjuvant oil containing 97% highly refined mineral oil

Use with

Asulox	Dow Shield + Goltix WG
Benlate	Goltix WG
Betanal E	Goltix WG + Laser
Basagran	Laser
Checkmate	Pilot
Checkmate + Goltix WG	Roundup

Label precautions — 21, 28, 29, 54, 63, 65

Contact	Interagro (UK) Ltd	0191

Type — An emulsifiable adjuvant oil containing 92% w/w paraffinic oil

Use with — Atrazine, Beetomax + metamitron, fenoxaprop-p-ethyl, metamitron, phenmedipham and other approved pesticides which have a label recommendation for use with authorised adjuvant mineral oils

Label precautions — A, C/6a, 6b, 14, 18, 21, 28, 29, 36, 37, 53, 63, 66

Cropoil	Chiltern	0137

Type — Adjuvant oil containing 99% highly refined mineral oil

	PRODUCT	SUPPLIER	ADJ. NO.

Use with	Any approved pesticide for which addition of a spraying oil is recommended
Label precautions	A, C/28, 29, 53, 60, 63, 66

	Cutinol	Techsol	0151

Type	Vegetable oil adjuvant containing 95% emulsifiable rapeseed oil
Use with	All approved pesticides for which addition of a wetter is recommended
Label precautions	A, C/6a, 6b, 18, 21, 29, 36, 37, 53, 63, 65

	Du Pont Adjuvant	Du Pont	0119

Type	Non-ionic wetter/spreader containing 900 g/l ethylene oxide condensate
Use with	Any spray for which additional wetter is approved and recommended
Label precautions	A, C/6a, 6b, 12c, 18, 21, 30, 36, 37, 54, 63, 65

	Emerald	Intracrop	0031

Type	Extender and antitranspirant containing 96% di-1-p-menthene
Use with	Alone or with most approved insecticides, fungicides or growth regulators. See label for details. Do not use in mixture with adjuvant oils or surfactants
Label precautions	28, 30, 53, 63, 65

	Enhance	Stefes	0147
	Enhance	Techsol	0147

Type	Non-ionic wetter/spreader containing 900 g/l phenol ethylene oxide condensate
Use with	Any spray for which additional wetter is approved and recommended
Label precautions	A, C/6a, 6b, 18, 21, 29, 36, 37, 53, 63, 65

	Enhance Low Foam	Techsol	0148

PRODUCT	SUPPLIER	ADJ. NO.

Type	Non-ionic wetter/spreader containing 900 g/l alkyl phenol ethylene oxide condensate and silicone anti-foam agent
Use with	Any spray for which additional wetter is approved and recommended through low-volume spraying systems or where low foam is required
Label precautions	A, C/6a, 6b, 18, 21, 29, 36, 37, 53, 63, 65

Ethokem	Techsol	0146

Type	Cationic surfactant containing 870 g/l polyoxyethylene tallow amine
Use with	A wide range of fungicides, herbicides, insecticides, growth regulators and micronutrients. See leaflet for details
Label precautions	A, C/5c, 6a, 6b, 12c, 18, 21, 25, 28, 29, 36, 37, 53, 63, 65

Ethokem C/12	Techsol	0149

Type	Cationic surfactant containing bis-2 hydroxyethyl coco-amine
Use with	glyphosate, for difficult to kill weeds including rhododendron, bracken and foxglove, especially in forestry
Label precautions	A, C/5c, 6a, 6b, 12c, 18, 21, 25, 28, 29, 36, 37, 53, 63, 65

Exell	Truchem	0017

Type	Wetter/spreader containing 64% polyethoxylated tallow amine	
Use with	diquat	glyphosate
Label precautions	A, C/6a, 6b, 18, 21, 28, 29, 36, 37, 53, 63, 65, 70, 78	

Frigate	ISK Biosciences	0128

Type	Wetter/spreader containing 800 g/l tallow amine ethoxylate
Use with	glyphosate
Label precautions	A, C/5c, 6a, 6b, 6c, 12c, 18, 21, 25, 29, 36, 37, 53, 63, 65, 78

Fyzol 11E	Schering	0072
Fyzol 11E	AgrEvo	0172

PRODUCT	SUPPLIER	ADJ. NO.

Type	Adjuvant oil containing 99% highly refined mineral oil
Use with	Betanal E Betanal E + Goltix WG Pilot
Label precautions	28, 29, 53, 63, 66, 70

Galion	Intracrop	0162

Type	Wetter/spreader containing 60% polyoxyalkylene glycol
Use with	diclofop-methyl mecoprop, salt formulations fluazifop-P-butyl fenoxaprop-ethyl
Label precautions	A, C/6a, 6b, 12c, 18, 21, 30, 36, 37, 54, 63, 65

GS 800	Stefes	0150
GS 800	Techsol	0150

Type	Cationic surfactant containing 800 g/l polyoxyethylene tallow amine
Use with	A wide range of pesticides. See label for details
Label precautions	A, C/5c, 6a, 6b, 12c, 18, 21, 25, 28, 29, 36, 37, 53, 63, 65

Headland Guard	Headland	0073

Type	Sticker/spreader containing 550 g/l organic co-polymer and surfactants
Use with	A wide range of agricultural chemicals and micronutrients
Label precautions	18, 29, 36, 37, 63

Headland Intake	Headland	0074

Type	Penetrant containing 450 g/l organic acids and surfactants
Use with	A wide range of fungicides, herbicides, desiccants and growth regulators
Label precautions	A, C/6a, 6b, 18, 21, 28, 29, 36, 37, 53, 63, 65

Headland Quilt	Headland	0188

Type	Wetter/spreader containing 870 g/l (85% w/w) bis-2 hydroxyethyl cocoamine

	PRODUCT	SUPPLIER	ADJ. NO.

Use with	glyphosate
Label precautions	A, C/5c, 6a, 6b, 14, 18, 21, 25, 28, 29, 36, 37, 53, 63, 65, 70, 78

	Hyspray	Fine	0020
Type	Cationic surfactant containing 800 g/l polyethoxylated tallow amine		
Use with	glyphosate		
Label precautions	A, C/5c, 6a, 6b, 12c, 18, 21, 25, 28, 29, 36, 37, 53, 63, 66, 70		

	Intracrop BLA	Intracrop	0125
Type	Sticker containing 52% synthetic latex and 20% alkyl phenol ethylene oxide condensate		
Use with	Any fungicides, herbicides, insecticides or trace elements where addition of an authorised extender/sticker is recommended		
Label precautions	A, C/14, 18, 21, 28, 31, 36, 37, 52, 63, 65, 70		

	Intracrop Non-ionic Wetter	Intracrop	0009
Type	Wetter spreader containing 900 g/l alkyl phenol ethoxylate		
Use with	Any spray for which additional wetter is approved and recommended		
Label precautions	A, C/6a, 6b, 12c, 14, 18, 21, 28, 29, 36, 37, 53, 63, 65		

	Jogral	Ideal	0109
Type	Cationic surfactant containing 800 g/l tallow amine ethoxylate		
Use with	glyphosate		
Label precautions	A, C/5c, 6a, 6b, 12c, 18, 21, 28, 29, 36, 37, 53, 63, 65, 70, 78		

	LI-700	Newman	0038
Type	Penetrating, acidifying surfactant containing 750 g/l soyal phospholids		
Use with	chlormequat and a wide range of systemic pesticides and trace elements. Contact supplier for details		
Label precautions	A, C/6a, 6b, 18, 21, 28, 29, 36, 37, 53, 63, 65		

	PRODUCT	SUPPLIER	ADJ. NO.

	Lightning	Batson	0177

Type | Adjuvant oil containing 97% highly refined mineral oil

Use with

Asulox	Dow Shield + Goltix WG
Basagran	Fusilade 5
Benlate	Gesaprim 500 FW
Betanal E	Goltix WG
Checkmate	Goltix WG + Laser
Checkmate + Goltix WG	Pilot
	Roundup

Label precautions | 21, 28, 29, 54, 63, 65

	Lo-Dose	ISK Biosciences	0129
	Lo-Dose	Quadrangle	0059

Type | Cationic surfactant containing 800 g/l tallow amine ethoxylate

Use with | Roundup | | Roundup Four 80

Label precautions | A, C/5c, 6a, 6b, 12c, 18, 21, 28, 29, 36, 37, 53, 63, 65, 70, 78

	Luxan	Luxan	0139
	Non-Ionic		
	Wetter		

Type | Non-ionic wetter/spreader containing 900 g/l alkyl phenol ethylene oxide condensate

Use with | Any spray for which additional wetter/spreader is approved and recommended

Label precautions | A, C/6a, 6b, 12c, 18, 21, 30, 36, 37, 53, 63, 66

	Lyrol	Batson	0165

Type | Self-emulsifying adjuvant oil containing 97% highly refined mineral oil

Use with

Asulox	Goltix WG + Checkmate
Basagran	Goltix WG + Dow Shield
Benlate	Goltix WG + Laser
Betanal E	Laser
Checkmate	Pilot
Gesaprim 500SC	Roundup
Goltix WG	
Goltix WG + Betanal E	

	PRODUCT	SUPPLIER	ADJ. NO.

Label precautions	21, 28, 29, 53, 63, 65

	Minder	Stoller	0087
Type	Rape oil adjuvant containing 94% vegetable oil
Use with	glyphosate
Label precautions	29, 37, 54, 63, 67

	Mixture B	Service	0161
Type	Non-ionic wetter/spreader containing 500 g/l nonyl phenol ethylene oxide condensate and 500 g/l primary alcohol ethylene oxide condensate
Use with	glyphosate in forestry, amenity and non-crop situations
Label precautions	A, C/5a, 5c, 6a, 6b, 18, 22, 36, 37, 53, 63, 65, 70, 78

	Non-ionic Wetter Service		0009
Type	Wetter/spreader containing 900 g/l alkyl phenol ethoxylate
Use with	Any spray for which additional wetter is approved and recommended
Label precautions	A, C/6a, 6b, 12c, 14, 18, 21, 28, 29, 36, 37, 53, 63, 65

	Nu-Film P	Intracrop	0039
Type	Spray deposit protector containing 96% poly-l-p-menthene
Use with	glyphosate and many other pesticides and growth regulators for which a protectant is recommended. Do not use in mixture with adjuvant oils or surfactants
Label precautions	28, 30, 53, 63, 65

	Output	Zeneca	0163
Type	Adjuvant oil containing 60% mineral oil and 40% surfactants
Use with	Grasp
Label precautions	A, C/6b, 18, 21, 28, 29, 36, 37, 53, 63, 66

PRODUCT	SUPPLIER	ADJ. NO.

Planet	Ideal	0140

Type	Non-ionic wetter/spreader containing 85% alkyl polyglycol ether and fatty acid
Use with	Any spray for which additional wetter is recommended
Label precautions	5c, 6a, 6b, 12c, 21, 29, 36, 37, 63, 65, 78

Quadrangle Q 900	Quadrangle	0112

Type	Wetter/spreader containing 90% alkylphenol ethylene oxide concentrate
Use with	Any spray for which additional wetter is approved and recommended
Label precautions	30, 54, 63, 65, 70

Quadrangle Quad-Fast	Quadrangle	0054

Type	Coating agent/surfactant containing di-1-p-menthene
Use with	glyphosate and other pesticides. Growth regulators and nutrients for which a coating agent is approved and recommended
Label precautions	30, 54, 63, 65

Rapide	Intracrop	0116

Type	Surfactant and penetrant containing 40% propionic acid
Use with	A wide range of pesticides and growth regulators, especially chlormequat
Label precautions	A, C/6a, 6b, 7, 8, 14, 18, 21, 28, 29, 36, 37, 63, 65

Ryda	Interagro (UK) Ltd	0168

Type	A cationic surfactant containing 800 g/litre polyethoxylated tallow amine
Use with	glyphosate
Label precautions	A, C/5c, 6a, 6b, 14, 18, 21, 25, 28, 29, 36, 37, 53, 63, 66, 70, 78

	PRODUCT	SUPPLIER	ADJ. NO.
	Slippa	Interagro (UK) Ltd	0206
Type	A liquid concentrate formulation containing 655 g/litre polyalkyleneoxide modified heptamethyltrisiloxane + non-ionic wetters		
Use with	Cereal fungicides and a range of other pesticides and trace elements		
Label precautions	A, C/5a, 5c, 6b, 6c, 9, 10a, 14, 18, 21, 24, 28, 29, 36, 37, 53, 63, 65, 70, 78		

	SM99	Newman	0127/0134
Type	Adjuvant oil containing 99% w/w highly refined paraffinic oil		
Use with	A wide range of pesticides in sugar beet, cereals, oilseed rape, maize and sweetcorn. Contact supplier for details		
Label precautions	28, 29, 54, 63, 66		

	Solar	Ideal	0111
Type	Non-ionic spreader/activator containing 75% polypropoxy propanol and 15% alkyl polyglycol ether		
Use with	Foliar applied plant growth regulators		
Label precautions	A, C/16, 21, 29, 36, 37, 54, 63, 66, 70, 78		

	Sprayfast	Mandops	0131
Type	Coating agent/surfactant containing di-1-p-menthene and nonyl phenol ethylene oxide condensate		
Use with	glyphosate and other pesticides, growth regulators or nutrients for which a coating agent is approved and recommended		
Label precautions	30, 54, 63, 65		

	Sprayfix	Newman	0145
Type	A sticker/extender containing 450 g/l synthetic latex		
Use with	Any spray where the addition of a sticker/extender is approved and recommended		
Label precautions	A, C/6a, 6b, 14, 18, 21, 28, 29, 36, 37, 53, 63, 65, 78		

PRODUCT SUPPLIER ADJ. NO.

Spraymac Newman 0144

Type	A non-ionic acidifying surfactant containing 350 g/l propionic acid and 100 g/l alkylphenyl-hydroxypolyoxyethylene
Use with	Any spray for which the addition of a wetter/spreader or self-emulsifying oil is approved and recommended
Label precautions	A, C/6a, 6b, 14, 16, 18, 23, 24, 28, 29, 36, 37, 53, 63, 67, 78

Sprayprover Fine 0027

Type	Adjuvant oil containing 92% highly refined mineral oil
Use with	Any spray for which the addition of a mineral oil is approved and recommended
Label precautions	A, C/6a, 6b, 18, 21, 28, 29, 36, 37, 53, 63, 66

Spreader PBI 0034

Type	Non-ionic wetter/spreader
Use with	Any spray for which additional wetter is approved and recommended
Label precautions	30, 54, 63, 65, 70

Stamina Interagro (UK) Ltd 0202

Type:	A liquid concentrate formulation containing 100% w/w alkoxylated fatty amine
Use with:	glyphosate and a range of other pesticides
Label precautions:	A, C/5c, 6b, 14, 18, 29, 36, 37, 52, 54a, 63, 70, 78

Stefes CAT 80 Stefes 0182

Type	Cationic surfactant containing 800 g/l (78% w/w) polyoxyethylene tallow amine
Use with	glyphosate
Label precautions	A, C/5, 18, 21, 25, 28, 29, 36, 37, 53, 63, 65, 70, 78

Stefes Cover Stefes 0187

	PRODUCT	SUPPLIER	ADJ. NO.

Type — Wetting and spreading agent containing 921 g/l (90% w/w) alkyl phenol ethoxylate

Use with — Any spray for which additional wetter or spreader is approved and recommended

Label precautions — A, C/6a, 6b, 18, 21, 28, 29, 36, 37, 53, 63, 65, 70

	Stefes Spread and Seal	Stefes	0122

Type — Wetter/sticker/extender and anti-transpirant containing di-l-p-menthene and nonyl phenol ethylene oxide condensate

Use with — glyphosate and other pesticides, growth regulators and nutrients for which a coating agent is approved and recommended

Label precautions — 30, 35, 53, 63, 65

	Stik-It	Quadrangle	0071

Type — Non-ionic wetter/spreader

Use with — A wide range of fungicides and insecticides for which additional wetter is approved and recommended

Label precautions — 30, 54, 63, 65, 70

	Swirl	Cyanamid	0167

Type — Adjuvant oil containing 590 g/l highly refined mineral oil

Use with — Commando

Label precautions — 12c, 21, 29, 52, 63, 66, 70

	TopUp Surfactant	FCC	0080

Type — Cationic surfactant containing 800 g/l ethoxylated tallow amine

Use with — A wide range of crop protection products. See label for details

Label precautions — A, C/5c, 6a, 6b, 12c, 14, 18, 21, 28, 29, 36, 37, 53, 63, 65, 70, 78

	Tripart Acer	Tripart	0097

	PRODUCT	SUPPLIER	ADJ. NO.

Type	Penetrating, acidifying surfactant containing 750 g/l soyal phospholipids
Use with	chlormequat and a wide range of systemic pesticides and trace elements. Contact supplier for details
Label precautions	A, C/6a, 6b, 18, 21, 28, 29, 36, 37, 53, 63, 65

	Tripart Lentus Tripart	0117

Type	Sticker and extender containing 450 g/l synthetic latex
Use with	A wide range of contact herbicides, fungicides and insecticides. Contact supplier for further details
Label precautions	A, C/6a, 6b, 18, 21, 28, 29, 36, 37, 53, 63, 65

	Tripart Minax Tripart	0016 0108

Type	Non-ionic wetter/spreader containing alkyl alcohol ethoxylate
Use with	Any spray for which additional wetter or spreader is approved and recommended
Label precautions	63, 65

	Tripart Orbis Tripart	0204

Type	Adjuvant oil containing 99% w/w highly refined paraffinic oil
Use with	A wide range of approved pesticides on sugar beet, cereals, oilseed rape, maize, sweet corn, fodder beet, mangels and red beet (including aerial use where recommended). See label for details
Label precautions	28, 29, 53, 54, 60, 63, 66

	Vassgro Spreader Vass	0035

Type	Non-ionic wetter/spreader containing nonyl phenol-ethylene oxide condensates
Use with	Any spray for which additional wetter or spreader is approved and recommended
Label precautions	29, 54, 63, 65, 70

PRODUCT	SUPPLIER	ADJ. NO.
Wayfarer	Hortichem	0045
Wayfarer	Service	0045

Type	Cationic wetter/spreader containing 80% tallow amine ethoxylate
Use with	glyphosate in agriculture
Label precautions	A, C/6a, 6b, 12c, 18, 21, 30, 36, 37, 54, 63, 65

POISONING BY PESTICIDES — FIRST AID MEASURES

If pesticides are handled in accordance with the required safety precautions, as given on the container label, poisoning should not occur. It is difficult however, to guard completely against the occasional accidental exposure. Thus, if a person handling, or exposed to, pesticides becomes ill, it is a wise precaution to apply first aid measures appropriate to pesticide poisoning even though the cause of illness may eventually prove to have been quite different. An employer has a legal duty to make adequate first aid provision for employees. Regular pesticide users should consider appointing a trained first aider, even if numbers of employees are not large, since there is a specific hazard.

The first essential in a case of suspected poisoning is for the person involved to stop work, to be moved away from any area of possible contamination and for a doctor to be called at once. If no doctor is available the patient should be taken to hospital as quickly as possible. In either event it is most important that the name of the chemical being used should be recorded and preferably the whole product label or leaflet should be shown to the doctor or hospital concerned.

Some pesticides which are unlikely to cause poisoning in normal use are extremely toxic if swallowed accidentally or deliberately. In such cases get the patient to hospital as quickly as possible, with all the information you have.

General Measures
Measures appropriate in all cases of suspected poisoning include:

- Remove any protective or other contaminated clothing (taking care to avoid personal contamination)
- Wash any contaminated areas carefully with water or with soap and water if available
- In cases of eye contamination, flush with plenty of clean water for at least 15 min
- Lay the patient down, keep at rest and under shelter. Cover with one clean blanket or coat etc. Avoid overheating
- Monitor level of consciousness, breathing and pulse-rate
- If consciousness is lost, place the casualty in the recovery position (on his/her side with head down and tongue forward to prevent inhalation of vomit)
- If breathing ceases or weakens commence mouth to mouth resuscitation. Ensure that the mouth is clear of obstructions such as false teeth, that the breathing passages are clear, and that tight clothing around the neck, chest and waist has been loosened. If a poisonous chemical has been swallowed, it is essential that the first aider is protected by the use of a resuscitation device (several types are available on the market)

Specific measures
In case of poisoning with particular chemical groups, the following measures may be taken before transfer to hospital:

Dinitro compounds* (dinoseb, DNOC, dinocap etc.)
Keep the patient at rest and cool.

Organophosphorus and carbamate insecticides
(aldicarb, azinphos-methyl, benfuracarb, carbofuran, chlorfenvinphos, chlorpyrifos, dichlorvos, heptenophos, malathion, methiocarb, mevinphos, propoxur, quinalphos, thiometon, triazophos etc.)
Keep the patient at rest. The patient may suddenly stop breathing so be ready to give artificial respiration.

Organochlorine compounds (aldrin[†], dienochlor, endosulfan, gamma-HCH)
If convulsions occur, do not interfere unless the patient is in danger of injury; if so any restraint must be gentle. When convulsions cease, place the casualty on his or her side with head down, tongue forward (recovery position).

Paraquat, diquat
Irrigate skin and eye splashes copiously with water. If any chemical has been swallowed, take to hospital for tests.

Cyanide (including sodium cyanide)
Send for medical aid. Remove casualty to fresh air, if necessary using breathing apparatus and protective clothing. Remove casualty's contaminated clothing. Gently brush solid particles from the skin, making sure you protect your own skin from contamination. Wash the skin and eyes copiously with water. Transfer casualty to nearest accident and emergency hospital by the quickest possible means together with the first aid cyanide antidote, if held on the premises.

Reporting of Pesticide Poisoning
Any cases of poisoning by pesticides must be reported without delay to an HM Agricultural Inspector of the Health and Safety Executive. In addition any cases of poisoning by substances named in schedule 2 of The Reporting of Injuries, Diseases and Dangerous Occurrences Regulations 1985, must also be reported to HM Agricultural Inspectorate (this includes organophosphorus chemicals, mercury and some fumigants).

Cases of pesticide poisoning should also be reported to the manufacturer concerned.

Additional information
General advice on the safe use of pesticides is given in a range of Health and Safety Executive leaflets available from HSE Books (see Appendix 2)

A useful booklet produced by GIFAP (International Group of National Associations of Manufacturers of Agrochemical Products) titled 'Guidelines for emergency measures in cases of pesticide poisoning' is available from GIFAP, Avenue Hamoir 12, 1180 Brussels, Belgium.

---

* Approvals for use of dinoseb, dinoseb acetate, dinoseb amine, dinoterb and binapacryl were revoked on 22 Jan 1988. Approval for storage of these materials was withdrawn on 30 June 1988. Approvals for supply of DNOC were revoked in Dec 1988 and for its use in Dec 1989.

† Approvals for supply, storage and use of aldrin were revoked in May 1989.

The major agrochemical companies are able to provide authoritative medical advice about their own pesticide products. Detailed advice is also available to doctors, from :
- The National Poisons Information Service, New Cross Hospital, London SE14 5ER (0171 635 9191)

and from regional centres in :
- Belfast, Royal Victoria Hospital (01232 240503)
- Birmingham, Dudley Road Hospital (0121 554 3801)
- Cardiff, Llandough Hospital (01222 709901)
- Edinburgh, Royal Infirmary (0131 229 2477)
- Leeds, Leeds General Infirmary (0113 243 2799)
- Newcastle, Royal Victoria Hospital (0191 232 5131)

SECTION 2
CROP/PEST GUIDE

CROP INDEX

GROUP	PAGE	CROP
Arable crops	71	Beans, field/broad
	72	Beet crops
	74	Cereals
	81	Cereals undersown
	81	Clovers
	82	Fodder brassica seed crops
	82	Fodder brassicas
	83	General
	84	Grass seed crops
	85	Linseed/flax
	86	Lucerne
	87	Lupins
	87	Maize/sweetcorn
	88	Miscellaneous field crops
	88	Oilseed rape
	91	Peas
	93	Potatoes
	95	Seed brassicas/mustard
Field vegetables	96	Asparagus
	96	Beans, french/runner
	97	Brassica seed crops
	98	Brassicas
	100	Carrots/parsnips/parsley
	101	Celery
	102	Cucurbits
	103	General
	103	Herb crops
	104	Lettuce, outdoor
	105	Miscellaneous field vegetables
	106	Onions/leeks
	107	Peas, mange-tout
	108	Red beet
	109	Rhubarb
	109	Root brassicas
	111	Root crops
	111	Spinach
	111	Tomatoes, outdoor
	111	Watercress
Flowers and ornamentals	112	Annuals/biennials
	112	Bedding plants
	113	Bulbs/corms
	114	Camellias

GROUP	PAGE	CROP
	114	Carnations
	114	Chrysanthemums
	115	Container-grown stock
	115	Dahlias
	115	General
	117	Hardy ornamental nursery stock
	118	Hedges
	118	Perennials
	118	Roses
	119	Standing ground
	119	Trees/shrubs
	121	Water lily
Forestry	121	Broadleaved trees
	121	Cut logs/timber
	121	Farm woodland
	121	Forest nursery beds
	122	Forestry plantations
	123	Transplant lines
Fruit and hops	124	Apples/pears
	126	Apricots/peaches/nectarines
	126	Blueberries/cranberries
	126	Cane fruit
	128	Currants
	129	Fruit nursery stock
	129	General
	130	Gooseberries
	131	Grapevines
	131	Hops
	132	Plums/cherries/damsons
	133	Quinces
	133	Strawberries
	135	Tree fruit
	136	Tree nuts
Grain/crop store uses	136	Food storage
	136	Stored cabbages
	136	Stored grain/rapeseed/linseed
	137	Stored products
Grassland	137	Grassland
	139	Leys
Non-crop pest control	140	Farm buildings/yards
	141	Farmland
	141	Food storage areas

GROUP	PAGE	CROP
	141	Manure heaps
	141	Miscellaneous situations
	141	Refuse tips
	142	Stored plant material
Protected crops	142	Aubergines
	142	Beans
	142	Cucumbers
	143	General
	145	Herbs
	145	Lettuce
	146	Mushrooms
	146	Mustard and cress
	146	Onions, leeks and garlic
	146	Peppers
	147	Pot plants
	148	Protected cucurbits
	148	Protected cut flowers
	149	Seed potatoes under protection
	149	Tomatoes
Setaside	150	Setaside
Total vegetation control	151	Non-crop areas
	152	Weeds in or near water
Turf/amenity grass	153	Turf/amenity grass

GROUP	PAGE	CROP

CROP/PEST GUIDE

Important note: For convenience, some crops and pests have been grouped into generic units in this guide, eg "cereals", "annual grasses". It is essential to check the profile entry in Section 3 *and* the label to ensure that a product is approved for a specific crop/pest combination, eg winter wheat/blackgrass.

Arable crops

Beans, field/broad

Diseases	Ascochyta	benomyl *(seed treatment)*, metalaxyl + thiabendazole + thiram *(seed treatment)*, thiabendazole + thiram
	Chocolate spot	benomyl, carbendazim (MBC), carbendazim + chlorothalonil, chlorothalonil, iprodione, iprodione + thiophanate-methyl, tebuconazole, thiophanate-methyl, vinclozolin
	Damping off	metalaxyl + thiabendazole + thiram *(seed treatment)*, thiabendazole + thiram, thiram *(seed treatment)*
	Downy mildew	chlorothalonil + metalaxyl, fosetyl-aluminium, metalaxyl + thiabendazole + thiram *(seed treatment)*
	Rust	fenpropimorph, tebuconazole
Pests	Aphids	demeton-S-methyl, dimethoate, disulfoton, fatty acids, heptenophos, malathion, nicotine, phorate, pirimicarb, resmethrin
	Bean beetle	triazophos
	Birds	aluminium ammonium sulfate
	Capsids	phorate
	Damaging mammals	aluminium ammonium sulfate
	Leaf miners	cypermethrin *(protected crops, off-label)*, oxamyl *(off-label)*, triazophos *(off-label)*
	Mealy bugs	fatty acids
	Pea and bean weevils	cypermethrin, deltamethrin, esfenvalerate, lambda-cyhalothrin, phorate, triazophos
	Red spider mites	fatty acids
	Scale insects	fatty acids
	Whitefly	fatty acids
Weeds	Annual and perennial weeds	glyphosate (agricultural) *(pre-harvest)*

	Annual dicotyledons	bentazone, carbetamide, chlorpropham + fenuron, cyanazine, cyanazine *(Scotland only)*, fomesafen + terbutryn, glufosinate-ammonium *(pre-harvest)*, pendimethalin + prometryn, prometryn + terbutryn, propyzamide, simazine, simazine + trietazine, terbuthylazine + terbutryn, terbutryn + trietazine, trifluralin
	Annual grasses	carbetamide, chlorpropham + fenuron, cyanazine, cyanazine *(Scotland only)*, cycloxydim, diclofop-methyl, fluazifop-P-butyl, glufosinate-ammonium *(pre-harvest)*, prometryn + terbutryn, propaquizafop, propyzamide, simazine, terbuthylazine + terbutryn, trifluralin
	Annual meadow grass	pendimethalin + prometryn, terbutryn + trietazine, tri-allate
	Blackgrass	cycloxydim, diclofop-methyl, fluazifop-P-butyl, tri-allate
	Canary grass	diclofop-methyl
	Chickweed	chlorpropham + fenuron
	Couch	cycloxydim, glyphosate (agricultural) *(pre-harvest)*
	Creeping bent	cycloxydim, glyphosate (agricultural) *(pre-harvest)*
	Perennial dicotyledons	glyphosate (agricultural) *(pre-harvest)*
	Perennial grasses	cycloxydim, diclofop-methyl, glyphosate (agricultural) *(pre-harvest)*, propaquizafop, propyzamide, tri-allate
	Rough meadow grass	diclofop-methyl
	Ryegrass	diclofop-methyl
	Volunteer cereals	carbetamide, cycloxydim, fluazifop-P-butyl, propyzamide
	Wild oats	cycloxydim, diclofop-methyl, fluazifop-P-butyl, propyzamide, tri-allate
Crop control	Pre-harvest desiccation	diquat, glufosinate-ammonium
Plant growth regulation	Increasing yield	chlormequat with di-1-p-menthene
	Lodging control	chlormequat with di-1-p-menthene

Beet crops

Diseases	Black leg	hymexazol
	Damping off	hymexazol
	Powdery mildew	copper sulfate + sulfur, propiconazole, sulfur, triadimefon, triadimenol
	Ramularia leaf spots	propiconazole
	Rust	propiconazole
	Seed-borne diseases	thiram *(seed soak)*

FOR FULL CONDITIONS OF USE ALWAYS READ THE PRODUCT LABEL

Pests	Aphids	aldicarb, carbosulfan, deltamethrin + heptenophos, demeton-S-methyl, dimethoate, dimethoate *(excluding Myzus persicae)*, disulfoton, imidacloprid, oxamyl, phorate, pirimicarb
	Birds	aluminium ammonium sulfate
	Capsids	phorate
	Cutworms	cypermethrin, gamma-HCH, triazophos
	Damaging mammals	aluminium ammonium sulfate
	Docking disorder vectors	aldicarb, carbofuran, oxamyl
	Flea beetles	carbofuran, carbosulfan, deltamethrin, gamma-HCH, imidacloprid, trichlorfon
	Leaf miners	aldicarb, pirimiphos-methyl, triazophos *(off-label)*
	Leatherjackets	chlorpyrifos, gamma-HCH
	Mangold fly	carbofuran, carbosulfan, demeton-S-methyl, dimethoate, disulfoton, imidacloprid, oxamyl, trichlorfon
	Millipedes	aldicarb, bendiocarb, carbofuran, carbosulfan, gamma-HCH, imidacloprid, oxamyl, tefluthrin
	Nematodes	aldicarb, carbofuran, carbosulfan
	Pygmy beetle	aldicarb, bendiocarb, carbofuran, carbosulfan, chlorpyrifos, gamma-HCH, imidacloprid, oxamyl, tefluthrin
	Springtails	bendiocarb, carbofuran, carbosulfan, gamma-HCH, imidacloprid, tefluthrin
	Symphylids	bendiocarb, carbosulfan, gamma-HCH, imidacloprid, tefluthrin
	Tortrix moths	carbofuran
	Wireworms	bendiocarb, carbofuran, carbosulfan, gamma-HCH
Weeds	Annual dicotyledons	carbetamide, chloridazon, chloridazon + chlorpropham + fenuron + propham, chloridazon + ethofumesate, chloridazon + lenacil, chlorpropham + fenuron + propham, clopyralid, ethofumesate, ethofumesate + phenmedipham, glufosinate-ammonium *(pre-emergence)*, lenacil, lenacil + phenmedipham, metamitron, paraquat, phenmedipham, propyzamide, trifluralin
	Annual grasses	carbetamide, chloridazon + chlorpropham + fenuron + propham, chlorpropham + fenuron + propham, cycloxydim, diclofop-methyl, fluazifop-P-butyl, glufosinate-ammonium *(pre-emergence)*, metamitron, paraquat, propaquizafop, propyzamide, quizalofop-ethyl, trifluralin
	Annual meadow grass	chloridazon, chloridazon + ethofumesate, chloridazon + lenacil, ethofumesate, lenacil, metamitron, tri-allate
	Annual weeds	glyphosate (agricultural) *(pre-drilling/pre-emergence)*
	Blackgrass	chloridazon + ethofumesate, cycloxydim, diclofop-methyl, ethofumesate, tri-allate
	Canary grass	diclofop-methyl
	Corn marigold	clopyralid

73

Couch	cycloxydim, quizalofop-ethyl, sethoxydim
Creeping bent	cycloxydim, sethoxydim
Creeping thistle	clopyralid
Fat-hen	metamitron
Mayweeds	clopyralid
Perennial dicotyledons	clopyralid
Perennial grasses	cycloxydim, diclofop-methyl, fluazifop-P-butyl, propaquizafop, propyzamide, quizalofop-ethyl, sethoxydim, tri-allate
Rough meadow grass	diclofop-methyl
Ryegrass	diclofop-methyl
Volunteer cereals	carbetamide, cycloxydim, fluazifop-P-butyl, glyphosate (agricultural) *(pre-drilling/pre-emergence)*, paraquat, quizalofop-ethyl, sethoxydim
Weed beet	glyphosate (agricultural) *(wiper application)*
Wild oats	cycloxydim, diclofop-methyl, fluazifop-P-butyl, sethoxydim, tri-allate

Cereals

Diseases	Alternaria	iprodione
	Blue mould	fuberidazole + triadimenol
	Botrytis	carbendazim + chlorothalonil + maneb, fenpropidin + tebuconazole, iprodione, tebuconazole + tridemorph
	Brown foot rot and ear blight	carbendazim + chlorothalonil + maneb, carboxin + imazalil + thiabendazole, carboxin + thiabendazole, epoxiconazole + fenpropimorph *(reduction)*, epoxiconazole + tridemorph *(reduction)*, epoxiconazole *(reduction)*, ethirimol + flutriafol + thiabendazole, fenpropidin + tebuconazole, fludioxonil, fuberidazole + triadimenol, guazatine + imazalil *(reduction)*, guazatine *(reduction)*, maneb + zinc, tebuconazole, tebuconazole + triadimenol, tebuconazole + tridemorph
	Covered smut	carboxin + imazalil + thiabendazole, ethirimol + flutriafol + thiabendazole, fludioxonil, fuberidazole + triadimenol
	Crown rust	fuberidazole + triadimenol, triadimenol + tridemorph

---

FOR FULL CONDITIONS OF USE ALWAYS READ THE PRODUCT LABEL

Eyespot	benomyl, carbendazim (MBC), carbendazim + chlorothalonil + maneb, carbendazim + cyproconazole, carbendazim + flusilazole, carbendazim + flusilazole *(reduction)*, carbendazim + flutriafol, carbendazim + mancozeb, carbendazim + maneb, carbendazim + maneb + tridemorph, carbendazim + prochloraz, carbendazim + propiconazole, chlorothalonil + cyproconazole, cyproconazole, cyproconazole + prochloraz, cyproconazole + tridemorph, epoxiconazole + fenpropimorph *(reduction)*, epoxiconazole + tridemorph *(reduction)*, epoxiconazole *(reduction)*, fenpropidin + prochloraz, fenpropimorph + prochloraz, flusilazole, maneb, maneb *(late infections)*, prochloraz, propiconazole *(low levels only)*
Foot rot	bitertanol + fuberidazole, fludioxonil, guazatine + imazalil
Late ear diseases	maneb
Leaf stripe	carboxin + imazalil + thiabendazole, ethirimol + flutriafol + thiabendazole, fludioxonil *(partial control)*, fuberidazole + triadimenol, guazatine + imazalil, tebuconazole + triazoxide
Loose smut	bitertanol + fuberidazole *(partial control)*, carboxin + imazalil + thiabendazole, carboxin + thiabendazole, ethirimol + flutriafol + thiabendazole, fuberidazole + triadimenol, tebuconazole + triazoxide
Net blotch	carbendazim + chlorothalonil + maneb, carbendazim + cyproconazole, carbendazim + flusilazole, carbendazim + flutriafol, carbendazim + flutriafol *(moderate control)*, carbendazim + mancozeb, carbendazim + maneb, carbendazim + maneb + sulfur, carbendazim + prochloraz, carbendazim + propiconazole, carboxin + imazalil + thiabendazole, chlorothalonil + cyproconazole, chlorothalonil + propiconazole, cyproconazole, cyproconazole + prochloraz, cyproconazole + tridemorph, epoxiconazole, epoxiconazole + fenpropimorph, epoxiconazole + tridemorph, ethirimol + flutriafol + thiabendazole, fenpropidin + prochloraz, fenpropidin + propiconazole, fenpropidin + tebuconazole, fenpropimorph + flusilazole, fenpropimorph + flusilazole + tridemorph, fenpropimorph + flusilazole + tridemorph *(reduction)*, fenpropimorph + prochloraz, fenpropimorph + propiconazole, flusilazole, flusilazole + tridemorph, flutriafol, guazatine + imazalil, iprodione, mancozeb, mancozeb *(moderate control)*, mancozeb *(reduction)*, maneb, nuarimol, prochloraz, propiconazole, propiconazole + tebuconazole, propiconazole + tridemorph, tebuconazole, tebuconazole + triadimenol, tebuconazole + tridemorph, triforine

Powdery mildew

carbendazim (MBC) *(partial control)*, carbendazim +
chlorothalonil + maneb, carbendazim + cyproconazole,
carbendazim + flusilazole, carbendazim + flutriafol,
carbendazim + mancozeb, carbendazim + maneb, carbendazim
+ maneb + sulfur, carbendazim + maneb + tridemorph,
carbendazim + prochloraz, carbendazim + propiconazole,
chlorothalonil + cyproconazole, chlorothalonil + fenpropimorph,
chlorothalonil + flutriafol, chlorothalonil + propiconazole,
copper oxychloride + maneb + sulfur, cyproconazole,
cyproconazole + prochloraz, cyproconazole + tridemorph,
difenzoquat *(reduction)*, epoxiconazole, epoxiconazole +
fenpropimorph, epoxiconazole + tridemorph, ethirimol +
flutriafol + thiabendazole, fenbuconazole + fenpropimorph,
fenpropidin, fenpropidin + prochloraz, fenpropidin +
propiconazole, fenpropidin + tebuconazole, fenpropimorph,
fenpropimorph + flusilazole, fenpropimorph + flusilazole +
tridemorph, fenpropimorph + prochloraz, fenpropimorph +
propiconazole, fenpropimorph + tridemorph, flusilazole,
flusilazole + tridemorph, flutriafol, fuberidazole + triadimenol,
maneb + zinc, maneb *(partial suppression)*, nuarimol,
prochloraz, prochloraz *(protection)*, propiconazole,
propiconazole + tebuconazole, propiconazole + tridemorph,
sulfur, tebuconazole, tebuconazole + triadimenol, tebuconazole
+ tridemorph, triadimefon, triadimenol, triadimenol +
tridemorph, tridemorph, triforine

Pyrenophora leaf spot

carboxin + imazalil + thiabendazole, fuberidazole +
triadimenol, guazatine + imazalil

Rhynchosporium

benomyl, carbendazim (MBC), carbendazim + chlorothalonil,
carbendazim + chlorothalonil + maneb, carbendazim +
cyproconazole, carbendazim + flusilazole, carbendazim +
flutriafol, carbendazim + mancozeb, carbendazim + maneb,
carbendazim + maneb + sulfur, carbendazim + maneb +
tridemorph, carbendazim + prochloraz, carbendazim +
propiconazole, chlorothalonil, chlorothalonil + cyproconazole,
chlorothalonil + propiconazole, copper oxychloride + maneb +
sulfur, cyproconazole, cyproconazole + prochloraz,
cyproconazole + tridemorph, epoxiconazole, epoxiconazole +
fenpropimorph, epoxiconazole + tridemorph, ethirimol +
flutriafol + thiabendazole, fenbuconazole + fenpropimorph,
fenpropidin, fenpropidin + prochloraz, fenpropidin +
propiconazole, fenpropidin + tebuconazole, fenpropimorph,
fenpropimorph + flusilazole, fenpropimorph + flusilazole +
tridemorph, fenpropimorph + prochloraz, fenpropimorph +
propiconazole, fenpropimorph + tridemorph, flusilazole,
flusilazole + tridemorph, flutriafol, fuberidazole + triadimenol,
mancozeb, maneb, maneb + zinc, nuarimol, prochloraz,
propiconazole, propiconazole + tebuconazole, propiconazole +
tridemorph, tebuconazole, tebuconazole + triadimenol,
tebuconazole + tridemorph, triadimefon, triadimenol,
triadimenol + tridemorph

FOR FULL CONDITIONS OF USE ALWAYS READ THE PRODUCT LABEL

Rust	carbendazim + chlorothalonil + maneb, carbendazim + cyproconazole, carbendazim + flusilazole, carbendazim + flutriafol, carbendazim + mancozeb, carbendazim + maneb, carbendazim + maneb + sulfur, carbendazim + maneb + tridemorph, carbendazim + propiconazole, chlorothalonil + cyproconazole, chlorothalonil + fenpropimorph, chlorothalonil + flutriafol, chlorothalonil + propiconazole, cyproconazole, cyproconazole + prochloraz, cyproconazole + tridemorph, difenoconazole, epoxiconazole, epoxiconazole + fenpropimorph, epoxiconazole + tridemorph, ethirimol + flutriafol + thiabendazole, fenbuconazole + fenpropimorph, fenpropidin, fenpropidin + prochloraz, fenpropidin + propiconazole, fenpropidin + tebuconazole, fenpropimorph, fenpropimorph + flusilazole, fenpropimorph + flusilazole + tridemorph, fenpropimorph + prochloraz, fenpropimorph + propiconazole, flusilazole, flusilazole + tridemorph, flutriafol, fuberidazole + triadimenol, mancozeb, mancozeb *(moderate control)*, mancozeb *(reduction)*, maneb, maneb + zinc, oxycarboxin, propiconazole, propiconazole + tebuconazole, propiconazole + tridemorph, tebuconazole, tebuconazole + triadimenol, tebuconazole + tridemorph, triadimefon, triadimenol, triadimenol + tridemorph
Septoria diseases	anilazine, bitertanol + fuberidazole *(reduction)*, carbendazim (MBC) *(partial control)*, carbendazim + chlorothalonil, carbendazim + chlorothalonil + maneb, carbendazim + cyproconazole, carbendazim + flusilazole, carbendazim + flutriafol, carbendazim + mancozeb, carbendazim + maneb, carbendazim + maneb + sulfur, carbendazim + prochloraz, carbendazim + propiconazole, chlorothalonil, chlorothalonil + cyproconazole, chlorothalonil + fenpropimorph, chlorothalonil + flutriafol, chlorothalonil + propiconazole, copper oxychloride + maneb + sulfur, cyproconazole, cyproconazole + prochloraz, cyproconazole + tridemorph, difenoconazole, epoxiconazole, epoxiconazole + fenpropimorph, epoxiconazole + tridemorph, fenbuconazole + fenpropimorph, fenpropidin, fenpropidin + prochloraz, fenpropidin + propiconazole, fenpropidin + tebuconazole, fenpropimorph + flusilazole, fenpropimorph + flusilazole + tridemorph, fenpropimorph + prochloraz, fenpropimorph + propiconazole, flusilazole, flusilazole + tridemorph, flutriafol, fuberidazole + triadimenol, guazatine, iprodione, mancozeb, maneb, maneb + zinc, nuarimol, prochloraz, propiconazole, propiconazole + tebuconazole, propiconazole + tridemorph, tebuconazole, tebuconazole + triadimenol, tebuconazole + tridemorph, triadimenol, triadimenol + tridemorph
Snow mould	fludioxonil
Snow rot	propiconazole, triadimefon, triadimenol, triadimenol + tridemorph

	Sooty moulds	carbendazim (MBC), carbendazim + chlorothalonil + maneb, carbendazim + cyproconazole, carbendazim + mancozeb, carbendazim + maneb, carbendazim + maneb + tridemorph, chlorothalonil + flutriafol, cyproconazole + tridemorph, epoxiconazole + fenpropimorph *(reduction)*, epoxiconazole + tridemorph *(reduction)*, epoxiconazole *(reduction)*, fenpropidin + tebuconazole, fenpropimorph + propiconazole, mancozeb, maneb, maneb + zinc, propiconazole, propiconazole + tridemorph, tebuconazole, tebuconazole + triadimenol, tebuconazole + tridemorph
	Stinking smut	bitertanol + fuberidazole, carboxin + thiabendazole, fludioxonil, fuberidazole + triadimenol, guazatine
	Stripe smut	fuberidazole + triadimenol
Pests	Aphids	alpha-cypermethrin, chlorpyrifos, cypermethrin, deltamethrin, demeton-S-methyl, dimethoate, esfenvalerate, heptenophos, lambda-cyhalothrin, pirimicarb
	Barley yellow dwarf virus vectors	bifenthrin, cypermethrin, deltamethrin, lambda-cyhalothrin
	Birds	aluminium ammonium sulfate
	Cutworms	gamma-HCH
	Damaging mammals	aluminium ammonium sulfate, sulfonated cod liver oil
	Frit fly	chlorpyrifos, fonofos, pirimiphos-methyl, triazophos
	Leatherjackets	chlorpyrifos, fenitrothion, gamma-HCH, methiocarb *(reduction)*, quinalphos, triazophos
	Saddle gall midge	fenitrothion
	Slugs and snails	metaldehyde, metaldehyde *(seed admixture)*, methiocarb, methiocarb *(seed admixture)*, thiodicarb
	Thrips	chlorpyrifos, fenitrothion
	Wheat bulb fly	chlorfenvinphos, chlorpyrifos, dimethoate, fonofos, pirimiphos-methyl
	Wheat-blossom midges	chlorpyrifos, fenitrothion, triazophos
	Wireworms	fonofos, gamma-HCH
	Yellow cereal fly	chlorfenvinphos, cypermethrin, deltamethrin
Weeds	Annual and perennial weeds	glyphosate (agricultural) *(pre-harvest)*

FOR FULL CONDITIONS OF USE ALWAYS READ THE PRODUCT LABEL

Annual dicotyledons	2,4-D, 2,4-DB + MCPA, MCPA, MCPA + MCPB, amidosulfuron, benazolin + bromoxynil + ioxynil, bifenox + dicamba, bromoxynil + clopyralid, bromoxynil + clopyralid + fluroxypyr, bromoxynil + fluroxypyr, bromoxynil + fluroxypyr + ioxynil, bromoxynil + ioxynil, bromoxynil + ioxynil + mecoprop-P, bromoxynil + ioxynil + triasulfuron, chlorotoluron, chlorotoluron + pendimethalin, clodinafop-propargyl + diflufenican, clopyralid, clopyralid + fluroxypyr + ioxynil, cyanazine, cyanazine + terbuthylazine, dicamba + MCPA + mecoprop, dicamba + MCPA + mecoprop-P, dicamba + mecoprop, dicamba + mecoprop-P, dichlorprop, dichlorprop + MCPA, diflufenican + isoproturon, diflufenican + terbuthylazine, diflufenican + trifluralin, fluoroglycofen-ethyl + mecoprop-P, fluroxypyr, glufosinate-ammonium *(pre-harvest)*, imazamethabenz-methyl, isoproturon, isoproturon + isoxaben, isoproturon + pendimethalin, isoproturon + simazine, isoproturon + trifluralin, linuron, linuron + trifluralin, mecoprop, mecoprop-P, methabenzthiazuron, metoxuron, metsulfuron-methyl, metsulfuron-methyl + thifensulfuron-methyl, pendimethalin, pendimethalin + simazine, pyridate, terbutryn, terbutryn + trifluralin, thifensulfuron-methyl + tribenuron-methyl, triasulfuron, tribenuron-methyl, trifluralin
Annual grasses	chlorotoluron, chlorotoluron + pendimethalin, cyanazine, diclofop-methyl, diflufenican + isoproturon, glufosinate-ammonium *(pre-harvest)*, isoproturon, isoproturon + isoxaben, isoproturon + pendimethalin, isoproturon + simazine, isoproturon + trifluralin, linuron + trifluralin, metoxuron, pendimethalin, terbutryn + trifluralin, trifluralin
Annual meadow grass	cyanazine + terbuthylazine, diflufenican + terbuthylazine, diflufenican + trifluralin, linuron, linuron + trifluralin, methabenzthiazuron, pendimethalin, terbutryn, terbutryn + trifluralin, tri-allate
Annual weeds	glyphosate (agricultural) *(pre-drilling/pre-emergence)*
Barren brome	metoxuron
Black bindweed	dichlorprop, dichlorprop + MCPA, fluroxypyr, linuron
Blackgrass	chlorotoluron, chlorotoluron + pendimethalin, clodinafop-propargyl, clodinafop-propargyl + diflufenican, clodinafop-propargyl + trifluralin, diclofop-methyl, diclofop-methyl + fenoxaprop-P-ethyl, diflufenican + isoproturon, fenoxaprop-P-ethyl, fenoxaprop-ethyl, imazamethabenz-methyl, isoproturon, isoproturon + isoxaben, isoproturon + pendimethalin, isoproturon + simazine, isoproturon + trifluralin, methabenzthiazuron, metoxuron, pendimethalin, terbutryn, terbutryn + trifluralin, tralkoxydim, tri-allate
Canary grass	diclofop-methyl, fenoxaprop-P-ethyl
Chickweed	bromoxynil + clopyralid + fluroxypyr, bromoxynil + fluroxypyr + ioxynil, clopyralid + fluroxypyr + ioxynil, dicamba + MCPA + mecoprop, dicamba + MCPA + mecoprop-P, dicamba + mecoprop, dicamba + mecoprop-P, fluroxypyr, linuron, mecoprop, mecoprop-P, metsulfuron-methyl, terbutryn + trifluralin, thifensulfuron-methyl + tribenuron-methyl, triasulfuron, tribenuron-methyl

Cleavers	amidosulfuron, bifenox + dicamba, bromoxynil + clopyralid + fluroxypyr, bromoxynil + fluroxypyr + ioxynil, clopyralid + fluroxypyr + ioxynil, dicamba + MCPA + mecoprop, dicamba + MCPA + mecoprop-P, dicamba + mecoprop, dicamba + mecoprop-P, fluoroglycofen-ethyl + mecoprop-P, fluroxypyr, mecoprop, mecoprop-P, metsulfuron-methyl + thifensulfuron-methyl, pendimethalin, pyridate
Corn marigold	clopyralid, linuron
Couch	amitrole *(direct-drilled)*, glyphosate (agricultural) *(pre-harvest)*
Creeping bent	glyphosate (agricultural) *(pre-harvest)*
Creeping thistle	clopyralid
Docks	dicamba + MCPA + mecoprop, fluroxypyr
Fat-hen	MCPA, linuron
Field pansy	bifenox + dicamba, metsulfuron-methyl + thifensulfuron-methyl
Hemp-nettle	MCPA, bromoxynil + clopyralid + fluroxypyr, bromoxynil + fluroxypyr + ioxynil, clopyralid + fluroxypyr + ioxynil, dichlorprop + MCPA, fluroxypyr
Loose silky bent	fenoxaprop-P-ethyl, imazamethabenz-methyl
Mayweeds	bromoxynil + clopyralid + fluroxypyr, clopyralid, clopyralid + fluroxypyr + ioxynil, dicamba + MCPA + mecoprop, dicamba + MCPA + mecoprop-P, dicamba + mecoprop, dicamba + mecoprop-P, metsulfuron-methyl, terbutryn + trifluralin, thifensulfuron-methyl + tribenuron-methyl, triasulfuron, tribenuron-methyl
Perennial dicotyledons	2,4-D, 2,4-DB + MCPA, MCPA, MCPA + MCPB, clopyralid, dicamba + MCPA + mecoprop, dicamba + MCPA + mecoprop-P, dicamba + mecoprop, dicamba + mecoprop-P, dichlorprop, dichlorprop + MCPA, glyphosate (agricultural) *(pre-harvest)*, mecoprop, mecoprop-P
Perennial grasses	diclofop-methyl, glyphosate (agricultural) *(pre-harvest)*, imazamethabenz-methyl, pendimethalin + simazine, tri-allate
Perennial ryegrass	linuron + trifluralin, terbutryn
Polygonums	2,4-DB + MCPA, bifenox + dicamba, dicamba + MCPA + mecoprop, dicamba + MCPA + mecoprop-P, dicamba + mecoprop, dicamba + mecoprop-P, dichlorprop, dichlorprop + MCPA, linuron, metsulfuron-methyl + thifensulfuron-methyl
Rough meadow grass	chlorotoluron, clodinafop-propargyl, clodinafop-propargyl + diflufenican, clodinafop-propargyl + trifluralin, diclofop-methyl, fenoxaprop-P-ethyl, isoproturon, linuron + trifluralin, methabenzthiazuron, terbutryn
Ryegrass	diclofop-methyl, diclofop-methyl + fenoxaprop-P-ethyl

FOR FULL CONDITIONS OF USE ALWAYS READ THE PRODUCT LABEL

	Speedwells	bifenox + dicamba, bromoxynil + clopyralid + fluroxypyr, bromoxynil + fluroxypyr + ioxynil, clopyralid + fluroxypyr + ioxynil, fluoroglycofen-ethyl + mecoprop-P, metsulfuron-methyl + thifensulfuron-methyl, pendimethalin, pyridate, terbutryn + trifluralin
	Timothy	fenoxaprop-P-ethyl
	Volunteer cereals	glyphosate (agricultural) *(pre-drilling/pre-emergence)*
	Volunteer oilseed rape	imazamethabenz-methyl
	Volunteer potatoes	amitrole *(stubble)*, fluroxypyr
	Wild oats	chlorotoluron, chlorotoluron + pendimethalin, clodinafop-propargyl, clodinafop-propargyl + diflufenican, clodinafop-propargyl + trifluralin, diclofop-methyl, diclofop-methyl + fenoxaprop-P-ethyl, difenzoquat, diflufenican + isoproturon, fenoxaprop-P-ethyl, fenoxaprop-ethyl, flamprop-M-isopropyl, glyphosate (agricultural) *(wiper glove)*, imazamethabenz-methyl, isoproturon, isoproturon + pendimethalin, pendimethalin, tralkoxydim, tri-allate
Crop control	Pre-harvest desiccation	diquat, glufosinate-ammonium, glyphosate (agricultural)
Plant growth regulation	Increasing yield	2-chloroethylphosphonic acid + mepiquat chloride *(low lodging situations)*, chlormequat, chlormequat + choline chloride, chlormequat + choline chloride + imazaquin, sulfur
	Lodging control	2-chloroethylphosphonic acid, 2-chloroethylphosphonic acid + mepiquat chloride, 2-chloroethylphosphonic acid + mepiquat chloride *(some cultivars only)*, chlormequat, chlormequat + 2-chloroethylphosphonic acid, chlormequat + choline chloride, chlormequat + choline chloride + imazaquin

Cereals undersown

Weeds	Annual dicotyledons	2,4-DB, 2,4-DB + MCPA, 2,4-DB + linuron + MCPA, MCPA + MCPB, MCPB, benazolin + 2,4-DB + MCPA, bentazone + MCPB, bentazone + cyanazine + 2,4-DB, bromoxynil + ioxynil, dichlorprop
	Black bindweed	dichlorprop
	Chickweed	benazolin + 2,4-DB + MCPA, bentazone + cyanazine + 2,4-DB
	Cleavers	benazolin + 2,4-DB + MCPA
	Perennial dicotyledons	2,4-DB, 2,4-DB + MCPA, MCPA + MCPB, MCPB, dichlorprop
	Polygonums	2,4-DB + MCPA, benazolin + 2,4-DB + MCPA, dichlorprop

Clovers

Weeds	Annual dicotyledons	2,4-DB + MCPA, 2,4-DB + linuron + MCPA, MCPB, benazolin + 2,4-DB + MCPA, bentazone + cyanazine + 2,4-DB, carbetamide, propyzamide

	Annual grasses	carbetamide, propyzamide
	Annual meadow grass	tri-allate
	Blackgrass	sethoxydim *(off-label)*, tri-allate
	Chickweed	benazolin + 2,4-DB + MCPA, bentazone + cyanazine + 2,4-DB
	Cleavers	benazolin + 2,4-DB + MCPA
	Docks	asulam *(off-label)*
	Perennial dicotyledons	2,4-DB + MCPA, MCPB
	Perennial grasses	propyzamide
	Polygonums	2,4-DB + MCPA, benazolin + 2,4-DB + MCPA
	Volunteer cereals	carbetamide, sethoxydim *(off-label)*
	Wild oats	sethoxydim *(off-label)*, tri-allate
Crop control	Pre-harvest desiccation	diquat

Fodder brassica seed crops

Diseases	Alternaria	iprodione
	Botrytis	iprodione
Pests	Aphids	dimethoate *(excluding Myzus persicae)*
Weeds	Annual dicotyledons	benazolin + clopyralid *(off-label)*, carbetamide
	Annual grasses	carbetamide
	Chickweed	benazolin + clopyralid *(off-label)*
	Cleavers	benazolin + clopyralid *(off-label)*
	Mayweeds	benazolin + clopyralid *(off-label)*
	Volunteer cereals	carbetamide

Fodder brassicas

Diseases	Alternaria	iprodione
	Damping off	gamma-HCH + thiram, thiabendazole + thiram
	Seed-borne diseases	thiabendazole + thiram
Pests	Aphids	alpha-cypermethrin, bifenthrin, chlorpyrifos, chlorpyrifos + dimethoate, cypermethrin, malathion, pirimicarb
	Cabbage root fly	carbofuran, chlorpyrifos, chlorpyrifos + dimethoate
	Cabbage stem flea beetle	cypermethrin
	Cabbage stem weevil	carbofuran

FOR FULL CONDITIONS OF USE ALWAYS READ THE PRODUCT LABEL

	Caterpillars	alpha-cypermethrin, bifenthrin, carbofuran, chlorpyrifos, cypermethrin, deltamethrin, esfenvalerate
	Cutworms	chlorpyrifos, chlorpyrifos + dimethoate
	Flea beetles	alpha-cypermethrin, carbofuran, gamma-HCH + thiram
	Leatherjackets	chlorpyrifos, chlorpyrifos + dimethoate
	Whitefly	bifenthrin
	Wireworms	chlorpyrifos + dimethoate
Weeds	Annual dicotyledons	clopyralid, desmetryn, desmetryn *(off-label)*, propachlor, sodium monochloroacetate, tebutam, trifluralin
	Annual grasses	fluazifop-P-butyl *(stockfeed only)*, propachlor, tebutam, trifluralin
	Annual meadow grass	tri-allate
	Blackgrass	tri-allate
	Corn marigold	clopyralid
	Creeping thistle	clopyralid
	Fat-hen	desmetryn, desmetryn *(off-label)*
	Mayweeds	clopyralid
	Perennial dicotyledons	clopyralid
	Perennial grasses	fluazifop-P-butyl *(stockfeed only)*
	Volunteer cereals	fluazifop-P-butyl *(stockfeed only)*, tebutam
	Wild oats	fluazifop-P-butyl *(stockfeed only)*, tri-allate

General

Diseases	Soil-borne diseases	dazomet, methyl bromide with amyl acetate, methyl bromide with chloropicrin
Pests	Birds	aluminium ammonium sulfate, quinalbarbitone-sodium, ziram
	Damaging mammals	aluminium ammonium sulfate, ziram
	Nematodes	dazomet, methyl bromide with amyl acetate, methyl bromide with chloropicrin
	Slugs and snails	aluminium sulfate, metaldehyde, methiocarb
	Soil pests	dazomet, methyl bromide with amyl acetate, methyl bromide with chloropicrin
Weeds	Annual and perennial weeds	glyphosate (agricultural), glyphosate (agricultural) *(stubble treatment)*, glyphosate (agricultural) *(sward destruction/direct drilling)*
	Annual dicotyledons	diquat + paraquat *(autumn stubble, pre-planting/sowing, sward destruction)*, glufosinate-ammonium *(pre-cropping situations)*, glufosinate-ammonium *(sward destruction)*, paraquat *(minimum cultivation)*, paraquat *(stubble treatment)*

83

Annual grasses	diquat + paraquat *(autumn stubble, pre-planting/sowing, sward destruction)*, fluazifop-P-butyl *(off-label)*, glufosinate-ammonium *(pre-cropping situations)*, glufosinate-ammonium *(sward destruction)*, glyphosate (agricultural) *(stubble treatment)*, paraquat *(minimum cultivation)*, paraquat *(stubble treatment)*, paraquat *(sward destruction/direct drilling)*
Annual weeds	amitrole, amitrole *(pre-sowing, autumn stubble)*, glufosinate-ammonium, glyphosate (agricultural) *(pre-drilling/pre-emergence)*
Barren brome	fluazifop-P-butyl *(off-label)*, paraquat *(stubble treatment)*
Couch	amitrole, amitrole *(pre-sowing, autumn stubble)*, glyphosate (agricultural) *(stubble treatment)*, glyphosate (agricultural) *(sward destruction/direct drilling)*
Creeping bent	glyphosate (agricultural) *(stubble treatment)*, paraquat *(minimum cultivation)*, paraquat *(stubble treatment)*, paraquat *(sward destruction/direct drilling)*
Docks	amitrole, amitrole *(pre-sowing, autumn stubble)*
General weed control	dazomet, methyl bromide with amyl acetate, methyl bromide with chloropicrin
Perennial dicotyledons	glufosinate-ammonium *(pre-cropping situations)*, glufosinate-ammonium *(sward destruction)*, glyphosate (agricultural) *(stubble treatment)*
Perennial grasses	diquat + paraquat *(autumn stubble, pre-planting/sowing, sward destruction)*, glufosinate-ammonium *(pre-cropping situations)*, glufosinate-ammonium *(sward destruction)*, glyphosate (agricultural) *(stubble treatment)*
Perennial ryegrass	paraquat *(sward destruction/direct drilling)*
Perennial weeds	amitrole, amitrole *(pre-sowing, autumn stubble)*, glufosinate-ammonium
Rough meadow grass	paraquat *(sward destruction/direct drilling)*
Volunteer cereals	diquat + paraquat *(autumn stubble, pre-planting/sowing, sward destruction)*, fluazifop-P-butyl *(off-label)*, glyphosate (agricultural) *(pre-drilling/pre-emergence)*, glyphosate (agricultural) *(stubble treatment)*, glyphosate (agricultural) *(sward destruction/direct drilling)*, paraquat *(minimum cultivation)*, paraquat *(stubble treatment)*
Volunteer potatoes	glyphosate (agricultural) *(stubble treatment)*
Wild oats	paraquat *(stubble treatment)*

Grass seed crops

Diseases	Crown rust	propiconazole
	Drechslera leaf spot	propiconazole

FOR FULL CONDITIONS OF USE ALWAYS READ THE PRODUCT LABEL

	Powdery mildew	propiconazole
	Rhynchosporium	propiconazole
Pests	Aphids	deltamethrin *(off-label)*, demeton-S-methyl, dimethoate
Weeds	Annual dicotyledons	2,4-D, 2,4-D + mecoprop, MCPA, bromoxynil + ethofumesate + ioxynil, clopyralid *(off-label)*, dicamba + MCPA + mecoprop, mecoprop-P, methabenzthiazuron
	Annual grasses	ethofumesate
	Annual meadow grass	bromoxynil + ethofumesate + ioxynil, methabenzthiazuron
	Blackgrass	ethofumesate, sethoxydim *(off-label)*
	Chickweed	dicamba + MCPA + mecoprop, ethofumesate, mecoprop-P
	Cleavers	dicamba + MCPA + mecoprop, ethofumesate, mecoprop-P
	Docks	dicamba + MCPA + mecoprop
	Fat-hen	MCPA
	Hemp-nettle	MCPA
	Mayweeds	dicamba + MCPA + mecoprop
	Perennial dicotyledons	2,4-D, 2,4-D + mecoprop, MCPA, clopyralid *(off-label)*, dicamba + MCPA + mecoprop, mecoprop-P
	Polygonums	dicamba + MCPA + mecoprop
	Rough meadow grass	methabenzthiazuron
	Ryegrass	chlorpropham *(off-label)*
	Volunteer cereals	ethofumesate, sethoxydim *(off-label)*
	Wild oats	difenzoquat, sethoxydim *(off-label)*

Linseed/flax

Diseases	Alternaria	iprodione *(seed treatment)*
	Damping off	thiabendazole + thiram
	Seed-borne diseases	thiabendazole + thiram
Weeds	Annual and perennial weeds	glyphosate (agricultural) *(pre-harvest)*
	Annual dicotyledons	bentazone, bromoxynil + clopyralid, clopyralid, cyanazine, glufosinate-ammonium *(pre-harvest)*, metsulfuron-methyl, trifluralin, trifluralin *(off-label)*
	Annual grasses	cyanazine, cycloxydim, diclofop-methyl, glufosinate-ammonium *(pre-harvest)*, propaquizafop, sethoxydim, trifluralin, trifluralin *(off-label)*
	Annual meadow grass	tri-allate
	Blackgrass	cycloxydim, diclofop-methyl, tri-allate
	Canary grass	diclofop-methyl
	Chickweed	metsulfuron-methyl

	Corn marigold	clopyralid
	Couch	cycloxydim, glyphosate (agricultural) *(pre-harvest)*
	Creeping bent	cycloxydim
	Creeping thistle	clopyralid
	Mayweeds	clopyralid, metsulfuron-methyl
	Perennial dicotyledons	clopyralid
	Perennial grasses	cycloxydim, diclofop-methyl, propaquizafop
	Rough meadow grass	diclofop-methyl
	Ryegrass	diclofop-methyl
	Volunteer cereals	cycloxydim, sethoxydim
	Wild oats	cycloxydim, diclofop-methyl, tri-allate
Crop control	Pre-harvest desiccation	diquat, glufosinate-ammonium, sulfuric acid (commodity substance)
Plant growth regulation	Lodging control	chlormequat + choline chloride
	Stem shortening	chlormequat + choline chloride

Lucerne

Pests	Weevils	deltamethrin *(off-label)*
Weeds	Annual dicotyledons	2,4-DB, carbetamide, chlorpropham, paraquat *(off-label)*, propyzamide
	Annual grasses	carbetamide, chlorpropham, diclofop-methyl, paraquat *(off-label)*, propyzamide
	Annual meadow grass	tri-allate
	Blackgrass	diclofop-methyl, sethoxydim *(off-label)*, tri-allate
	Canary grass	diclofop-methyl
	Chickweed	chlorpropham
	Perennial dicotyledons	2,4-DB
	Perennial grasses	diclofop-methyl, propyzamide
	Polygonums	chlorpropham
	Rough meadow grass	diclofop-methyl
	Ryegrass	diclofop-methyl
	Volunteer cereals	carbetamide, sethoxydim *(off-label)*
	Wild oats	diclofop-methyl, sethoxydim *(off-label)*, tri-allate

FOR FULL CONDITIONS OF USE ALWAYS READ THE PRODUCT LABEL

Lupins

Diseases	Damping off	thiabendazole + thiram
	Seed-borne diseases	thiabendazole + thiram
Weeds	Annual dicotyledons	terbuthylazine + terbutryn
	Annual grasses	diclofop-methyl, sethoxydim, terbuthylazine + terbutryn
	Blackgrass	diclofop-methyl
	Canary grass	diclofop-methyl
	Perennial grasses	diclofop-methyl
	Rough meadow grass	diclofop-methyl
	Ryegrass	diclofop-methyl
	Volunteer cereals	sethoxydim
	Wild oats	diclofop-methyl
Crop control	Pre-harvest desiccation	diquat

Maize/sweetcorn

Diseases	Damping off	thiram *(seed treatment)*
Pests	Aphids	pirimicarb
	Cutworms	gamma-HCH
	Frit fly	aldicarb *(off-label)*, bendiocarb, carbofuran, chlorfenvinphos, chlorpyrifos, fenitrothion, phorate, triazophos
	General insect control	dimethoate *(off-label)*
	Leatherjackets	gamma-HCH
	Thrips	fenitrothion *(off-label)*
	Wireworms	bendiocarb, gamma-HCH
Weeds	Annual dicotyledons	atrazine, bromoxynil, clopyralid, cyanazine, fluroxypyr, pendimethalin, pyridate, simazine
	Annual grasses	atrazine, cyanazine, pendimethalin *(off-label)*, simazine
	Annual meadow grass	pendimethalin, tri-allate
	Black bindweed	fluroxypyr
	Blackgrass	tri-allate
	Chickweed	fluroxypyr
	Cleavers	fluroxypyr, pyridate
	Corn marigold	clopyralid
	Creeping thistle	clopyralid
	Docks	fluroxypyr
	Hemp-nettle	fluroxypyr
	Mayweeds	clopyralid

Perennial dicotyledons	clopyralid
Speedwells	pendimethalin, pyridate
Volunteer potatoes	fluroxypyr
Wild oats	difenzoquat, tri-allate

Miscellaneous field crops

Diseases	Fungus diseases	iprodione *(off-label)*, propiconazole *(off-label)*
Pests	Aphids	pirimicarb *(off-label)*
	Flea beetles	deltamethrin *(off-label)*
	Leaf miners	oxamyl *(off-label)*
	Pollen beetles	deltamethrin *(off-label)*
Weeds	Annual dicotyledons	MCPA + MCPB, benazolin + clopyralid *(off-label)*, bentazone *(off-label)*, carbetamide, clopyralid + propyzamide *(off-label)*, cyanazine *(off-label)*, metazachlor *(off-label)*, pendimethalin *(off-label)*, propyzamide *(off-label)*, trifluralin *(off-label)*
	Annual grasses	carbetamide, clopyralid + propyzamide *(off-label)*, pendimethalin *(off-label)*, propyzamide *(off-label)*, trifluralin *(off-label)*
	Annual meadow grass	tri-allate
	Blackgrass	sethoxydim *(off-label)*, tri-allate
	Chickweed	benazolin + clopyralid *(off-label)*
	Cleavers	benazolin + clopyralid *(off-label)*
	Mayweeds	benazolin + clopyralid *(off-label)*, clopyralid + propyzamide *(off-label)*, clopyralid *(off-label)*
	Perennial dicotyledons	MCPA + MCPB, clopyralid *(off-label)*
	Perennial grasses	propyzamide *(off-label)*, tri-allate
	Volunteer cereals	carbetamide, sethoxydim *(off-label)*
	Wild oats	sethoxydim *(off-label)*, tri-allate
Crop control	Pre-harvest desiccation	diquat, diquat *(off-label)*

Oilseed rape

Diseases	Alternaria	carbendazim + iprodione, carbendazim + prochloraz, carbendazim + vinclozolin *(reduction)*, fenpropimorph + gamma-HCH + thiram, iprodione, iprodione + thiophanate-methyl, iprodione *(seed treatment)*, maneb + zinc, prochloraz, propiconazole, tebuconazole, vinclozolin

FOR FULL CONDITIONS OF USE ALWAYS READ THE PRODUCT LABEL

	Black scurf and stem canker	iprodione + thiophanate-methyl
	Botrytis	benomyl, carbendazim (MBC), carbendazim + iprodione, carbendazim + maneb, carbendazim + prochloraz, carbendazim + vinclozolin *(reduction)*, chlorothalonil, iprodione, iprodione + thiophanate-methyl, prochloraz, vinclozolin
	Canker	carbendazim + flusilazole, carbendazim + prochloraz, carboxin + gamma-HCH + thiram, fenpropimorph + gamma-HCH + thiram, gamma-HCH + thiabendazole + thiram, prochloraz
	Damping off	carboxin + gamma-HCH + thiram, fenpropimorph + gamma-HCH + thiram, gamma-HCH + thiabendazole + thiram, gamma-HCH + thiram, thiabendazole + thiram, thiram *(seed treatment)*
	Downy mildew	benalaxyl + mancozeb, carbendazim + mancozeb, carbendazim + maneb, carbendazim + maneb + sulfur, chlorothalonil, chlorothalonil + metalaxyl, mancozeb, maneb + zinc
	Light leaf spot	benomyl, carbendazim (MBC), carbendazim + flusilazole, carbendazim + flusilazole *(reduction)*, carbendazim + iprodione, carbendazim + maneb, carbendazim + maneb + sulfur, carbendazim + prochloraz, carbendazim + vinclozolin *(reduction)*, flusilazole, iprodione + thiophanate-methyl, prochloraz, propiconazole, tebuconazole, tebuconazole *(reduction)*
	Phoma leaf spot	carbendazim + flusilazole
	Ring spot	tebuconazole *(reduction)*
	Sclerotinia	carbendazim + vinclozolin, tebuconazole *(reduction)*
	Sclerotinia stem rot	carbendazim + iprodione, carbendazim + prochloraz, iprodione, iprodione + thiophanate-methyl, prochloraz, tebuconazole, vinclozolin
	Seed-borne diseases	thiabendazole + thiram
	White leaf spot	carbendazim + prochloraz, prochloraz
Pests	Aphids	chlorpyrifos + dimethoate, deltamethrin, lambda-cyhalothrin, pirimicarb
	Birds	aluminium ammonium sulfate
	Cabbage root fly	carbofuran, chlorpyrifos + dimethoate, phorate
	Cabbage seed weevil	alpha-cypermethrin, deltamethrin, malathion
	Cabbage stem flea beetle	alpha-cypermethrin, bifenthrin, carbofuran, cypermethrin, deltamethrin, gamma-HCH, lambda-cyhalothrin, phorate, pirimiphos-methyl
	Cabbage stem weevil	deltamethrin, gamma-HCH, malathion
	Cutworms	chlorpyrifos + dimethoate
	Damaging mammals	aluminium ammonium sulfate, sulfonated cod liver oil
	Flea beetles	carboxin + gamma-HCH + thiram, esfenvalerate, fenpropimorph + gamma-HCH + thiram, gamma-HCH + thiabendazole + thiram, gamma-HCH + thiram

	Leatherjackets	chlorpyrifos + dimethoate
	Millipedes	gamma-HCH + thiabendazole + thiram
	Pod midge	alpha-cypermethrin, cypermethrin *(restricted area)*, deltamethrin, endosulfan, triazophos
	Pollen beetles	alpha-cypermethrin, cypermethrin *(restricted area)*, deltamethrin, endosulfan, esfenvalerate, gamma-HCH, malathion
	Rape winter stem weevil	alpha-cypermethrin, bifenthrin, carbofuran, cypermethrin, deltamethrin, phorate
	Seed weevil	alpha-cypermethrin, cypermethrin *(restricted area)*, endosulfan, gamma-HCH, triazophos
	Slugs and snails	metaldehyde, methiocarb, thiodicarb
	Virus vectors	deltamethrin, lambda-cyhalothrin
	Weevils	gamma-HCH
	Wireworms	chlorpyrifos + dimethoate
Weeds	Annual and perennial weeds	glyphosate (agricultural) *(pre-harvest)*
	Annual dicotyledons	benazolin + clopyralid, benazolin + clopyralid *(off-label)*, carbetamide, chlorthal-dimethyl, clopyralid, clopyralid + propyzamide, cyanazine, glufosinate-ammonium *(pre-harvest)*, metazachlor, napropamide, propachlor, propyzamide, pyridate, tebutam, trifluralin
	Annual grasses	carbetamide, clopyralid + propyzamide, cyanazine, cycloxydim, diclofop-methyl, fluazifop-P-butyl, glufosinate-ammonium *(pre-harvest)*, napropamide, propachlor, propaquizafop, propyzamide, quizalofop-ethyl, sethoxydim, tebutam, trifluralin
	Annual meadow grass	metazachlor, tri-allate
	Barren brome	clopyralid + propyzamide
	Blackgrass	cycloxydim, diclofop-methyl, metazachlor, tri-allate
	Canary grass	diclofop-methyl
	Chickweed	benazolin + clopyralid
	Cleavers	benazolin + clopyralid, napropamide, pyridate
	Corn marigold	clopyralid
	Couch	cycloxydim, glyphosate (agricultural) *(pre-harvest)*, quizalofop-ethyl
	Creeping bent	cycloxydim, glyphosate (agricultural) *(pre-harvest)*
	Creeping thistle	clopyralid
	Groundsel	napropamide

FOR FULL CONDITIONS OF USE ALWAYS READ THE PRODUCT LABEL

	Groundsel, triazine resistant	metazachlor
	Mayweeds	benazolin + clopyralid, clopyralid, clopyralid + propyzamide
	Perennial dicotyledons	clopyralid, glyphosate (agricultural) *(pre-harvest)*
	Perennial grasses	cycloxydim, diclofop-methyl, fluazifop-P-butyl, glyphosate (agricultural) *(pre-harvest)*, propaquizafop, propyzamide, quizalofop-ethyl
	Rough meadow grass	diclofop-methyl
	Ryegrass	diclofop-methyl
	Speedwells	pyridate
	Volunteer cereals	carbetamide, cycloxydim, fluazifop-P-butyl, propyzamide, quizalofop-ethyl, sethoxydim, tebutam
	Wild oats	cycloxydim, diclofop-methyl, fluazifop-P-butyl, propyzamide, tri-allate
Crop control	Pre-harvest desiccation	diquat, glufosinate-ammonium, glyphosate (agricultural)
Plant growth regulation	Increasing yield	chlormequat with di-1-p-menthene, sulfur
	Lodging control	chlormequat + choline chloride, chlormequat with di-1-p-menthene
	Stem shortening	chlormequat + choline chloride

Peas

Diseases	Ascochyta	benomyl *(seed treatment)*, carbendazim + chlorothalonil, carbendazim + iprodione, chlorothalonil, fosetyl-aluminium *(off-label)*, iprodione + thiophanate-methyl, metalaxyl + thiabendazole + thiram *(seed treatment)*, thiabendazole + thiram, vinclozolin
	Botrytis	benomyl, carbendazim + chlorothalonil, carbendazim + iprodione, chlorothalonil, iprodione + thiophanate-methyl, vinclozolin
	Damping off	metalaxyl + thiabendazole + thiram *(seed treatment)*, thiabendazole + thiram, thiram *(seed treatment)*
	Downy mildew	carbendazim + chlorothalonil, fosetyl-aluminium *(off-label)*, metalaxyl + thiabendazole + thiram *(seed treatment)*
	Mycosphaerella	carbendazim + iprodione, chlorothalonil, vinclozolin
	Sclerotinia stem rot	carbendazim + iprodione
	Stem rot	iprodione + thiophanate-methyl
Pests	Aphids	bifenthrin, cypermethrin, deltamethrin, deltamethrin + heptenophos, demeton-S-methyl, dimethoate, disulfoton, fatty acids, fenitrothion, heptenophos, malathion, phorate, pirimicarb, triazophos
	Birds	aluminium ammonium sulfate

	Capsids	phorate
	Damaging mammals	aluminium ammonium sulfate
	Leatherjackets	chlorpyrifos
	Mealy bugs	fatty acids
	Pea and bean weevils	cypermethrin, deltamethrin, esfenvalerate, fenitrothion, lambda-cyhalothrin, oxamyl, phorate, triazophos
	Pea cyst nematode	oxamyl
	Pea midge	demeton-S-methyl, dimethoate, disulfoton, fenitrothion, triazophos
	Pea moth	bifenthrin, cypermethrin, deltamethrin, deltamethrin + heptenophos, esfenvalerate, fenitrothion, lambda-cyhalothrin, triazophos
	Red spider mites	fatty acids
	Scale insects	fatty acids
	Thrips	dimethoate, disulfoton, fenitrothion
	Whitefly	fatty acids
Weeds	Annual and perennial weeds	glyphosate (agricultural) *(pre-harvest)*
	Annual dicotyledons	MCPA + MCPB, MCPB, bentazone, bentazone + MCPB, chlorpropham + fenuron, cyanazine, fomesafen + terbutryn *(spring sown)*, glufosinate-ammonium *(pre-harvest)*, isoxaben + terbuthylazine, pendimethalin, pendimethalin + prometryn, prometryn, prometryn + terbutryn, simazine + trietazine, terbuthylazine + terbutryn, terbutryn + trietazine, trifluralin *(off-label)*
	Annual grasses	chlorpropham + fenuron, cyanazine, cycloxydim, diclofop-methyl, fluazifop-P-butyl, glufosinate-ammonium *(pre-harvest)*, pendimethalin, prometryn, prometryn + terbutryn, propaquizafop, sethoxydim, terbuthylazine + terbutryn, trifluralin *(off-label)*
	Annual meadow grass	pendimethalin, pendimethalin + prometryn, terbutryn + trietazine, tri-allate
	Annual weeds	glyphosate (agricultural) *(pre-drilling/pre-emergence)*
	Blackgrass	cycloxydim, diclofop-methyl, pendimethalin, tri-allate
	Canary grass	diclofop-methyl
	Chickweed	chlorpropham + fenuron
	Cleavers	pendimethalin
	Couch	cycloxydim, glyphosate (agricultural) *(pre-harvest)*
	Creeping bent	cycloxydim, glyphosate (agricultural) *(pre-harvest)*
	Perennial dicotyledons	MCPA + MCPB, MCPB, glyphosate (agricultural) *(pre-harvest)*

FOR FULL CONDITIONS OF USE ALWAYS READ THE PRODUCT LABEL

	Perennial grasses	cycloxydim, diclofop-methyl, fluazifop-P-butyl, glyphosate (agricultural) *(pre-harvest)*, propaquizafop, tri-allate
	Rough meadow grass	diclofop-methyl
	Ryegrass	diclofop-methyl
	Speedwells	pendimethalin
	Volunteer cereals	cycloxydim, fluazifop-P-butyl, glyphosate (agricultural) *(pre-drilling/pre-emergence)*, sethoxydim
	Wild oats	cycloxydim, diclofop-methyl, fluazifop-P-butyl, pendimethalin, tri-allate
Crop control	Pre-harvest desiccation	diquat, glufosinate-ammonium
Plant growth regulation	Increasing yield	chlormequat with di-1-p-menthene
	Lodging control	chlormequat with di-1-p-menthene

Potatoes

Diseases	Black leg	dichlorophen *(in store)*
	Black scurf and stem canker	imazalil + pencycuron, imazalil + pencycuron *(reduction)*, imazalil + thiabendazole *(reduction)*, iprodione *(seed treatment)*, pencycuron, prochloraz + tolclofos-methyl, thiabendazole *(pre-planting)*, tolclofos-methyl
	Blight	Bordeaux mixture, benalaxyl + mancozeb, chlorothalonil, chlorothalonil + cymoxanil, chlorothalonil + propamocarb hydrochloride, copper oxychloride, copper sulfate + sulfur, cymoxanil + mancozeb, cymoxanil + mancozeb + oxadixyl, dimethomorph + mancozeb, fentin acetate + maneb, fentin hydroxide, ferbam + maneb + zineb, fluazinam, mancozeb, mancozeb + metalaxyl, mancozeb + ofurace, mancozeb + oxadixyl, mancozeb + propamocarb hydrochloride, maneb, maneb + zinc, maneb + zinc oxide, zineb
	Dry rot	carbendazim + tecnazene, imazalil, imazalil + thiabendazole *(reduction)*, tecnazene, tecnazene + thiabendazole, thiabendazole *(post-harvest)*
	Gangrene	2-aminobutane, carbendazim + tecnazene, dichlorophen *(in store)*, imazalil, imazalil + thiabendazole *(reduction)*, prochloraz + tolclofos-methyl *(reduction)*, tecnazene + thiabendazole, thiabendazole *(post-harvest)*
	Powdery scab	maneb + zinc oxide
	Rhizoctonia	tolclofos-methyl *(off-label)*
	Silver scurf	2-aminobutane, carbendazim + tecnazene, dichlorophen *(in store)*, imazalil, imazalil + pencycuron *(reduction)*, imazalil + thiabendazole *(reduction)*, prochloraz + tolclofos-methyl *(reduction)*, tecnazene + thiabendazole, thiabendazole *(post-harvest)*

	Skin spot	2-aminobutane, carbendazim + tecnazene, imazalil, imazalil + thiabendazole *(reduction)*, prochloraz + tolclofos-methyl *(reduction)*, tecnazene + thiabendazole, thiabendazole *(post-harvest)*, thiabendazole *(pre-planting)*
	Soft rot	dichlorophen *(in store)*
Pests	Aphids	aldicarb, deltamethrin + heptenophos, demeton-S-methyl, dimethoate *(excluding Myzus persicae)*, disulfoton, malathion, oxamyl, phorate, pirimicarb
	Capsids	phorate
	Colorado beetle	chlorfenvinphos *(off-label)*, chlorpyrifos *(off-label)*, deltamethrin *(off-label)*, endosulfan *(off-label)*, permethrin *(off-label)*, triazophos *(off-label)*
	Cutworms	chlorpyrifos, cypermethrin, triazophos
	Leaf miners	cypermethrin *(protected crops, off-label)*, triazophos *(off-label)*
	Leaf roll virus vectors	deltamethrin + heptenophos
	Leafhoppers	phorate
	Nematodes	1,3-dichloropropene, aldicarb, oxamyl
	Potato cyst nematode	1,3-dichloropropene, aldicarb, carbofuran, ethoprophos, oxamyl
	Potato virus vectors	deltamethrin + heptenophos, nicotine
	Slugs and snails	metaldehyde, methiocarb
	Spraing vectors	oxamyl
	Wireworms	ethoprophos, phorate
Weeds	Annual dicotyledons	bentazone, cyanazine, diquat + paraquat, glufosinate-ammonium *(pre-emergence)*, glufosinate-ammonium *(pre-harvest)*, linuron, metribuzin, monolinuron, monolinuron + paraquat, paraquat, pendimethalin, pendimethalin + prometryn, prometryn, prometryn + terbutryn, terbuthylazine + terbutryn, terbutryn + trietazine
	Annual grasses	cyanazine, cycloxydim, diclofop-methyl, diquat + paraquat, glufosinate-ammonium *(pre-emergence)*, glufosinate-ammonium *(pre-harvest)*, metribuzin, monolinuron, monolinuron + paraquat, paraquat, pendimethalin, prometryn, prometryn + terbutryn, terbuthylazine + terbutryn
	Annual meadow grass	linuron, monolinuron, monolinuron + paraquat, pendimethalin, pendimethalin + prometryn, terbutryn + trietazine
	Black bindweed	linuron
	Blackgrass	cycloxydim, diclofop-methyl, monolinuron + paraquat, pendimethalin
	Canary grass	diclofop-methyl
	Chickweed	linuron

FOR FULL CONDITIONS OF USE ALWAYS READ THE PRODUCT LABEL

	Cleavers	pendimethalin
	Corn marigold	linuron
	Couch	cycloxydim, sethoxydim
	Creeping bent	cycloxydim, sethoxydim
	Fat-hen	linuron, monolinuron
	Perennial grasses	cycloxydim, diclofop-methyl, diquat + paraquat, sethoxydim
	Polygonums	linuron, monolinuron
	Rough meadow grass	diclofop-methyl
	Ryegrass	diclofop-methyl
	Speedwells	pendimethalin
	Volunteer cereals	cycloxydim, diquat + paraquat, paraquat, sethoxydim
	Volunteer oilseed rape	metribuzin
	Wild oats	cycloxydim, diclofop-methyl, monolinuron + paraquat, pendimethalin, sethoxydim
Crop control	Haulm destruction	sulfuric acid (commodity substance)
	Pre-harvest desiccation	diquat, glufosinate-ammonium
Plant growth regulation	Increasing yield	sulfur
	Sprout suppression	carbendazim + tecnazene, chlorpropham, chlorpropham + propham, maleic hydrazide, tecnazene, tecnazene + thiabendazole
	Volunteer suppression	maleic hydrazide

Seed brassicas/mustard

Diseases	Alternaria	iprodione
	Botrytis	iprodione
Pests	Cabbage seed weevil	deltamethrin, malathion
	Cabbage stem weevil	deltamethrin, malathion
	Pod midge	deltamethrin, endosulfan
	Pollen beetles	deltamethrin, endosulfan, malathion
	Seed weevil	endosulfan
Weeds	Annual and perennial weeds	glyphosate (agricultural) *(pre-harvest)*
	Annual dicotyledons	benazolin + clopyralid *(off-label)*, chlorthal-dimethyl, chlorthal-dimethyl + propachlor, propachlor, propyzamide, trifluralin
	Annual grasses	chlorthal-dimethyl + propachlor, diclofop-methyl, propachlor, propaquizafop, propyzamide, quizalofop-ethyl, sethoxydim, trifluralin

	Annual meadow grass	tri-allate
	Blackgrass	diclofop-methyl, tri-allate
	Canary grass	diclofop-methyl
	Chickweed	benazolin + clopyralid *(off-label)*
	Cleavers	benazolin + clopyralid *(off-label)*
	Couch	glyphosate (agricultural) *(pre-harvest)*, quizalofop-ethyl
	Mayweeds	benazolin + clopyralid *(off-label)*
	Perennial grasses	diclofop-methyl, propaquizafop, propyzamide, quizalofop-ethyl
	Rough meadow grass	diclofop-methyl
	Ryegrass	diclofop-methyl
	Volunteer cereals	quizalofop-ethyl, sethoxydim
	Wild oats	diclofop-methyl, tri-allate
Crop control	Pre-harvest desiccation	diquat *(off-label)*, glyphosate (agricultural)

Field vegetables

Asparagus

Diseases	Fungus diseases	thiabendazole *(off-label)*
Pests	Asparagus beetle	cypermethrin *(off-label)*
	Leaf miners	cypermethrin *(protected crops, off-label)*, triazophos *(off-label)*
Weeds	Annual and perennial weeds	glyphosate (agricultural) *(off-label)*
	Annual dicotyledons	MCPA, diquat *(off-label)*, simazine, terbacil
	Annual grasses	simazine, terbacil
	Couch	terbacil
	Perennial dicotyledons	MCPA
	Perennial grasses	terbacil

Beans, french/runner

Diseases	Anthracnose	carbendazim (MBC)
	Ascochyta	metalaxyl + thiabendazole + thiram *(seed treatment)*
	Botrytis	carbendazim (MBC), thiophanate-methyl, vinclozolin
	Damping off	metalaxyl + thiabendazole + thiram *(seed treatment)*, thiram *(seed treatment)*

FOR FULL CONDITIONS OF USE ALWAYS READ THE PRODUCT LABEL

	Downy mildew	metalaxyl + thiabendazole + thiram *(seed treatment)*
Pests	Aphids	demeton-S-methyl, dimethoate, disulfoton, heptenophos, malathion, nicotine, pirimicarb
	Caterpillars	Bacillus thuringiensis *(off-label)*, cypermethrin, lambda-cyhalothrin *(off-label)*
	General insect control	nicotine *(off-label)*
	Red spider mites	tetradifon
	Whitefly	cypermethrin
Weeds	Annual dicotyledons	bentazone, chlorthal-dimethyl, diquat *(off-label)*, monolinuron, monolinuron + paraquat, simazine *(off-label)*, trifluralin
	Annual grasses	cycloxydim, diclofop-methyl, monolinuron, monolinuron + paraquat, simazine *(off-label)*, trifluralin
	Annual meadow grass	monolinuron, monolinuron + paraquat, tri-allate
	Blackgrass	cycloxydim, diclofop-methyl, monolinuron + paraquat, tri-allate
	Canary grass	diclofop-methyl
	Couch	cycloxydim
	Creeping bent	cycloxydim
	Fat-hen	monolinuron
	Perennial grasses	cycloxydim, diclofop-methyl
	Polygonums	monolinuron
	Rough meadow grass	diclofop-methyl
	Ryegrass	diclofop-methyl
	Volunteer cereals	cycloxydim
	Wild oats	cycloxydim, diclofop-methyl, monolinuron + paraquat, tri-allate
Crop control	Pre-harvest desiccation	diquat *(off-label)*

Brassica seed crops

Diseases	Alternaria	iprodione
	Botrytis	iprodione
Pests	Cabbage stem flea beetle	gamma-HCH
	Cabbage stem weevil	gamma-HCH
	Pollen beetles	gamma-HCH
	Seed weevil	gamma-HCH
Weeds	Annual dicotyledons	carbetamide, propyzamide, trifluralin *(off-label)*
	Annual grasses	carbetamide, propyzamide, trifluralin *(off-label)*
	Volunteer cereals	carbetamide

Brassicas

Diseases	Alternaria	chlorothalonil, fenpropimorph, iprodione, maneb + zinc, tebuconazole, triadimenol
	Botrytis	chlorothalonil
	Canker	gamma-HCH + thiabendazole + thiram
	Clubroot	cresylic acid
	Damping off	fosetyl-aluminium *(off-label)*, gamma-HCH + thiabendazole + thiram, gamma-HCH + thiram, propamocarb hydrochloride, thiabendazole + thiram, thiram *(seed treatment)*
	Damping off and wirestem	chlorothalonil, tolclofos-methyl
	Downy mildew	chlorothalonil, chlorothalonil *(limited control)*, copper oxychloride + metalaxyl *(off-label)*, dichlofluanid, fosetyl-aluminium *(off-label)*, maneb + zinc, propamocarb hydrochloride
	Fungus diseases	mancozeb + metalaxyl *(off-label)*
	Light leaf spot	benomyl, fenpropimorph, tebuconazole, triadimenol
	Powdery mildew	copper sulfate + sulfur, tebuconazole, triadimefon, triadimenol
	Ring spot	benomyl, chlorothalonil, fenpropimorph, tebuconazole, triadimenol
	Seed-borne diseases	thiabendazole + thiram
	Spear rot	copper oxychloride *(off-label)*
	White blister	chlorothalonil + metalaxyl, mancozeb + metalaxyl *(off-label)*
	Wirestem	quintozene
Pests		cypermethrin
	Aphids	aldicarb, alpha-cypermethrin, bifenthrin, chlorpyrifos, chlorpyrifos + dimethoate, chlorpyrifos + disulfoton, cypermethrin, deltamethrin *(off-label)*, demeton-S-methyl, dimethoate, disulfoton, disulfoton + quinalphos, fatty acids, heptenophos, heptenophos *(off-label)*, nicotine, permethrin, phorate, pirimicarb, resmethrin
	Birds	aluminium ammonium sulfate
	Cabbage root fly	aldicarb, carbofuran, carbosulfan, chlorfenvinphos, chlorpyrifos, chlorpyrifos + dimethoate, chlorpyrifos + disulfoton, disulfoton + quinalphos, fonofos, trichlorfon
	Cabbage stem flea beetle	cypermethrin, gamma-HCH
	Cabbage stem weevil	carbofuran, carbosulfan, gamma-HCH

FOR FULL CONDITIONS OF USE ALWAYS READ THE PRODUCT LABEL

	Caterpillars	Bacillus thuringiensis, alpha-cypermethrin, bifenthrin, carbofuran, chlorpyrifos, cypermethrin, deltamethrin, deltamethrin *(off-label)*, diflubenzuron, esfenvalerate, lambda-cyhalothrin, permethrin, pirimiphos-methyl, resmethrin, triazophos, trichlorfon
	Cutworms	chlorpyrifos, chlorpyrifos + dimethoate, gamma-HCH, permethrin
	Damaging mammals	aluminium ammonium sulfate, sulfonated cod liver oil
	Flea beetles	aldicarb, alpha-cypermethrin, carbofuran, carbosulfan, deltamethrin, gamma-HCH, gamma-HCH + thiabendazole + thiram, gamma-HCH + thiram
	General insect control	deltamethrin + heptenophos *(off-label)*
	Leaf miners	aldicarb, cypermethrin *(protected crops, off-label)*, dichlorvos *(off-label)*, permethrin, trichlorfon, trichlorfon *(off-label)*
	Leatherjackets	chlorpyrifos, chlorpyrifos + dimethoate, gamma-HCH
	Mealy bugs	fatty acids
	Millipedes	gamma-HCH + thiabendazole + thiram
	Pollen beetles	gamma-HCH
	Red spider mites	fatty acids
	Scale insects	fatty acids
	Slugs and snails	metaldehyde, metaldehyde *(seed admixture)*, methiocarb
	Weevils	gamma-HCH
	Western flower thrips	dichlorvos *(off-label)*
	Whitefly	bifenthrin, cypermethrin, fatty acids, permethrin, pirimiphos-methyl
	Wireworms	chlorpyrifos + dimethoate, gamma-HCH
Weeds	Annual dicotyledons	carbetamide, chlorthal-dimethyl, chlorthal-dimethyl + propachlor, clopyralid, cyanazine *(off-label)*, desmetryn, desmetryn *(off-label)*, metazachlor, pendimethalin, propachlor, pyridate, sodium monochloroacetate, sodium monochloroacetate *(off-label)*, tebutam, trifluralin
	Annual grasses	carbetamide, chlorthal-dimethyl + propachlor, cyanazine *(off-label)*, cycloxydim, diclofop-methyl, pendimethalin, propachlor, sethoxydim, tebutam, trifluralin
	Annual meadow grass	metazachlor, tri-allate
	Blackgrass	cycloxydim, diclofop-methyl, metazachlor, tri-allate
	Canary grass	diclofop-methyl
	Cleavers	pyridate
	Corn marigold	clopyralid
	Couch	cycloxydim
	Creeping bent	cycloxydim
	Creeping thistle	clopyralid

	Fat-hen	desmetryn, desmetryn *(off-label)*
	Groundsel, triazine resistant	metazachlor
	Mayweeds	clopyralid
	Perennial dicotyledons	clopyralid
	Perennial grasses	cycloxydim, diclofop-methyl
	Rough meadow grass	diclofop-methyl
	Ryegrass	diclofop-methyl
	Speedwells	pyridate
	Volunteer cereals	carbetamide, cycloxydim, sethoxydim, tebutam
	Wild oats	cycloxydim, diclofop-methyl, tri-allate
Plant growth regulation	Increasing yield	sulfur

Carrots/parsnips/parsley

Diseases	Alternaria	iprodione + thiophanate-methyl *(off-label)*
	Cavity spot	mancozeb + metalaxyl
	Crown rot	fenpropimorph *(off-label)*, iprodione + thiophanate-methyl *(off-label)*, metalaxyl + thiram *(off-label)*
	Damping off	thiabendazole + thiram, thiram *(seed treatment)*
	Fungus diseases	triadimenol *(off-label)*
	Powdery mildew	sulfur *(off-label)*, triadimefon
	Seed-borne diseases	thiabendazole + thiram, thiram *(seed soak)*
Pests	Aphids	aldicarb, carbofuran, carbosulfan, demeton-S-methyl, dimethoate, disulfoton, malathion, phorate, pirimicarb
	Birds	aluminium ammonium sulfate
	Carrot fly	carbofuran, carbosulfan, chlorfenvinphos, disulfoton, lambda-cyhalothrin *(off-label)*, phorate, pirimiphos-methyl, pirimiphos-methyl *(off-label)*, quinalphos, triazophos
	Celery fly	malathion
	Cutworms	chlorpyrifos, cypermethrin, quinalphos, triazophos
	Damaging mammals	aluminium ammonium sulfate
	General insect control	lambda-cyhalothrin *(off-label)*, pyrazophos *(off-label)*
	Nematodes	aldicarb, carbofuran, carbosulfan

FOR FULL CONDITIONS OF USE ALWAYS READ THE PRODUCT LABEL

Weeds	Annual dicotyledons	chlorpropham, chlorpropham + pentanochlor, linuron, metoxuron, pendimethalin, pendimethalin *(off-label)*, pentanochlor, prometryn, trifluralin
	Annual grasses	chlorpropham, cycloxydim, diclofop-methyl, fluazifop-P-butyl, metoxuron, pendimethalin, pendimethalin *(off-label)*, prometryn, propaquizafop, sethoxydim *(off-label)*, trifluralin
	Annual meadow grass	chlorpropham + pentanochlor, linuron, pentanochlor, tri-allate
	Black bindweed	linuron
	Blackgrass	cycloxydim, diclofop-methyl, tri-allate
	Canary grass	diclofop-methyl
	Chickweed	chlorpropham, linuron
	Corn marigold	linuron
	Couch	cycloxydim
	Creeping bent	cycloxydim
	Docks	asulam *(off-label)*
	Fat-hen	linuron
	Mayweeds	metoxuron
	Perennial grasses	cycloxydim, diclofop-methyl, fluazifop-P-butyl, propaquizafop
	Polygonums	chlorpropham, linuron
	Rough meadow grass	diclofop-methyl
	Ryegrass	diclofop-methyl
	Volunteer cereals	cycloxydim, fluazifop-P-butyl, sethoxydim *(off-label)*
	Wild oats	cycloxydim, diclofop-methyl, fluazifop-P-butyl, tri-allate

Celery

Diseases	Botrytis	carbendazim (MBC) *(off-label)*
	Celery leaf spot	Bordeaux mixture, carbendazim (MBC), chlorothalonil, copper ammonium carbonate, copper oxychloride, zineb
	Damping off	propamocarb hydrochloride
	Seed-borne diseases	thiram *(seed soak)*
Pests	Aphids	cypermethrin, deltamethrin *(off-label)*, demeton-S-methyl, dimethoate, heptenophos, malathion, pirimicarb
	Carrot fly	disulfoton, phorate
	Caterpillars	cypermethrin, deltamethrin *(off-label)*
	Celery fly	malathion, quinalphos
	Cutworms	chlorpyrifos, cypermethrin, triazophos
	General insect control	dimethoate *(off-label)*
	Leaf miners	cypermethrin *(protected crops, off-label)*, nicotine, triazophos *(off-label)*

	Whitefly	permethrin
Weeds	Annual dicotyledons	chlorpropham, chlorpropham + pentanochlor, linuron, pentanochlor, prometryn
	Annual grasses	chlorpropham, diclofop-methyl, prometryn
	Annual meadow grass	chlorpropham + pentanochlor, linuron, pentanochlor
	Black bindweed	linuron
	Blackgrass	diclofop-methyl
	Canary grass	diclofop-methyl
	Chickweed	chlorpropham, linuron
	Corn marigold	linuron
	Fat-hen	linuron
	Perennial grasses	diclofop-methyl
	Polygonums	chlorpropham, linuron
	Rough meadow grass	diclofop-methyl
	Ryegrass	diclofop-methyl
	Wild oats	diclofop-methyl
Plant growth regulation	Increasing germination	gibberellins
	Increasing yield	gibberellins

Cucurbits

Diseases	Black spot	benomyl *(off-label)*
	Botrytis	iprodione *(off-label)*
	Damping off	propamocarb hydrochloride
	Gummosis	benomyl *(off-label)*
	Powdery mildew	bupirimate, imazalil *(off-label)*, iprodione *(off-label)*, quinomethionate
Pests	Aphids	disulfoton, fatty acids, heptenophos
	General insect control	dimethoate *(off-label)*
	Leaf miners	cypermethrin *(protected crops, off-label)*, oxamyl *(off-label)*, triazophos *(off-label)*, trichlorfon *(off-label)*
	Mealy bugs	fatty acids
	Red spider mites	fatty acids, quinomethionate
	Scale insects	fatty acids

FOR FULL CONDITIONS OF USE ALWAYS READ THE PRODUCT LABEL

	Whitefly	fatty acids

General

Diseases	Damping off	etridiazole, propamocarb hydrochloride
	Soil-borne diseases	dazomet, methyl bromide with chloropicrin
Pests	Aphids	gamma-HCH, nicotine, rotenone
	Birds	aluminium ammonium sulfate
	Capsids	gamma-HCH, nicotine
	Cutworms	trichlorfon
	Damaging mammals	aluminium ammonium sulfate
	Leaf miners	gamma-HCH, nicotine
	Leafhoppers	gamma-HCH, nicotine
	Nematodes	dazomet, methyl bromide with chloropicrin
	Sawflies	nicotine
	Slugs and snails	aluminium sulfate, metaldehyde
	Soil pests	dazomet, methyl bromide with chloropicrin
	Thrips	nicotine
	Woolly aphid	nicotine
Weeds	Annual dicotyledons	diquat, diquat + paraquat, glufosinate-ammonium *(pre-emergence)*, paraquat
	Annual grasses	diquat + paraquat, glufosinate-ammonium *(pre-emergence)*, paraquat
	Annual weeds	ammonium sulfamate *(pre-planting)*
	General weed control	dazomet, methyl bromide with chloropicrin
	Perennial grasses	diquat + paraquat
	Perennial weeds	ammonium sulfamate *(pre-planting)*
	Volunteer cereals	diquat + paraquat, paraquat
Plant growth regulation	Increasing yield	sulfur

Herb crops

Diseases	Crown rot	fenpropimorph *(off-label)*
	Fungus diseases	triadimenol *(off-label)*
Pests	Leafhoppers	deltamethrin *(off-label)*

Weeds	Annual dicotyledons	chlorpropham + pentanochlor *(off-label)*, chlorthal-dimethyl, chlorthal-dimethyl *(off-label)*, clopyralid *(off-label)*, diquat *(off-label)*, lenacil, lenacil *(off-label)*, metamitron *(off-label)*, paraquat *(off-label)*, pendimethalin *(off-label)*, pentanochlor *(off-label)*, prometryn *(off-label)*, propachlor, propachlor *(off-label)*, propyzamide *(off-label)*, simazine *(off-label)*, terbacil *(off-label)*, trifluralin *(off-label)*
	Annual grasses	ethofumesate *(off-label)*, lenacil *(off-label)*, paraquat *(off-label)*, pendimethalin *(off-label)*, prometryn *(off-label)*, propachlor, propachlor *(off-label)*, propyzamide *(off-label)*, simazine *(off-label)*, terbacil *(off-label)*, trifluralin *(off-label)*
	Annual meadow grass	chlorpropham + pentanochlor *(off-label)*, lenacil, pentanochlor *(off-label)*
	Annual weeds	linuron *(off-label)*
	Chickweed	ethofumesate *(off-label)*
	Cleavers	ethofumesate *(off-label)*
	Couch	terbacil *(off-label)*
	Docks	asulam *(off-label)*
	Perennial dicotyledons	clopyralid *(off-label)*
	Perennial grasses	propyzamide *(off-label)*, terbacil *(off-label)*

Lettuce, outdoor

Diseases	Botrytis	iprodione, iprodione *(off-label)*, quintozene, thiram
	Damping off	thiram *(seed treatment)*
	Downy mildew	fosetyl-aluminium *(off-label)*, mancozeb, metalaxyl + thiram, zineb
	Rhizoctonia	quintozene, tolclofos-methyl
	Ring spot	prochloraz *(off-label)*
	Sclerotinia	iprodione *(off-label)*, quintozene
Pests	Aphids	cypermethrin, deltamethrin + heptenophos *(off-label)*, deltamethrin *(off-label)*, demeton-S-methyl, dimethoate, fatty acids, gamma-HCH, heptenophos, malathion, nicotine, phorate, pirimicarb, resmethrin
	Caterpillars	cypermethrin, deltamethrin, deltamethrin *(off-label)*
	Cutworms	chlorpyrifos *(outdoor only)*, cypermethrin, deltamethrin
	Damaging mammals	sulfonated cod liver oil
	Leaf miners	cypermethrin *(protected crops, off-label)*, trichlorfon *(off-label)*
	Lettuce root aphid	phorate, phorate *(higher rate, off-label)*
	Mealy bugs	fatty acids

FOR FULL CONDITIONS OF USE ALWAYS READ THE PRODUCT LABEL

	Red spider mites	dicofol, fatty acids
	Scale insects	fatty acids
	Sciarid flies	resmethrin
	Whitefly	fatty acids, resmethrin
Weeds	Annual dicotyledons	chlorpropham, chlorpropham + diuron + propham, propachlor *(off-label)*, propyzamide, trifluralin
	Annual grasses	chlorpropham, chlorpropham + diuron + propham, diclofop-methyl, propachlor *(off-label)*, propyzamide, trifluralin
	Blackgrass	diclofop-methyl
	Canary grass	diclofop-methyl
	Chickweed	chlorpropham
	Perennial grasses	diclofop-methyl, propyzamide
	Polygonums	chlorpropham
	Rough meadow grass	diclofop-methyl
	Ryegrass	diclofop-methyl
	Wild oats	diclofop-methyl

Miscellaneous field vegetables

Diseases	Botrytis	iprodione *(off-label)*
	Celery leaf spot	chlorothalonil *(off-label)*
	Crown rot	fenpropimorph *(off-label)*
	Fungus diseases	triadimenol *(off-label)*
	General insect control	etridiazole
	Root rot	iprodione *(off-label)*
	Sclerotinia	iprodione *(off-label)*, quintozene *(forcing)*
	Septoria diseases	chlorothalonil *(off-label)*
Pests	Aphids	deltamethrin *(off-label)*, heptenophos *(off-label)*, pirimicarb *(off-label)*
	Carrot fly	chlorfenvinphos *(off-label)*
	Caterpillars	deltamethrin *(off-label)*
	Flea beetles	deltamethrin *(off-label)*
	General insect control	deltamethrin + heptenophos *(off-label)*, nicotine *(off-label)*, pyrazophos *(off-label)*
	Leaf miners	cypermethrin *(protected crops, off-label)*, triazophos *(off-label)*, trichlorfon *(off-label)*
Weeds	Annual dicotyledons	chlorpropham + pentanochlor, pentanochlor, propyzamide *(off-label)*, trifluralin *(off-label)*
	Annual grasses	propyzamide *(off-label)*, trifluralin *(off-label)*

	Annual meadow grass	chlorpropham + pentanochlor, pentanochlor
	Annual weeds	linuron *(off-label)*
	Perennial grasses	propyzamide *(off-label)*
Crop control	Pre-harvest desiccation	diquat *(off-label)*

Onions/leeks

Diseases	Botrytis	carbendazim (MBC) *(off-label)*, chlorothalonil, iprodione
	Cladosporium leaf blotch	propiconazole
	Collar rot	iprodione
	Damping off	propamocarb hydrochloride, thiabendazole + thiram, thiram *(seed treatment)*
	Downy mildew	chlorothalonil + metalaxyl, mancozeb + metalaxyl *(off-label)*, propamocarb hydrochloride
	Fungus diseases	ferbam + maneb + zineb *(off-label)*
	Rust	fenpropimorph, propiconazole, triadimefon
	Seed-borne diseases	thiabendazole + thiram
	White tip	chlorothalonil + metalaxyl, mancozeb + metalaxyl *(off-label)*
Pests	Cutworms	chlorpyrifos, triazophos
	General insect control	deltamethrin *(off-label)*, dimethoate *(off-label)*
	Leaf miners	cypermethrin *(protected crops, off-label)*, oxamyl *(off-label)*, triazophos *(off-label)*
	Onion fly	carbofuran
	Stem and bulb nematodes	oxamyl
	Stem nematodes	aldicarb, aldicarb *(off-label)*, carbofuran, oxamyl *(off-label)*
	Thrips	fenitrothion *(off-label)*, malathion
Weeds	Annual dicotyledons	bentazone *(off-label)*, chloridazon + propachlor, chloridazon + propachlor *(off-label, post-emergence)*, chloridazon + propachlor *(off-label)*, chloridazon *(off-label)*, chlorpropham, chlorpropham + cresylic acid + fenuron, chlorpropham + fenuron, chlorthal-dimethyl, chlorthal-dimethyl + propachlor, clopyralid, cyanazine, cyanazine *(organic fen soils)*, fluroxypyr *(off-label)*, ioxynil, ioxynil *(off-label)*, monolinuron, pendimethalin, pendimethalin *(off-label)*, prometryn, prometryn *(off-label)*, propachlor, sodium monochloroacetate

FOR FULL CONDITIONS OF USE ALWAYS READ THE PRODUCT LABEL

	Annual grasses	chloridazon + propachlor, chloridazon + propachlor *(off-label, post-emergence)*, chloridazon + propachlor *(off-label)*, chlorpropham, chlorpropham + cresylic acid + fenuron, chlorpropham + fenuron, chlorthal-dimethyl + propachlor, cyanazine, cycloxydim, diclofop-methyl, ethofumesate *(off-label)*, fluazifop-P-butyl, monolinuron, pendimethalin, pendimethalin *(off-label)*, prometryn, prometryn *(off-label)*, propachlor, propaquizafop
	Annual meadow grass	chloridazon *(off-label)*, monolinuron, tri-allate
	Annual weeds	glyphosate (agricultural) *(pre-drilling/pre-emergence)*, linuron *(off-label)*
	Blackgrass	cycloxydim, diclofop-methyl, tri-allate
	Canary grass	diclofop-methyl
	Chickweed	chlorpropham, chlorpropham + cresylic acid + fenuron, chlorpropham + fenuron, ethofumesate *(off-label)*
	Corn marigold	clopyralid
	Couch	cycloxydim, sethoxydim
	Creeping bent	cycloxydim, sethoxydim
	Creeping thistle	clopyralid
	Fat-hen	monolinuron
	Mayweeds	bentazone *(off-label)*, clopyralid
	Perennial dicotyledons	clopyralid
	Perennial grasses	cycloxydim, diclofop-methyl, fluazifop-P-butyl, propaquizafop, sethoxydim
	Polygonums	chlorpropham, monolinuron
	Rough meadow grass	diclofop-methyl
	Ryegrass	diclofop-methyl
	Volunteer cereals	cycloxydim, fluazifop-P-butyl, glyphosate (agricultural) *(pre-drilling/pre-emergence)*, sethoxydim
	Volunteer potatoes	fluroxypyr *(off-label)*
	Wild oats	cycloxydim, diclofop-methyl, fluazifop-P-butyl, sethoxydim, tri-allate
Crop control	Pre-harvest desiccation	sulfuric acid (commodity substance)
Plant growth regulation	Sprout suppression	maleic hydrazide

Peas, mange-tout

Diseases	Ascochyta	metalaxyl + thiabendazole + thiram *(seed treatment)*
	Damping off	metalaxyl + thiabendazole + thiram *(seed treatment)*
	Downy mildew	metalaxyl + thiabendazole + thiram *(seed treatment)*

Pests	Aphids	demeton-S-methyl *(off-label)*
	Leaf miners	triazophos *(off-label)*
	Thrips	demeton-S-methyl *(off-label)*
Weeds	Annual dicotyledons	MCPB *(off-label)*, simazine + trietazine *(off-label)*, simazine *(off-label)*, trifluralin *(off-label)*
	Annual grasses	simazine *(off-label)*, trifluralin *(off-label)*
	Annual meadow grass	tri-allate *(off-label)*
	Blackgrass	tri-allate *(off-label)*
	Perennial dicotyledons	MCPB *(off-label)*
	Wild oats	tri-allate *(off-label)*

Red beet

Diseases	Aphanomyces cochlioides	hymexazol *(off-label)*
	Seed-borne diseases	thiram *(seed soak)*
Pests	Aphids	dimethoate *(excluding Myzus persicae)*, malathion
	Cutworms	chlorpyrifos
	Flea beetles	carbofuran, trichlorfon
	General insect control	pyrazophos *(off-label)*
	Leaf miners	oxamyl *(off-label)*
	Mangold fly	dimethoate, trichlorfon
Weeds	Annual dicotyledons	chlorpropham + fenuron + propham, clopyralid, ethofumesate, lenacil, metamitron, phenmedipham
	Annual grasses	chlorpropham + fenuron + propham, diclofop-methyl, metamitron, quizalofop-ethyl
	Annual meadow grass	ethofumesate, lenacil, metamitron
	Blackgrass	diclofop-methyl, ethofumesate, tri-allate
	Canary grass	diclofop-methyl
	Corn marigold	clopyralid
	Couch	quizalofop-ethyl
	Creeping thistle	clopyralid
	Fat-hen	metamitron
	Mayweeds	clopyralid
	Perennial dicotyledons	clopyralid
	Perennial grasses	diclofop-methyl, quizalofop-ethyl, tri-allate

FOR FULL CONDITIONS OF USE ALWAYS READ THE PRODUCT LABEL

	Rough meadow grass	diclofop-methyl
	Ryegrass	diclofop-methyl
	Volunteer cereals	quizalofop-ethyl
	Wild oats	diclofop-methyl, tri-allate

Rhubarb

Pests	Leaf miners	cypermethrin *(protected crops, off-label)*, triazophos *(off-label)*
	Rosy rustic moth	chlorpyrifos
Weeds	Annual dicotyledons	chlorpropham + cresylic acid + fenuron, propyzamide, propyzamide *(outdoor)*, simazine
	Annual grasses	chlorpropham + cresylic acid + fenuron, propyzamide, propyzamide *(outdoor)*, simazine
	Annual weeds	dichlobenil *(off-label)*
	Chickweed	chlorpropham + cresylic acid + fenuron
	Perennial dicotyledons	dichlobenil *(off-label)*
	Perennial grasses	dichlobenil *(off-label)*, propyzamide, propyzamide *(outdoor)*
Plant growth regulation	Increasing yield	gibberellins

Root brassicas

Diseases	Canker	gamma-HCH + thiabendazole + thiram
	Damping off	gamma-HCH + thiabendazole + thiram, gamma-HCH + thiram, thiabendazole + thiram, thiram *(seed treatment)*
	Downy mildew	propamocarb hydrochloride *(off-label)*
	Fungus diseases	mancozeb + metalaxyl *(off-label)*
	Powdery mildew	copper sulfate + sulfur, sulfur, tebuconazole, triadimefon, triadimenol, tridemorph
	Seed-borne diseases	thiabendazole + thiram
	White blister	mancozeb + metalaxyl *(off-label)*
Pests	Aphids	carbofuran, chlorpyrifos, chlorpyrifos + dimethoate, cypermethrin, malathion, pirimicarb
	Cabbage root fly	aldicarb, carbofuran, carbofuran *(off-label)*, carbosulfan, chlorfenvinphos, chlorfenvinphos *(off-label)*, chlorpyrifos, chlorpyrifos + dimethoate, chlorpyrifos *(off-label)*, fonofos, trichlorfon *(off-label)*
	Cabbage stem flea beetle	cypermethrin
	Cabbage stem weevil	carbofuran, carbosulfan
	Caterpillars	chlorpyrifos, cypermethrin, deltamethrin

	Cutworms	chlorpyrifos, chlorpyrifos + dimethoate, cypermethrin
	Flea beetles	aldicarb, carbofuran, carbosulfan, deltamethrin, gamma-HCH + thiabendazole + thiram, gamma-HCH + thiram
	General insect control	chlorfenvinphos *(off-label)*, deltamethrin + heptenophos *(off-label)*, dimethoate *(off-label)*
	Leatherjackets	chlorpyrifos, chlorpyrifos + dimethoate
	Millipedes	gamma-HCH + thiabendazole + thiram
	Nematodes	aldicarb
	Soil pests	carbofuran *(off-label)*
	Turnip root fly	carbofuran
	Whitefly	cypermethrin
	Wireworms	chlorpyrifos + dimethoate
Weeds	Annual dicotyledons	chlorthal-dimethyl, clopyralid, metazachlor, prometryn *(off-label)*, propachlor, tebutam, trifluralin
	Annual grasses	cycloxydim, fluazifop-P-butyl *(stockfeed only)*, prometryn *(off-label)*, propachlor, tebutam, trifluralin
	Annual meadow grass	metazachlor
	Annual weeds	glyphosate (agricultural) *(pre-drilling/pre-emergence)*
	Blackgrass	cycloxydim, metazachlor
	Corn marigold	clopyralid
	Couch	cycloxydim
	Creeping bent	cycloxydim
	Creeping thistle	clopyralid
	Groundsel, triazine resistant	metazachlor
	Mayweeds	clopyralid
	Perennial dicotyledons	clopyralid
	Perennial grasses	cycloxydim, fluazifop-P-butyl *(stockfeed only)*
	Volunteer cereals	cycloxydim, fluazifop-P-butyl *(stockfeed only)*, glyphosate (agricultural) *(pre-drilling/pre-emergence)*, tebutam
	Wild oats	cycloxydim, fluazifop-P-butyl *(stockfeed only)*
Plant growth regulation	Increasing yield	sulfur

FOR FULL CONDITIONS OF USE ALWAYS READ THE PRODUCT LABEL

Root crops

Pests	Cutworms	cypermethrin

Spinach

Pests	Flea beetles	trichlorfon
	Leaf miners	cypermethrin *(protected crops, off-label)*, trichlorfon *(off-label)*
	Mangold fly	trichlorfon
Weeds	Annual dicotyledons	chlorpropham + fenuron, lenacil *(off-label)*, phenmedipham *(off-label)*
	Annual grasses	chlorpropham + fenuron, lenacil *(off-label)*
	Chickweed	chlorpropham + fenuron

Tomatoes, outdoor

Diseases	Blight	copper ammonium carbonate, copper oxychloride
	Botrytis	iprodione
	Leaf mould	copper ammonium carbonate
	Soil-borne diseases	metam-sodium *(Jersey)*
Pests	Nematodes	metam-sodium *(Jersey)*
	Soil pests	metam-sodium *(Jersey)*
Weeds	General weed control	metam-sodium *(Jersey)*

Watercress

Diseases	Damping off	mancozeb + metalaxyl *(off-label)*
	Phytophthora	etridiazole *(off-label)*
	Pythium	etridiazole *(off-label)*
	Rhizoctonia	benomyl *(off-label)*
Pests	Aphids	dimethoate *(off-label)*, fatty acids *(off-label)*
	Flea beetles	malathion *(off-label)*
	Midges	malathion *(off-label)*
	Mustard beetle	malathion *(off-label)*
Plant growth regulation	Growth regulation	gibberellins *(off-label)*

Flowers and ornamentals

Annuals/biennials

Weeds		
	Annual dicotyledons	chlorpropham, pentanochlor
	Annual grasses	chlorpropham
	Annual meadow grass	pentanochlor
	Chickweed	chlorpropham
	Polygonums	chlorpropham

Bedding plants

Diseases		
	Botrytis	carbendazim (MBC), carbendazim (MBC) (off-label), quintozene
	Clubroot	cresylic acid
	Damping off	furalaxyl, propamocarb hydrochloride
	Phytophthora	propamocarb hydrochloride
	Powdery mildew	carbendazim (MBC), carbendazim (MBC) (off-label)
	Rhizoctonia	quintozene
	Sclerotinia	quintozene
Pests	Browntail moth	teflubenzuron
	Caterpillars	teflubenzuron
	Sciarid flies	teflubenzuron
	Western flower thrips	teflubenzuron
	Whitefly	teflubenzuron
Weeds	Annual dicotyledons	pentanochlor
	Annual grasses	sethoxydim (off-label)
	Annual meadow grass	pentanochlor
	Perennial grasses	sethoxydim (off-label)
Plant growth regulation	Increasing flowering	paclobutrazol
	Stem shortening	chlormequat, chlormequat + choline chloride, daminozide, paclobutrazol

FOR FULL CONDITIONS OF USE ALWAYS READ THE PRODUCT LABEL

Bulbs/corms

Diseases	Botrytis	carbendazim (MBC), thiram, zineb
	Fire	carbendazim (MBC), dichlofluanid, mancozeb, thiram, zineb
	Fungus diseases	captan *(off-label)*, formaldehyde (commodity substance) *(dip)*
	Fusarium diseases	carbendazim (MBC), carbendazim (MBC) *(off-label)*, prochloraz *(off-label)*, thiabendazole
	Ink disease	chlorothalonil
	Penicillium rot	carbendazim (MBC), prochloraz *(off-label)*
	Phytophthora	etridiazole, propamocarb hydrochloride
	Pythium	etridiazole, propamocarb hydrochloride
	Rhizoctonia	quintozene
	Sclerotinia	carbendazim (MBC), quintozene
	Stagonospora	carbendazim (MBC)
Pests	Aphids	nicotine
	Birds	aluminium ammonium sulfate
	Bulb scale mite	endosulfan
	Capsids	nicotine
	Damaging mammals	aluminium ammonium sulfate
	Large narcissus fly	carbofuran *(off-label)*
	Leaf miners	nicotine
	Leafhoppers	nicotine
	Sawflies	nicotine
	Stem and bulb nematodes	1,3-dichloropropene
	Thrips	nicotine
	Woolly aphid	nicotine
Weeds	Annual dicotyledons	bentazone, chlorpropham, chlorpropham + cresylic acid + fenuron, chlorpropham + linuron, chlorpropham + pentanochlor, cyanazine, diquat + paraquat, lenacil, paraquat, pentanochlor
	Annual grasses	chlorpropham, chlorpropham + cresylic acid + fenuron, chlorpropham + linuron, cyanazine, cycloxydim, diquat + paraquat, paraquat, sethoxydim *(off-label)*
	Annual meadow grass	chlorpropham + pentanochlor, lenacil, pentanochlor
	Blackgrass	cycloxydim
	Chickweed	chlorpropham, chlorpropham + cresylic acid + fenuron
	Couch	cycloxydim
	Creeping bent	cycloxydim, paraquat
	Perennial grasses	cycloxydim, diquat + paraquat, sethoxydim *(off-label)*

	Polygonums	chlorpropham
	Volunteer cereals	cycloxydim
	Wild oats	cycloxydim
Crop control	Pre-harvest desiccation	sulfuric acid (commodity substance)
Plant growth regulation	Increasing flowering	paclobutrazol
	Stem shortening	2-chloroethylphosphonic acid, paclobutrazol

Camellias

Plant growth regulation	Stem shortening	chlormequat, chlormequat + choline chloride

Carnations

Diseases	Rust	zineb

Chrysanthemums

Diseases	Botrytis	carbendazim (MBC) *(off-label)*, iprodione, quintozene, thiram
	Powdery mildew	bupirimate, copper ammonium carbonate, dinocap
	Ray blight	mancozeb
	Rhizoctonia	quintozene
	Rust	oxycarboxin, propiconazole *(off-label)*, thiram
	Sclerotinia	quintozene
	Soil-borne diseases	metam-sodium
Pests	Aphids	demeton-S-methyl, nicotine, permethrin, propoxur, resmethrin
	Browntail moth	teflubenzuron
	Capsids	permethrin + thiram
	Caterpillars	permethrin, permethrin + thiram, teflubenzuron
	Earwigs	permethrin + thiram
	Leaf miners	nicotine, permethrin, permethrin + thiram *(partial control)*
	Nematodes	metam-sodium
	Red spider mites	demeton-S-methyl
	Sciarid flies	resmethrin, teflubenzuron
	Soil pests	metam-sodium

FOR FULL CONDITIONS OF USE ALWAYS READ THE PRODUCT LABEL

	Western flower thrips	teflubenzuron
	Whitefly	permethrin, propoxur, resmethrin, teflubenzuron
Weeds	Annual dicotyledons	chlorpropham, chlorpropham + pentanochlor, pentanochlor
	Annual grasses	chlorpropham
	Annual meadow grass	chlorpropham + pentanochlor, pentanochlor
	Chickweed	chlorpropham
	General weed control	metam-sodium
	Polygonums	chlorpropham
Plant growth regulation	Increasing flowering	paclobutrazol
	Stem shortening	daminozide, paclobutrazol

Container-grown stock

Diseases	Fungus diseases	prochloraz
	Phytophthora	etridiazole, propamocarb hydrochloride
Weeds	Annual dicotyledons	isoxaben, napropamide, oxadiazon
	Annual grasses	napropamide, oxadiazon
	Cleavers	napropamide
	Groundsel	napropamide
Plant growth regulation	Stem shortening	chlormequat, chlormequat + choline chloride

Dahlias

Diseases	Botrytis	quintozene
	Rhizoctonia	quintozene
	Sclerotinia	quintozene

General

Diseases	Black spot	bupirimate + triforine
	Botrytis	dichlofluanid *(off-label)*, prochloraz
	Damping off	copper ammonium carbonate, propamocarb hydrochloride
	Downy mildew	propamocarb hydrochloride
	Foot rot	tolclofos-methyl
	Leaf spots	bupirimate + triforine
	Phytophthora	propamocarb hydrochloride

	Powdery mildew	bupirimate, bupirimate + triforine, fenarimol, imazalil, sulfur
	Root rot	tolclofos-methyl
	Soil-borne diseases	metam-sodium, methyl bromide with chloropicrin
	Verticillium wilt	chloropicrin
Pests	Adelgids	gamma-HCH
	Aphids	cypermethrin, deltamethrin, demeton-S-methyl, dimethoate, gamma-HCH, malathion, nicotine, permethrin, pirimicarb, pyrethrins + resmethrin, rotenone
	Birds	aluminium ammonium sulfate, ziram
	Browntail moth	Bacillus thuringiensis
	Capsids	cypermethrin, deltamethrin, gamma-HCH, nicotine
	Caterpillars	cypermethrin, diflubenzuron, permethrin
	Chrysanthemum midge	gamma-HCH
	Cutworms	cypermethrin
	Damaging mammals	aluminium ammonium sulfate, bone oil, ziram
	General insect control	cypermethrin *(off-label)*, endosulfan *(off-label)*
	Leaf miners	abamectin, cypermethrin, cypermethrin *(protected crops, off-label)*, dimethoate, gamma-HCH, heptenophos *(off-label)*, nicotine, oxamyl *(off-label)*, permethrin, triazophos *(off-label)*, trichlorfon *(off-label)*
	Leafhoppers	gamma-HCH, nicotine
	Nematodes	metam-sodium, methyl bromide with chloropicrin
	Red spider mites	abamectin, bifenthrin, demeton-S-methyl, dicofol, dimethoate, tetradifon
	Rhododendron bug	gamma-HCH
	Sawflies	nicotine
	Scale insects	deltamethrin
	Sciarid flies	fonofos
	Slugs and snails	aluminium sulfate, metaldehyde, methiocarb
	Soil pests	metam-sodium, methyl bromide with chloropicrin
	Springtails	gamma-HCH
	Symphylids	gamma-HCH
	Thrips	cypermethrin, deltamethrin, malathion, nicotine, pyrethrins + resmethrin
	Vine weevil	fonofos
	Whitefly	cypermethrin, permethrin, pyrethrins + resmethrin

FOR FULL CONDITIONS OF USE ALWAYS READ THE PRODUCT LABEL

	Woolly aphid	nicotine
Weeds	Annual dicotyledons	chlorpropham + fenuron, chlorthal-dimethyl, diquat + paraquat, metazachlor, propachlor, trifluralin *(off-label)*
	Annual grasses	chlorpropham + fenuron, diquat + paraquat, propachlor, trifluralin *(off-label)*
	Annual meadow grass	metazachlor
	Annual weeds	ammonium sulfamate *(pre-planting)*
	Blackgrass	metazachlor
	Chickweed	chlorpropham + fenuron
	General weed control	metam-sodium, methyl bromide with chloropicrin
	Groundsel, triazine resistant	metazachlor
	Perennial grasses	diquat + paraquat
	Perennial weeds	ammonium sulfamate *(pre-planting)*
Plant growth regulation	Increasing branching	dikegulac
	Rooting of cuttings	2-(1-naphthyl)acetic acid, 4-indol-3-yl-butyric acid, 4-indol-3-yl-butyric acid + 2-(1-naphthyl)acetic acid with dichlorophen, indol-3-ylacetic acid
	Stem shortening	chlormequat, chlormequat + choline chloride

Hardy ornamental nursery stock

Diseases	Fungus diseases	prochloraz
	Phytophthora	etridiazole, fosetyl-aluminium, furalaxyl, propamocarb hydrochloride
	Pythium	furalaxyl, propamocarb hydrochloride
	Soil-borne diseases	methyl bromide with chloropicrin
Pests	Capsids	deltamethrin
	Nematodes	methyl bromide with chloropicrin
	Scale insects	deltamethrin
	Soil pests	methyl bromide with chloropicrin
	Thrips	deltamethrin
	Vine weevil	carbofuran, chlorpyrifos
Weeds	Annual dicotyledons	chlorpropham + cresylic acid + fenuron, diuron, diuron + paraquat, glufosinate-ammonium, isoxaben, lenacil, metazachlor, pentanochlor
	Annual grasses	chlorpropham + cresylic acid + fenuron, diuron, diuron + paraquat, glufosinate-ammonium
	Annual meadow grass	lenacil, metazachlor, pentanochlor

	Blackgrass	metazachlor
	Chickweed	chlorpropham + cresylic acid + fenuron
	General weed control	methyl bromide with chloropicrin
	Groundsel, triazine resistant	metazachlor
	Perennial dicotyledons	diuron + paraquat, glufosinate-ammonium
	Perennial grasses	diuron + paraquat, glufosinate-ammonium

Hedges

Pests

	Browntail moth	trichlorfon
	Small ermine moth	trichlorfon

Weeds

	Annual and perennial weeds	glyphosate (horticulture, forestry, amenity etc.) *(directed spray)*
	Annual weeds	dalapon + dichlobenil
	Bracken	dalapon + dichlobenil
	Perennial grasses	dalapon + dichlobenil
	Rushes	dalapon + dichlobenil

Plant growth regulation

	Growth retardation	dikegulac, maleic hydrazide

Perennials

Weeds

	Annual dicotyledons	lenacil, pentanochlor, propachlor
	Annual grasses	propachlor, sethoxydim *(off-label)*
	Annual meadow grass	lenacil, pentanochlor
	Perennial grasses	sethoxydim *(off-label)*

Plant growth regulation

	Increasing branching	dikegulac

Roses

Diseases

	Black spot	captan, dichlofluanid, dodine, mancozeb, myclobutanil, triforine
	Powdery mildew	bupirimate, carbendazim (MBC), dichlofluanid, dinocap, dodemorph, fenarimol, imazalil, myclobutanil, pyrazophos
	Rust	bupirimate + triforine, mancozeb, myclobutanil, oxycarboxin, penconazole

FOR FULL CONDITIONS OF USE ALWAYS READ THE PRODUCT LABEL

	Verticillium wilt	cresylic acid
Pests	Aphids	dimethoate, nicotine
	Browntail moth	teflubenzuron
	Caterpillars	teflubenzuron, trichlorfon
	Leaf miners	dimethoate
	Red spider mites	fenpropathrin
	Scale insects	malathion
	Sciarid flies	teflubenzuron
	Slug sawflies	rotenone
	Western flower thrips	teflubenzuron
	Whitefly	teflubenzuron
Weeds	Annual dicotyledons	atrazine, chlorthal-dimethyl, pentanochlor, propyzamide, simazine
	Annual grasses	atrazine, propyzamide, simazine
	Annual meadow grass	pentanochlor
	Annual weeds	dalapon + dichlobenil, dichlobenil
	Perennial dicotyledons	dichlobenil
	Perennial grasses	dalapon + dichlobenil, dichlobenil, propyzamide
Plant growth regulation	Increasing flowering	paclobutrazol
	Stem shortening	paclobutrazol

Standing ground

Weeds	Annual dicotyledons	chlorpropham + cresylic acid + fenuron
	Annual grasses	chlorpropham + cresylic acid + fenuron
	Chickweed	chlorpropham + cresylic acid + fenuron

Trees/shrubs

Diseases	Canker	cresylic acid, octhilinone
	Crown gall	cresylic acid
	Dutch elm disease	metam-sodium *(off-label)*, thiabendazole *(injection -off-label)*
	Honey fungus	cresylic acid
	Powdery mildew	penconazole
	Pruning wounds	octhilinone
	Scab	penconazole
	Silver leaf	octhilinone

Pests	Adelgids	gamma-HCH
	Aphids	dimethoate, fatty acids, gamma-HCH
	Browntail moth	permethrin
	Capsids	deltamethrin
	Caterpillars	diflubenzuron
	Leaf miners	dimethoate
	Leafhoppers	gamma-HCH
	Mealy bugs	fatty acids
	Red spider mites	dimethoate, fatty acids
	Rhododendron bug	gamma-HCH
	Scale insects	deltamethrin, fatty acids
	Thrips	deltamethrin
	Whitefly	fatty acids
Weeds	Annual and perennial weeds	glyphosate (horticulture, forestry, amenity etc.), glyphosate (horticulture, forestry, amenity etc.) *(directed spray)*, glyphosate (horticulture, forestry, amenity etc.) *(pre-planting/sowing)*, glyphosate (horticulture, forestry, amenity etc.) *(wiper application)*
	Annual dicotyledons	clopyralid, diuron, diuron + glyphosate, diuron + paraquat, glufosinate-ammonium, isoxaben, napropamide, oxadiazon, paraquat, propyzamide, simazine
	Annual grasses	diuron, diuron + glyphosate, diuron + paraquat, glufosinate-ammonium, napropamide, oxadiazon, paraquat, propyzamide, sethoxydim *(off-label)*, simazine
	Annual weeds	dalapon + dichlobenil, dichlobenil
	Bindweeds	oxadiazon
	Bracken	glyphosate (horticulture, forestry, amenity etc.)
	Cleavers	napropamide, oxadiazon
	Corn marigold	clopyralid
	Creeping bent	paraquat
	Creeping thistle	clopyralid
	Groundsel	napropamide
	Mayweeds	clopyralid
	Perennial dicotyledons	clopyralid, dichlobenil, diuron + glyphosate, diuron + paraquat, glufosinate-ammonium
	Perennial grasses	dalapon + dichlobenil, dichlobenil, diuron + glyphosate, diuron + paraquat, glufosinate-ammonium, propyzamide, sethoxydim *(off-label)*

FOR FULL CONDITIONS OF USE ALWAYS READ THE PRODUCT LABEL

	Polygonums	oxadiazon
Plant growth regulation	Growth retardation	dikegulac
	Sucker inhibition	maleic hydrazide

Water lily

Diseases	Crown rot	carbendazim + metalaxyl *(off-label)*

Forestry

Broadleaved trees

Plant growth regulation	Increasing germination	gibberellins

Cut logs/timber

Pests	Ambrosia beetle	chlorpyrifos, gamma-HCH
	Elm bark beetle	chlorpyrifos, gamma-HCH
	Great spruce bark beetle	gamma-HCH
	Larch shoot beetle	chlorpyrifos
	Pine shoot beetle	chlorpyrifos

Farm woodland

Weeds	Annual dicotyledons	cyanazine *(off-label)*, metazachlor *(off-label)*, pendimethalin *(off-label)*, propyzamide
	Annual grasses	cyanazine *(off-label)*, fluazifop-P-butyl *(off-label)*, metazachlor *(off-label)*, pendimethalin *(off-label)*, propaquizafop, propyzamide
	Perennial grasses	fluazifop-P-butyl *(off-label)*, propaquizafop, propyzamide

Forest nursery beds

Pests	Aphids	pirimicarb
	Birds	aluminium ammonium sulfate
	Black pine beetle	permethrin *(off-label)*
	Clay-coloured weevil	gamma-HCH
	Cutworms	gamma-HCH

	Damaging mammals	aluminium ammonium sulfate
	Pine weevil	permethrin *(off-label)*
	Poplar leaf beetles	gamma-HCH
	Strawberry root weevil	gamma-HCH
Weeds	Annual dicotyledons	paraquat *(stale seedbed)*, pentanochlor, simazine
	Annual grasses	paraquat *(stale seedbed)*, simazine
	Annual meadow grass	pentanochlor

Forestry plantations

Pests	Adelgids	gamma-HCH
	Birds	aluminium ammonium sulfate, ziram
	Black pine beetle	permethrin *(off-label)*
	Browntail moth	diflubenzuron
	Caterpillars	diflubenzuron
	Damaging mammals	aluminium ammonium sulfate, aluminium phosphide, sulfonated cod liver oil, ziram
	Grey squirrels	warfarin
	Large pine weevil	carbosulfan, gamma-HCH
	Pine looper	diflubenzuron, resmethrin
	Pine weevil	permethrin *(off-label)*
	Rodents	aluminium phosphide
	Vine weevil	chlorpyrifos
	Winter moth	diflubenzuron
Weeds	Annual and perennial weeds	glyphosate (agricultural) *(directed spray)*, glyphosate (agricultural) *(pre-planting)*, glyphosate (horticulture, forestry, amenity etc.) *(directed spray)*, glyphosate (horticulture, forestry, amenity etc.) *(pre-planting)*
	Annual dicotyledons	2,4-D, 2,4-D + dicamba + triclopyr, atrazine, clopyralid *(off-label)*, diquat + paraquat, isoxaben, paraquat, propyzamide
	Annual grasses	atrazine, diquat + paraquat, paraquat, propyzamide
	Annual weeds	ammonium sulfamate, dalapon + dichlobenil, glufosinate-ammonium, glyphosate (agricultural), imazapyr
	Bracken	asulam, asulam *(off-label)*, dalapon + dichlobenil, dicamba, glyphosate (agricultural) *(directed spray)*, glyphosate (agricultural) *(overall dormant spray)*, glyphosate (horticulture, forestry, amenity etc.) *(directed spray)*, glyphosate (horticulture, forestry, amenity etc.) *(overall dormant spray)*

FOR FULL CONDITIONS OF USE ALWAYS READ THE PRODUCT LABEL

	Brambles	2,4-D + dicamba + triclopyr, glyphosate (agricultural) *(overall dormant spray)*, glyphosate (horticulture, forestry, amenity etc.) *(overall dormant spray)*, triclopyr
	Broom	triclopyr
	Couch	atrazine, glyphosate (agricultural) *(directed spray)*, glyphosate (horticulture, forestry, amenity etc.) *(directed spray)*
	Creeping bent	paraquat
	Docks	2,4-D + dicamba + triclopyr, triclopyr
	Firebreak desiccation	paraquat
	General weed control	atrazine *(off-label)*
	Gorse	2,4-D + dicamba + triclopyr, triclopyr
	Heather	2,4-D
	Horsetails	propyzamide
	Japanese knotweed	2,4-D + dicamba + triclopyr
	Perennial dicotyledons	2,4-D, 2,4-D + dicamba + triclopyr, ammonium sulfamate, clopyralid *(off-label)*, glyphosate (agricultural) *(wiper application)*, glyphosate (horticulture, forestry, amenity etc.) *(wiper application)*, imazapyr, triclopyr
	Perennial grasses	ammonium sulfamate, atrazine, dalapon + dichlobenil, diquat + paraquat, glyphosate (agricultural) *(overall dormant spray)*, glyphosate (horticulture, forestry, amenity etc.) *(overall dormant spray)*, imazapyr, propyzamide
	Perennial weeds	glufosinate-ammonium, glyphosate (agricultural)
	Rhododendrons	2,4-D + dicamba + triclopyr, ammonium sulfamate, triclopyr
	Rushes	dalapon + dichlobenil, glyphosate (agricultural) *(directed spray)*, glyphosate (horticulture, forestry, amenity etc.) *(directed spray)*
	Sedges	propyzamide
	Stinging nettle	2,4-D + dicamba + triclopyr, triclopyr
	Volunteer cereals	diquat + paraquat
	Woody weeds	2,4-D, 2,4-D + dicamba + triclopyr, ammonium sulfamate, fosamine-ammonium, glyphosate (agricultural), glyphosate (agricultural) *(directed spray)*, glyphosate (agricultural) *(overall dormant spray)*, glyphosate (horticulture, forestry, amenity etc.) *(directed spray)*, glyphosate (horticulture, forestry, amenity etc.) *(overall dormant spray)*, triclopyr
Crop control	Chemical thinning	glyphosate (agricultural), glyphosate (horticulture, forestry, amenity etc.)

Transplant lines

Pests	Black pine beetle	chlorpyrifos
	Pine weevil	chlorpyrifos

	Vine weevil	chlorpyrifos
Weeds	Annual dicotyledons	paraquat, simazine
	Annual grasses	paraquat, simazine
	Creeping bent	paraquat

Fruit and hops

Apples/pears

Diseases	Blossom wilt	vinclozolin
	Botrytis	iprodione *(off-label)*
	Botrytis fruit rot	captan
	Brown rot	captan
	Canker	Bordeaux mixture, carbendazim (MBC), carbendazim (MBC) *(partial control)*, copper oxychloride, octhilinone, thiophanate-methyl
	Collar rot	copper oxychloride + metalaxyl, copper oxychloride *(off-label)*, fosetyl-aluminium
	Crown rot	fosetyl-aluminium
	Gloeosporium	captan, carbendazim (MBC)
	Phytophthora fruit rot	captan, mancozeb + metalaxyl *(off-label)*
	Powdery mildew	bupirimate, captan + penconazole, carbendazim (MBC), dinocap, dithianon + penconazole, fenarimol, myclobutanil, penconazole, pyrazophos, pyrifenox, sulfur, thiophanate-methyl, triadimefon, triforine
	Pruning wounds	octhilinone
	Scab	Bordeaux mixture, captan, captan + penconazole, carbendazim (MBC), carbendazim (MBC) *(off-label)*, dithianon, dithianon + penconazole, dodine, fenarimol, fenbuconazole, mancozeb, maneb + zinc, myclobutanil, pyrifenox, sulfur, thiophanate-methyl, thiram, triforine
	Silver leaf	Trichoderma viride, octhilinone
	Storage rots	captan, carbendazim (MBC), carbendazim (MBC) *(off-label)*, carbendazim + metalaxyl, thiophanate-methyl
Pests	Aphids	chlorpyrifos, cypermethrin, deltamethrin, demeton-S-methyl, dimethoate, fenitrothion, heptenophos, lambda-cyhalothrin, malathion, nicotine, permethrin, pirimicarb, pirimiphos-methyl, tar oils
	Apple blossom weevil	chlorpyrifos, fenitrothion, gamma-HCH

FOR FULL CONDITIONS OF USE ALWAYS READ THE PRODUCT LABEL

	Bryobia mites	demeton-S-methyl, dicofol + tetradifon, dimethoate, malathion
	Capsids	carbaryl, chlorpyrifos, cypermethrin, deltamethrin, dimethoate, fenitrothion, pirimiphos-methyl, triazophos
	Caterpillars	chlorpyrifos, cypermethrin, deltamethrin, diflubenzuron, fenitrothion, fenpropathrin, permethrin, pirimiphos-methyl
	Cherry bark tortrix	trichlorfon
	Codling moth	carbaryl, chlorpyrifos, cypermethrin, deltamethrin, diflubenzuron, fenitrothion, malathion, pirimiphos-methyl, triazophos
	Earwigs	carbaryl, diflubenzuron
	Leaf midges	lambda-cyhalothrin
	Leaf miners	diflubenzuron
	Leafhoppers	demeton-S-methyl, malathion
	Red spider mites	amitraz, bifenthrin, chlorpyrifos, clofentezine, demeton-S-methyl, dicofol, dicofol + tetradifon, dimethoate, fenpropathrin, fenpyroximate, malathion, tebufenpyrad, tetradifon, triazophos
	Rust mite	amitraz, diflubenzuron, pirimiphos-methyl
	Sawflies	chlorpyrifos, cypermethrin, deltamethrin, demeton-S-methyl, diflubenzuron, dimethoate, fenitrothion, gamma-HCH, pirimiphos-methyl
	Scale insects	tar oils
	Slug sawflies	rotenone
	Suckers	amitraz, chlorpyrifos, cypermethrin, deltamethrin, demeton-S-methyl, diflubenzuron, dimethoate, fenitrothion, gamma-HCH, lambda-cyhalothrin, malathion, permethrin, pirimiphos-methyl, tar oils
	Tortrix moths	carbaryl, chlorpyrifos, cypermethrin, deltamethrin, diflubenzuron, fenitrothion, permethrin, pirimiphos-methyl, triazophos
	Winter moth	carbaryl, chlorpyrifos, cypermethrin, diflubenzuron, fenitrothion, tar oils
	Woolly aphid	chlorpyrifos, demeton-S-methyl, heptenophos, malathion
Weeds	Annual and perennial weeds	glyphosate (agricultural), glyphosate (horticulture, forestry, amenity etc.)
	Annual dicotyledons	2,4-D + dichlorprop + MCPA + mecoprop, dicamba + MCPA + mecoprop, diuron, isoxaben, oxadiazon, pendimethalin, pentanochlor, propyzamide, simazine, sodium monochloroacetate
	Annual grasses	diuron, oxadiazon, pendimethalin, propyzamide, simazine
	Annual meadow grass	pentanochlor
	Annual weeds	amitrole, dichlobenil, glufosinate-ammonium
	Bindweeds	oxadiazon
	Chickweed	2,4-D + dichlorprop + MCPA + mecoprop

	Cleavers	2,4-D + dichlorprop + MCPA + mecoprop, oxadiazon
	Couch	amitrole
	Creeping thistle	amitrole
	Docks	amitrole, asulam, dicamba + MCPA + mecoprop
	Perennial dicotyledons	2,4-D + dichlorprop + MCPA + mecoprop, dicamba + MCPA + mecoprop, dichlobenil
	Perennial grasses	dichlobenil, propyzamide
	Perennial weeds	amitrole, glufosinate-ammonium
	Polygonums	oxadiazon
Crop control	Control of water shoots	2-(1-naphthyl)acetic acid
	Sucker control	2-(1-naphthyl)acetic acid, glyphosate (agricultural)
Plant growth regulation	Controlling vigour	paclobutrazol
	Fruit thinning	carbaryl
	Increasing fruit set	gibberellins, gibberellins *(off-label)*, paclobutrazol
	Reducing fruit russeting	gibberellins

Apricots/peaches/nectarines

Diseases	Leaf curl	Bordeaux mixture, copper ammonium carbonate, copper oxychloride
Pests	Aphids	malathion, tar oils
	Red spider mites	tetradifon
	Scale insects	tar oils
	Winter moth	tar oils
Weeds	Docks	asulam *(off-label)*
	General weed control	amitrole *(off-label)*

Blueberries/cranberries

Weeds	Docks	asulam *(off-label)*

Cane fruit

Diseases	Botrytis	carbendazim (MBC), carbendazim (MBC) *(off-label)*, chlorothalonil, dichlofluanid, iprodione, thiram
	Cane blight	dichlofluanid

FOR FULL CONDITIONS OF USE ALWAYS READ THE PRODUCT LABEL

	Cane spot	Bordeaux mixture, carbendazim (MBC), carbendazim (MBC) *(off-label)*, chlorothalonil, copper ammonium carbonate, copper oxychloride, dichlofluanid, thiram
	Powdery mildew	bupirimate, carbendazim (MBC), carbendazim (MBC) *(off-label)*, chlorothalonil, dichlofluanid, fenarimol, triadimefon *(off-label)*
	Root rot	mancozeb + metalaxyl *(off-label)*, mancozeb + oxadixyl *(off-label)*
	Spur blight	Bordeaux mixture, carbendazim (MBC), carbendazim (MBC) *(off-label)*, thiram
Pests	Aphids	chlorpyrifos, demeton-S-methyl, dimethoate, malathion, pirimicarb, tar oils
	Birds	aluminium ammonium sulfate
	Blackberry mite	endosulfan
	Capsids	dimethoate
	Caterpillars	Bacillus thuringiensis
	Damaging mammals	aluminium ammonium sulfate
	Leafhoppers	dimethoate, malathion
	Nematodes	1,3-dichloropropene
	Overwintering pests	cresylic acid
	Raspberry beetle	chlorpyrifos, deltamethrin, fenitrothion, malathion, rotenone
	Raspberry cane midge	chlorpyrifos, fenitrothion, gamma-HCH
	Red spider mites	chlorpyrifos, clofentezine *(off-label)*, demeton-S-methyl, dimethoate, tetradifon
	Scale insects	tar oils
	Virus vectors	1,3-dichloropropene
	Winter moth	tar oils
Weeds	Annual dicotyledons	atrazine, bromacil, chlorthal-dimethyl, isoxaben, lenacil, napropamide, oxadiazon, paraquat, propyzamide, propyzamide *(England only)*, simazine, trifluralin
	Annual grasses	atrazine, bromacil, fluazifop-P-butyl, napropamide, oxadiazon, paraquat, propyzamide, propyzamide *(England only)*, simazine, trifluralin
	Annual meadow grass	lenacil
	Annual weeds	dichlobenil, glufosinate-ammonium
	Bindweeds	oxadiazon
	Cleavers	napropamide, oxadiazon
	Couch	bromacil, sethoxydim
	Creeping bent	paraquat, sethoxydim
	Docks	asulam *(off-label)*
	Groundsel	napropamide

	Perennial dicotyledons	bromacil, dichlobenil
	Perennial grasses	bromacil, dichlobenil, fluazifop-P-butyl, propyzamide, propyzamide *(England only)*, sethoxydim
	Perennial ryegrass	paraquat
	Perennial weeds	glufosinate-ammonium
	Polygonums	oxadiazon
	Rough meadow grass	paraquat
	Volunteer cereals	fluazifop-P-butyl, sethoxydim
	Wild oats	fluazifop-P-butyl, sethoxydim
Crop control	Sucker control	2-(1-naphthyl)acetic acid, sodium monochloroacetate *(off-label)*

Currants

Diseases	Botrytis	carbendazim (MBC), chlorothalonil *(suppression)*, dichlofluanid
	Currant leaf spot	Bordeaux mixture, carbendazim (MBC), chlorothalonil, copper ammonium carbonate, dodine, mancozeb, quinomethionate, triforine, zineb
	Powdery mildew	bupirimate, carbendazim (MBC), chlorothalonil *(suppression)*, fenarimol, myclobutanil, penconazole *(restricted area)*, quinomethionate, triadimefon *(off-label)*, triforine
	Rust	copper oxychloride, thiram
Pests	Aphids	chlorpyrifos, demeton-S-methyl, dimethoate, malathion, pirimicarb, tar oils
	Big-bud mite	endosulfan
	Capsids	chlorpyrifos, cypermethrin, fenitrothion
	Caterpillars	chlorpyrifos, fenpropathrin
	Earwigs	trichlorfon
	Gall mite	sulfur
	Overwintering pests	cresylic acid
	Red spider mites	bifenthrin, chlorpyrifos, clofentezine *(off-label)*, demeton-S-methyl, dicofol + tetradifon, dimethoate, fenpropathrin, tetradifon
	Sawflies	cypermethrin, diflubenzuron, fenitrothion
	Scale insects	tar oils
	Winter moth	diflubenzuron, tar oils
Weeds	Annual dicotyledons	MCPB, chlorpropham, chlorthal-dimethyl, isoxaben, lenacil, napropamide, oxadiazon, paraquat, pendimethalin, pentanochlor, propyzamide, simazine, sodium monochloroacetate

FOR FULL CONDITIONS OF USE ALWAYS READ THE PRODUCT LABEL

Annual grasses	chlorpropham, fluazifop-P-butyl, napropamide, oxadiazon, paraquat, pendimethalin, propyzamide, simazine
Annual meadow grass	lenacil, pentanochlor
Annual weeds	dichlobenil, glufosinate-ammonium
Bindweeds	oxadiazon
Chickweed	chlorpropham
Cleavers	napropamide, oxadiazon
Creeping bent	paraquat
Docks	asulam, asulam *(off-label)*
Groundsel	napropamide
Perennial dicotyledons	MCPB, dichlobenil
Perennial grasses	dichlobenil, fluazifop-P-butyl, propyzamide
Perennial ryegrass	paraquat
Perennial weeds	glufosinate-ammonium
Polygonums	chlorpropham, oxadiazon
Rough meadow grass	paraquat
Volunteer cereals	fluazifop-P-butyl
Wild oats	fluazifop-P-butyl

Fruit nursery stock

Weeds

Annual dicotyledons	metazachlor, trifluralin *(off-label)*
Annual grasses	trifluralin *(off-label)*
Annual meadow grass	metazachlor
Blackgrass	metazachlor
Groundsel, triazine resistant	metazachlor

General

Diseases

Powdery mildew	fenpropimorph *(off-label)*

Pests

Aphids	nicotine, resmethrin, rotenone
Birds	aluminium ammonium sulfate
Capsids	nicotine
Damaging mammals	aluminium ammonium sulfate
Leaf miners	nicotine
Leafhoppers	nicotine
Sawflies	nicotine
Slugs and snails	aluminium sulfate, metaldehyde, methiocarb

	Thrips	nicotine
	Woolly aphid	nicotine
Weeds	Annual dicotyledons	diquat + paraquat, trifluralin *(off-label)*
	Annual grasses	diquat + paraquat, trifluralin *(off-label)*
	Perennial grasses	diquat + paraquat
	Volunteer cereals	diquat + paraquat
Plant growth regulation	Increasing yield	sulfur

Gooseberries

Diseases	Botrytis	carbendazim (MBC), dichlofluanid
	Currant leaf spot	carbendazim (MBC), chlorothalonil, dodine, mancozeb, quinomethionate
	Powdery mildew	bupirimate, carbendazim (MBC), fenarimol, myclobutanil, quinomethionate, sulfur, triadimefon *(off-label)*
	Septoria diseases	chlorothalonil
Pests	Aphids	chlorpyrifos, demeton-S-methyl, dimethoate, malathion, pirimicarb, tar oils
	Capsids	chlorpyrifos, cypermethrin, dimethoate, fenitrothion
	Caterpillars	chlorpyrifos
	Red spider mites	chlorpyrifos, demeton-S-methyl, dimethoate, malathion, quinomethionate, tetradifon
	Sawflies	cypermethrin, fenitrothion, malathion, nicotine, rotenone
	Scale insects	tar oils
	Winter moth	tar oils
Weeds	Annual dicotyledons	chlorpropham, chlorthal-dimethyl, isoxaben, lenacil, napropamide, oxadiazon, paraquat, pendimethalin, pentanochlor, propyzamide, simazine, sodium monochloroacetate
	Annual grasses	chlorpropham, fluazifop-P-butyl, napropamide, oxadiazon, paraquat, pendimethalin, propyzamide, simazine
	Annual meadow grass	lenacil, pentanochlor
	Annual weeds	dichlobenil
	Bindweeds	oxadiazon
	Chickweed	chlorpropham
	Cleavers	napropamide, oxadiazon
	Creeping bent	paraquat

FOR FULL CONDITIONS OF USE ALWAYS READ THE PRODUCT LABEL

Docks	asulam *(off-label)*	
Groundsel	napropamide	
Perennial dicotyledons	dichlobenil	
Perennial grasses	dichlobenil, fluazifop-P-butyl, propyzamide	
Perennial ryegrass	paraquat	
Polygonums	chlorpropham, oxadiazon	
Rough meadow grass	paraquat	
Volunteer cereals	fluazifop-P-butyl	
Wild oats	fluazifop-P-butyl	

Grapevines

Diseases	Botrytis	dichlofluanid, iprodione *(off-label)*
	Downy mildew	chlorothalonil, copper oxychloride, copper oxychloride + metalaxyl *(off-label)*
	Powdery mildew	chlorothalonil, copper sulfate + sulfur, fenarimol, sulfur, triadimefon *(off-label)*
Pests	General insect control	dimethoate *(off-label)*
	Mealy bugs	petroleum oil
	Red spider mites	petroleum oil, tetradifon
	Scale insects	petroleum oil, tar oils
Weeds	Annual and perennial weeds	glyphosate (agricultural) *(off-label)*
	Annual dicotyledons	isoxaben, oxadiazon, paraquat, simazine *(off-label)*
	Annual grasses	oxadiazon, paraquat, simazine *(off-label)*
	Annual weeds	glufosinate-ammonium
	Bindweeds	oxadiazon
	Cleavers	oxadiazon
	Perennial weeds	glufosinate-ammonium
	Polygonums	oxadiazon
Plant growth regulation	Increasing yield	sulfur

Hops

Diseases	Canker	benomyl *(off-label)*
	Downy mildew	Bordeaux mixture, chlorothalonil, copper oxychloride, copper oxychloride + metalaxyl, copper sulfate + sulfur, fosetyl-aluminium, zineb

	Powdery mildew	bupirimate, copper sulfate + sulfur, myclobutanil *(off-label)*, penconazole, pyrazophos, sulfur, triadimefon, triforine
Pests	Aphids	bifenthrin, cypermethrin, deltamethrin, dimethoate *(Fuggles only)*, endosulfan, fenpropathrin, imidacloprid, lambda-cyhalothrin, mephosfolan, tebufenpyrad
	Caterpillars	fenpropathrin
	Nematodes	1,3-dichloropropene
	Red spider mites	amitraz, bifenthrin, dicofol, dicofol + tetradifon, dimethoate *(Fuggles only)*, fenpropathrin, lambda-cyhalothrin, tebufenpyrad, tetradifon
	Virus vectors	1,3-dichloropropene
Weeds	Annual dicotyledons	diquat + paraquat, isoxaben, oxadiazon, paraquat, pendimethalin, simazine
	Annual grasses	diquat + paraquat, fluazifop-P-butyl, oxadiazon, paraquat, pendimethalin, simazine
	Bindweeds	oxadiazon
	Blackgrass	fluazifop-P-butyl
	Cleavers	oxadiazon
	Creeping bent	paraquat
	Docks	asulam
	Perennial grasses	diquat + paraquat
	Perennial ryegrass	paraquat
	Polygonums	oxadiazon
	Rough meadow grass	paraquat
	Volunteer cereals	diquat + paraquat, fluazifop-P-butyl
	Wild oats	fluazifop-P-butyl
Crop control	Chemical stripping	anthracene oil, diquat, diquat + paraquat, paraquat, sodium monochloroacetate, sodium monochloroacetate *(off-label)*
Plant growth regulation	Increasing quality	gibberellins *(off-label)*
	Increasing yield	sulfur

Plums/cherries/damsons

Diseases	Bacterial canker	Bordeaux mixture, copper oxychloride
	Brown rot	carbendazim (MBC) *(off-label)*
	Pruning wounds	octhilinone

FOR FULL CONDITIONS OF USE ALWAYS READ THE PRODUCT LABEL

	Silver leaf	Trichoderma viride, octhilinone
Pests	Aphids	chlorpyrifos, cypermethrin, deltamethrin, demeton-S-methyl, dimethoate, fenitrothion, malathion, pirimicarb, tar oils
	Caterpillars	cypermethrin, deltamethrin, fenitrothion
	Cherry fruit moth	dimethoate
	Plum fruit moth	deltamethrin, diflubenzuron
	Red spider mites	chlorpyrifos, clofentezine, demeton-S-methyl, dimethoate, malathion, tetradifon
	Sawflies	deltamethrin, demeton-S-methyl, dimethoate
	Scale insects	tar oils
	Tortrix moths	chlorpyrifos, cypermethrin, diflubenzuron
	Winter moth	chlorpyrifos, cypermethrin, diflubenzuron, tar oils
Weeds	Annual and perennial weeds	glyphosate (agricultural), glyphosate (horticulture, forestry, amenity etc.)
	Annual dicotyledons	isoxaben, pendimethalin, pentanochlor, propyzamide, sodium monochloroacetate
	Annual grasses	pendimethalin, propyzamide
	Annual meadow grass	pentanochlor
	Annual weeds	glufosinate-ammonium
	Docks	asulam, asulam (off-label)
	General weed control	amitrole (off-label)
	Perennial grasses	propyzamide
	Perennial weeds	glufosinate-ammonium
Crop control	Sucker control	2-(1-naphthyl)acetic acid, glyphosate (agricultural)

Quinces

Weeds	Docks	asulam (off-label)
	General weed control	amitrole (off-label)

Strawberries

Diseases	Botrytis	carbendazim (MBC), chlorothalonil, dichlofluanid, iprodione, thiram
	Crown rot	chloropicrin
	Phytophthora	propamocarb hydrochloride
	Powdery mildew	bupirimate, carbendazim (MBC), dinocap, fenarimol, myclobutanil, sulfur, triadimefon (off-label)
	Red core	chloropicrin, copper oxychloride + metalaxyl, etridiazole (Scotland only), fosetyl-aluminium, fosetyl-aluminium (spring treatment, off-label)

	Soil-borne diseases	methyl bromide with chloropicrin
	Verticillium wilt	chloropicrin
Pests	Aphids	chlorpyrifos, demeton-S-methyl, dimethoate, disulfoton, heptenophos, malathion, nicotine, phorate, pirimicarb
	Birds	aluminium ammonium sulfate
	Capsids	cypermethrin
	Caterpillars	cypermethrin
	Chafer grubs	gamma-HCH
	Damaging mammals	aluminium ammonium sulfate
	Leatherjackets	gamma-HCH
	Mites	fenbutatin oxide
	Nematodes	1,3-dichloropropene, chloropicrin, methyl bromide with chloropicrin
	Red spider mites	bifenthrin, chlorpyrifos, clofentezine *(off-label)*, demeton-S-methyl, dicofol, dicofol + tetradifon, dimethoate, fenbutatin oxide, fenpropathrin, quinomethionate, tetradifon, triazophos
	Soil pests	methyl bromide with chloropicrin
	Stem nematodes	1,3-dichloropropene
	Strawberry seed beetle	methiocarb
	Tarsonemid mites	dicofol, dicofol + tetradifon, endosulfan
	Tortrix moths	Bacillus thuringiensis, chlorpyrifos, cypermethrin, fenitrothion, triazophos, trichlorfon
	Vine weevil	carbofuran, chlorpyrifos
	Virus vectors	1,3-dichloropropene
	Wireworms	gamma-HCH
Weeds	Annual dicotyledons	chlorpropham, chlorpropham + fenuron, chlorthal-dimethyl, clopyralid, diquat + paraquat, isoxaben, lenacil, napropamide, pendimethalin, phenmedipham, propachlor, propyzamide, simazine, trifluralin
	Annual grasses	chlorpropham, chlorpropham + fenuron, cycloxydim, diquat + paraquat, fluazifop-P-butyl, napropamide, pendimethalin, propachlor, propyzamide, simazine, trifluralin
	Annual meadow grass	lenacil
	Annual weeds	glufosinate-ammonium
	Blackgrass	cycloxydim
	Chickweed	chlorpropham, chlorpropham + fenuron
	Cleavers	napropamide

FOR FULL CONDITIONS OF USE ALWAYS READ THE PRODUCT LABEL

	Corn marigold	clopyralid
	Couch	cycloxydim, sethoxydim
	Creeping bent	cycloxydim, sethoxydim
	Creeping thistle	clopyralid
	Docks	asulam *(off-label)*
	General weed control	methyl bromide with chloropicrin
	Groundsel	napropamide
	Mayweeds	clopyralid
	Perennial dicotyledons	clopyralid
	Perennial grasses	cycloxydim, diquat + paraquat, fluazifop-P-butyl, propyzamide, sethoxydim
	Perennial weeds	glufosinate-ammonium
	Polygonums	chlorpropham
	Volunteer cereals	cycloxydim, diquat + paraquat, fluazifop-P-butyl, sethoxydim
	Wild oats	cycloxydim, fluazifop-P-butyl, sethoxydim
Crop control	Runner desiccation	paraquat

Tree fruit

Diseases	Canker	octhilinone
	Pruning wounds	octhilinone
	Silver leaf	octhilinone
	Soil-borne diseases	methyl bromide with chloropicrin
	Verticillium wilt	chloropicrin
Pests	Aphids	fatty acids, rotenone
	Birds	aluminium ammonium sulfate, ziram
	Damaging mammals	aluminium ammonium sulfate, ziram
	Mealy bugs	fatty acids
	Nematodes	methyl bromide with chloropicrin
	Overwintering pests	cresylic acid
	Red spider mites	fatty acids
	Scale insects	fatty acids
	Soil pests	methyl bromide with chloropicrin
	Whitefly	fatty acids
Weeds	Annual dicotyledons	paraquat
	Annual grasses	paraquat
	Annual weeds	glyphosate (agricultural)
	Creeping bent	paraquat

	General weed control	methyl bromide with chloropicrin
	Perennial dicotyledons	glyphosate (agricultural) *(wiper application)*, glyphosate (horticulture, forestry, amenity etc.) *(wiper application)*
	Perennial ryegrass	paraquat
	Perennial weeds	glyphosate (agricultural)
	Rough meadow grass	paraquat
Plant growth regulation	Increasing yield	sulfur

Tree nuts

Diseases	Gloeosporium	benomyl *(off-label)*
Weeds	Annual and perennial weeds	glyphosate (agricultural) *(off-label)*
	Annual weeds	glufosinate-ammonium
	Perennial weeds	glufosinate-ammonium

Grain/crop store uses

Food storage

Pests	Food storage pests	methyl bromide with amyl acetate

Stored cabbages

Diseases	Alternaria	iprodione
	Botrytis	carbendazim + metalaxyl *(off-label)*, iprodione
	Phytophthora	carbendazim + metalaxyl *(off-label)*

Stored grain/rapeseed/linseed

Pests	Birds	aluminium ammonium sulfate
	Damaging mammals	aluminium ammonium sulfate
	Grain storage pests	aluminium phosphide, chlorpyrifos-methyl, deltamethrin, etridiazole, etrimfos, fenitrothion, fenitrothion + permethrin + resmethrin, gamma-HCH *(empty)*, iodofenphos, methyl bromide, methyl bromide with chloropicrin, permethrin, pirimiphos-methyl

FOR FULL CONDITIONS OF USE ALWAYS READ THE PRODUCT LABEL

Stored products

Pests   Stored product pests   aluminium phosphide

Grassland

Grassland

Diseases   Crown rust   propiconazole, triadimefon

                 Damping off   thiabendazole + thiram, thiram *(seed treatment)*

                 Drechslera leaf spot   propiconazole

                 Powdery mildew   propiconazole, triadimefon

                 Rhynchosporium   propiconazole, triadimefon

                 Seed-borne diseases   benomyl *(off-label)*, thiabendazole + thiram

Pests   Aphids   pirimicarb

                 Birds   aluminium ammonium sulfate

                 Cutworms   gamma-HCH

                 Damaging mammals   aluminium ammonium sulfate, strychnine hydrochloride
                                             (commodity substance) *(areas of restricted public access)*

                 Frit fly   chlorpyrifos, triazophos

                 Leatherjackets   chlorpyrifos, gamma-HCH, triazophos

                 Slugs and snails   metaldehyde, methiocarb

                 Wireworms   gamma-HCH

Weeds   Annual and perennial   glyphosate (agricultural) *(pre-cut/graze)*, glyphosate
                 weeds   (agricultural) *(sward destruction)*

                 Annual dicotyledons   2,4-D, 2,4-D + dicamba + mecoprop, 2,4-D + dicamba +
                                             triclopyr, 2,4-D + mecoprop, 2,4-DB + linuron + MCPA,
                                             MCPA, MCPA + MCPB, benazolin + bromoxynil + ioxynil,
                                             bentazone + MCPA + MCPB, bentazone + cyanazine + 2,4-
                                             DB, bromoxynil + ioxynil + mecoprop-P, clopyralid, dicamba +
                                             MCPA + mecoprop, dicamba + MCPA + mecoprop-P, dicamba
                                             + mecoprop, dicamba + mecoprop + triclopyr, fluroxypyr,
                                             glyphosate (agricultural) *(sward destruction)*, mecoprop,
                                             mecoprop-P

                 Annual grasses   ethofumesate, glyphosate (agricultural) *(sward destruction)*

                 Annual weeds   glyphosate (agricultural)

                 Black bindweed   fluroxypyr

                 Blackgrass   ethofumesate

                 Bracken   asulam, glyphosate (agricultural) *(sward destruction)*

                 Brambles   2,4-D + dicamba + mecoprop, 2,4-D + dicamba + triclopyr,
                                             clopyralid + triclopyr, dicamba + mecoprop + triclopyr,
                                             triclopyr

Broom	clopyralid + triclopyr, dicamba + mecoprop + triclopyr, triclopyr
Buttercups	MCPA
Chickweed	bentazone + cyanazine + 2,4-DB, dicamba + MCPA + mecoprop, dicamba + MCPA + mecoprop-P, ethofumesate, fluroxypyr, mecoprop, mecoprop-P
Cleavers	dicamba + MCPA + mecoprop, dicamba + MCPA + mecoprop-P, ethofumesate, fluroxypyr, mecoprop, mecoprop-P
Corn marigold	clopyralid
Creeping thistle	clopyralid, clopyralid *(off-label, spot treatment)*
Destruction of short term leys	glyphosate (agricultural)
Docks	2,4-D + dicamba + triclopyr, MCPA, asulam, clopyralid + fluroxypyr + triclopyr, clopyralid + triclopyr, dicamba + MCPA + mecoprop, dicamba + mecoprop, dicamba + mecoprop + triclopyr, fluroxypyr, fluroxypyr + triclopyr, thifensulfuron-methyl, triclopyr
Fat-hen	MCPA
Gorse	2,4-D + dicamba + triclopyr, clopyralid + triclopyr, dicamba + mecoprop + triclopyr, triclopyr
Hemp-nettle	MCPA, fluroxypyr
Japanese knotweed	2,4-D + dicamba + triclopyr
Mayweeds	clopyralid, dicamba + MCPA + mecoprop, dicamba + MCPA + mecoprop-P
Perennial dicotyledons	2,4-D, 2,4-D + dicamba + mecoprop, 2,4-D + dicamba + triclopyr, 2,4-D + mecoprop, MCPA, MCPA + MCPB, clopyralid, clopyralid + fluroxypyr + triclopyr, clopyralid + triclopyr, clopyralid + triclopyr *(weed wiper -off-label)*, dicamba + MCPA + mecoprop, dicamba + MCPA + mecoprop-P, dicamba + mecoprop, dicamba + mecoprop + triclopyr, dicamba *(wiper application)*, glyphosate (agricultural) *(sward destruction)*, glyphosate (agricultural) *(wiper application)*, glyphosate (horticulture, forestry, amenity etc.) *(wiper application)*, mecoprop, mecoprop-P, triclopyr
Perennial grasses	glyphosate (agricultural) *(sward destruction)*
Polygonums	dicamba + MCPA + mecoprop, dicamba + MCPA + mecoprop-P
Rhododendrons	2,4-D + dicamba + triclopyr
Rushes	glyphosate (agricultural) *(sward destruction)*, triclopyr
Stinging nettle	2,4-D + dicamba + mecoprop, 2,4-D + dicamba + triclopyr, clopyralid + fluroxypyr + triclopyr, clopyralid + triclopyr, dicamba + mecoprop + triclopyr, triclopyr

FOR FULL CONDITIONS OF USE ALWAYS READ THE PRODUCT LABEL

	Volunteer cereals	ethofumesate
	Volunteer potatoes	fluroxypyr
	Woody weeds	2,4-D + dicamba + mecoprop, 2,4-D + dicamba + triclopyr, dicamba + mecoprop + triclopyr, triclopyr
Plant growth regulation	Increasing yield	sulfur

Leys

Pests	Frit fly	cypermethrin
Weeds	Annual dicotyledons	2,4-DB + MCPA, MCPA, MCPA + MCPB, MCPB, benazolin + 2,4-DB + MCPA, bromoxynil + ethofumesate + ioxynil, dicamba + MCPA + mecoprop, dicamba + mecoprop, dicamba + mecoprop + triclopyr, fluroxypyr, mecoprop, mecoprop-P, methabenzthiazuron
	Annual grasses	ethofumesate
	Annual meadow grass	bromoxynil + ethofumesate + ioxynil, methabenzthiazuron
	Black bindweed	fluroxypyr
	Blackgrass	ethofumesate
	Brambles	dicamba + mecoprop + triclopyr
	Broom	dicamba + mecoprop + triclopyr
	Chickweed	benazolin + 2,4-DB + MCPA, dicamba + MCPA + mecoprop, ethofumesate, fluroxypyr, mecoprop, mecoprop-P
	Cleavers	benazolin + 2,4-DB + MCPA, dicamba + MCPA + mecoprop, ethofumesate, fluroxypyr, mecoprop, mecoprop-P
	Docks	dicamba + MCPA + mecoprop, dicamba + mecoprop, dicamba + mecoprop + triclopyr, fluroxypyr
	Fat-hen	MCPA
	Gorse	dicamba + mecoprop + triclopyr
	Hemp-nettle	MCPA, fluroxypyr
	Mayweeds	dicamba + MCPA + mecoprop
	Perennial dicotyledons	2,4-DB + MCPA, MCPA, MCPA + MCPB, MCPB, dicamba + MCPA + mecoprop, dicamba + mecoprop, dicamba + mecoprop + triclopyr, mecoprop, mecoprop-P
	Polygonums	2,4-DB + MCPA, benazolin + 2,4-DB + MCPA, dicamba + MCPA + mecoprop
	Rough meadow grass	methabenzthiazuron
	Stinging nettle	dicamba + mecoprop + triclopyr
	Volunteer cereals	ethofumesate
	Volunteer potatoes	fluroxypyr
	Woody weeds	dicamba + mecoprop + triclopyr

Non-crop pest control

Farm buildings/yards

Diseases	Fungus diseases	formaldehyde (commodity substance)
Pests	Ants	fenitrothion
	Beetles	dichlorvos, fenitrothion
	Birds	aluminium ammonium sulfate
	Bugs	fenitrothion
	Cockroaches	fenitrothion
	Crickets	fenitrothion
	Damaging mammals	aluminium ammonium sulfate, bone oil, sodium cyanide
	Earwigs	fenitrothion
	Ectoparasites	iodofenphos
	Fleas	fenitrothion
	Flies	azamethiphos, bioallethrin + permethrin, cyromazine, dichlorvos, fenitrothion, iodofenphos, methomyl + (Z)-9-tricosene, phenothrin, phenothrin + tetramethrin, pyrethrins, tetramethrin
	General insect control	bioresmethrin
	Hide beetle	alpha-cypermethrin, iodofenphos
	Mealworms	alpha-cypermethrin, iodofenphos
	Mites	fenitrothion
	Mosquitoes	dichlorvos, phenothrin + tetramethrin
	Moths	fenitrothion
	Poultry ectoparasites	dichlorvos
	Poultry house pests	alpha-cypermethrin, fenitrothion
	Red mite	carbaryl
	Rodents	bromadiolone, calciferol + difenacoum, chloralose, chlorophacinone, coumatetralyl, difenacoum, diphacinone, sodium cyanide, warfarin, zinc phosphide
	Silverfish	fenitrothion
	Wasps	phenothrin + tetramethrin
	Wax moths	Bacillus thuringiensis

FOR FULL CONDITIONS OF USE ALWAYS READ THE PRODUCT LABEL

Farmland

Pests	Damaging mammals	aluminium phosphide, bone oil
	Rodents	aluminium phosphide
Weeds	Volunteer potatoes	dichlobenil *(blight prevention)*

Food storage areas

Pests	Ants	fenitrothion
	Beetles	fenitrothion
	Bugs	fenitrothion
	Cockroaches	cypermethrin, fenitrothion
	Crickets	fenitrothion
	Earwigs	fenitrothion
	Fleas	fenitrothion
	Flies	fenitrothion
	Food storage pests	methyl bromide
	Grain storage pests	cypermethrin
	Mites	fenitrothion
	Moths	cypermethrin, fenitrothion
	Silverfish	fenitrothion

Manure heaps

Pests	Flies	cyromazine, fenitrothion, permethrin, trichlorfon
	General insect control	bioresmethrin

Miscellaneous situations

Pests	Birds	carbon dioxide (commodity substance), paraffin oil (commodity substance) *(egg treatment)*
	Damaging mammals	strychnine hydrochloride (commodity substance) *(areas of restricted public access)*
	Rodents	carbon dioxide (commodity substance)
	Wasps	resmethrin + tetramethrin

Refuse tips

Pests	Ants	fenitrothion
	Beetles	fenitrothion
	Bugs	fenitrothion

	Cockroaches	fenitrothion
	Crickets	fenitrothion, iodofenphos
	Earwigs	fenitrothion
	Fleas	fenitrothion
	Flies	chlorpyrifos-methyl, fenitrothion, iodofenphos, pyrethrins, trichlorfon *(off-label)*
	General insect control	bioresmethrin
	Mites	fenitrothion
	Moths	fenitrothion
	Silverfish	fenitrothion

Stored plant material

Pests	General insect control	methyl bromide

Protected crops

Aubergines

Diseases	Botrytis	dichlofluanid *(off-label)*, iprodione *(off-label)*
	Damping off	propamocarb hydrochloride
	Root rot	propamocarb hydrochloride
Pests	Aphids	Verticillium lecanii, permethrin, resmethrin
	Leaf miners	cypermethrin *(protected crops, off-label)*, oxamyl *(off-label)*, trichlorfon *(off-label)*
	Sciarid flies	resmethrin
	Tomato fruitworm	permethrin
	Whitefly	Verticillium lecanii, buprofezin, permethrin, resmethrin

Beans

Pests	Aphids	Verticillium lecanii
	Whitefly	Verticillium lecanii

Cucumbers

Diseases	Black root rot	carbendazim (MBC)

FOR FULL CONDITIONS OF USE ALWAYS READ THE PRODUCT LABEL

	Botrytis	carbendazim (MBC), carbendazim (MBC) *(off-label)*, chlorothalonil, dicloran, iprodione
	Damping off	etridiazole, propamocarb hydrochloride, propamocarb hydrochloride *(off-label)*
	Downy mildew	copper oxychloride + metalaxyl *(off-label)*
	Powdery mildew	bupirimate, carbendazim (MBC), carbendazim (MBC) *(off-label)*, chlorothalonil, copper ammonium carbonate, fenarimol, imazalil, sulfur *(off-label)*, thiophanate-methyl
	Rhizoctonia	dicloran, quintozene
	Root rot	propamocarb hydrochloride, propamocarb hydrochloride *(off-label)*
Pests	Aphids	Verticillium lecanii, deltamethrin, demeton-S-methyl, fatty acids, permethrin, pirimicarb, propoxur, pyrethrins + resmethrin, resmethrin
	Caterpillars	Bacillus thuringiensis *(off-label)*, deltamethrin, permethrin
	Cutworms	cypermethrin
	French fly	pirimiphos-methyl
	Leaf miners	cypermethrin, cypermethrin *(protected crops, off-label)*, permethrin, trichlorfon *(off-label)*
	Leafhoppers	heptenophos
	Mealy bugs	deltamethrin, fatty acids, petroleum oil
	Mites	dicofol + tetradifon, fenbutatin oxide
	Red spider mites	demeton-S-methyl, dicofol, dicofol + tetradifon, fatty acids, fenbutatin oxide, petroleum oil, tetradifon
	Scale insects	deltamethrin, fatty acids, petroleum oil
	Sciarid flies	resmethrin
	Springtails	gamma-HCH
	Symphylids	gamma-HCH
	Tarsonemid mites	dicofol + tetradifon
	Thrips	gamma-HCH, heptenophos
	Tomato fruitworm	permethrin
	Whitefly	Verticillium lecanii, buprofezin, cypermethrin, deltamethrin, fatty acids, permethrin, propoxur, pyrethrins + resmethrin, resmethrin

General

Diseases	Botrytis	carbendazim (MBC), chlorothalonil, dicloran, iprodione *(off-label)*
	Fungus diseases	formaldehyde (commodity substance)
	Phytophthora	fosetyl-aluminium, propamocarb hydrochloride

	Powdery mildew	carbendazim (MBC), imazalil
	Rhizoctonia	dicloran
	Soil-borne diseases	cresylic acid, dazomet, formaldehyde (commodity substance), metam-sodium, methyl bromide with amyl acetate, methyl bromide with chloropicrin
Pests	Ants	cresylic acid, gamma-HCH
	Aphids	Verticillium lecanii, cypermethrin, deltamethrin, gamma-HCH, heptenophos *(off-label)*, nicotine, permethrin, rotenone
	Capsids	cypermethrin, gamma-HCH, nicotine
	Caterpillars	cypermethrin, deltamethrin, diflubenzuron, permethrin
	Cutworms	cypermethrin
	Earwigs	gamma-HCH
	General insect control	chlorpyrifos *(off-label)*, cypermethrin *(off-label)*, endosulfan *(off-label)*
	Leaf miners	abamectin, cypermethrin, deltamethrin *(off-label)*, dichlorvos *(off-label)*, gamma-HCH, heptenophos *(off-label)*, nicotine, pyrethrins + resmethrin *(off-label)*, resmethrin *(off-label)*, trichlorfon *(off-label)*
	Leafhoppers	nicotine
	Mealy bugs	deltamethrin
	Mites	dicofol + tetradifon, fenbutatin oxide
	Nematodes	dazomet, metam-sodium, methyl bromide with amyl acetate, methyl bromide with chloropicrin
	Red spider mites	abamectin, dicofol + tetradifon, dienochlor, fenbutatin oxide
	Rodents	chloralose
	Sawflies	nicotine
	Scale insects	deltamethrin
	Sciarid flies	gamma-HCH
	Slugs and snails	aluminium sulfate, cresylic acid, metaldehyde, methiocarb *(off-label)*
	Soil pests	dazomet, metam-sodium, methyl bromide with amyl acetate, methyl bromide with chloropicrin
	Tarsonemid mites	dicofol + tetradifon
	Thrips	cypermethrin, gamma-HCH, nicotine
	Western flower thrips	deltamethrin *(off-label)*, dichlorvos *(off-label)*
	Whitefly	Verticillium lecanii, buprofezin, cypermethrin, deltamethrin, gamma-HCH, nicotine, permethrin
	Woodlice	cresylic acid, gamma-HCH

FOR FULL CONDITIONS OF USE ALWAYS READ THE PRODUCT LABEL

	Woolly aphid	nicotine
Weeds	Algae	benzalkonium chloride
	Annual dicotyledons	chlorpropham + fenuron, isoxaben (off-label)
	Annual grasses	chlorpropham + fenuron
	Chickweed	chlorpropham + fenuron
	General weed control	dazomet, metam-sodium, methyl bromide with amyl acetate, methyl bromide with chloropicrin
	Mosses	benzalkonium chloride

Herbs

Diseases	Powdery mildew	sulfur (off-label)
Pests	Leaf miners	deltamethrin (off-label), dichlorvos (off-label)
	Western flower thrips	deltamethrin (off-label), dichlorvos (off-label)
Weeds	Annual dicotyledons	prometryn (off-label), propyzamide (off-label), terbacil (off-label), trifluralin (off-label)
	Annual grasses	prometryn (off-label), propyzamide (off-label), terbacil (off-label), trifluralin (off-label)
	Couch	terbacil (off-label)
	Perennial grasses	propyzamide (off-label), terbacil (off-label)

Lettuce

Diseases	Botrytis	dicloran, iprodione, propamocarb hydrochloride (off-label), thiram
	Downy mildew	fosetyl-aluminium, mancozeb, metalaxyl + thiram, propamocarb hydrochloride (off-label), thiram, zineb
	Rhizoctonia	dicloran, thiram, tolclofos-methyl
	Ring spot	prochloraz (off-label)
Pests	Aphids	Verticillium lecanii, deltamethrin + heptenophos (off-label), dimethoate, malathion, pirimicarb
	Cutworms	cypermethrin
	Leaf miners	cypermethrin
	Leafhoppers	malathion
	Mealy bugs	malathion
	Scale insects	malathion
	Thrips	malathion
	Whitefly	Verticillium lecanii, cypermethrin, malathion, permethrin
Weeds	Annual dicotyledons	chlorpropham with cetrimide
	Annual grasses	chlorpropham with cetrimide

	Chickweed	chlorpropham with cetrimide
	Polygonums	chlorpropham with cetrimide

Mushrooms

Diseases	Cobweb	carbendazim (MBC), prochloraz
	Dry bubble	carbendazim (MBC), chlorothalonil, prochloraz, tecnazene
	Fungus diseases	dichlorophen, formaldehyde (commodity substance)
	Trichoderma	carbendazim (MBC), carbendazim (MBC) *(spawn treatment - off-label)*
	Wet bubble	carbendazim (MBC), chlorothalonil, prochloraz, tecnazene
Pests	Caterpillars	deltamethrin *(off-label)*
	Leaf miners	deltamethrin *(off-label)*
	Mushroom flies	permethrin, pyrethrins + resmethrin *(Off-label)*
	Sciarid flies	chlorfenvinphos, diazinon, dichlorvos, diflubenzuron, malathion, methoprene, nicotine, permethrin, pyrethrins + resmethrin *(Off-label)*, resmethrin

Mustard and cress

Diseases	Damping off	etridiazole

Onions, leeks and garlic

Pests	Stem nematodes	oxamyl *(off-label)*
	Thrips	deltamethrin *(off-label)*

Peppers

Diseases	Botrytis	carbendazim (MBC) *(off-label)*, dichlofluanid *(off-label)*, iprodione *(off-label)*
	Damping off	etridiazole, propamocarb hydrochloride
	Powdery mildew	carbendazim (MBC) *(off-label)*, fenarimol *(off-label)*
	Root rot	propamocarb hydrochloride
Pests	Aphids	Verticillium lecanii, deltamethrin, fatty acids, permethrin, pirimicarb, resmethrin
	Caterpillars	Bacillus thuringiensis, deltamethrin, diflubenzuron, permethrin
	Leaf miners	oxamyl *(off-label)*, permethrin, trichlorfon *(off-label)*
	Mealy bugs	deltamethrin, fatty acids

FOR FULL CONDITIONS OF USE ALWAYS READ THE PRODUCT LABEL

	Red spider mites	fatty acids, tetradifon
	Scale insects	deltamethrin, fatty acids
	Sciarid flies	resmethrin
	Tomato fruitworm	permethrin
	Whitefly	Verticillium lecanii, buprofezin, deltamethrin, fatty acids, permethrin, resmethrin

Pot plants

Diseases	Botrytis	carbendazim (MBC), carbendazim (MBC) *(off-label)*, iprodione, quintozene, thiram
	Damping off	propamocarb hydrochloride
	Phytophthora	fosetyl-aluminium, furalaxyl, propamocarb hydrochloride
	Powdery mildew	carbendazim (MBC), carbendazim (MBC) *(off-label)*, pyrazophos
	Pythium	furalaxyl, propamocarb hydrochloride
	Rhizoctonia	quintozene
	Rust	mancozeb, oxycarboxin
	Sclerotinia	quintozene
Pests	Aphids	deltamethrin, heptenophos, malathion, permethrin, resmethrin
	Browntail moth	teflubenzuron
	Capsids	permethrin + thiram
	Caterpillars	deltamethrin, permethrin, permethrin + thiram, teflubenzuron
	Earwigs	permethrin + thiram
	Leaf miners	abamectin, permethrin, permethrin + thiram *(partial control)*
	Leafhoppers	malathion
	Mealy bugs	deltamethrin, malathion, petroleum oil
	Red spider mites	abamectin, petroleum oil
	Scale insects	deltamethrin, malathion, petroleum oil
	Sciarid flies	resmethrin, teflubenzuron
	Thrips	heptenophos, malathion
	Vine weevil	carbofuran
	Western flower thrips	teflubenzuron
	Whitefly	deltamethrin, malathion, permethrin, resmethrin, teflubenzuron
Plant growth regulation	Flower induction	2-chloroethylphosphonic acid
	Flower life prolongation	sodium silver thiosulfate
	Improving colour	paclobutrazol

147

Increasing branching	2-chloroethylphosphonic acid, dikegulac
Increasing flowering	paclobutrazol
Stem shortening	chlormequat, chlormequat + choline chloride, daminozide, paclobutrazol

Protected cucurbits

Pests

Bean seed flies	bendiocarb *(off-label seed treatment)*
Leaf miners	cypermethrin *(protected crops, off-label)*, heptenophos *(off-label)*, trichlorfon *(off-label)*
Red spider mites	tetradifon

Protected cut flowers

Diseases

Botrytis	quintozene, thiram
Fusarium wilt	carbendazim (MBC)
Powdery mildew	imazalil
Rhizoctonia	quintozene
Rust	mancozeb, oxycarboxin, thiram
Sclerotinia	quintozene
Soil-borne diseases	metam-sodium
Verticillium wilt	carbendazim (MBC)

Pests

Aphids	Verticillium lecanii, demeton-S-methyl, dimethoate, malathion, pirimicarb, propoxur, resmethrin
Leaf miners	dimethoate
Leafhoppers	malathion
Mealy bugs	malathion
Nematodes	metam-sodium
Red spider mites	demeton-S-methyl, tebufenpyrad
Scale insects	malathion
Sciarid flies	resmethrin
Soil pests	metam-sodium
Thrips	malathion
Whitefly	Verticillium lecanii, malathion, propoxur, resmethrin

Weeds

Annual dicotyledons	pentanochlor
Annual meadow grass	pentanochlor
General weed control	metam-sodium

FOR FULL CONDITIONS OF USE ALWAYS READ THE PRODUCT LABEL

Plant growth regulation	Basal bud stimulation	2-chloroethylphosphonic acid
	Flower life prolongation	sodium silver thiosulfate

Seed potatoes under protection

Diseases	Blight	cymoxanil + mancozeb *(off-label)*

Tomatoes

Diseases	Blight	Bordeaux mixture *(outdoor)*, chlorothalonil, copper ammonium carbonate, copper sulfate + sulfur, maneb + zinc, zineb
	Botrytis	carbendazim (MBC), chlorothalonil, dichlofluanid, dicloran, iprodione, quintozene, thiram
	Damping off	copper oxychloride, etridiazole, propamocarb hydrochloride, propamocarb hydrochloride *(off-label)*
	Didymella stem rot	carbendazim (MBC), maneb *(off-label)*
	Foot rot	copper oxychloride
	Fusarium wilt	carbendazim (MBC)
	Leaf mould	carbendazim (MBC), chlorothalonil, copper ammonium carbonate, dichlofluanid, maneb + zinc, zineb
	Phytophthora	copper oxychloride, propamocarb hydrochloride
	Powdery mildew	bupirimate *(off-label)*, fenarimol *(off-label)*, sulfur *(off-label)*
	Rhizoctonia	dicloran, quintozene
	Root diseases	etridiazole *(off-label)*
	Root rot	propamocarb hydrochloride, propamocarb hydrochloride *(off-label)*, zineb
	Sclerotinia	quintozene
	Soil-borne diseases	metam-sodium
	Verticillium wilt	carbendazim (MBC)
Pests	Aphids	deltamethrin, demeton-S-methyl, dimethoate, fatty acids, gamma-HCH, heptenophos, malathion, nicotine, permethrin, pirimicarb, propoxur, pyrethrins + resmethrin, resmethrin
	Capsids	gamma-HCH
	Caterpillars	Bacillus thuringiensis *(off-label)*, deltamethrin, deltamethrin *(off-label)*, diflubenzuron, permethrin
	Leaf miners	deltamethrin *(off-label)*, gamma-HCH, oxamyl *(off-label)*, permethrin, trichlorfon *(off-label)*
	Leafhoppers	gamma-HCH, heptenophos, malathion
	Mealy bugs	deltamethrin, fatty acids, malathion, petroleum oil
	Mites	dicofol + tetradifon, fenbutatin oxide

149

	Nematodes	metam-sodium
	Red spider mites	demeton-S-methyl, dicofol, dicofol + tetradifon, dimethoate, fatty acids, fenbutatin oxide, petroleum oil, tetradifon
	Scale insects	deltamethrin, fatty acids, malathion, petroleum oil
	Sciarid flies	resmethrin
	Soil pests	metam-sodium
	Springtails	gamma-HCH
	Symphylids	gamma-HCH
	Tarsonemid mites	dicofol + tetradifon
	Thrips	gamma-HCH, malathion
	Tomato fruitworm	permethrin
	Tomato moth	Bacillus thuringiensis
	Whitefly	buprofezin, deltamethrin, fatty acids, malathion, permethrin, propoxur, pyrethrins + resmethrin, resmethrin
Weeds	Annual dicotyledons	pentanochlor
	Annual meadow grass	pentanochlor
	General weed control	metam-sodium
Plant growth regulation	Fruit ripening	2-chloroethylphosphonic acid
	Increasing fruit set	(2-naphthyloxy)acetic acid

Setaside

Setaside

Weeds	Annual and perennial weeds	glyphosate (agricultural)
	Annual dicotyledons	glufosinate-ammonium
	Annual grasses	fluazifop-P-butyl *(off-label)*, glufosinate-ammonium
	Barren brome	fluazifop-P-butyl *(off-label)*
	Green cover	cycloxydim, glyphosate (agricultural)
	Perennial dicotyledons	glufosinate-ammonium
	Perennial grasses	fluazifop-P-butyl *(off-label)*, glufosinate-ammonium
	Volunteer cereals	fluazifop-P-butyl *(off-label)*

FOR FULL CONDITIONS OF USE ALWAYS READ THE PRODUCT LABEL

Total vegetation control

Non-crop areas

Diseases	Fungus diseases	dichlorophen
Weeds	Algae	benzalkonium chloride, cresylic acid, dichlorophen
	Annual and perennial weeds	glyphosate (agricultural), glyphosate (horticulture, forestry, amenity etc.)
	Annual dicotyledons	2,4-D + dicamba + mecoprop, 2,4-D + dicamba + triclopyr, MCPA, chlorpropham + cresylic acid + fenuron, dicamba + mecoprop + triclopyr, diquat + paraquat, diuron + glyphosate, paraquat, picloram
	Annual grasses	chlorpropham + cresylic acid + fenuron, diquat + paraquat, diuron + glyphosate, paraquat
	Annual weeds	diuron, glufosinate-ammonium, glyphosate (agricultural)
	Bracken	asulam, dicamba, glyphosate (horticulture, forestry, amenity etc.), imazapyr, picloram
	Brambles	2,4-D + dicamba + mecoprop, 2,4-D + dicamba + triclopyr, dicamba + mecoprop + triclopyr, triclopyr
	Broom	dicamba + mecoprop + triclopyr, triclopyr
	Chickweed	chlorpropham + cresylic acid + fenuron
	Couch	amitrole
	Creeping bent	amitrole, paraquat
	Creeping thistle	MCPA, amitrole
	Daisies	MCPA
	Docks	2,4-D + dicamba + triclopyr, MCPA, amitrole, dicamba + mecoprop + triclopyr, triclopyr
	Gorse	2,4-D + dicamba + triclopyr, dicamba + mecoprop + triclopyr, triclopyr
	Green cover	glyphosate (agricultural)
	Japanese knotweed	2,4-D + dicamba + triclopyr, picloram
	Lichens	cresylic acid
	Liverworts	cresylic acid
	Mosses	benzalkonium chloride, cresylic acid, dichlorophen
	Perennial dicotyledons	2,4-D + dicamba + mecoprop, 2,4-D + dicamba + triclopyr, MCPA, dicamba + mecoprop + triclopyr, diuron + glyphosate, picloram, triclopyr
	Perennial grasses	amitrole, diquat + paraquat, diuron + glyphosate
	Perennial ryegrass	paraquat
	Perennial weeds	diuron, glufosinate-ammonium, glyphosate (agricultural)

	Rhododendrons	2,4-D + dicamba + triclopyr
	Rough meadow grass	paraquat
	Rushes	triclopyr
	Stinging nettle	2,4-D + dicamba + mecoprop, 2,4-D + dicamba + triclopyr, dicamba + mecoprop + triclopyr, triclopyr
	Total vegetation control	amitrole, amitrole + 2,4-D + diuron, amitrole + bromacil + diuron, bromacil, bromacil + diuron, bromacil + picloram, dalapon + dichlobenil, dichlobenil, diuron + paraquat, glyphosate (agricultural) *(amenity situations)*, glyphosate (horticulture, forestry, amenity etc.) *(amenity situations)*, imazapyr, sodium chlorate
	Volunteer cereals	diquat + paraquat
	Volunteer potatoes	dichlobenil
	Woody weeds	2,4-D + dicamba + mecoprop, 2,4-D + dicamba + triclopyr, dicamba + mecoprop + triclopyr, fosamine-ammonium, picloram, triclopyr
Crop control	Green cover	glyphosate (agricultural)
Plant growth regulation	Growth suppression	maleic hydrazide

Weeds in or near water

Weeds	Aquatic weeds	2,4-D, dichlobenil, diquat, glyphosate (agricultural), glyphosate (horticulture, forestry, amenity etc.), terbutryn
	Bracken	asulam
	Docks	asulam
	Perennial dicotyledons	2,4-D
	Perennial grasses	glyphosate (agricultural), glyphosate (horticulture, forestry, amenity etc.)
	Rushes	glyphosate (agricultural), glyphosate (horticulture, forestry, amenity etc.)
	Sedges	glyphosate (agricultural), glyphosate (horticulture, forestry, amenity etc.)
	Waterlilies	glyphosate (agricultural), glyphosate (horticulture, forestry, amenity etc.)
	Woody weeds	fosamine-ammonium
Plant growth regulation	Growth suppression	maleic hydrazide

FOR FULL CONDITIONS OF USE ALWAYS READ THE PRODUCT LABEL

Turf/amenity grass

Turf/amenity grass

Diseases	Anthracnose	carbendazim + chlorothalonil, carbendazim + iprodione, chlorothalonil
	Brown patch	chlorothalonil, iprodione
	Cladosporium leaf spot	carbendazim + iprodione
	Dollar spot	carbendazim (MBC), carbendazim + chlorothalonil, fenarimol, iprodione, quintozene, thiabendazole, thiophanate-methyl
	Fairy rings	oxycarboxin, triforine
	Fungus diseases	dichlorophen
	Fusarium patch	carbendazim (MBC), carbendazim + chlorothalonil, carbendazim + iprodione, chlorothalonil, fenarimol, iprodione, quintozene, thiabendazole, thiophanate-methyl
	Grey snow mould	chlorothalonil, iprodione
	Melting out	chlorothalonil, iprodione
	Red thread	carbendazim (MBC), carbendazim + chlorothalonil, carbendazim + iprodione, chlorothalonil, dichlorophen, fenarimol, iprodione, quintozene, thiabendazole, thiophanate-methyl
	Take-all patch	chlorothalonil
Pests	Chafer grubs	gamma-HCH
	Earthworms	carbaryl, carbendazim (MBC), gamma-HCH + thiophanate-methyl
	Frit fly	chlorpyrifos
	Leatherjackets	chlorpyrifos, gamma-HCH, gamma-HCH + thiophanate-methyl
	Millipedes	gamma-HCH
	Nematodes	methyl bromide with chloropicrin
	Soil pests	methyl bromide with chloropicrin
	Wireworms	gamma-HCH
Weeds	Algae	dichlorophen
	Annual and perennial weeds	glyphosate (horticulture, forestry, amenity etc.) *(directed spray)*, glyphosate (horticulture, forestry, amenity etc.) *(pre-planting/sowing)*, glyphosate (horticulture, forestry, amenity etc.) *(wiper application)*

Annual dicotyledons	2,4-D, 2,4-D + dicamba, 2,4-D + dicamba + ferrous sulfate, 2,4-D + dicamba + mecoprop, 2,4-D + mecoprop, 2,4-D + picloram, MCPA, MCPA + mecoprop, MCPA + mecoprop-P, chlormequat + choline chloride + imazaquin, clopyralid + diflufenican + MCPA, dicamba + MCPA + mecoprop, dicamba + MCPA + mecoprop-P, dicamba + dichlorprop + MCPA, dicamba + maleic hydrazide + MCPA, dicamba + paclobutrazol, dichlorprop + MCPA, dichlorprop + ferrous sulfate + MCPA, fluroxypyr + mecoprop-P, ioxynil, isoxaben, mecoprop, picloram
Annual grasses	ethofumesate
Blackgrass	ethofumesate
Bracken	dicamba, picloram
Brambles	2,4-D + dicamba + mecoprop, 2,4-D + picloram, triclopyr
Broom	triclopyr
Buttercups	clopyralid + diflufenican + MCPA, dichlorprop + MCPA
Chickweed	dicamba + MCPA + mecoprop-P, ethofumesate, mecoprop
Cleavers	dicamba + MCPA + mecoprop-P, ethofumesate
Clovers	dichlorprop + MCPA, mecoprop
Corn marigold	clopyralid + diflufenican + MCPA
Creeping thistle	2,4-D + picloram, MCPA, clopyralid + diflufenican + MCPA
Daisies	MCPA, dichlorprop + MCPA
Docks	2,4-D + picloram, MCPA, triclopyr
General weed control	methyl bromide with chloropicrin
Gorse	triclopyr
Japanese knotweed	2,4-D + picloram, picloram
Mayweeds	clopyralid + diflufenican + MCPA, dicamba + MCPA + mecoprop-P
Mosses	2,4-D + dicamba + ferrous sulfate, cresylic acid, dichlorophen, dichlorophen + ferrous sulfate, dichlorprop + ferrous sulfate + MCPA, ferrous sulfate
Perennial dicotyledons	2,4-D, 2,4-D + dicamba, 2,4-D + dicamba + ferrous sulfate, 2,4-D + dicamba + mecoprop, 2,4-D + mecoprop, 2,4-D + picloram, MCPA, MCPA + mecoprop, MCPA + mecoprop-P, chlormequat + choline chloride + imazaquin, clopyralid + diflufenican + MCPA, dicamba + MCPA + mecoprop, dicamba + MCPA + mecoprop-P, dicamba + dichlorprop + MCPA, dicamba + maleic hydrazide + MCPA, dicamba + paclobutrazol, dicamba *(wiper application)*, dichlorprop + MCPA, dichlorprop + ferrous sulfate + MCPA, fluroxypyr + mecoprop-P, mecoprop, picloram, triclopyr

FOR FULL CONDITIONS OF USE ALWAYS READ THE PRODUCT LABEL

Polygonums	dicamba + MCPA + mecoprop-P
Ragwort	2,4-D + picloram
Slender speedwell	chlorthal-dimethyl, fluroxypyr + mecoprop-P
Stinging nettle	2,4-D + dicamba + mecoprop, triclopyr
Volunteer cereals	ethofumesate
Weed grasses	amitrole *(off-label)*
Woody weeds	2,4-D + dicamba + mecoprop, 2,4-D + picloram, picloram, triclopyr

Plant growth regulation

Growth retardation	dicamba + maleic hydrazide + MCPA, dicamba + paclobutrazol
Growth suppression	maleic hydrazide, mefluidide
Increasing yield	sulfur

SECTION 3
PESTICIDE PROFILES

1   abamectin

A selective acaricide for use in ornamentals

| Products | Dynamec | MSD AGVET | 18 g/l | EC | 06804 |

Uses   American serpentine leaf miner, two-spotted spider mite in FLOWERS, ORNAMENTALS, PROTECTED FLOWERS, PROTECTED ORNAMENTALS.

Notes   **Efficacy**
* Treat at first sign of infestation. Repeat sprays may be required
* For effective control total cover of all plant surfaces is essential, but avoid run-off
* Mites quickly become immobilised but 3-5 d may be required for maximum mortality

Crop Safety/Restrictions
* Maximum concentration must not exceed 50 ml per 100 l water
* Do not use on ferns (*Adiantum* spp)
* Some spotting or staining may occur on carnation, kalanchoe and begonia foliage
* Consult manufacturer for list of plant varieties tested for safety
* There is insufficient evidence to support product compatibility with integrated and biological pest control programmes

Special precautions/Environmental safety
* Harmful if swallowed, in contact with skin and by inhalation
* Irritating to eyes
* Extremely dangerous to fish or other aquatic life. Do not contaminate surface waters or ditches with chemical or used container
* Extremely dangerous to bees. Do not apply to crops in flower or to those in which bees are actively foraging. Do not apply when flowering weeds are present
* Unprotected persons must be kept out of treated areas until the spray has dried
* Keep in original container, tightly closed, in a safe place, under lock and key

Protective clothing/Label precautions
* A, C, D, H, K, M
* 5a, 5b, 5c, 6a, 12c, 14, 18, 22, 28, 29, 34, 36, 37, 38, 48a, 51, 61a, 64, 66, 70, 79

2   alachlor

A soil-acting herbicide approvals for which expired on 30 June 1992

3   aldicarb

A soil-applied, systemic carbamate insecticide and nematicide

Products					
1 Standon Aldicarb 10G	Standon	10% w/w	GR	05915
2 Temik 10G	RP Agric.	10% w/w	GR	06210

Uses   Aphids, cabbage root fly, flea beetles, leaf miners in BRASSICAS [1, 2]. Aphids, docking disorder vectors, free-living nematodes, leaf miners, millipedes, pygmy beetle in SUGAR BEET [1, 2]. Aphids, free-living nematodes, potato cyst nematode in POTATOES [1, 2]. Aphids, free-living nematodes in CARROTS, PARSNIPS [1, 2]. Cabbage root fly, flea beetles, free-living nematodes in ROOT BRASSICAS [1, 2]. Frit fly in SWEETCORN *(off-label)* [2]. Stem nematodes in BULB ONIONS [1, 2]. Stem nematodes in LEEKS *(off-label)* [2].

Notes

Efficacy
* Must be incorporated into soil by physical means. See label for details of application rates, timing, suitable applicators and techniques of incorporation
* Persistence and activity may be reduced in very wet soils or where pH exceeds 8.0. Do not use within 14 d of liming
* Use in potatoes reduces incidence of spraing disease

Crop Safety/Restrictions
* Maximum number of treatments 1 per crop. No edible crops other than those listed (see label) should be planted into treated soil for at least 8 wk after application

Special precautions/Environmental safety
* Aldicarb is subject to the Poisons Rules 1982 and the Poisons Act 1972. See notes in Section 1
* This product contains an anticholinesterase carbamate compound. Do not use if under medical advice not to work with such compounds
* Keep in original container, tightly closed, in a safe place, under lock and key
* Toxic in contact with skin, by inhalation and if swallowed
* Dangerous to game, wild birds and animals. Cover granules completely and immediately after application. Bury spillages. Failure to bury granules immediately and completely is hazardous to wildlife
* Dangerous to livestock
* Dangerous to fish or other aquatic life. Do not contaminate surface waters or ditches with chemical or used container
* Keep unprotected persons out of treated glasshouses for at least 1 d
* Wear gloves when handling pots treated in past 4 wk

Protective clothing/Label precautions
* A, B, C or D+E, H, K, M
* 2, 4a, 4b, 4c, 14, 16, 18, 22, 25, 28, 29, 36, 37, 38, 40, 45, 52, 64, 67, 70, 79

Withholding period
* Keep all livestock out of treated areas for at least 13 wk. Bury or remove spillages

Latest application/Harvest Interval(HI)
* At planting/sowing for sugar beet, potatoes, onions.
* HI potatoes 8 wk; brassicas 10 wk; carrots, parsnips 12 wk; spring sown bulb onions do not harvest until mature bulb stage

Approval
* Off-label approval unlimited for use on leeks, sweetcorn (OLA 0159, 0160/92)[2]

4   aldrin

A persistent organochlorine insecticide approvals for sale, supply, storage and use of which were revoked in May 1989

FOR FULL CONDITIONS OF USE ALWAYS READ THE PRODUCT LABEL

5   alpha-cypermethrin

A contact and ingested pyrethroid insecticide for use in arable crops

Products

1 Acquit	DuPont	100 g/l	EC	07000
2 Contest	Cyanamid	100 g/l	EC	07001
3 Fastac	Cyanamid	100 g/l	EC	07008
4 Littac	Sorex	15 g/l	SC	H5176

Uses

Brassica pod midge, cabbage seed weevil, cabbage stem flea beetle, pollen beetles, rape winter stem weevil in WINTER OILSEED RAPE [1, 3]. Brassica pod midge, cabbage stem flea beetle, pollen beetles, rape winter stem weevil, seed weevil in OILSEED RAPE [2]. Cabbage aphid, caterpillars, flea beetles in BROCCOLI, BRUSSELS SPROUTS, CABBAGES, CALABRESE, CAULIFLOWERS, KALE [1, 3]. Cabbage seed weevil, pollen beetles in SPRING OILSEED RAPE [1, 3]. Cereal aphids in CEREALS [1-3]. Hide beetle, lesser mealworm, poultry house pests in POULTRY HOUSES [4].

Notes

Efficacy

- For cabbage stem flea beetle control spray oilseed rape when adult or larval damage first seen and about 1 mth later [1-3]
- For flowering pests on oilseed rape apply at any time during flowering, on pollen beetle best results achieved at green to yellow bud stage (GS 3,3-3,7), on seed weevil between 20 pods set stage and 80% petal fall (GS 4,7-5,8)
- Spray cereals in autumn for control of cereal aphids, in spring/summer for grain aphids. (See label for details) [1-3]
- For flea beetle, caterpillar and cabbage aphid control on brassicas apply when the pest or damage first seen or as a preventive spray. Repeat if necessary [1-3]
- For lesser mealworm control in poultry houses apply a coarse, low-pressure spray as routine treatment after clean-out and before each new crop. Spray vertical surfaces and ensure an overlap onto ceilings. It is not necessary to treat the floor [4]

Crop Safety/Restrictions

- Apply up to 4 sprays per crop on edible brassicas, up to 3 sprays per crop on winter oilseed rape, up to 2 on spring oilseed rape (only 1 after yellow bud stage - GS 3,7)
- Apply up to 2 sprays on cereals in autumn and spring, 1 in summer between 1 Apr and 31 Aug. See label for details of rates
- Only 1 aphicide treatment may be applied in cereals between 1 Apr and 31 Aug in any one year
- Treatment presents minimal hazard to bees but on flowering crops spray in evening, early morning or in dull weather as a precaution

Special precautions/Environmental safety

- Harmful in contact with skin or if swallowed
- Irritating to skin. Risk of serious damage to eyes
- Flammable
- Dangerous to bees. Do not apply at flowering stage except as directed on oilseed rape and cereals. Keep down flowering weeds in all crops
- Extremely dangerous to fish or other aquatic life. Do not contaminate surface waters or ditches with chemical or used container
- Do not allow spray from ground sprayers to fall within 6 m or the direct spray from hand-held sprayers to fall within 2 m of surface waters or ditches
- For summer cereal application do not spray within 6 m from edge of crop
- Reduced volume spraying must not be used

- Do not apply directly to poultry; collect eggs before application [4]

Protective clothing/Label precautions
- A, C, H
- 5a, 5c, 6b, 9, 12c, 14, 16, 18, 36, 48, 60, 66, 70, 78 [1, 3]; 22 [2]; 28, 29, 37, 51, 63 [1-3]

Latest application/Harvest Interval(HI)
- Before the end of flowering for oilseed rape, before 31 Mar in year of harvest for cereals (autumn and spring application), before early dough stage (GS 83) for cereals (summer application).
- HI brassicas 7.d

6   aluminium ammonium sulfate
An inorganic bird and animal repellent

Products

1	Curb Crop Spray Powder	Sphere	88% w/w	WP	02480
2	Guardsman	Chiltern	83 g/l	SL	05494
3	Guardsman STP Seed Dressing Powder	Sphere	88% w/w	DP	03606
4	Liquid Curb Crop Spray	Sphere		SC	03164
5	Scoot	Garotta	88% w/w	SP	07388
6	Stay Off	Vitax	88% w/w	SP	02019

Uses

Birds, damaging mammals, rabbits in BRASSICAS, BUSH FRUIT, CEREALS, OILSEED RAPE, PEAS [2, 6]. Birds, damaging mammals in FIELD CROPS, FRUIT CROPS, ORNAMENTALS, VEGETABLES [1, 4, 5]. Birds, damaging mammals in CORMS, FLOWER BULBS, SEEDS [3, 6]. Birds, damaging mammals in BEANS, CANE FRUIT, CARROTS, FARM BUILDINGS, FOREST NURSERY BEDS, FORESTRY PLANTATIONS, GRAIN STORES, GRASSLAND, STRAWBERRIES, SUGAR BEET, TOP FRUIT [2]. Birds, rabbits in FRUIT TREES [6].

Notes   **Efficacy**
- Apply as overall spray to growing crops before damage starts or mix powder with seed depending on type of protection required
- Spray deposit protects growth present at spraying but gives little protection to new growth
- Product must be sprayed onto dry foliage to be effective and must dry completely before dew or frost forms. In winter this may require some wind
- Use of additional sticker (coating agent) recommended to prolong activity [6]

Protective clothing/Label precautions
- 17, 23, 66, 80 [2]; 29, 54, 63, 67 [1-6]; 73, 74 [2, 3, 6]

Latest application/Harvest Interval(HI)
- HI fruit crops 6 wk

FOR FULL CONDITIONS OF USE ALWAYS READ THE PRODUCT LABEL

7   aluminium phosphide

A phosphine generating compound used against vertebrates and grain store pests

Products				
1 Amos Talunex	Luxan	57% w/w	GE	06567
2 Luxan Talunex	Luxan	57% w/w	GE	06563
3 Phostek	Killgerm	57% w/w	GE	05115
4 Phostek	Killgerm	57% w/w	GE	05116
5 Phostoxin	Rentokil	57% w/w	GE	01775

Uses

Grain storage pests in GRAIN STORES [4]. Mice, moles, rabbits, rats in FARMLAND, FORESTRY [1, 2, 4, 5]. Stored product pests in STORED PRODUCTS [3].

Notes

Efficacy
* Product releases poisonous hydrogen phosphide gas in contact with moisture
* Place pellets in burrows or runs and seal hole by heeling in or covering with turf. Do not cover pellets with soil. Inspect daily and treat any new or re-opened holes
* Apply pellets by means of Luxan Topex Applicator [1, 2]
* See label for details of fumigation of grain in silos and commodities not stored in bulk [3, 4]

Special precautions/Environmental safety
* Aluminium phosphide is subject to the Poisons Rules 1982 and the Poisons Act 1972. See notes in Section 1
* Only to be used by farmers, growers, foresters or other qualified users in the course of their business. See label for full precautions
* Keep in original container, tighty closed, in a safe place, under lock and key
* Very toxic by inhalation, in contact with skin and if swallowed
* Product liberates toxic, highly flammable gas
* Highly flammable
* Spontaneous combustion can arise due to sudden release of phosphine gas if container, having been opened once, is then re-opened
* Wear suitable protective gloves (synthetic rubber/plastics) when handling product
* Only open container outdoors [1, 2, 5], in well ventilated space [3, 4], and for immediate use. Keep away from liquid or water as this causes immediate release of gas. Do not use in wet weather
* Do not use within 10 m of human or animal habitation [1, 2, 5], before application ensure that no humans or domestic animals are in adjacent buildings or structures [3, 4]
* Dangerous to fish or other aquatic life. Do not contaminate surface waters or ditches with chemical or used container
* Pellets must not be placed or allowed to remain on ground surface
* Do not use adjacent to watercourses
* Dust remaining after decomposition is harmless and of no environmental hazard

Protective clothing/Label precautions
* A
* 3a, 3b, 3c, 12b, 19, 22, 23, 25, 26, 27, 28, 29, 37, 61, 64, 65

8   aluminium sulfate

An inorganic salt for slug and snail control

Products					
Growing Success Slug Killer	Growing Success	99.5% w/w	SG	04386

Uses   Slugs, snails in FIELD CROPS, FRUIT CROPS, ORNAMENTALS, PROTECTED CROPS, VEGETABLES.

Notes   **Efficacy**
* Apply granules to soil surface. Product has contact effect
* Best results achieved during mild, damp weather when slugs and snails active

9   amidosulfuron

A post-emergence sulfonylurea herbicide for cleavers and other broad-leaved weed control in cereals

Products				
1 Eagle	AgrEvo	75% w/w	WG	07318
2 Eagle	Hoechst	75% w/w	WG	06980

Uses   Annual dicotyledons, cleavers in BARLEY, DURUM WHEAT, OATS, RYE, TRITICALE, WHEAT.

Notes   **Efficacy**
* For optimum results apply in spring (from 1 Feb) in warm weather when soil moist and weeds growing actively
* Weed kill is slow, especially under cool, dry conditions. Weeds may sometimes only be stunted but will have little or no competitive effect on crop
* May be used on all soil types
* Spray is rainfast after 1 h
* Cleavers controlled from emergence to flower bud stage. If present at application charlock (up to flower bud), shepherds purse (up to flower bud) and field forget-me-not (up to 6 leaves) will also be controlled

Crop Safety/Restrictions
* Maximum number of treatments 1 per crop
* Apply from 2 leaf stage of crop up to and including first awns visible (GS 12-49)
* Do not apply to crops undersown or due to be undersown with clover or lucerne
* Broadcast crops should be sprayed post-emergence after plants have a well established root system
* Do not spray crops under stress, suffering drought, waterlogged, grazed, lacking nutrients or if soil compacted
* Do not spray if frost expected
* Do not roll or harrow within 1 wk of spraying

FOR FULL CONDITIONS OF USE ALWAYS READ THE PRODUCT LABEL

- Only cereals, winter oilseed rape, mustard, winter field beans, vetches may be sown in the same year as treatment
- Do not spray in tank mixture, or in sequence, with a product containing any other sulfonylurea
- If crop fails cereals may be sown after 15 d
- Avoid drift onto neighbouring broad-leaved plants or onto surface waters or ditches
- Take care to wash out sprayers thoroughly. See label for details

Special precautions/Environmental safety
- Dangerous to fish or other aquatic life. Do not contaminate surface waters or ditches with chemical or used container

Protective clothing/Label precautions
- 29, 52, 65

Latest application/Harvest Interval(HI)
- Before first spikelets just visible (GS 51)

10   2-aminobutane

A fumigant alkylamine fungicide for use in stored seed potatoes

Products					
Hortichem 2-Aminobutane	Hortichem	720 g/l	VP	06147

Uses Gangrene, silver scurf, skin spot in SEED POTATOES.

Notes

Efficacy
- Treatment must only be carried out by trained operators in suitable fumigation chambers under licence from the British Technology Group
- Fumigate within 21 d of lifting

Crop Safety/Restrictions
- Maximum number of treatments 1 per batch
- Do not treat immature tubers. Allow period of healing before treating damaged tubers

Special precautions/Environmental safety
- Keep in original container, tightly closed, in a safe place, under lock and key
- Harmful by inhalation. Irritating to skin, eyes and respiratory system
- Highly flammable. Keep away from sources of ignition. No smoking
- Do not empty into drains
- Do not contaminate surface waters or ditches with chemical or used container
- Do not supply treated potatoes for consumption by humans or lactating dairy cows
- Use must be in accordance with approved Code of Practice for the Control of Substances Hazardous to Health: Fumigation Operations
- The quantity of potatoes to be fumigated in a single stack must not exceed 2000 tonnes

Protective clothing/Label precautions
- A, C
- 5a, 6a, 6b, 6c, 12b, 14, 16, 18, 24, 25, 28, 29, 36, 37, 54, 64, 65, 78

Maximum Residue Levels (mg residue/kg food)
- citrus fruits 5; potatoes 1

11 amitraz

An amidine acaricide and insecticide for use in top fruit and hops

Products Mitac HF Promark 200 g/l EC 07358

Uses Pear sucker, red spider mites in PEARS. Red spider mites, rust mite in APPLES. Two-spotted spider mite in HOPS.

Notes **Efficacy**
- For red spider mites on apples and pears spray at 60-80% egg hatch and repeat 3 wk later, on hops when numbers build up and repeat every 3 wk as necessary
- For pear sucker control spray when significant numbers of nymphs have hatched but before there is significant contamination of the fruit with honeydew, normally June/July
- Best results achieved in dry conditions, do not spray if rain imminent

Crop Safety/Restrictions
- Maximum total dose 7 l/ha per yr

Special precautions/Environmental safety
- Harmful in contact with skin and if swallowed. Flammable
- Harmful to fish. Do not contaminate surface waters or ditches with chemical or used container
- Keep in original container, tightly closed, in a safe place, under lock and key

Protective clothing/Label precautions
- A, C, H
- 5a, 5c, 14, 16, 18, 21, 25, 28, 29, 35 (2 wk apples/pears), 36, 37, 53, 58, 60, 64, 66, 70, 78

Latest application/Harvest Interval(HI)
- HI apples, pears 2 wk; hops 7 wk

12 amitrole

A translocated, foliar-acting, non-selective triazole herbicide

Products
1 MSS Aminotriazole Technical Mirfield 98% w/w TC 04645
2 Weedazol-TL Bayer 225 g/l SL 02979

Uses Annual weeds, couch, creeping thistle, docks, perennial weeds in APPLES, PEARS [2]. Annual weeds, couch, docks, perennial weeds in FALLOWS, FIELD CROPS *(pre-sowing, autumn stubble)*, HEADLANDS [2]. Bent grasses, couch, creeping bent, creeping thistle, docks, total vegetation control in NON-CROP AREAS [1]. Contaminant grasses in AMENITY TURF - AMITROLE RESISTANT *(off-label)* [2]. Couch in WINTER WHEAT *(direct-drilled)* [2]. General weed control in APRICOTS *(off-label)*, CHERRIES *(off-label)*, PEACHES *(off-label)*, PLUMS *(off-label)*, QUINCES *(off-label)* [2]. Volunteer potatoes in BARLEY *(stubble)* [2].

FOR FULL CONDITIONS OF USE ALWAYS READ THE PRODUCT LABEL

Notes

Efficacy
* In non-crop land may be applied at any time from Apr to Oct. Best results achieved in spring or early summer when weeds growing actively. For coltsfoot, hogweed and horsetail summer and autumn applications are preferred [1]
* In cropland apply when couch in active growth and foliage at least 7.5 cm high [2]
* In fallows and stubble plough 3-6 wk after application to depth of 20 cm, taking care to seal the furrow [2]

Crop Safety/Restrictions
* Maximum number of treatments 1 per yr for stone fruit, 2 per yr for amenity turf
* Keep off suckers or foliage of desirable trees or shrubs [1]
* Do not spray areas in which the roots of adjacent trees or shrubs extend [1]
* Do not spray on sloping ground when rain imminent and run-off may occur [1]
* Allow specified interval between treatment and sowing crops (see label) [2]
* Apply in autumn on land to be used for spring barley [2]
* Do not sow direct-drilled winter wheat less than 2 wk after application [2]
* Apply round base of established fruit trees taking care to avoid contaminating trees, especially where bark is damaged. Keep off trees [2]

Special precautions/Environmental safety
* Harmful to fish. Do not contaminate surface waters or ditches with chemical or used container

Protective clothing/Label precautions
* A, C [2]; A [1]
* 21, 29, 53 [1]; 22, 36, 54 [2]; 28, 63, 66 [1, 2]

Latest application/Harvest Interval(HI)
* After harvest but before 30 Jun around fruit trees [2]

Approval
* Off-label approval unlimited for use on amenity turf (OLA 0693/91) and around stone fruit trees (OA 0333/92) [2]

Maximum Residue Levels (mg residue/kg food)
* tea, hops 0.1; fruits, vegetables, pulses, oilseeds, potatoes 0.05

---

13   amitrole + bromacil + diuron

A total herbicide mixture of translocated and residual chemicals

Products   BR Destral   RP Environ.   52.8:17.8:17.8%   WP   05184
                                                     w/w

Uses   Total vegetation control in NON-CROP AREAS, RAILWAY TRACKS.

Notes   **Efficacy**
* Apply as foliage spray at medium to high volume at any time during growing season
* Ensure continuous mechanical or hydraulic agitation during spraying to prevent settling

Crop Safety/Restrictions
* Do not apply or drain or flush equipment on or near young trees, shrubs or other desirable plants or over areas where their roots may extend

- Do not use where chemical may come into contact with roots of desirable plants

Special precautions/Environmental safety
- Irritating to skin, eyes and respiratory system
- Harmful to fish. Do not contaminate surface waters or ditches with chemical or used container

Protective clothing/Label precautions
- A, E, F, H
- 6a, 6b, 6c, 16, 18, 23, 24, 28, 29, 36, 37, 43, 53, 63, 67

Withholding period
- Keep livestock out of treated areas until poisonous weeds such as ragwort has died and become unpalatable

Maximum Residue Levels (mg residue/kg food)
- see amitrole entry

---

14   amitrole + 2,4-D + diuron
A total herbicide mixture of translocated and residual chemicals

Products	Trik	Mirfield	26.6:11.0:46.4% w/w	WP	02182

Uses   Total vegetation control in NON-CROP AREAS.

Notes   **Efficacy**
- Apply in spring or late summer/early autumn when weeds are growing actively and have sufficient leaf area to absorb chemical
- Apply maintenance treatment if necessary at lower rate when weeds 7-10 cm high
- Increase rate on areas of peat or high carbon content

Crop Safety/Restrictions
- Do not use on ground under which roots of valuable trees or shrubs are growing

Special precautions/Environmental safety
- Harmful to fish. Do not contaminate surface waters or ditches with chemical or used container

Protective clothing/Label precautions
- A
- 28, 29, 43, 53, 63, 65

Withholding period
- Keep livestock out of treated areas until poisonous weeds such as ragwort has died and become unpalatable

Maximum Residue Levels (mg residue/kg food)
- see amitrole entry

FOR FULL CONDITIONS OF USE ALWAYS READ THE PRODUCT LABEL

15   ammonium sulfamate

A non-selective, inorganic, general purpose herbicide and tree-killer

Products					
1 Amcide	B H & B	99.5% w/w	CR	04246
2 Root-Out	Dax	98.5% w/w	CR	03510

Uses   Annual weeds, perennial dicotyledons, perennial grasses, rhododendrons, woody weeds in FORESTRY. Annual weeds, perennial weeds in ORNAMENTALS *(pre-planting)*, VEGETABLES *(pre-planting)*.

Notes   **Efficacy**
* Apply as spray to low scrub and herbaceous weeds from Apr to Sep in dry weather when rain unlikely and cultivate after 3-8 wk
* Apply as crystals in frills or notches in trunks of standing trees at any time of year
* Apply as concentrated solution or crystals to stump surfaces within 48 h of cutting. Rhododendrons must be cut level with ground and sprayed to cover cut surface, bark and immediate root area
* Stainless steel or plastic sprayers are recommended. Solutions are corrosive to mild steel, galvanised iron, brass and copper

Crop Safety/Restrictions
* Allow 8-12 wk after treatment before replanting
* Keep spray at least 30 cm from growing plants. Low doses may be used under mature trees with undamaged bark

Special precautions/Environmental safety
* Harmful to fish. Do not contaminate surface waters or ditches with chemical or used container

Protective clothing/Label precautions
* 24, 26, 27, 28, 29, 37, 53, 63

16   anilazine

A triazine fungicide for cereals

Products				
Dyrene	Bayer	480 g/l	SC	06982

Uses   Septoria leaf spot in SPRING WHEAT, WINTER WHEAT.

Notes   **Efficacy**
* Treat normally from third node detectable (GS 33) when conditions favour disease development
* Improved control can be achieved by tank mixtures. See label for details
* A follow-up application alone or in mixture should be made if reinfection occurs within permitted timing

Crop Safety/Restrictions
* Maximum total dose 4 l/ha per crop
* Product must be sprayed immediately after mixing and agitation maintained at all times

Special precautions/Environmental safety
* Irritating to eyes and skin
* May cause sensitisation by skin contact
* Dangerous to fish or other aquatic life. Do not contaminate surface waters or ditches with chemical or used container
* Do not allow direct spray from vehicle-mounted/drawn hydraulic sprayers to fall within 6 m of surface waters or ditches

Protective clothing/Label precautions
* A, C, H
* 6a, 6b, 10a, 14, 18, 23, 24, 26, 27, 29, 36, 37, 52, 54a, 63, 67, 70

Latest application/Harvest Interval(HI)
* Before anthesis complete (GS 65)

17 anthracene oil
A crop desiccant

| Products | Sterilite Hop Defoliant | Coventry Chemicals | 63.6% w/w | EC | 05060 |

Uses Chemical stripping in HOPS.

Notes **Efficacy**
* Spray when hop bines 1.2 m high and direct spray downward at 45° onto area to be defoliated. Repeat as necessary until cones are formed

Crop Safety/Restrictions
* Do not spray if temperature is above 21°C or after cones have formed
* Do not drench rootstocks
* Do not spray on windy, wet or frosty days

Special precautions/Environmental safety
* Harmful if swallowed. Irritating to eyes, skin and respiratory system
* Dangerous to fish. Do not contaminate surface waters or ditches with chemical or used container

Protective clothing/Label precautions
* A, C, H
* 5c, 6a, 6b, 6c, 18, 21, 28, 29, 35, 36, 51, 59, 60, 62, 63, 69, 77

18 asulam
A translocated carbamate herbicide for control of docks and bracken

Products					
1 Asulox	RP Agric.	400 g/l	SL	06124
2 Asulox	RP Environ.	400 g/l	SL	05235

FOR FULL CONDITIONS OF USE ALWAYS READ THE PRODUCT LABEL

Uses   Bracken, docks in WATERSIDE AREAS [1, 2]. Bracken in FORESTRY, NON-CROP AREAS, PERMANENT PASTURE, ROUGH GRAZING [1, 2]. Bracken in FORESTRY PLANTATIONS *(off-label)* [1]. Docks in APPLES, BLACKCURRANTS, CHERRIES, GRASSLAND, HOPS, PEARS, PLUMS [1, 2]. Docks in BLUEBERRIES *(off-label)*, CANE FRUIT *(off-label)*, CRANBERRIES *(off-label)*, CURRANTS *(off-label)*, DAMSONS *(off-label)*, GOOSEBERRIES *(off-label)*, MINT *(off-label)*, NECTARINES *(off-label)*, PARSLEY *(off-label)*, QUINCES *(off-label)*, STRAWBERRIES *(off-label)*, TARRAGON *(off-label)*, WHITE CLOVER SEED CROPS *(off-label)* [1].

Notes   **Efficacy**
* Spray bracken when fronds fully expanded but not senescent, usually Jul-Aug, docks in full leaf before flower stem emergence
* Bracken fronds must not be damaged by stock, frost or cutting before treatment
* Do not apply in drought or hot, dry conditions
* Uptake and reliability of bracken control may be improved by use of specified additives - see label. Additives not recommended on forestry land
* To allow adequate translocation do not cut or admit stock for 14 d after spraying bracken or 7 d after spraying docks
* Complete bracken control rarely achieved by one treatment. Survivors should be sprayed when they recover to full green frond, which may be in the ensuing year but more likely in the second year following initial application

Crop Safety/Restrictions
* Maximum number of treatments 1 per yr (2 per yr for some crops - see OLA notice for details)
* In forestry areas some young trees may be checked if sprayed directly (see label)
* Allow at least 6 wk between spraying and planting any crop
* Do not use in pasture before mowing for hay
* In fruit crops apply as a directed spray
* Do not treat blackcurrant cuttings, hop sets or weak hills
* Cocksfoot and Yorkshire fog may be severely checked, bents, fescues, meadow-grasses and timothy may be checked temporarily
* Apply as spot treatment in parsley, mint and tarragon, not directly to crop [1]

Protective clothing/Label precautions
* A, C, D, H, M (for ULVA application)
* 30, 43, 54, 63, 65 (+28 for ULVA application)

Withholding period
* Keep livestock out of treated areas until foliage of poisonous weeds such as ragwort has died and become unpalatable. An interval of at least 4-6 wk is recommended

Latest application/Harvest Interval(HI)
* 6 wk before harvest for parsley and mint; 10 wk before harvest for tarragon [1]

Approval
* May be applied through CDA equipment. See notes in Section 1
* Approved for aerial application on bracken. See notes in Section 1
* Cleared for aquatic weed control. See notes in Section 1
* Off-label Approval unlimited for use on white clover seed crops (OLA 0931/92), strawberries (OLA 0932/92), cane fruit, currants, gooseberries, strawberries, blueberries, cranberries, damsons, nectarines, quinces (OLA 0930/92) [1]; unlimited for use in forestry areas (OLA 1001/92) [1], as spot treatment in parsley, mint, tarragon (OLA 0097/93) [1]

19 atrazine
A triazine herbicide with residual and foliar activity

Products					
1 Atlas Atrazine	Atlas	500 g/l	SC	03097
2 Gesaprim 500SC	Ciba Agric.	500 g/l	SC	05845
3 MSS Atrazine 50 FL	Mirfield	500 g/l	SC	01398
4 MSS Atrazine 80 WP	Mirfield	80% w/w	WP	04360
5 Unicrop Flowable Atrazine	Unicrop	500 g/l	SC	05446

Uses

Annual dicotyledons, annual grasses, couch, perennial grasses in CONIFER PLANTATIONS [1, 5]. Annual dicotyledons, annual grasses in ROSES [1, 5]. Annual dicotyledons, annual grasses in MAIZE [1-5]. Annual dicotyledons, annual grasses in RASPBERRIES [1]. Annual dicotyledons, annual grasses in SWEETCORN [2-5]. General weed control in BROADLEAVED TREES *(off-label)* [5].

Notes

Efficacy
* May be used pre- or early post-weed emergence in maize and sweetcorn
* Root activity enhanced by rainfall soon after application and reduced on high organic soils. Foliar activity effective on weeds up to 3 cm high
* Not recommended for use on soils with more than 10% organic matter
* In conifers apply as overall spray in Feb-Apr. May be used in first spring after planting
* Apply to raspberries in spring before new cane emergence, not in season of planting
* Application rates vary with crop, soil type and weed problem. See label for details
* Resistant weed strains may develop with repeated use of atrazine or other triazines

Crop Safety/Restrictions
* Maximum number of applications 1 per crop for maize and sweetcorn [1, 2], 2 per crop for maize and sweetcorn [3-5], 1 per yr for roses and conifers
* Do not apply to raspberries in season of planting
* On slopes heavy rainfall soon after application may cause surface run-off
* To reduce soil run-off, especially from forest plantations, users are advised to plant grass strips 6 m wide between treated areas and surface waters
* Do not use on Christmas trees
* After annual weed control only maize or sweetcorn should be sown for at least 7 mth after application. Do not sow oats in autumn following spring treatment
* After perennial grass control only maize or sweetcorn may be sown for at least 18 mth

Special precautions/Environmental safety
* Dangerous to fish or other aquatic life and aquatic higher plants. Do not contaminate surface waters or ditches with chemical or used container
* Do not allow direct spray from vehicle-mounted/drawn hydraulic sprayers to fall within 6 m, or from hand-held sprayers to within 2 m, of surface waters or ditches. Direct spray away from water
* Use must be restricted to one product containing atrazine or simazine, and either to a single application at the maximum approved rate or (subject to any existing maximum permitted number of treatments) to several applications at lower doses up to the maximum approved rate for a single application

FOR FULL CONDITIONS OF USE ALWAYS READ THE PRODUCT LABEL

Protective clothing/Label precautions
- A, B, C, D, H, M [2-5]; A, C, D, H, M [1]
- 29 [1, 3]; 30 [2, 4, 5]; 37, 66 [2]; 52, 54a, 63 [1-5]; 60 [1]; 65 [1, 4, 5]; 67 [3]

Latest application/Harvest Interval(HI)
- Before new cane emerges for raspberries

Approval
- Off-label Approval unlimited for use on broad-leaved trees (OLA 1339/94) [5]

Maximum Residue Levels (mg residue/kg food)
- fruits, vegetables, pulses, oilseeds, potatoes, tea, hops 0.1

20   azamethiphos
A residual organophosphorus insecticide for fly control

Products   Alfacron 10 WP   Ciba Agric.   10% w/w   WP   02832

Uses   Flies in LIVESTOCK HOUSES.

Notes   **Efficacy**
- Add water as directed to produce paint consistency and paint onto 2.5% of total wall and ceiling surface area at a minimum of 5 points in building or apply as spray to 30% of surface area

Crop Safety/Restrictions
- Only apply to areas out of reach of children and animals

Special precautions/Environmental safety
- Irritating to eyes and skin. May cause sensitization by skin contact
- This product contains an anticholinesterase organophosphorus compound. Do not use if under medical advice not to work with such compounds
- Do not apply directly to livestock and poultry
- Do not apply to surfaces on which food or feed is stored, prepared or eaten. Cover feedtuffs and remove exposed milk and eggs before application
- Dangerous to fish. Do not contaminate surface waters or ditches with chemical or used container

Protective clothing/Label precautions
- A, C
- 1, 6a, 6b, 10a, 14, 18, 23, 25, 26, 27, 28, 29, 36, 37, 52, 63, 67

21   Bacillus thuringiensis
A bacterial insecticide for control of caterpillars

Products
1 Bactospeine WP	Koppert		WP	02913
2 Certan	Steele & Brodie	3500 GMU/mg	SC	00465
3 Dipel	English Woodland	16000 IU/mg	WP	03214
4 Thuricide HP	Fargro	16000 IU/mg	WP	02136

Uses	Browntail moth in ORNAMENTALS [1, 3, 4]. Cabbage moth, cabbage white butterfly, diamond-back moth in BROCCOLI, BRUSSELS SPROUTS, CABBAGES, CAULIFLOWERS [1, 3, 4]. Caterpillars in PEPPERS, RASPBERRIES [4]. Caterpillars in CUCUMBERS *(off-label)*, FRENCH BEANS *(off-label)*, TOMATOES *(off-label)* [1]. Strawberry tortrix in STRAWBERRIES [1, 4]. Tomato moth in TOMATOES [1, 3, 4]. Wax moths in BEEHIVES [2].

Notes

Efficacy
* Product affects gut of larvae and must be eaten to be effective. Caterpillars cease feeding and die in 2-3 d (3-5 d for large caterpillars)
* Apply as soon as larvae appear on crop and repeat every 3-14 d for outdoor crops, every 3 wk under glass
* Addition of a wetter recommended for use on brassicas
* Good coverage is essential, especially of undersides of leaves. Use a drop-leg sprayer in field crops

Crop Safety/Restrictions
* Maximum number of treatments 1 per frame per yr in beehives

Protective clothing/Label precautions
* 29, 54, 63, 65

Latest application/Harvest Interval(HI)
* Before treated empty combs are replaced on hives [2]
* HI zero for crops [1, 3, 4]

Approval
* Off-label approval unlimited for use on French beans, tomatoes and cucumbers (OLA 0029/92)[1]

22 benalaxyl
A phenylamide (acylalanine) fungicide available only in mixtures

23 benalaxyl + mancozeb
A systemic and protectant fungicide mixture

Products					
1 Barclay Bezant	Barclay	8:65% w/w	WP	05914
2 Galben M	DowElanco	8:65% w/w	WP	05092
3 Galben M	Isagro	8:65% w/w	WP	05904

Uses	Blight in EARLY POTATOES, MAINCROP POTATOES. Downy mildew in WINTER OILSEED RAPE.

Notes

Efficacy
* Apply to potatoes at blight warning prior to crop becoming infected and repeat at 10-21 d intervals depending on risk of infection

FOR FULL CONDITIONS OF USE ALWAYS READ THE PRODUCT LABEL

- Spray irrigated potatoes after irrigation and at 14 d intervals, crops in polythene tunnels at 10 d intervals
- When active potato growth ceases use a non-systemic fungicide to end of season, starting not more than 10 d after last application
- Do not treat potatoes showing active blight infection
- For reduction of downy mildew in oilseed rape apply at seedling to 7-leaf stage, before end Nov as soon as infection seen

Crop Safety/Restrictions
- Maximum number of treatments 5 per crop (including other phenylamide fungicides) for potatoes, 1 per crop for winter oilseed rape
- Do not use on potatoes after mid-Aug in Northern Britain or early-Jul in Southern Britain and Northern Ireland [1], after end of Aug [2]
- Other phenylamide fungicides should not be used following a programme based on these products

Special precautions/Environmental safety
- Irritating to eyes
- Dangerous to fish or other aquatic life. Do not contaminate surface waters or ditches with chemical or used container

Protective clothing/Label precautions
- A
- 6a, 18, 21, 29, 35, 36, 37, 52, 62, 63, 67

Latest application/Harvest Interval(HI)
- HI 7 d

Approval
- Approved for aerial application on potatoes. See notes in Section 1

Maximum Residue Levels (mg residue/kg food)
- (benalaxyl) grapes, onions, tomatoes, peppers 0.2; tea, hops 0.1; fruits (except grapes), vegetables (except onions, tomatoes, peppers), pulses, oilseed, potatoes, cereals, animal products 0.05. See also mancozeb entry

---

24   benazolin

A translocated arylacetic acid herbicide available only in mixtures

---

25   benazolin + bromoxynil + ioxynil

A post-emergence, HBN herbicide mixture for cereal crops and grass

Products	1 Asset	AgrEvo	50:125:62.5 g/l	EC	07243
	2 Asset	Schering	50:125:62.5 g/l	EC	03824

Uses	Annual dicotyledons in BARLEY, DURUM WHEAT, NEWLY SOWN GRASS, OATS, WHEAT.

Notes	**Efficacy**
	• Commonly used in mixture with mecoprop and other cereal herbicides to extend the range of weeds controlled. Sequential treatments also recommended

- Best results achieved when weeds small and actively growing and crop competitive
- Weeds should be dry when sprayed

Crop Safety/Restrictions
- Maximum number of treatments 2 per crop or yr
- Use in cereals from 2-leaf stage to first node detectable (GS 12-31), in grass from 2-leaf stage when crop growing vigorously (other times may be advised for mixtures)
- Use on oats only in spring, on other cereals in autumn or spring
- Do not apply to crops undersown with legumes
- Do not use on crops affected by pests, disease, waterlogging or prolonged frost
- Severe frost within 3-4 wk of spraying may scorch crop
- Do not roll or harrow crops for 3 d before or after spraying
- Do not spray grass seed crops later than 5 wk before heading

Special precautions/Environmental safety
- Harmful if swallowed. Irritating to skin and eyes
- Do not apply by knapsack sprayer or at less than recommended volume
- Flammable
- Dangerous to fish or other aquatic life. Do not contaminate surface waters or ditches with chemical or used container
- Do not allow direct spray from vehicle-mounted/drawn hydraulic sprayers to fall within 6 m of surface waters or ditches
- Direct spray away from water
- Harmful to bees. Do not apply to crops in flower or to those in which bees are actively foraging. Do not apply when flowering weeds are present
- Do not apply to crops in flower or to those in which bees are actively foraging. Do not apply when flowering weeds are present

Protective clothing/Label precautions
- A, C
- 5c, 6a, 6b, 12c, 18, 21, 25, 28, 29, 36, 37, 43, 50, 52, 63, 66, 70, 78

Withholding period
- Keep livestock out of treated areas for at least 6 wk and until foliage of poisonous plants such as ragwort has died and become unpalatable

Latest application/Harvest Interval(HI)
- Before second node detectable for cereals (GS 32), 6 wk before grazing or harvest for grass
- HI 6 wk (before grazing)

Maximum Residue Levels (mg residue/kg food)
- see ioxynil entry

---

26   benazolin + clopyralid
A post-emergence herbicide mixture for use mainly in winter oilseed rape

Products					
1	Benazalox	AgrEvo	30:5% w/w	WP	07246
2	Benazalox	Schering	30:5% w/w	WP	03858

FOR FULL CONDITIONS OF USE ALWAYS READ THE PRODUCT LABEL

Uses   Annual dicotyledons, chickweed, cleavers, mayweeds in EVENING PRIMROSE *(off-label)*, SWEDE SEED CROPS *(off-label)*, WHITE MUSTARD *(off-label)*, WINTER OILSEED RAPE. Annual dicotyledons in HONESTY *(off-label)*, SPRING OILSEED RAPE *(off-label)*.

Notes   **Efficacy**
- Weeds are controlled from cotyledon stage to early flowerbud or up to 15 cm high or across in winter oilseed rape
- For best results apply during mild, moist weather when weeds are growing actively and still visible in young crop. Do not spray if rain expected within 4 h
- Do not spray if frost present on foliage; frost after application will not reduce effectiveness
- For grass weed control various tank mixtures are recommended or sequential treatments, allowing specified interval after application of grass-killer. See label for details

Crop Safety/Restrictions
- Maximum number of treatments 1 per crop
- May be used on winter oilseed rape when crop between stages of 3 fully developed leaves to flower buds hidden beneath leaves (GS 1,3-3,1), on spring oilseed rape before green bud
- Oilseed rape may be replanted in the event of crop failure. Wheat, barley or oats may be sown after 1 mth, other crops in the following autumn after ploughing. Winter beans should not be planted in the same year
- Chop and incorporate straw and trash in early autumn to release any clopyralid residues. Ensure remains of plants completely decayed before planting susceptible crops

Special precautions/Environmental safety
- Irritating to eyes and skin
- Harmful to fish or other aquatic life. Do not contaminate surface waters or ditches with chemical or used container

Protective clothing/Label precautions
- A, C
- 6a, 6b, 18, 22, 25, 28, 29, 36, 37, 43, 53, 63, 67

Withholding period
- Keep livestock out of treated areas until foliage of any poisonous weeds such as ragwort has died and become unpalatable

Latest application/Harvest Interval(HI)
- Oilseed rape before green bud stage (GS 3,1) (before end Jan if in tank mix with Butisan S), honesty up to 4-5 pairs true leaves

Approval
- Off-label approval unlimited for use on evening primrose, white mustard and swedes grown for seed (OLA 0871, 0872/92); on honesty (OLA 0610/93); on spring oilseed rape (OLA 0654/93)

27   **benazolin + 2,4-DB + MCPA**

A post-emergence herbicide for undersown cereals, grass and clover

Products   Setter 33   DowElanco   50:237:43 g/l   SL   05623

Uses   Annual dicotyledons, chickweed, cleavers, knotgrass in CLOVERS, DIRECT-SOWN SEEDLING CLOVERS, SEEDLING LEYS, UNDERSOWN BARLEY, UNDERSOWN OATS, UNDERSOWN WHEAT.

Notes

Efficacy
- Spray when weeds are young and in active growth
- Spray perennial weeds when well developed but before flowering. Higher rates may be used against perennials in established grassland
- Do not spray when rain is imminent or during drought

Crop Safety/Restrictions
- Maximum number of treatments 1 per crop or yr
- Spray undersown cereals before 1st node detectable (GS 31), grass seedlings should have at least 3 leaves
- Spray winter cereals in spring from leaf sheath erect exceeding 5 cm, spring wheat after 5 leaves unfolded (GS 15), spring barley and oats after 2 leaves unfolded (GS 12)
- Spray clovers after the 1-trifoliate leaf stage and before red clover has more than 3 trifoliate leaves. Do not spray clover seed crops in the yr seed is to be taken
- Spray new swards before end Oct when grasses have at least 3 leaves and clovers at stages as above
- Clovers may be damaged if frost occurs soon after spraying
- Do not spray legumes other than clover
- Do not roll, harrow or cut crops for 3 d before or after spraying

Special precautions/Environmental safety
- Harmful to fish or other aquatic life. Do not contaminate surface waters or ditches with chemical or used container

Protective clothing/Label precautions
- 21, 29, 43, 53, 60, 63, 66

Withholding period
- Keep livestock out of treated areas for at least 2 wk and until foliage of poisonous weeds such as ragwort has died and become unpalatable

Latest application/Harvest Interval(HI)
- Before first node detectable (GS 31) for undersown cereals

28   bendiocarb

A contact, ingested and systemic carbamate insecticide

Products					
1	Garvox 3G	AgrEvo	3% w/w	GR	07256
2	Garvox 3G	Schering	3% w/w	GR	03866
3	Seedox SC	Uniroyal	500 g/l	FS	07591

Uses

Bean seed flies in PROTECTED COURGETTES *(off-label seed treatment)*, PROTECTED GHERKINS *(off-label seed treatment)*, PROTECTED MARROWS *(off-label seed treatment)*, PROTECTED MELONS *(off-label seed treatment)*, PROTECTED PUMPKINS *(off-label seed treatment)*, PROTECTED SQUASHES *(off-label seed treatment)* [3]. Frit fly, wireworms in MAIZE, SWEETCORN [1-3]. Millipedes, pygmy beetle, springtails, symphylids, wireworms in SUGAR BEET [1, 2].

FOR FULL CONDITIONS OF USE ALWAYS READ THE PRODUCT LABEL

Notes

Efficacy
- Apply in the furrow with seed using a suitable applicator [1, 2] or mixed with DowElanco Adjuvant Oil using suitable seed treatment machinery [3]
- Depending on climatic conditions soil pests are controlled for 6-8 wk [1, 2]
- Calibration of applicator should be checked in the field before drilling [1, 2]
- Use of suitable adjuvant advised to ensure effectiveness. See label for details [3]

Crop Safety/Restrictions
- Maximum number of treatments 1 per crop or batch of seed

Special precautions/Environmental safety
- This product contains an anticholinesterase carbamate compound. Do not use if under medical advice not to work with such compounds
- Harmful in contact with skin or if swallowed [3]
- Ensure adequate ventilation in confined spaces [3]
- Dangerous to fish or other aquatic life. Do not contaminate surface waters or ditches with chemical or used container
- Dangerous to game, wild birds and animals. Bury spillages
- Treated seed harmful to game and wildlife [3]
- Treated seed not to be used as food or feed. Do not re-use sack [3]

Protective clothing/Label precautions
- A, C
- 2, 22, 25, 28, 29, 45, 52, 63, 70 [1-3]; 18, 36, 37, 65, 72, 73, 74, 75, 76, 77, 79 [3]; 67 [1, 2]

Latest application/Harvest Interval(HI)
- At drilling [1, 2]; pre-drilling [3]

---

29 benomyl
A systemic, MBC fungicide with protectant and eradicant activity

Products					
Benlate Fungicide	DuPont	50% w/w	WP	00229

Uses

Black spot, gummosis in COURGETTES *(off-label)*. Botrytis, light leaf spot in OILSEED RAPE. Botrytis in PEAS. Canker in HOPS *(off-label)*. Chocolate spot in FIELD BEANS. Eyespot, rhynchosporium in CEREALS. Gloeosporium in HAZEL NUTS *(off-label)*. Leaf and pod spot in FIELD BEANS *(seed treatment)*, PEAS *(seed treatment)*. Light leaf spot, ring spot in BRUSSELS SPROUTS. Rhizoctonia in WATERCRESS *(off-label)*. Seed-borne diseases in GRASS SEED *(off-label)*.

Notes

Efficacy
- Apply as overall spray. One spray effective for some diseases, repeat sprays at 1-4 wk intervals needed for others. Recommended spray programmes vary with disease and crop. See label for details of timing and recommended tank mixes
- Addition of non-ionic wetter recommended for certain uses. See label for details
- To delay appearance of resistant strains in diseases needing more than 2 applications per season, use in programme with fungicide of different mode of action
- Product may lose effectiveness if allowed to become damp during storage

Crop Safety/Restrictions
• Maximum number of treatments (including applications of any product containing benomyl, carbendazim or thiophanate-methyl) 3 per crop for oilseed rape; 2 per crop for cereals, Brussels sprouts, field and broad beans, combining and vining peas; 1 per crop for mangetout peas, dwarf French beans

Protective clothing/Label precautions
• A, C, D, H, M
• 30, 54, 63, 67

Latest application/Harvest Interval(HI)
• At full bloom for mange-tout peas, dwarf French beans; up to and including grain watery ripe stage (GS 71) for cereals.
• HI courgettes 2 d, hops 7 d, oilseed rape 3 wk, Brussels sprouts (with Actipron or non-ionic wetter) combining and vining peas, field and broad beans 3 wk, hazel nuts 28 d

Approval
• Approved for aerial application on wheat, barley, oilseed rape. See notes in Section 1. Consult firm for details
• Off-label approval to Dec 1998 for use on watercress (OLA 1208/93); to Feb 1997 for use on hops, hazel nuts (OLA 0101/92) and grass seed (OLA 0100/92); to Oct 1996 for use on outdoor courgettes (OLA 0694/91)

---

30 bentazone
A post-emergence contact diazinone herbicide

Products					
1 Barclay Dingo	Barclay	480 g/l	SL	07642
2 Basagran	BASF	480 g/l	SL	00188
3 I T Bentazone 480	I T Agro	480 g/l	SL	07458
4 Standon Bentazone	Standon	480 g/l	SL	06895

Uses Annual dicotyledons, mayweeds in BULB ONIONS *(off-label)*, SALAD ONIONS *(off-label)* [2]. Annual dicotyledons in BROAD BEANS [2, 4]. Annual dicotyledons in DWARF BEANS, NARCISSI, NAVY BEANS, PEAS, POTATOES, RUNNER BEANS, SPRING FIELD BEANS, WINTER FIELD BEANS [2-4]. Annual dicotyledons in EVENING PRIMROSE *(off-label)*, OUTDOOR DWARF BEANS *(off-label)* [2]. Annual dicotyledons in LINSEED [1-4]. Annual dicotyledons in COMBINING PEAS, FIELD BEANS [1].

Notes Efficacy
• Most effective control obtained when weeds are growing actively and less than 5 cm high or across. Good spray cover is essential
• Split dose application may be made in beans, linseed, potatoes and narcissi and generally gives better weed control. See label for details
• Do not apply if rain or frost expected, if unseasonably cold, if foliage wet or in drought
• A minimum of 6 h free from rain is required after application
• Various recommendations are made for use in spray programmes with other herbicides or in tank mixes. See label for details

---

FOR FULL CONDITIONS OF USE ALWAYS READ THE PRODUCT LABEL

- Addition of Actipron, Adder or Cropspray 11E adjuvant oils recommended for use in dwarf beans and potatoes to improve fat hen control. Do not use under hot or humid conditions or on field beans

Crop Safety/Restrictions

- Maximum number of treatments 1 per crop (2 with repeat low dose treatment) for beans, evening primrose, linseed, narcissi and potatoes, 1 per crop for peas, 3 per crop for bulb onions (1 per crop using higher dose for mayweed control), 2 per crop for salad onions
- Crops must be treated at correct stage of growth to avoid danger of scorch. See label for details and for varietal tolerances. Apply to broad beans from 3 to 4 leaf pairs only
- Do not use on crops which have been affected by drought, waterlogging, frost or other stress conditions
- Do not spray at temp above 21°C. Delay spraying until evening if necessary
- Consult processor before using on crops for processing
- A satisfactory wax test must be carried out before use on peas
- Not all varieties are fully tolerant. Do not use on forage peas and other named varieties of peas, beans and potatoes. See label [3]
- May be used on selected varieties of maincrop and second early potatoes (see label for details), not on seed crops or first earlies
- Do not treat narcissi during flower bud initiation

Special precautions/Environmental safety

- Irritant. May cause sensitization by skin contact

Protective clothing/Label precautions

- A
- 6 [1, 3, 4]; 10a, 18, 21, 28, 29, 36, 37, 54, 66 [1-4]; 63 [1-3]

Latest application/Harvest Interval(HI)

- Before shoots exceed 15 cm high for potatoes and spring field beans (or 6 leaf pairs), 4 leaf pairs (6 pairs or 15 cm high with split dose) for broad beans, before flower bud visible for French, navy, runner and winter field beans and linseed, before flower bud can be found enclosed in terminal shoot for peas [2].
- Before crop exceeds 15 cm and 7 leaf pairs for spring sown field beans; before shoots exceed 15 cm for potatoes; before flower buds visible for winter sown field beans, navy beans, runner beans, linseed; before flower buds can be found enclosed in terminal shoot for peas [1, 3, 4]
- HI onions 3 wk

Approval

- Off-label approval unlimited for use on bulb and salad onions (OLA 0132/92)[2], and evening primrose (OLA 0596/92)[2], and for use of 3 split dose treatments on outdoor dwarf beans up to permitted maximum (OLA 0611/93)[2]

---

31   bentazone + cyanazine + 2,4-DB

A post-emergence herbicide for undersown cereals and grass

Products					
Topshot	Cyanamid	225:500:235 g/l	KL	07178

Uses	
Annual dicotyledons, chickweed in CLOVERS, NEWLY SOWN GRASS, UNDERSOWN SPRING BARLEY.	

Notes

Efficacy
- Apply at any time in spring, summer or autumn when growth is occurring provided crop at correct stage
- Best results achieved when weeds small and actively growing
- Do not use during drought, waterlogging, extremes of temperature or when frosts are expected
- Do not spray when rainfall imminent or if crop wet

Crop Safety/Restrictions
- Maximum number of treatments 1 per crop
- Apply to cereals from 2-fully expanded leaf stage to first node detectable (GS 12 or 15-31) provided clovers have reached 1-trifoliate leaf stage
- Do not treat red clover after 3-trifoliate leaf stage
- Apply to newly sown leys after grass has reached 2-leaf stage provided clovers have at least 1 trifoliate leaf and red clover has not passed 3-trifoliate leaf stage
- Do not use on seed crops or cereals undersown with lucerne
- Do not use on crops affected by pest, disease or herbicide damage
- Do not roll or harrow for 7 d before or after spraying
- Growth of clovers may be reduced but effects are normally outgrown

Special precautions/Environmental safety
- Harmful if swallowed. Irritating to eyes and skin
- Harmful to fish or other aquatic life. Do not contaminate surface waters or ditches with chemical or used container

Protective clothing/Label precautions
- A, C
- 5c, 6a, 6b, 18, 21, 28, 36, 37, 43, 53, 63, 66, 70, 78

Withholding period
- Keep livestock out of treated areas for at least 4 wk and until poisonous weeds such as ragwort has died and become unpalatable

Latest application/Harvest Interval(HI)
- Before first node detectable (GS 31) for spring cereals
- HI grass 4 wk

---

32   bentazone + MCPA + MCPB

A post-emergence herbicide for undersown cereals and grass

Products   Acumen   BASF   200:80:200 g/l   SL   00028

Uses   Annual dicotyledons in NEWLY SOWN GRASS, UNDERSOWN SPRING BARLEY, UNDERSOWN SPRING OATS, UNDERSOWN SPRING WHEAT.

Notes   **Efficacy**
- Best results when weeds small and actively growing provided crop at correct stage
- A minimum of 6 h free from rain is required after treatment

---

FOR FULL CONDITIONS OF USE ALWAYS READ THE PRODUCT LABEL

- Do not apply if frost expected, if crop wet or in high humidity at above 21°C
- On first year grass leys use alone or in mixture with cyanazine on seedling stage weed before end of Sep, use mixture with cyanazine on older chickweed in Oct-Nov

Crop Safety/Restrictions

- Maximum number of treatments 1 per crop on undersown cereals, 1 per yr on grassland
- Apply to cereals from 2-fully expanded leaf stage but before first node detectable (GS 12-31) provided clover has reached 1-trifoliate leaf stage
- Do not treat red clover after 3-trifoliate leaf stage
- Apply to newly sown leys after grass has reached 2-leaf stage provided clovers have at least 1 trifoliate leaf and red clover has not passed 3-trifoliate leaf stage. Grasses must have at least 3 leaves before treating with cyanazine mixture
- Do not use on crops suffering from herbicide damage or physical stress
- Do not use on seed crops or on cereals undersown with lucerne
- Do not roll or harrow for 7 d before or after spraying
- Clovers may be scorched and undersown crop checked but effects likely to be outgrown

Special precautions/Environmental safety

- Harmful if swallowed. Irritating to eyes and skin. May cause sensitization by skin contact
- Harmful to fish or other aquatic life. Do not contaminate surface waters or ditches with chemical or used container

Protective clothing/Label precautions

- A, C
- 5c, 6a, 6b, 10a, 18, 21, 28, 29, 36, 37, 43, 53, 63, 66, 70, 78

Withholding period

- Keep livestock out of treated areas for at least 2 wk and until poisonous weeds such as ragwort has died and become unpalatable

Latest application/Harvest Interval(HI)

- Before 1st node detectable for undersown cereals (GS 31), 2 wk before grazing for grassland (+ Fortrol 4 wk)

---

33   bentazone + MCPB

A post-emergence herbicide mixture for tank mixing with cyanazine

Products					
	Pulsar	BASF	200:200 g/l	SL	04002

Uses Annual dicotyledons in PEAS.

Notes

Efficacy

- To be used in tank mix with cyanazine (Fortrol)
- Best results achieved when weeds small and actively growing provided crop at correct stage. Good spray cover is essential
- May be applied as single treatment or as split dose treatment applying first spray when susceptible weeds are not beyond the 2 leaf stage
- Do not apply if rain or frost expected or if foliage wet. A minimum of 6 h free from rain required after treatment
- Do not apply during drought or unseasonably cold weather

Crop Safety/Restrictions
* Maximum number of treatments 1 per crop, 2 per crop with split dose
* Apply only to listed cultivars (see label) from 3 fully expanded leaf to before flower bud can be found enclosed in terminal shoot (GS 103 to before GS 201)
* Do not treat forage pea cultivars or mange-tout peas
* Apply after a satisfactory wax test. Early drilled crops or crops affected by frost or abrasion may not have sufficiently waxy cuticle
* Do not use as tank mix with any other product than cyanazine, nor after use of TCA
* Do not apply to any crop that may have been subjected to stress conditions, where foliage damaged or under hot, sunny conditions when temperature exceeds 21°C
* Do not apply insecticides or grass herbicides within 7 d of treatment

Special precautions/Environmental safety
* Harmful if swallowed or in contact with skin. Irritating to eyes and skin. May cause sensitization by skin contact
* Harmful to fish or other aquatic life. Do not contaminate surface waters or ditches with chemical or used container

Protective clothing/Label precautions
* A, C
* 5a, 5c, 6a, 6b, 10a, 18, 21, 28, 29, 36, 37, 43, 53, 63, 66, 70, 78

Withholding period
* Keep livestock out of treated areas until foliage of any poisonous weeds such as ragwort has died and become unpalatable

Latest application/Harvest Interval(HI)
* Before enclosed bud stage of crop (GS 10x)

---

34   benzalkonium chloride
A quaternary ammonium algicide and moss killer for paths, pots etc

Products	Paramos	Chemsearch	23.6 g/l	SL	H4524

Uses   Algae, mosses in FLOWER POTS, GLASSHOUSE BENCHES, PATHS, ROOFS, WALLS.

Notes

Efficacy
* Spray, sprinkle or brush on paths, walls and stonework under dry conditions. Do not use if it has rained within 3-4 d or if rain anticipated within 24 h
* Apply to plant beds or benches before plants are put in place. If applied to pots, plants should be removed during spraying or spray directed at base of pots
* Regrowth of moss normally prevented for whole growing season

Crop Safety/Restrictions
* Avoid contact with leaves of growing plants

Special precautions/Environmental safety
* Irritating to eyes and skin

FOR FULL CONDITIONS OF USE ALWAYS READ THE PRODUCT LABEL

- Dangerous to fish. Do not contaminate surface waters or ditches with chemical or used container

Protective clothing/Label precautions
- A, C, H
- 6a, 6b, 14, 16, 18, 21, 29, 36, 37, 52, 63, 67

---

35   bifenox
A diphenyl ether herbicide available only in mixtures

36   bifenox + dicamba
A contact and translocated herbicide mixture for winter cereals

Products	Quickstep	Sandoz	400:40 g/l	SC	07389

Uses   Annual dicotyledons, cleavers, field pansy, polygonums, speedwells in WINTER BARLEY, WINTER WHEAT.

Notes

Efficacy
- Best results achieved by application at early stages of weed growth in warm moist conditions

Crop Safety/Restrictions
- Maximum number of treatments 1 per crop
- Apply from 5 fully expanded leaves (GS 15) until before second node detectable stage (GS 32)
- Do not roll or harrow within 7 d of treatment
- Do not treat crops undersown or to be undersown
- Do not mix with any pesticide formulated as an emulsifiable concentrate nor in a 3-way mixture with mecoprop and isoproturon
- Do not treat crops suffering from any kind of stress or when rain is imminent or falling
- Initial leaf spotting often seen after treatment but is normally outgrown within a few wk without reducing yield

Special precautions/Environmental safety
- Dangerous to fish or other aquatic life. Do not contaminate surface waters or ditches with chemical or used container

Protective clothing/Label precautions
- 21, 29, 43, 52, 63, 66, 70

Withholding period
- Keep livestock out of treated areas for at least 2 wk and until poisonous weeds such as ragwort has died and become unpalatable

Latest application/Harvest Interval(HI)
- Before second node detectable stage (GS 32)

37   **bifenthrin**

A contact and residual pyrethroid acaricide/insecticide for use in agricultural and horticultural crops

Products	1 Talstar	DowElanco	100 g/l	EC	04916
	2 Talstar	PBI	100 g/l	EC	06913

Uses   Aphids, caterpillars, whitefly in BROCCOLI, BRUSSELS SPROUTS, CABBAGES, CALABRESE, CAULIFLOWERS [1, 2]. Aphids, caterpillars, whitefly in KALE [2]. Barley yellow dwarf virus vectors in WINTER BARLEY, WINTER OATS, WINTER WHEAT [1, 2]. Cabbage stem flea beetle, rape winter stem weevil in WINTER OILSEED RAPE [1, 2]. Damson-hop aphid, two-spotted spider mite in HOPS [1, 2]. Fruit tree red spider mite in APPLES, PEARS [1, 2]. Pea aphid, pea moth in PEAS [1, 2]. Two-spotted spider mite in ORNAMENTALS, STRAWBERRIES [1, 2]. Two-spotted spider mite in BLACKCURRANTS [2].

Notes   **Efficacy**
* Timing of application varies with crop and pest. See label for details
* Treatment may be used on all varieties of peas including vining, combining, forage and edible-podsded varieties

Crop Safety/Restrictions
* Maximum number of treatments 2 per yr for winter cereals, oilseed rape, peas, leaf brassicas, apples, pears and strawberries, 5 per yr for hops
* Minimum interval between applications on apples, pears, strawberries 14 d, on hops 10 d
* On ornamentals small-scale test treatment advised before applying overall treatment. Do not spray plants in blossom or under stress

Special precautions/Environmental safety
* Keep in original container, tightly closed, in a safe place under lock and key
* Harmful if swallowed and by inhalation
* Irritating to skin and eyes
* Flammable
* Dangerous [1], extremely dangerous [2] to bees. Do not apply to crops in flower or to those in which bees are actively foraging. Do not apply when flowering weeds are present
* Extremely dangerous to fish or other aquatic life. Do not contaminate surface waters or ditches with chemical or used container

Protective clothing/Label precautions
* A, C, H [1, 2]
* 5b, 5c, 6a, 6b, 12c, 14, 16, 18, 21, 28, 29, 36, 37, 51, 64, 66, 70, 78 [1, 2]; 48a, 54a [2]; 49 [1]

Latest application/Harvest Interval(HI)
* Before 31 Mar in year of harvest for winter wheat, winter barley and winter oats; before 30 Nov in year of sowing for winter oilseed rape.
* HI nil for peas, broccoli, Brussels sprouts, cabbages, calabrese, cauliflowers

---

FOR FULL CONDITIONS OF USE ALWAYS READ THE PRODUCT LABEL

38   bioallethrin

A pyrethroid insecticide available only in mixtures

39   bioallethrin + permethrin

A residual pyrethroid insecticide mixture for surface application

Products   Insektigun   Spraydex   0.05:0.238% w/w   RH   H3002

Uses   Flies in AGRICULTURAL PREMISES.

Notes   **Efficacy**
- Apply spray to doors, windows and other surfaces covering about 20% of area
- Repeat every 3 wk for continuous control

Crop Safety/Restrictions
- Not recommended for use in intensive animal houses as resistance can develop
- Do not spray onto food, grain or animals
- Remove exposed milk and collect eggs before spraying
- Protect milking machinery and containers from contamination

Special precautions/Environmental safety
- Harmful to bees. Do not apply to crops in flower or to those in which bees are actively foraging. Do not apply when flowering weeds are present
- Extremely dangerous to fish or other aquatic life. Do not contaminate surface waters or ditches with chemical or used container

Protective clothing/Label precautions
- 28, 29, 36, 37, 50, 51, 63

Maximum Residue Levels (mg residue/kg food)
- see permethrin entry

40   bioresmethrin

A contact acting pyrethroid insecticide for control of flying insects

Products   Blade   Chemsearch   0.3% w/w   ME   H4839

Uses   Insect pests in AGRICULTURAL PREMISES, MANURE HEAPS, REFUSE TIPS.

Notes   **Efficacy**
- Formulated as a non-staining, water-based emulsion for application as a space spray

Special precautions/Environmental safety
- For use only by professional operators
- Harmful to fish. Remove or cover fish tanks and bowls before application
- Do not apply to surfaces on which food is stored, prepared or eaten
- Cover food preparing equipment and eating utensils before application

Protective clothing/Label precautions
* A, H
* 28, 53, 67

41 bitertanol
A conazole fungicide available only in mixtures

42 bitertanol + fuberidazole
A broad spectrum fungicide mixture for seed treatment in wheat

Products

1 Sibutol	Bayer	375:23 g/l	FS	06983
2 Sibutol	Zeneca	375:23 g/l	LS	07487

Uses   Bunt, foot rot in WHEAT. Loose smut in WHEAT *(partial control)*. Septoria seedling blight in WHEAT *(reduction)*.

Notes   **Efficacy**
* Must be applied simultaneously with water in the ratio 1 part product to 2 parts water in a recommended seed treatment machine
* Treated seed should preferably be drilled in the same season
* Control of loose smut may be inadequate for use on seed for multiplication

Crop Safety/Restrictions
* Maximum number of treatments 1 per batch of seed
* Any delay of field emergence may be accentuated by treatment
* Do not use on seed with more than 16% moisture content, or on sprouted, cracked or skinned seed

Special precautions/Environmental safety
* Treated seed not to be used as food or feed, nor to be applied from the air
* Dangerous to fish or other aquatic life. Do not contaminate surface waters or ditches with chemical or used container
* Product supplied in returnable (1000 l) container. See label for guidance on handling, storage and protective clothing [2]

Protective clothing/Label precautions
* A, H [1, 2]
* 20a, 29, 69, 70, 71a [2]; 52, 63, 73, 74, 76, 77 [1, 2]; 65 [1]

Latest application/Harvest Interval(HI)
* Before drilling

Maximum Residue Levels (mg residue/kg food)
* (bitertanol) pome fruits, apricots, peaches, nectarines, plums 1; bananas 0.5

FOR FULL CONDITIONS OF USE ALWAYS READ THE PRODUCT LABEL

43   bone oil

A ready-to-use animal repellent

Products   Renardine   Roebuck Eyot   33.3% w/w   AL   04402

Uses   Badgers, cats, dogs, foxes, moles, rabbits in AMENITY AREAS, FARMLAND. Cats, dogs in AGRICULTURAL PREMISES.

Notes   **Efficacy**
* Soak pieces of stick, rags or sand in product and distribute around area to be protected as directed
* Repeat treatment weekly or after heavy rainfall

Crop Safety/Restrictions
* Do not apply directly to animals or crops
* Used as directed product is harmless to animals

44   Bordeaux mixture

A protectant copper sulfate/lime complex fungicide

Products   Wetcol 3   Ford Smith   30 g/l (copper)   SC   02360

Uses   Bacterial canker in CHERRIES. Blight in POTATOES, TOMATOES *(outdoor)*. Cane spot, spur blight in RASPBERRIES. Cane spot in LOGANBERRIES. Canker, scab in APPLES, PEARS. Celery leaf spot in CELERY. Currant leaf spot in BLACKCURRANTS. Downy mildew in HOPS. Leaf curl in APRICOTS, NECTARINES, PEACHES.

Notes   **Efficacy**
* Spray interval normally 7-14 d but varies with disease and crop. See label for details
* Commence spraying potatoes before crop meets in row or immediately first blight period occurs
* For canker control spray monthly from Aug to Oct
* For peach leaf curl control spray at leaf fall in autumn and again in Feb
* Spray when crop foliage dry. Do not spray if rain imminent

Crop Safety/Restrictions
* Do not use on copper sensitive cultivars, including Doyenne du Comice pears

Special precautions/Environmental safety
* Harmful if swallowed. Irritating to eyes, skin and respiratory system
* Harmful to fish or other aquatic life. Do not contaminate surface waters or ditches with chemical or used container
* Harmful to livestock

Protective clothing/Label precautions
* 5c, 6a, 6b, 6c, 25, 27, 29, 41, 53, 63, 67

Withholding period
* Keep all livestock out of treated areas for at least 3 wk

45 bromacil

A soil acting uracil herbicide for non-crop areas and cane fruit

See also amitrole + bromacil + diuron

Products	Hyvar X	DuPont	80% w/w	WP	01105

Uses Annual dicotyledons, annual grasses, couch, perennial dicotyledons, perennial grasses in BLACKBERRIES, LOGANBERRIES, RASPBERRIES, TAYBERRIES. Total vegetation control in NON-CROP AREAS.

Notes **Efficacy**
• Apply to established cane fruit as soon as possible in spring after cultivation and before bud break. Avoid further soil disturbance for as long as possible. If inter-row areas are cultivated, only a 30 cm band either side of row need be treated
• Best results achieved when soil is moist at time of application
• For total vegetation control apply to bare ground or standing vegetation. Adequate rainfall is needed to carry chemical into root zone
• Against existing vegetation best results achieved in late winter to early spring but application also satisfactory at other times

Crop Safety/Restrictions
• Maximum number of treatments 1 per yr
• May be used on cane fruit established for at least 2 yr
• In Scotland only may be used on newly planted raspberries at a reduced rate immediately after planting followed by light ridging
• Do not use in last 2 yr before grubbing crop to avoid injury to subsequent crops. Carrots, lettuce, beet, leeks and brassicas are extremely sensitive
• When used non-selectively, take care not to apply where chemical can be washed into root zone of desirable plants
• Treated land should not be cropped within 3 full yr of treatment and then only after obtaining advice from manufacturer

Special precautions/Environmental safety
• Irritating to eyes, skin and respiratory system

Protective clothing/Label precautions
• A, C
• 6a, 6b, 6c, 18, 22, 28, 29, 36, 37, 54, 63, 67

Latest application/Harvest Interval(HI)
• April for cane fruit

FOR FULL CONDITIONS OF USE ALWAYS READ THE PRODUCT LABEL

46   bromacil + diuron

A root-absorbed residual total herbicide mixture

Products   Borocil K   RP Environ.   0.88:0.88% w/w   GR   05183

Uses   Total vegetation control in NON-CROP AREAS.

Notes   **Efficacy**
- Spray or apply granules in early stage of weed growth at any time of year, provided adequate moisture to activate chemical is supplied by rainfall
- Use higher rates on adsorptive soils or established weed growth
- Do not apply when ground frozen

Crop Safety/Restrictions
- Do not apply on or near trees, shrubs, crops or other desirable plants
- Do not apply where roots of desirable plants may extend or where chemical may be washed into contact with their roots
- Do not use on ground intended for subsequent cultivation

Special precautions/Environmental safety
- Irritating to eyes, skin and respiratory system
- Harmful to fish or other aquatic life. Do not contaminate surface waters or ditches with chemical or used container

Protective clothing/Label precautions
- 29, 54, 63, 67

---

47   bromacil + picloram

A persistent residual and translocated herbicide mixture

Products   Hydon   Nomix-Chipman   1.08:0.33% w/w   GR   01088

Uses   Total vegetation control in NON-CROP AREAS.

Notes   **Efficacy**
- Apply at any time from spring to autumn. Best results achieved in Mar-Apr when weeds 50-75 mm high
- Do not apply in very dry weather as moisture needed to carry chemical to roots
- May be used on high fire risk sites

Crop Safety/Restrictions
- Do not apply near crops, cultivated plants and trees
- Do not apply on slopes where run-off to cultivated plants or water courses may occur

Special precautions/Environmental safety
- Harmful to fish or other aquatic life. Do not contaminate surface waters or ditches with chemical or used container

Protective clothing/Label precautions
- 28, 29, 43, 53, 63, 67

Withholding period
• Keep livestock out of treated area until foliage of any poisonous weeds such as ragwort has died and become unpalatable

48 bromadiolone

An anti-coagulant coumarin-derivative rodenticide

Products	1 Biotrol Plus Outdoor Rat Killer	Rentokil	0.005% ww	RB	03707
	2 Slaymor	Ciba Agric.	0.005% w/w	RB	01958
	3 Slaymor Bait Bags	Ciba Agric.	0.005% w/w	RB	03183

Uses   Mice, rats in FARM BUILDINGS, FARMYARDS.

Notes   **Efficacy**
• Ready-to-use baits are formulated on a mould-resistant, whole-wheat base
• Use in baiting programme. Place baits in protected situations, sufficient for continuous feeding between treatments
• Chemical is effective against warfarin- and coumatetralyl-resistant rats and mice and does not induce bait shyness
• Use bait bags where loose baiting inconvenient (eg behind ricks, silage clamps etc)

Special precautions/Environmental safety
• Access to baits by children, birds and animals, particularly cats, dogs, pigs and poultry, must be prevented
• Baits must not be placed where food, feed or water could become contaminated
• Remains of bait and bait containers must be removed after treatment and burned or buried
• Rodent bodies must be searched for and burned or buried. They must not be placed in refuse bins or on rubbish tips

Protective clothing/Label precautions
• 25, 29, 63, 65, 79, 81, 82, 84

49 bromoxynil

A contact acting HBN herbicide
See also benazolin + bromoxynil + ioxynil

Products	Alpha Bromotril P	Makhteshim	250 g/l	LI	07099

Uses   Annual dicotyledons in MAIZE.

Notes   **Efficacy**
• Apply to spring sown maize from 2 fully expanded leaves up to 9 fully expanded leaves
• Spray when main weed flush has germinated and the largest are at the 4 leaf stage

FOR FULL CONDITIONS OF USE ALWAYS READ THE PRODUCT LABEL

- Weed control can be enhanced by using a split treatment spraying each application when the weeds are seedling to 2 true leaves. Apply the second treatment before the crop canopy covers the ground
- Tank mixture with atrazine recommended for enhanced weed control - see label

Crop Safety/Restrictions
- Maximum total dose 2.4 l/ha/crop
- Foliar scorch, which rapidly disappears without affecting growth, will occur if treatment made in hot weather or during rapid growth
- Do not apply with oils or other adjuvants
- Do not apply during frosty weather, drought, when soil is water-logged, when rain expected within 4 hr or to crops under any stress
- Take particular care to avoid drift onto neighbouring susceptible crops or open water surfaces

Special precautions/Environmental safety
- Harmful if swallowed
- Irritating to eyes
- Dangerous to fish or other aquatic life. Do not contaminate surface waters or ditches with chemical or used container
- Harmful to bees. Do not apply to crops in flower or to those in which bees are actively foraging. Do not apply when flowering weeds are present

Protective clothing/Label precautions
- A, C
- 5c, 6a, 14, 18, 21, 25, 28, 29, 36, 37, 50, 52, 63, 66, 70, 78

Latest application/Harvest Interval(HI)
- Before 10 fully expanded leaf stage of crop

---

50   bromoxynil + clopyralid
A post-emergence contact and translocated herbicide mixture

Products					
Vindex	DowElanco	240:50 g/l	LI	05470

Uses   Annual dicotyledons in BARLEY, LINSEED, OATS, RYE, WHEAT.

Notes

Efficacy
- Recommended for tank mixing with approved formulations of MCPA, mecoprop, fluroxypyr, isoproturon and bentazone
- Best results achieved when weeds small and growing actively in warm, moist weather
- Do not spray during drought, waterlogging, frost or if rain is imminent
- Application rate, recommended crops and weed spectrum vary according to other components of tank mixture. See label for details

Crop Safety/Restrictions
- Maximum number of treatments 1 per crop
- Safe times for spraying crops vary according to other components of tank mixtures. See label for details
- Do not spray cereals undersown or to be undersown

- Spraying in frosty weather or when hard frost occurs within 3-4 wk may result in leaf scorch. This effect normally outgrown but may reduce yield of barley of low vigour or under stress on light soils
- Straw from treated crops must not be used in compost or manure for growing tomatoes under glass, but may be used for strawing down strawberries
- Do not roll, harrow or graze crops within 7 d before or after spraying
- Do not apply any other product within 7 d of application to linseed (10 d if crop is stressed)

Special precautions/Environmental safety
- Harmful in contact with skin or if swallowed. Irritating to skin and eyes
- Flammable
- Do not apply by knapsack sprayer or at concentrations higher than those recommended
- Dangerous to fish or other aquatic life. Do not contaminate surface waters or ditches with chemical or used container
- Harmful to bees. Do not apply to crops in flower or to those in which bees are actively foraging. Do not apply when flowering weeds are present

Protective clothing/Label precautions
- A, C
- 5a, 5c, 6a, 6b, 12c, 18, 21, 25, 28, 29, 36, 37, 50, 52, 63, 66, 70, 78

Withholding period
- Keep livestock out of treated areas for at least 6 wk after treatment

Latest application/Harvest Interval(HI)
- Before second node detectable (GS 32) for cereals, before first flower and before crop exceed 30 cm tall for linseed.
- HI 6 wk (for animal consumption)

---

51   bromoxynil + clopyralid + fluroxypyr
A contact and translocated herbicide mixture for use in cereals

Products	Crusader S	DowElanco	75:30:90 g/l	EC	05174

Uses   Annual dicotyledons, chickweed, cleavers, hemp-nettle, mayweeds, speedwells in WINTER BARLEY, WINTER WHEAT.

Notes

Efficacy
- Best results achieved by application in good growing conditions, when weeds small and growing actively in a strongly competitive crop
- Do not spray if night temperatures are low

Crop Safety/Restrictions
- Maximum number of treatments 1 per crop
- Apply to winter or spring cereals from 3-leaf stage to first node detectable (GS 13-31)
- Crops undersown with grass may be treated if the grasses are tillering. Do not treat cereals undersown with clover or other legumes

---

FOR FULL CONDITIONS OF USE ALWAYS READ THE PRODUCT LABEL

- Do not roll or harrow 10 d before or 7 d after treatment
- Do not spray during prolonged frosty weather or when crop is under stress from drought, waterlogging, nutrient deficiency or pest attack
- Straw from treated crops may contain residues which could damage susceptible crops. See label for detailed guidance on straw disposal or use
- Do not apply with hand-held equipment or at concentrations higher than those recommended

Special precautions/Environmental safety

- Harmful if swallowed. Irritating to eyes and skin
- Do not apply by knapsack sprayer or at concentrations higher than recommended
- Flammable
- Dangerous to fish or other aquatic life. Do not contaminate surface waters or ditches with chemical or used container
- Harmful to bees. Do not apply to crops in flower or to those in which bees are actively foraging. Do not apply when flowering weeds are present
- Do not harvest crops for human or animal consumption for at least 6 wk after last application

Protective clothing/Label precautions

- A, C
- 5c, 6a, 6b, 12c, 18, 21, 25, 28, 29, 35 (6 wk), 36, 37, 43 (6 wk), 50, 52, 63, 66, 70, 78

Withholding period

- Keep livestock out of treated areas for at least 6 wk

Latest application/Harvest Interval(HI)

- Before second node detectable stage (GS 32)

---

52   bromoxynil + ethofumesate + ioxynil

A post-emergence herbicide for new grass leys

Products	1 Leyclene	AgrEvo	50:200:25 g/l	EC	07263
	2 Leyclene	Schering	50:200:25 g/l	EC	05285

Uses   Annual dicotyledons, annual meadow grass in GRASS SEED CROPS, SEEDLING LEYS.

Notes   **Efficacy**
- Best results when weeds small and growing actively in a vigorous crop, soil moist and further rain within 10 d. Mid Oct to end Dec normally suitable
- Annual meadow grass controlled during early crop establishment. Spray weed grasses before fully tillered
- Do not spray in cold conditions or when heavy rain or frost imminent
- Do not use on soils with more than 10% organic matter
- Do not cut for 14 d after spraying or graze in Jan-Feb after spraying in Oct-Dec
- Ash or trash should be burned, buried or removed before spraying

Crop Safety/Restrictions
- Maximum number of treatments 2 per yr
- Apply to healthy ryegrasses or tall fescue after 2-3 leaf stage, to cocksfoot, timothy and meadow fescue at least 60 d after emergence and after 2-3 leaf stage

- Do not use on crops under stress, during periods of very dry weather, prolonged frost or waterlogging
- Do not use where clovers or other legumes are valued components of ley
- Do not roll for 7 d before or after spraying
- Any crop may be sown 5 mth after application following ploughing to at least 15 cm

Special precautions/Environmental safety
- Harmful if swallowed. Irritating to skin and eyes. Flammable
- Do not apply by knapsack sprayer or at concentrations higher than those recommended
- Dangerous to fish or other aquatic life. Do not contaminate surface waters or ditches with chemical or used container
- Harmful to bees. Do not apply to crops in flower or to those in which bees are actively foraging. Do not apply when flowering weeds are present

Protective clothing/Label precautions
- A, C
- 5c, 6a, 6b, 12c, 18, 22, 25, 28, 29, 36, 37, 50, 52, 63, 66, 70, 78

Withholding period
- Keep livestock out of treated areas for at least 6 wk after treatment

Latest application/Harvest Interval(HI)
- 6 wk before cutting or grazing

Maximum Residue Levels (mg residue/kg food)
- see ioxynil entry

---

53   bromoxynil + fluroxypyr
A post-emergence contact and translocated herbicide for cereals

Products	Sickle	DowElanco	300:150 g/l	EC	05187

Uses   Annual dicotyledons in BARLEY, WHEAT.

Notes

Efficacy
- Best results when weeds small and growing actively in a strongly competitive crop
- Spray is rainfast after 2 h

Crop Safety/Restrictions
- Maximum number of treatments 1 per crop
- Apply from 2-fully expanded leaf stage of crop to first node detectable (GS 12-31)
- Crops undersown with grass may be sprayed provided the grasses are tillering. Do not spray cereals undersown with clover or other legume mixtures
- Do not spray when crops are under stress from cold, drought, waterlogging etc, nor when frost imminent
- Do not roll or harrow for 10 d before or 7 d after spraying

---

FOR FULL CONDITIONS OF USE ALWAYS READ THE PRODUCT LABEL

Special precautions/Environmental safety
* Harmful if swallowed. Irritating to eyes and skin. Flammable
* Do not apply by knapsack sprayer or in less than recommended water volumes
* Dangerous to fish or other aquatic life. Do not contaminate surface waters or ditches with chemical or used container
* Harmful to bees. Do not apply to crops in flower or to those in which bees are actively foraging. Do not apply when flowering weeds are present

Protective clothing/Label precautions
* A, C
* 5c, 6a, 6b, 12c, 18, 21, 25, 28, 29, 35, 36, 37, 50, 52, 63, 66, 70, 78

Withholding period
* Keep livestock out of treated areas for at least 6 wk

Latest application/Harvest Interval(HI)
* Before second node detectable (GS 32).
* HI 6 wk

---

54   bromoxynil + fluroxypyr + ioxynil
A post-emergence contact and translocated herbicide for cereals

Products   Advance   DowElanco   100:90:100 g/l   LI   05173

Uses   Annual dicotyledons, chickweed, cleavers, hemp-nettle, speedwells in BARLEY, WHEAT.

Notes   **Efficacy**
* May be used in autumn or spring on winter or spring crops
* Best results when weeds small and growing actively in a highly competitive crop
* Do not spray when weed growth is hard from cold or drought

Crop Safety/Restrictions
* Maximum number of treatments 1 per crop
* Apply from 2-leaf stage of crop to first node detectable (GS 12-31)
* Do not spray undersown crops or crops to be undersown with clover or legume mixtures
* Do not spray crops stressed by frost, drought, mineral deficiency, pest or disease attack
* Do not roll or harrow for 7 d before or after spraying

Special precautions/Environmental safety
* Harmful if swallowed. Irritating to eyes and skin
* Flammable
* Do not apply by knapsack sprayer or in less than recommended water volumes
* Dangerous to fish or other aquatic life. Do not contaminate surface waters or ditches with chemical or used container
* Harmful to bees. Do not apply to crops in flower or to those in which bees are actively foraging. Do not apply when flowering weeds are present

Protective clothing/Label precautions
* A, C
* 5c, 6a, 12c, 18, 21, 25, 28, 29, 35, 36, 37, 50, 52, 60, 63, 66, 70, 78

Withholding period
• Keep livestock out of treated areas for at least 6 wk

Latest application/Harvest Interval(HI)
• Before second node detectable (GS 32).
• HI 6 wk

Maximum Residue Levels (mg residue/kg food)
• see ioxynil entry

55   bromoxynil + ioxynil

A contact acting post-emergence HBN herbicide for cereals

Products					
1 Briotril Plus 19/19	PBI	190:190 g/l	EC	04740
2 Deloxil	AgrEvo	190:190 g/l	EC	07313
3 Deloxil	Hoechst	190:190 g/l	EC	00664
4 Oxytril CM	RP Agric.	200:200 g/l	EC	06201

Uses   Annual dicotyledons in RYE, TRITICALE, UNDERSOWN CEREALS [3, 4]. Annual dicotyledons in BARLEY, OATS, WHEAT [1-4].

Notes   **Efficacy**
• Best results achieved on young weeds growing actively in a highly competitive crop
• Do not apply during periods of drought or when rain imminent (some labels say 'if likely within 4 or 6 h')
• Recommended for tank mixture with hormone herbicides to extend weed spectrum. See label for details

Crop Safety/Restrictions
• Maximum number of treatments 1 per crop
• Apply to winter or spring cereals from 2 fully expanded leaf stage [1-3], 1-fully expanded leaf [4], up to and including first node detectable (GS 31)
• Spray oats in spring when danger of frost past. Do not spray winter oats in autumn
• Apply to undersown cereals pre-sowing or pre-emergence of legume provided cover crop is at correct stage. Only spray trefoil pre-sowing [2-4]
• On crops undersown with grasses alone apply from 2-leaf stage of grass [2-4]
• Do not spray crops stressed by drought, waterlogging or other factors
• Do not roll or harrow for several days before or after spraying. Number of days specified varies with product, see label for details

Special precautions/Environmental safety
• Harmful if swallowed
• Irritating to skin and eyes [1, 4]
• Irritating to eyes [2, 3]
• Do not apply by knapsack sprayer or at concentrations higher than those recommended
• Dangerous to fish or other aquatic life. Do not contaminate surface waters or ditches with chemical or used container

FOR FULL CONDITIONS OF USE ALWAYS READ THE PRODUCT LABEL

- Direct spray from vehicle mounted sprayers must not be allowed to fall within 6 m of surface waters or ditches, from hand-held sprayers within 2 m. Spray must be directed away from water [2, 3]
- Harmful to bees. Do not apply to crops in flower or to those in which bees are actively foraging. Do not apply when flowering weeds are present

Protective clothing/Label precautions
- A, C
- 5a [2, 3]; 5c, 18, 21, 25, 28, 30, 36, 37, 50, 52, 63, 67, 70, 78 [1-4]; 6a, 6b [1, 4]

Withholding period
- Keep livestock out of treated areas for at least 6 wk [2, 3], 14 d [4]

Latest application/Harvest Interval(HI)
- Before 2nd node detectable stage (GS 31).
- HI (animal consumption) 6 wk [2, 3], 14 d [4]

Maximum Residue Levels (mg residue/kg food)
- see ioxynil entry

---

56   bromoxynil + ioxynil + mecoprop-P
A post-emergence contact and translocated herbicide

Products	Swipe P	Ciba Agric.	56:56:268 g/l	EC	07395

Uses   Annual dicotyledons in BARLEY, DURUM WHEAT, OATS, RYEGRASS, TRITICALE, WHEAT.

Notes   **Efficacy**
- Best results when weeds small and growing actively in a strongly competitive crop
- Do not spray in rain or when rain imminent. Control may be reduced by rain within 6 h
- Application to wet crops or weeds may reduce control
- To achieve optimum control of large over-wintered weeds or in advanced crops increase water volume to aid spray penetration and cover
- Recommended for tank-mixing with approved MCPA-amine for hemp nettle control

Crop Safety/Restrictions
- Maximum number of treatments 1 per crop
- Spray cereals from 3 leaves unfolded (GS 13) to before second node detectable (GS 32) for winter sown cereals, spring wheat and spring barley and before first node detectable (GS 31) for spring oats
- Apply to direct sown ryegrass or cereals undersown with ryegrass from 2-3 leaf stage of grass
- Do not spray crops undersown with legumes or use on winter oats or durum wheat in autumn
- Do not spray crops under stress from frost, waterlogging, drought or other causes
- Do not roll within 5 d after spraying
- Yield of barley may be reduced if frost occurs within 3-4 wk of treatment of low vigour crops on light soils or subject to stress
- Some crop yellowing may follow treatment but yield not normally affected
- Do not mix with manganese sulfate

- Avoid drift onto neighbouring susceptible crops

Special precautions/Environmental safety
- Harmful in contact with skin or if swallowed.
- Irritating to eyes and skin
- Do not apply by knapsack sprayer or in less than recommended volumes
- Dangerous to fish or other aquatic life. Do not contaminate surface waters or ditches with chemical or used container
- Do not allow direct spray from ground-based vehicle mounted/drawn sprayers to fall within 6 m, or from hand held sprayers to within 2 m, of surface waters or ditches. Direct spray away from water
- Harmful to bees. Do not apply to crops in flower or to those in which bees are actively foraging. Do not apply when flowering weeds are present
- Do not harvest crops for animal consumption for at least 6 wk after last application

Protective clothing/Label precautions
- A, C, H, M
- 5c, 6b, 9, 10a, 18, 21, 25, 26, 27, 28, 29, 35, 36, 37, 50, 52, 54a, 63, 66, 78

Withholding period
- Keep livestock out of treated areas for at least 6 wk and until foliage of poisonous weeds such as ragwort has died and become unpalatable

Latest application/Harvest Interval(HI)
- Before first node detectable (GS 31) for spring oats; before second node detectable (GS 32) for winter cereals, spring wheat, spring barley
- HI (animal consumption) 6 wk

Maximum Residue Levels (mg residue/kg food)
- see ioxynil entry

57   bromoxynil + ioxynil + triasulfuron
An HBN and sulfonylurea herbicide mixture for cereals

Products	Teal	Ciba Agric.	190:190 g/l:20% w/w	KK	06117

Uses   Annual dicotyledons in BARLEY, OATS, RYE, TRITICALE, WHEAT.

Notes   **Efficacy**
- For best results apply during warm, moist weather when weeds growing actively
- Do not spray when rain imminent. Rainfall within 6 h after spraying may reduce weed control

Crop Safety/Restrictions
- Maximum number of treatments 1 per crop
- Apply in spring, after 1 Feb, from 3 leaves unfolded but before second node detectable (GS 13-32)

FOR FULL CONDITIONS OF USE ALWAYS READ THE PRODUCT LABEL

- Do not spray undersown crops or those due to be undersown
- Do not spray winter barley of low vigour, on light soil and subject to stress during frosty weather
- Do not use on oats until risk of frost is over
- Do not use during frosty weather, when frost imminent, or on crops under stress from frost, waterlogging or drought
- Do not spray in tank mixture, or in sequence, with a product containing any other sulfonylurea
- See label for details of restrictions on sequential use of other herbicides and sowing of subsequent crops

Special precautions/Environmental safety
- Harmful if swallowed.
- Irritating to eyes
- Do not apply by knapsack sprayer or at concentrations higher than those recommended
- Dangerous to fish or other aquatic life and extremely dangerous to aquatic higher plants. Do not contaminate surface waters or ditches with chemical or used container
- Direct spray from vehicle-mounted sprayers must not be allowed to fall within 6 m, or from hand-held sprayers to within 2 m, of surface waters or ditches. Direct spray away from water
- Harmful to bees. Do not apply to crops in flower or to those in which bees are actively foraging. Do not apply when flowering weeds are present

Protective clothing/Label precautions
- A, C, H
- 5c, 6a, 18, 21, 25, 26, 27, 28, 29, 36, 37, 50, 52, 63, 67, 70, 78

Withholding period
- Keep livestock out of treated areas for at least 6 wk

Latest application/Harvest Interval(HI)
- From 1 Feb to before second node detectable (GS 32).
- HI (animal consumption) 6 wk

Maximum Residue Levels (mg residue/kg food)
- see ioxynil entry

58   bupirimate

A systemic pyrimidinol fungicide active against powdery mildew

Products	Nimrod	Zeneca	250 g/l	EC	06686

Uses Powdery mildew in APPLES, BLACKCURRANTS, CHRYSANTHEMUMS, COURGETTES, CUCUMBERS, GOOSEBERRIES, HOPS, MARROWS, ORNAMENTALS, PEARS, RASPBERRIES, ROSES, STRAWBERRIES, TOMATOES *(off-label)*.

Notes **Efficacy**
- Apply before or at first signs of disease and repeat at 5-14 d intervals. Timing and maximum dose vary with crop. See label for details
- On apples during periods that favour disease development lower doses applied weekly give better results than higher rates fortnightly

- Product has negligible effect on Phytoseiulus and Encarsia and may be used in conjunction with biological control of red spider mite

Crop Safety/Restrictions
- Maximum number of treatments for strawberries 3 per crop, for apples, pears 4-8 per crop (depending on dose), for gooseberries 5-10, for hops, cucurbits 6 per crop
- With apples, hops and ornamentals cultivars may vary in sensitivity to spray. See label for details
- If necessary to spray cucurbits in winter or early spring spray a few plants 10-14 d before spraying whole crop to test for likelihood of leaf spotting problem
- On roses some leaf puckering may occur on young soft growth in early spring or under low light intensity. Avoid use of high rates or wetter on such growth
- Never spray flowering begonias (or buds showing colour) as this can scorch petals
- Do not mix with other chemicals for application to begonias, cucumbers or gerberas

Special precautions/Environmental safety
- Irritating to eyes and skin.
- Flammable
- Harmful to fish or other aquatic life. Do not contaminate surface waters or ditches with chemical or used container

Protective clothing/Label precautions
- A, C
- 6a, 6b, 12c, 18, 29, 35, 36, 37, 51, 63, 66, 70

Latest application/Harvest Interval(HI)
- HI apples, pears, strawberries 1 d; cucurbits, tomatoes 2 d; blackcurrants 7 d; raspberries 8 d; gooseberries, hops 14 d

Approval
- Off-label approval unlimited for use on protected tomatoes (OLA 0822/94)

---

59   bupirimate + triforine

A systemic protectant and eradicant fungicide for ornamental crops

Products					
	1 Nimrod-T	ICI Professional	62.5:62.5 g/l	EC	01499
	2 Nimrod-T	Zeneca Prof.	62.5:62.5 g/l	EC	06859

Uses   Black spot, leaf spots, powdery mildew in ORNAMENTALS. Rust in ROSES.

Notes   **Efficacy**
- To prevent infection on roses spray in early May and repeat every 10-14 d. If infected with blackspot in previous season commence spraying at bud burst

Crop Safety/Restrictions
- Spraying in high glasshouse temperatures may cause temporary leaf damage
- Test varietal susceptibility of roses by spraying a few plants and allow 14 d for any symptoms to develop

---

FOR FULL CONDITIONS OF USE ALWAYS READ THE PRODUCT LABEL

Special precautions/Environmental safety
* Irritating to eyes
* Harmful to fish or other aquatic life. Do not contaminate surface waters or ditches with chemical or used container

Protective clothing/Label precautions
* C
* 6a, 18, 22, 29, 36, 37, 53, 63, 66

60   buprofezin

A moulting inhibitor, thiadiazine insecticide for whitefly control

Products   Applaud   Zeneca   250 g/l   SC   06900

Uses   Glasshouse whitefly, tobacco whitefly in AUBERGINES, CUCUMBERS, PEPPERS, PROTECTED ORNAMENTALS, TOMATOES.

Notes   **Efficacy**
* Product has contact, residual and some vapour activity
* Whitefly most susceptible at larval stages but residual effect can also kill nymphs emerging from treated eggs and application to pupae reduces emergence
* Adult whitefly not directly affected
* Product may be used either in IPM programme in association with *Encarsia formosa* or in All Chemical programme
* In IPM programme apply as single application and allow at least 60 d before re-applying
* In All Chemical programme apply twice at 7-14 d interval and allow at least 60 d before re-applying
* Do not leave spray liquid in sprayer for long periods
* Do not apply as fog or mist

Crop Safety/Restrictions
* Maximum number of treatments up to 4 programmes (IPM or All Chemical) per crop for tomatoes, cucumbers and ornamentals; up to 2 sprays (2 IPM or 1 All Chemical programme) per crop for aubergines and peppers
* See label for list of ornamentals successfully treated but small scale test advised to check varietal tolerance. This is especially important if spraying flowering ornamentals with buds showing colour
* Do not treat Dieffenbachia or Closmoplictrum
* Do not apply to crops under stress

Protective clothing/Label precautions
* 30, 54, 63, 67, 70

Latest application/Harvest Interval(HI)
* HI edible crops 3 d

61   calciferol

A hypercalcaemic rodenticide available only in mixtures

62   calciferol + difenacoum

A mixture of rodenticides with different modes of action

Products   Sorexa CD Ready to Use   Sorex   0.1:0.0025% w/w   RB   03514

Uses   Mice in FARM BUILDINGS.

Notes   **Efficacy**
* Lay baits in mouse runs in many locations throughout infested area
* It is important to lay many small baits as mice are sporadic feeders
* Cover baits to protect from moisture in outdoor situations
* Inspect baits frequently and replace or top up with fresh material as necessary

Special precautions/Environmental safety
* Protect baits from access by children, domestic or other animals

Protective clothing/Label precautions
* 25, 29, 63, 67, 81, 82, 83, 84

---

63   captafol

A protectant dicarboximide fungicide approvals for which expired on 31 December 1990

---

64   captan

A protectant dicarboximide fungicide with horticultural uses

Products   PP Captan 80 WG   Zeneca   80% w/w   SG   06696

Uses   Black spot in ROSES. Botrytis fruit rot, brown rot, gloeosporium rot, phytophthora fruit rot, scab, storage rots in APPLES, PEARS. Fungus diseases in FLOWER BULBS *(off-label)*.

Notes   **Efficacy**
* For control of scab, flyspeck and sooty blotch apply at bud burst and repeat at 7-14 d intervals until danger of scab infection ceased
* For suppression of fruit storage rots apply from late Jul and repeat at 2 wk intervals until 7 d before picking
* To reduce spread of fruit storage rots in store dip or drench fruit immediately after picking
* For black spot control in roses apply after pruning with 3 further applications at 14 d intervals or spray when spots appear and repeat at 7-10 d intervals
* Do not leave diluted material for more than 2 h. Agitate well before and during spraying
* Product is not compatible with adjuvant oils

FOR FULL CONDITIONS OF USE ALWAYS READ THE PRODUCT LABEL

Crop Safety/Restrictions
* Maximum number of treatments 1 per batch as post-harvest dip on apples and pears
* Do not use on apple cultivars Bramley, Monarch, Winston, King Edward, Kidd's Orange or Red Delicious or on pear cultivar D'Anjou

Special precautions/Environmental safety
* Irritating to eyes
* May cause sensitization by skin contact
* Harmful to fish or other aquatic life. Do not contaminate surface waters or ditches with chemical or used container

Protective clothing/Label precautions
* A, C, H
* 6a, 10a, 18, 30, 36, 37, 53, 63, 66

Latest application/Harvest Interval(HI)
* HI apples, pears 7 d

Approval
* Off-label approval unlimited for use on non-edible ornamental bulbs (OLA 1229/95)[1]

Maximum Residue Levels (mg residue/kg food)
* pome fruits, grapes, cane fruits, bilberries, cranberries, currants, gooseberries, tomatoes, peppers, aubergines 3; apricots, peaches, nectarines, plums, lettuce, beans, peas, leeks 2; citrus fruits, bananas, carrots, horseradish, parsnips, parsley root, salsify, swedes, turnips, garlic, onions, shallots, cucumbers, gherkins, courgettes, cauliflowers, Brussels sprouts, head cabbage, celery, rhubarb, mushrooms, potatoes 0.1

---

65   captan + penconazole
A protectant fungicide for use on apple trees

Products	Topas C 50 WP	Ciba Agric.	47.5:2.5% w/w	WP	03232

Uses   Powdery mildew, scab in APPLES.

Notes   **Efficacy**
* Use as a protective spray every 7-14 d from bud burst until extension growth ceases
* High antisporulant activity reduces development of primary mildew and controls spread of secondary mildew
* Does not affect beneficial insects so can be used in integrated control programmes
* Penconazole is recommended alone for mildew control if scab is not a problem

Crop Safety/Restrictions
* Maximum number of treatments 10 per yr

Special precautions/Environmental safety
* Irritating to eyes, skin and respiratory system
* Dangerous to fish or other aquatic life. Do not contaminate surface waters or ditches with chemical or used container

Protective clothing/Label precautions
* A, C
* 6a, 6b, 6c, 14, 18, 22, 29, 35, 36, 37, 52, 63, 70

Latest application/Harvest Interval(HI)
- HI apples 14 d

Maximum Residue Levels (mg residue/kg food)
- see captan entry

66   carbaryl

A contact carbamate insecticide, worm killer and fruit thinner

Products					
1	Microcarb Suspendable Powder	Grampian	50% w/w	WP	07183
2	Thinsec	Zeneca	450 g/l	SC	06710
3	Twister Flow	RP Environ.	240 g/l	SC	05712

Uses

Apple capsid, codling moth, earwigs, fruit thinning, tortrix moths, winter moth in APPLES [2]. Earthworms in AMENITY GRASS, TURF [3]. Red mite in POULTRY HOUSES [1].

Notes

Efficacy
- Application rate and timing vary with pest. See label for details
- Effective thinning of apples depends on thorough wetting of foliage and fruitlets. Dose and timing vary with cultivar
- Fruit from treated trees may need picking 5-7 d earlier than from untreated trees
- For earthworm control apply in autumn or spring when worms active. On dry soil apply a light watering after treatment
- For use in poultry houses spray when birds have been removed and the house has been thoroughly cleared out [1]

Crop Safety/Restrictions
- Maximum number of treatments 2 per yr for turf [3]
- Carbaryl is dangerous to pollinating insects and particular care must be taken when spraying close to blossom period or when orchard adjacent to other flowering crops
- Do not spray Laxton's Fortune between pink bud stage and end of first wk in Jun
- Do not use on apples between late pink bud and end of first wk in Jun except for purpose of fruit thinning
- Do not spray onto eggs and keep off birds [1]

Special precautions/Environmental safety
- This product contains an anticholinesterase carbamate compound. Do not use if under medical advice not to work with such compounds
- Harmful if swallowed
- Harmful to fish or other aquatic life. Do not contaminate surface waters or ditches with chemical or used container
- Dangerous to bees. Do not apply to crops in flower or to those in which bees are actively foraging. Do not apply when flowering weeds are present [2, 3]

Protective clothing/Label precautions
- A, C, D, H, J, M [1]; A, C [2]; A [3]
- 2, 5c, 48, 63, 65 [2, 3]; 5b, 28, 67 [1]; 18, 21, 29, 36, 37, 53, 70, 78 [1-3]; 35 [3]

FOR FULL CONDITIONS OF USE ALWAYS READ THE PRODUCT LABEL

Latest application/Harvest Interval(HI)
• HI apples 7 d

Maximum Residue Levels (mg residue/kg food)
• apricots, peaches, nectarines, plums, cane fruits, bilberries, cranberries, currants, gooseberries, lettuce 10; citrus fruits, strawberries 7; pome fruits, grapes, bananas, tomatoes, peppers, aubergines, head cabbage, beans, peas 5; cucumbers, gherkins, courgettes, celery, rhubarb 3; carrots, horseradish, parsnips, parsley root, salsify, swedes 2; turnips, garlic, onions, shallots, cauliflowers, Brussels sprouts, leeks, mushrooms 1; potatoes 0.2

67 carbendazim (MBC)

A systemic benzimidazole fungicide with curative and protectant activity

Products

1 Ashlade Carbendazim FL	Ashlade	500 g/l	SC	06213
2 Bavistin DF	BASF	50% w/w	WG	03848
3 Campbell's Carbendazim 50% Flowable	MTM Agrochem.	500 g/l	SC	02681
4 Carbate Flowable	PBI	500 g/l	SC	03341
5 Delsene 50 DF	DuPont	50% w/w	WG	02692
6 Derosal WDG	AgrEvo	80% w/w	WG	07316
7 Derosal WDG	Hoechst	80% w/w	WG	03404
8 Fisons Turfclear	Levington	500 g/l	SC	02253
9 Headland Addstem	Headland	511 g/l	SC	06755
10 Hinge	Quadrangle	500 g/l	SC	04929
11 Mascot Systemic	Rigby Taylor	500 g/l	SC	07654
12 Stefes C-Flo	Stefes	511 g/l	SC	07052
13 Stefes Carbendazim Flo	Stefes	500 g/l	SC	05677
14 Stefes Derosal WDG	Stefes	80% w/w	WG	07658
15 Tripart Defensor FL	Tripart	500 g/l	SC	02752
16 Turfclear	Levington	500 g/l	SC	07506
17 Turfclear WDG	Levington	80% w/w	SG	06275
18 Turfclear WDG	Levington	80% w/w	WG	07490

Uses

American gooseberry mildew, botrytis, currant leaf spot in BLACKCURRANTS, GOOSEBERRIES [1, 2, 6, 7, 9, 10, 12, 14, 15]. Anthracnose, botrytis in DWARF BEANS [1-3, 6, 7, 9, 12, 14]. Anthracnose, botrytis in NAVY BEANS [2, 3]. Black root rot, botrytis, powdery mildew in CUCUMBERS [6, 7, 9, 12, 14]. Botrytis, cane spot, powdery mildew, spur blight in BLACKBERRIES *(off-label)*, LOGANBERRIES *(off-label)*, RASPBERRIES *(off-label)* [2]. Botrytis, cane spot, powdery mildew, spur blight in CANE FRUIT [1, 6, 7, 15]. Botrytis, cane spot, powdery mildew, spur blight in RASPBERRIES [9, 10, 12, 14]. Botrytis, didymella, fusarium wilt, leaf mould, verticillium wilt in TOMATOES [2, 4, 6, 7, 9, 10, 12, 14]. Botrytis, fire, sclerotinia in TULIPS [14]. Botrytis, fusarium, penicillium rot, sclerotinia, stagonospora in CORMS, FLOWER BULBS [6, 7, 9, 12]. Botrytis, fusarium, penicillium rot, sclerotinia in BULBS/CORMS *(dip - off-label)* [2]. Botrytis, light leaf spot in OILSEED RAPE [1, 2, 4-7, 9, 10, 13-15]. Botrytis, powdery mildew in BEDDING PLANTS, POT PLANTS [2, 6, 7, 9]. Botrytis, powdery mildew in BEDDING PLANTS *(off-label)*, CUCUMBERS *(off-label)*, PEPPERS *(off-label)*, POT PLANTS *(off-label)*, PROTECTED ORNAMENTALS [2]. Botrytis, powdery mildew in STRAWBERRIES [1, 2, 6, 7, 9, 10, 12, 14, 15]. Botrytis, powdery mildew in PROTECTED POT PLANTS [14]. Botrytis, sclerotinia, stagonospora in BULBS/CORMS *(off-label)* [2]. Botrytis, sclerotinia in GLADIOLI, NARCISSI [14]. Botrytis in CELERY *(off-label)*, CHRYSANTHEMUMS *(off-label)*, ONIONS *(off-label)* [2]. Brown rot in CHERRIES *(off-label)*, PLUMS *(off-label)* [2]. Canker, eye rot, gloeosporium, powdery mildew, scab, storage rots in APPLES [1-3, 6, 7, 9, 10, 12, 14, 15]. Canker in APPLES *(partial control)* [10]. Celery leaf spot in

CELERY [6, 7, 9, 10, 12, 14]. Chocolate spot in BROAD BEANS [1, 3, 6, 7, 9, 10, 12, 14, 15]. Chocolate spot in FIELD BEANS [1-7, 9, 10, 12-15]. Dactylium, dry bubble, trichoderma, wet bubble in MUSHROOMS [2, 4]. Dollar spot, earthworms, fusarium patch, red thread in TURF [8, 11, 16-18]. Eye rot, gloeosporium, scab, storage rots in PEARS [3, 6, 7, 9, 14]. Eyespot, rhynchosporium, sooty moulds in BARLEY [1-7, 9, 12-15]. Eyespot, rhynchosporium in WINTER BARLEY [1-5, 9, 10, 12, 13, 15]. Eyespot, sooty moulds in WHEAT [6, 7, 14]. Eyespot in WINTER RYE [2, 4, 5, 13]. Eyespot in WINTER WHEAT [1-5, 9, 12, 13, 15]. Fusarium wilt, verticillium wilt in PROTECTED CARNATIONS [2, 4]. Fusarium in FREESIAS *(off-label)*[2]. Glume blotch, powdery mildew in WINTER WHEAT *(partial control)*[10]. Mildew in ROSES [2]. Powdery mildew in WINTER BARLEY *(partial control)* [10]. Scab, storage rots in PEARS *(off-label)* [2]. Trichoderma in MUSHROOMS *(spawn treatment - off-label)* [2].

Notes

Efficacy
- Products vary in the diseases listed as controlled for several crops. Labels must be consulted for full details and for rates and timings
- Mostly applied as spray or drench. Spray treatments normally applied at first sign of disease and repeated after 10-14 d if required
- Apply as a drench to control soil-borne diseases in cucumbers and tomatoes and as a pre-planting dip treatment for bulbs
- On apples, bush and cane fruit the addition of a non-ionic wetter aids penetration and improves disease control [14]
- Apply by incorporation into casing to control mushroom diseases [2, 4]
- Do not apply if rain or frost expected or crop wet
- To delay appearance of resistant strains alternate treatment with non-MBC fungicide. Eyespot in cereals and *Botrytis cinerea* in many crops is now widely resistant
- Not compatible with alkaline products such as lime sulfur

Crop Safety/Restrictions
- Maximum number of treatments (including applications of any product containing benomyl, carbendazim or thiophanate-methyl) 1 per crop for apples (post-harvest), lettuce (peat incorporation), dwarf beans, mushrooms, bulbs, corms; 2 per crop for onions, field beans, broad beans, navy beans, cereals, lettuce, ornamentals; 3 per crop for plums, cherries, strawberries, cane fruit, blackcurrants, gooseberries, oilseed rape; 6 per crop for apples (foliar spray), pears, tomatoes, peppers, cucumbers (12 per crop for apples and pears [14]); 8 per crop for celery
- Do not treat crops suffering from drought or other physical or chemical stress
- Do not use on strawberry runner beds
- Apply as drench rather than spray where red spider mite predators are being used
- Consult processors before using on crops for processing
- Use drench for tomatoes, cucumbers and peppers on soil-grown crops only

Special precautions/Environmental safety
- Harmful to fish or other aquatic life. Do not contaminate surface waters or ditches with chemical or used container [2, 5-9, 11-14]
- After use dipping suspension must not be discharged directly into ditches or drains. Preferred disposal method is via a soakaway [14]

Protective clothing/Label precautions
- A, B, C, D, H, K, M [14]; A, B, C, H, K, M [1-9, 12, 13, 15-18]; A, C, H, M [10, 11]

FOR FULL CONDITIONS OF USE ALWAYS READ THE PRODUCT LABEL

- 29 [10, 13]; 30 [1-9, 11-14, 16-18]; 36 [10]; 37 [1, 10, 15]; 51 [17, 18]; 53 [2, 5, 8-14, 16]; 54 [1, 3, 4, 6, 7]; 58 [11]; 63 [1-5, 7-13, 16-18]; 65 [11, 13, 15]; 66 [1, 3, 4, 7, 9, 12]; 67 [2, 5-8, 10, 14, 16-18]; 70 [3, 15]

Latest application/Harvest Interval(HI)
- Up to and including grain watery-ripe stage (GS 71) for cereals; at full bloom for dwarf beans.
- HI 2 d for strawberries, cane fruit, blackcurrants, gooseberries, tomatoes, peppers, cucumbers; 7 d for celery, lettuce; 7-14 d (depend on dose) for apples, pears; 14 d for plums, cherries, onions, mushrooms; 21 d for field beans, broad beans, navy beans, oilseed rape.

Approval
- Approved for aerial application on winter wheat, winter and spring barley [1, 2, 4-7, 15]; winter rye [2]; field beans [2, 4-7]; dwarf beans [2]; oilseed rape [1, 2, 4, 5, 15]; onions [2]. See notes in Section 1
- Off-label approval unlimited for use on mushrooms (spawn treatment), protected cucumbers, onions, celery, cucumbers, peppers, ornamental bulbs and corms, pears, plums, cherries, raspberries, loganberries, blackberries, freesias, chrysanthemums, pot and bedding plants (OLA 1470/94, 1002/95, 1144/95)[2]

Maximum Residue Levels (mg residue/kg food)
- peaches, nectarines, grapes 10; citrus fruits, strawberries, raspberries, currants, tomatoes, lettuce 5; potatoes 3; pome fruits, plums, onions, celery 2; bananas, cultivated mushrooms 1; aubergines, cucumbers, melons, squashes, Brussels sprouts 0.5; soya beans 0.2; tree nuts, bilberries, cranberries, wild berries, miscellaneous fruits (except bananas, olives), beetroot, horseradish, Jerusalem artichokes, parsnips, parsley root, radishes, sweet potatoes, yams, garlic, shallots, spring onions, courgettes, watermelons, sweet corn, Chinese cabbage, kale, kohlrabi, spinach, beet leaves, watercress, chervil, chives, parsley, celery leaves, stem vegetables (except celery), wild mushrooms, lentils, oilseeds (except soya beans), cereals, animal products 0.1

68   carbendazim + chlorothalonil
A systemic and protectant fungicide mixture

Products					
1 Bravocarb	ISK Biosciences	100:450 g/l	SC	05119
2 Greenshield	Zeneca Prof.	100:450 g/l	SC	06763

Uses   Anthracnose, dollar spot, fusarium patch, red thread in TURF [2]. Ascochyta, botrytis, downy mildew in PEAS [1]. Chocolate spot in FIELD BEANS [1]. Rhynchosporium in BARLEY [1]. Septoria in WINTER WHEAT [1].

Notes   **Efficacy**
- Apply to winter wheat and barley between leaf sheath lengthening and first node detectable stages (GS 30-31) for early Septoria control. Further application up to ear emergence (GS 51) may be needed for Rhynchosporium control [1]
- To protect wheat against Septoria spray immediately infection visible between flag leaf just visible and ear just fully emerged (GS 37-59) [1]
- Treatment gives good control of cereal ear diseases particularly when applied directly to the ear. It also suppresses rusts, mildew and net blotch [1]
- Tank-mix with a specific fungicide if rust/mildew levels become high and with prochloraz if MBC-resistant strains of eyespot are present [1]

* Spray peas and beans at flowering and 2-4 wk later [1]
* Apply at any time of year at the first sign of turf disease. Further treatments may be necessary if conditions remain favourable [2]
* Established or severe infections of Anthracnose will not be controlled [2]

Crop Safety/Restrictions
* Maximum number of treatments (including applications of any product containing benomyl, carbendazim or thiophanate-methyl) 2 per crop for cereals, peas, field beans [1]
* Do not mow or water within 24 h of treatment [2]
* Do not mix with other pesticides, surfactants or fertilisers [2]

Special precautions/Environmental safety
* Irritating to eyes, skin and respiratory system
* Dangerous to fish or other aquatic life. Do not contaminate surface waters or ditches with chemical or used container

Protective clothing/Label precautions
* A, C, H, M [2]; A, C [1]
* 6a, 6b, 6c, 18, 29, 36, 37, 52, 63, 65 [1, 2]; 21, 26, 60, 9 [1]; 22, 28 [2]

Latest application/Harvest Interval(HI)
* Up to and including grain watery-ripe stage (GS 71) for cereals [1]
* HI 14 d for peas, 21 d for field beans, oilseed rape [1]

Approval
* Approved for aerial application on winter wheat, barley. See notes in Section 1. Consult firm for details [1]

Maximum Residue Levels (mg residue/kg food)
* see carbendazim and chlorothalonil entries

69   carbendazim + chlorothalonil + maneb
A broad-spectrum protectant and eradicant fungicide for cereals

Products					
1 Ashlade Mancarb Plus	Ashlade	80:150:200 g/l	SC	06222
2 Tripart Victor	Tripart	80:150:200 g/l	SC	04359

Uses Botrytis, brown foot rot and ear blight, brown rust, eyespot, powdery mildew, septoria diseases, sooty moulds, yellow rust in SPRING WHEAT, WINTER WHEAT. Botrytis, brown rust, eyespot, net blotch, rhynchosporium, yellow rust in SPRING BARLEY, WINTER BARLEY.

Notes **Efficacy**
* Apply at any stage, dependent on disease pressure, up to ears fully emerged (GS 59)
* Highly effective against late diseases on flag leaf and ear of winter cereals

Crop Safety/Restrictions
* Maximum number of treatments (including application of any product containing benomyl, carbendazim or thiophanate-methyl) 2 per crop.

FOR FULL CONDITIONS OF USE ALWAYS READ THE PRODUCT LABEL

Special precautions/Environmental safety
* Irritating to eyes, skin and respiratory system
* Dangerous to fish or other aquatic life. Do not contaminate surface waters or ditches with chemical or used container

Protective clothing/Label precautions
* A, C
* 6a, 6b, 6c, 18, 21, 28, 29, 35, 36, 37, 52, 63, 67

Latest application/Harvest Interval(HI)
* Up to and including grain watery-ripe stage (GS 71)

Maximum Residue Levels (mg residue/kg food)
* see carbendazim, chlorothalonil and maneb entries

---

70   carbendazim + cyproconazole
A systemic protective and curative fungicide for cereals

Products	Alto Combi	Sandoz	300:160 g/l	SC	05066

Uses   Brown rust, eyespot, net blotch, powdery mildew, rhynchosporium, yellow rust in WINTER BARLEY. Brown rust, eyespot, powdery mildew, septoria diseases, sooty moulds, yellow rust in WINTER WHEAT. Brown rust, net blotch, powdery mildew, rhynchosporium, yellow rust in SPRING BARLEY.

Notes

Efficacy
* Apply at start of disease development or as preventive treatment
* Most effective timing of treatment varies with disease - see label for details
* For eradication of established mildew tank-mix with approved formulation of tridemorph or fenpropimorph
* For high *Septoria tritici* infection tank-mix with approved formulation of chlorothalonil
* When applied early, prior to GS 33, useful reduction of eyespot is obtained

Crop Safety/Restrictions
* Maximum number of treatments (including applications of any product containing benomyl, carbendazim or thiophanate-methyl) 2 per crop
* Application to winter wheat in spring at GS 30-33 may cause straw shortening but does not cause loss of yield

Special precautions/Environmental safety
* Harmful to fish or other aquatic life. Do not contaminate surface waters or ditches with chemical or used container

Protective clothing/Label precautions
* A, C, H
* 30, 53, 63, 66

Latest application/Harvest Interval(HI)
* Up to and including beginning of anthesis (GS 61) for winter wheat; up to and including emergence of ear complete (GS 59) for barley

Maximum Residue Levels (mg residue/kg food)
* see carbendazim entry

---

71   carbendazim + flusilazole
A broad-spectrum systemic and protectant fungicide for cereals

Products

1 Contrast	DuPont	125:250 g/l	SC	06150
2 Punch C	DuPont	125:250 g/l	SC	06801
3 Standon Flusilazole Plus	Standon	125:250 g/l	SC	07403

Uses

Brown rust, eyespot, mildew, net blotch, powdery mildew, rhynchosporium, septoria, yellow rust in SPRING BARLEY, WINTER BARLEY [1-3]. Brown rust, eyespot, mildew, powdery mildew, septoria, septoria diseases, yellow rust in WINTER WHEAT [1-3]. Brown rust, powdery mildew, septoria diseases, yellow rust in SPRING WHEAT [3]. Canker, light leaf spot, phoma leaf spot in OILSEED RAPE [1, 2]. Eyespot in SPRING WHEAT *(reduction)*, WINTER BARLEY *(reduction)*, WINTER WHEAT *(reduction)* [3]. Light leaf spot in OILSEED RAPE *(reduction)* [3].

Notes

Efficacy
* Apply at early stage of disease development or in routine preventive programme
* Most effective timing of treatment varies with disease - see label for details
* Higher rate active against both MBC-sensitive and MBC-resistant eyespot
* Rain occurring within 2-3 h of spraying may reduce effectiveness
* To prevent build-up of resistant strains of mildew tank mix with approved morpholine fungicide [1, 2]
* Treat oilseed rape in autumn when leaf lesions first appear and spring from the start of stem extension when disease appears

Crop Safety/Restrictions
* Maximum number of treatments (including applications of any product containing flusilazole) 3 per winter wheat crop, 2 per crop of winter barley, spring barley, spring wheat or oilseed rape
* Do not apply to crops under stress or during frosty weather

Special precautions/Environmental safety
* Irritating to eyes
* Harmful if swallowed
* Dangerous to fish or other aquatic life. Do not contaminate surface waters or ditches with chemical or used container

Protective clothing/Label precautions
* A, C, H, M [3]; A, C [1, 2]
* 5c, 78 [3]; 6a, 18, 24, 28, 29, 36, 37, 52, 63 [1-3]; 12c [1, 2]; 66 [2, 3]; 68 [1]; 70 [1, 3]

Latest application/Harvest Interval(HI)
* Up to and including grain watery ripe stage (GS 71) for wheat and barley; before first flower opened stage for oilseed rape

---

FOR FULL CONDITIONS OF USE ALWAYS READ THE PRODUCT LABEL

Maximum Residue Levels (mg residue/kg food)
* see carbendazim entry

72   carbendazim + flutriafol

A systemic fungicide for cereals with protectant and eradicant activity

Products	1 Early Impact	ICI Agrochem.	150:94 g/l	SC	02915
	2 Early Impact	Zeneca	150:94 g/l	SC	06659
	3 Pacer	Zeneca	150:94 g/l	SC	06690
	4 Palette	Zeneca	150:94 g/l	SC	06691

Uses   Brown rust, eyespot, glume blotch, leaf spot, powdery mildew, yellow rust in WINTER WHEAT [1-4]. Brown rust, eyespot, leaf blotch, net blotch, powdery mildew, yellow rust in SPRING BARLEY, WINTER BARLEY [1-4]. Net blotch in WINTER BARLEY *(moderate control)* [4].

Notes   **Efficacy**
* Apply at early stage of disease development or as a protectant in high risk situations
* For MBC sensitive eyespot control apply between leaf sheath erect stage and before second node detectable (GS 30-31). See label for timing details [1-3]
* When mildew is established at 5-10% level mix with a morpholine fungicide for improved control. See label for details of timing. Seed treatment may be preferable on spring barley

Crop Safety/Restrictions
* Maximum number of treatments (including applications of any product containing benomyl, carbendazim, thiophanate-methyl or flutriafol) 2 per crop
* Under conditions of stress some wheat varieties can exhibit flag leaf tip scorch, which may be increased by fungicide application

Special precautions/Environmental safety
* Irritating to eyes and skin
* Harmful to fish or other aquatic life. Do not contaminate surface waters or ditches with chemical or used container

Protective clothing/Label precautions
* A, C, H [3, 4]; A, C [1, 2]
* 6a, 6b, 14, 18, 28, 29, 36, 37, 53, 63, 66 [1-4]; 21, 70 [1-3]; 22 [4]

Latest application/Harvest Interval(HI)
* Up to and including grain watery-ripe stage (GS 71)

Maximum Residue Levels (mg residue/kg food)
* see carbendazim entry

73   carbendazim + iprodione

A systemic and contact fungicide mixture

Products	1 Calidan	RP Agric.	87.5:175 g/l	SC	06536
	2 Vitesse	RP Environ.	87.5:175 g/l	SL	06537

Uses	Alternaria, botrytis, light leaf spot, sclerotinia stem rot in OILSEED RAPE [1]. Anthracnose, fusarium patch, pink patch, red thread, timothy leaf spot in TURF [2]. Ascochyta, botrytis, mycosphaerella, sclerotinia stem rot in COMBINING PEAS, VINING PEAS [1].

Notes

Efficacy
- Best results obtained by application at first signs of disease, repeated monthly as necessary [2]
- Maximum efficacy against Anthracnose achieved by treatment at early disease development stage. Curative treatments for well-established Anthracnose are not recommended [2]
- May be used all year round, but is best suited for spring or late summer/early autumn application [2]
- Timing for oilseed rape varies according to disease. See label [1]
- Complete control of Botrytis and Alternaria in rape may require 2 treatments separated by not less than 3 wk [1]
- Treat peas at mid-flowering when first pods 2.5 cm long. On combining peas repeat 2-3 wk later if necessary [1]

Crop Safety/Restrictions
- Maximum number of treatments 2 per crop for oilseed rape and combining peas, 1 per crop for vining peas [1]. (NB Oilseed rape may be treated with 3 applications of products containing benomyl, carbendazim or thiophanate-methyl per season)
- Where grass is being mown, apply after mowing. Delay further mowing for at least 48 h after treatment [2]

Special precautions/Environmental safety
- Harmful to fish or other aquatic life. Do not contaminate surface waters or ditches with chemical or used container

Protective clothing/Label precautions
- A, C, H, M
- 30, 53, 63, 66

Latest application/Harvest Interval(HI)
- HI 3 wk for oilseed rape, peas [1]

Maximum Residue Levels (mg residue/kg food)
- see carbendazim and iprodione entries

---

74 carbendazim + mancozeb
A broad-spectrum systemic and protectant fungicide for cereals

Products	1 Kombat WDG	AgrEvo	12.4:63.3% w/w	WG	07328
	2 Kombat WDG	Hoechst	12.4:63.3% w/w	WG	04344

Uses	Downy mildew in WINTER OILSEED RAPE. Eyespot, mildew, septoria, sooty moulds, yellow rust in WINTER WHEAT. Mildew, net blotch, rhynchosporium, sooty moulds in WINTER BARLEY.

FOR FULL CONDITIONS OF USE ALWAYS READ THE PRODUCT LABEL

Notes

Efficacy
- Best results achieved by application at beginning of disease development
- Do not spray when crops wet, if rain imminent or if temperature exceeds 30°C
- Spray wheat between leaf sheath erect and first spikelets visible (GS 30-51) and barley between flag leaf ligule visible and complete ear emergence (GS 39-59)
- Will protect against mildew/rusts but tank-mix with a specific rust/mildew fungicide if these diseases are already present in the crop
- MBC resistant eyespot not controlled. If present use products containing prochloraz
- Spray winter oilseed rape pre-flowering up to yellow bud stage (GS 3,7)

Crop Safety/Restrictions
- Maximum number of treatments (including applications of any product containing benomyl, carbendazim or thiophanate-methyl) 2 per crop for cereals, 3 per crop for oilseed rape
- Do not spray crops suffering stress from drought

Special precautions/Environmental safety
- Irritating to respiratory system
- May cause sensitization by skin contact
- Harmful to fish or other aquatic life. Do not contaminate surface waters or ditches with chemical or used container

Protective clothing/Label precautions
- A, C, D, H
- 6c, 10a, 18, 21, 29, 36, 37, 53, 63, 67

Latest application/Harvest Interval(HI)
- Up to and including grain watery-ripe stage (GS 71).
- HI 21 d for oilseed rape.

Maximum Residue Levels (mg residue/kg food)
- see carbendazim and mancozeb entries

75   carbendazim + maneb
A broad-spectrum systemic and protectant fungicide

Products

1 Ashlade Mancarb FL	Ashlade	50:320 g/l	SC	06217
2 Campbell's MC Flowable	MTM Agrochem.	62:400 g/l	SC	03467
3 Headland Dual	Headland	62:400 g/l	SC	03782
4 MC Flowable	United Phosphorus	62:400 g/l	SC	06295
5 Multi-W FL	PBI	50:320 g/l	SC	04131
6 New Squadron	Quadrangle	50:320 g/l	SC	05981
7 Tripart Legion	Tripart	50:320 g/l	SC	02997
8 Tripart Legion	Tripart	50:320 g/l	SC	06113

Uses

Botrytis, downy mildew, light leaf spot in OILSEED RAPE [5, 6, 8]. Brown rust, eyespot, powdery mildew, septoria diseases, sooty moulds, yellow rust in WINTER WHEAT [1-6]. Brown rust, net blotch, powdery mildew, rhynchosporium, sooty moulds, yellow rust in BARLEY [1-8]. Brown rust, powdery mildew, septoria, sooty moulds, yellow rust in WHEAT [1-4, 7, 8]. Eyespot, sooty moulds in WINTER BARLEY [1-4, 6].

Notes

Efficacy
- Apply in cereals from flag leaf just visible (GS 37) until first ears visible (GS 51), best when flag leaf ligules visible (GS 39) or, if not possible to spray earlier, from ear emergence complete (GS 59) until grain watery ripe (GS 71)
- Treatment gives useful control of late attacks of rust or mildew but for early attacks or established infection tank-mix with specific mildew or rust fungicide
- Apply to oilseed rape as soon as disease appears in autumn or early spring and repeat if necessary between green bud and early flowering (GS 3,3-4,0) [5], when 25% of plants infected and disease spreading [6]
- Do not spray if frost or rain expected

Crop Safety/Restrictions
- Maximum number of treatments (including applications of any product containing benomyl, carbendazim or thiophanate-methyl) 2 per crop for cereals, 3 per crop for oilseed rape

Special precautions/Environmental safety
- Harmful if swallowed [2-4]. Irritating to eyes and skin [5, 6]
- Irritating to eyes, skin and respiratory system [1-4, 7, 8]
- Harmful to fish or other aquatic life. Do not contaminate surface waters or ditches with chemical or used container

Protective clothing/Label precautions
- A, C, H, M [6]; A, C [2-4]; A [1, 5, 7, 8]
- 5c, 6a, 6b, 6c, 14, 16, 18, 21, 28, 29, 35, 36, 37, 53, 54, 63, 66, 70, 78 [1-8]

Latest application/Harvest Interval(HI)
- Up to and including grain watery-ripe stage (GS 71).
- HI 21 d for oilseed rape

Approval
- Approved for aerial application on cereals [1]; wheat, barley [2, 3]; winter wheat, winter barley [7, 8]. See notes in Section 1
- Approval expiry: 30 Sep 95 [03467]; replaced by MAFF 06295. 31 Dec 95 [02997]; replaced by MAFF 06113

Maximum Residue Levels (mg residue/kg food)
- see carbendazim and maneb entries

76   carbendazim + maneb + sulfur

A broad spectrum protectant fungicide mixture for cereals and oilseed rape

Products   Bolda FL   Atlas   50:320:100 g/l   SC   07653

Uses   Brown rust, powdery mildew, septoria leaf spot, yellow rust in WHEAT. Downy mildew, light leaf spot in OILSEED RAPE. Net blotch, powdery mildew, rhynchosporium in BARLEY.

FOR FULL CONDITIONS OF USE ALWAYS READ THE PRODUCT LABEL

Notes
Efficacy
- Optimum timing for treatment of wheat and barley between flag leaf just visible stage and ear emergence (GS 39-59)
- Treat oilseed rape in autumn or spring when disease becomes active

Crop Safety/Restrictions
- Maximum number of treatments 2 per crop
- Under severe disease pressure or heavy infection in cereals specific fungicides should be used

Special precautions/Environmental safety
- Irritating to eyes, skin and respiratory system
- Harmful to fish or other aquatic life. Do not contaminate surface waters or ditches with chemical or used container

Protective clothing/Label precautions
- A, C, H
- 6a, 6b, 6c, 18, 30, 36, 37, 53, 63, 67, 70

Latest application/Harvest Interval(HI)
- Before grain milky ripe stage (GS 73) for cereals
- HI 21 d for oilseed rape

Maximum Residue Levels (mg residue/kg food)
- see carbendazim and maneb entries

---

77 carbendazim + maneb + tridemorph
A protectant and systemic fungicide for use in cereals

Products   Cosmic FL   BASF   40:320:90 g/l   SC   03473

Uses   Brown rust, eyespot, powdery mildew, rhynchosporium, sooty moulds, yellow rust in BARLEY. Brown rust, eyespot, powdery mildew, sooty moulds, yellow rust in WINTER WHEAT.

Notes   **Efficacy**
- Apply before leaf disease becomes established
- In winter barley autumn application is particularly recommended on disease susceptible cultivars but further treatment may be needed in spring
- Tank mixtures with other fungicides recommended for increased protection against net blotch, rusts, powdery mildew and Septoria. See label for details
- To delay appearance of resistant strains alternate treatment with non-MBC fungicide. Eyespot in cereals is now widely resistant to MBC fungicides
- Do not apply if rain expected or crop wet
- Systemic effect reduced under conditions of severe drought stress

Crop Safety/Restrictions
- Maximum number of treatments (including applications of any product containing benomyl, carbendazim or thiophanate-methyl) 2 per crop
- Do not apply to wheat during periods of temperature above 21°C or high light intensity. Under such conditions spray in late evening

Special precautions/Environmental safety
* Irritating to eyes, skin and respiratory system
* Harmful to livestock
* Harmful to fish or other aquatic life. Do not contaminate surface waters or ditches with chemical or used container

Protective clothing/Label precautions
* A, C
* 6a, 6b, 6c, 16, 18, 21, 29, 36, 37, 43, 53, 60, 63, 66

Withholding period
* Keep all livestock out of treated areas for at least 14 d

Latest application/Harvest Interval(HI)
* Up to and including grain watery-ripe stage (GS 71)

Approval
* Approved for aerial application on barley and winter wheat. See notes in Section 1

Maximum Residue Levels (mg residue/kg food)
* see carbendazim and maneb entries

78   carbendazim + metalaxyl
A protectant fungicide for use in fruit and cabbage storage

Products	Ridomil mbc 60 WP	Ciba Agric.	50:10% w/w	WP	01804

Uses   Botrytis, phytophthora in STORED CABBAGES *(off-label)*. Crown rot in WATER LILY *(off-label)*. Storage rots in APPLES, PEARS.

Notes   **Efficacy**
* Apply as post-harvest drench or dip to prevent spread of storage diseases
* Crop must be treated immediately after harvest
* Little curative activity on crop infected at or before harvest
* Best results achieved when crop stored in a controlled environment
* Use with calcium chloride for bitter pit control in susceptible apple cultivars

Crop Safety/Restrictions
* Maximum number of treatments (including applications of any product containing benomyl, carbendazim or thiophanate-methyl) 1 per batch. Allow crop to drain thoroughly

Special precautions/Environmental safety
* Irritating to eyes and skin
* Harmful to fish or other aquatic life. Do not contaminate surface waters or ditches with chemical or used container (remove fish before treating water-lily)

Protective clothing/Label precautions
* A
* 6a, 6b, 18, 22, 29, 35, 36, 37, 53, 63, 67, 70

FOR FULL CONDITIONS OF USE ALWAYS READ THE PRODUCT LABEL

Latest application/Harvest Interval(HI)
- Treated fruit must not be processed or sold for at least 4 wk, cabbage for 7 wk

Approval
- Off-label approval unlimited for use on water-lily (OLA 0912/92); unlimited for use on stored cabbages (OLA 0777/95)

Maximum Residue Levels (mg residue/kg food)
- see carbendazim and metalaxyl entries

---

79   carbendazim + prochloraz

A broad-spectrum systemic and contact fungicide for cereals and oilseed rape

Products					
1 Sportak Alpha	AgrEvo	100:267 g/l	SC	07222
2 Sportak Alpha	Schering	100:267 g/l	SC	03872

Uses

Botrytis, canker, dark leaf spot, light leaf spot, sclerotinia stem rot, white leaf spot in OILSEED RAPE. Eyespot, glume blotch, leaf spot, powdery mildew in WINTER WHEAT. Eyespot, net blotch, powdery mildew, rhynchosporium in WINTER BARLEY. Net blotch, powdery mildew, rhynchosporium in SPRING BARLEY.

Notes

Efficacy
- To protect against eyespot in high risk situations apply to winter barley from when leaf sheaths begin to become erect to first node detectable (GS 30-31). May also be used to control eyespot already in crop (up to 10% of tillers affected)
- Applied for eyespot control, product will also control rhynchosporium and mildew and protect against new infections of net blotch, mildew and Septoria. Where MBC resistance is not a problem product can be used against eyespot and leaf spot but prochloraz alone is preferable where these diseases are resistant
- Tank-mix with tridemorph, fenpropimorph or fenpropidin to control established mildew
- May be applied to winter barley up to full ear emergence (GS 59), but if used earlier in season, prochloraz alone or plus a non-MBC fungicide is preferred treatment
- Apply in oilseed rape at first signs of disease and repeat if necessary. Timing varies with disease - see label for details.
- A period of at least 3 h without rain should follow spraying

Crop Safety/Restrictions
- Maximum number of treatments (including applications of any product containing benomyl, carbendazim or thiophanate-methyl) 2 per crop for cereals; 2 at normal rate or 2 at split plus 1 at normal rate for oilseed rape

Special precautions/Environmental safety
- Harmful in contact with skin. Irritating to eyes and skin
- Flammable
- Dangerous to fish or other aquatic life. Do not contaminate surface waters or ditches with chemical or used container

Protective clothing/Label precautions
- A, C
- 5a, 6a, 6b, 12c, 18, 21, 29, 36, 37, 52, 63, 66, 70, 78

Latest application/Harvest Interval(HI)
- Up to full ear emergence (GS 59) for barley, before flowering (GS 60) for wheat. HI 6 wk for oilseed rape
- HI 21 d for oilseed rape

Maximum Residue Levels (mg residue/kg food)
- see carbendazim entry

---

80   carbendazim + propiconazole
A contact and systemic fungicide for winter cereals

Products					
1 Hispor 45 WP	Ciba Agric.	20:25% w/w	WP	01050
2 Sparkle 45 WP	Ciba Agric.	20:25% w/w	WP	04968

Uses   Brown rust, eyespot, net blotch, powdery mildew, rhynchosporium, yellow rust in WINTER BARLEY. Brown rust, eyespot, powdery mildew, septoria, yellow rust in WINTER WHEAT.

Notes   **Efficacy**
- Major benefit obtained from spring treatment. Control provided for about 30 d
- Apply in spring from stage when leaf sheath begins to lengthen until first node detectable (GS 30-31). Further sprays may be applied up to fully emerged ear (GS 59)

Crop Safety/Restrictions
- Maximum number of treatments (including applications of any product containing benomyl, carbendazim or thiophanate-methyl) 2 per crop

Special precautions/Environmental safety
- Irritating to eyes and skin
- Dangerous to fish or other aquatic life. Do not contaminate surface waters or ditches with chemical or used container
- Harmful to bees. Do not apply to crops in flower or to those in which bees are actively foraging. Do not apply when flowering weeds are present

Protective clothing/Label precautions
- A, C
- 6a, 6b, 18, 22, 28, 29, 35, 36, 37, 50, 52, 63

Latest application/Harvest Interval(HI)
- Up to and including grain watery-ripe stage (GS 71)

Approval
- Approved for aerial application in wheat and barley. See notes in Section 1

Maximum Residue Levels (mg residue/kg food)
- see carbendazim and propiconazole entries

FOR FULL CONDITIONS OF USE ALWAYS READ THE PRODUCT LABEL

81   carbendazim + tecnazene

A protectant fungicide and sprout suppressant for stored potatoes

Products

1 Hickstor 6 + MBC	Hickson & Welch	2:6% w/w	DP	04176
2 Hortag Tecnacarb Dust	Hortag	1.8:6% w/w	DP	02929
3 New Arena Plus	Hickson & Welch	2:6% w/w	DP	04598
4 Tripart Arena Plus	Tripart	2:6% w/w	DP	05602

Uses

Dry rot, gangrene, silver scurf, skin spot, sprout suppression in WARE POTATOES. Dry rot, gangrene, silver scurf, skin spot in SEED POTATOES.

Notes

Efficacy
- Potatoes should be dormant, have a mature skin and be dry and free from dirt
- Treat tubers as they go into store using a dusting machine (or automatic vibratory applicator [1])
- Ensure even cover of tubers. Cover clamps with straw etc to aid vapour-phase transmission of a.i. Pack boxes as tightly together as possible
- Effectiveness of treatment is reduced if ventilation in store is inadequate or excessive
- Treatment can give protection for 3-4 mth [2] or up to 6 mth [1, 4] but does not cure blemishes already present on tubers. It will not control sprouting if tubers have already broken dormancy

Crop Safety/Restrictions
- Maximum number of treatments (including applications of any product containing benomyl, carbendazim or thiophanate-methyl) 1 per batch
- Treated tubers must not be removed for sale or processing, including washing, for at least 6 wk after application
- Air seed potatoes for 6 wk and ensure that chitting has commenced before planting out. Treatment may delay emergence and possibly slightly reduce ware yield

Protective clothing/Label precautions
- 28, 29, 35, 54, 63, 67

Latest application/Harvest Interval(HI)
- 6 wk before sale or planting

Maximum Residue Levels (mg residue/kg food)
- see carbendazim and tecnazene entries

---

82   carbendazim + vinclozolin

A protectant and systemic fungicide for use in oilseed rape

Products

Konker	BASF	165:250 g/l	SC	03988

Uses

Alternaria, botrytis, light leaf spot in OILSEED RAPE *(reduction)*. Sclerotinia in OILSEED RAPE.

Notes

Efficacy
- Best results achieved if applied before any disease becomes well established

- Apply from start of rapid spring growth up to and including full flower
- Timing depends on diseases present and weather conditions. See label for details

Crop Safety/Restrictions
- Maximum number of treatments 2 per crop
- Do not apply if rain or frost is expected or when the crop is wet

Special precautions/Environmental safety
- Irritant. May cause sensitisation by skin contact
- Harmful to fish or other aquatic life. Do not contaminate surface waters or ditches with chemical or used container
- Must be applied only by vehicle mounted or trailed hydraulic sprayers
- Vehicles must be fitted with a cab and a forced air filtration unit with a pesticide filter complying with HSE Guidance Note PM74, or equivalent

Protective clothing/Label precautions
- A, C, H, K, M
- 6, 10a, 18, 21, 28, 29, 36, 37, 53, 63, 66

Latest application/Harvest Interval(HI)
- HI 7 wk

Maximum Residue Levels (mg residue/kg food)
- see carbendazim and vinclozolin entries

83   carbetamide

A residual pre- and post-emergence carbamate herbicide

Products					
Carbetamex		RP Agric.	70% w/w	WP	06186

Uses   Annual grasses, some annual dicotyledons, volunteer cereals in CABBAGE SEED CROPS, COLLARDS, FODDER RAPE SEED CROPS, KALE SEED CROPS, LUCERNE, RED CLOVER, SAINFOIN, SPRING CABBAGE, SUGAR BEET STECKLINGS, SWEDE SEED CROPS, TURNIP SEED CROPS, WHITE CLOVER, WINTER FIELD BEANS, WINTER OILSEED RAPE.

Notes   **Efficacy**
- Best results achieved pre- or early post-emergence of weed under cool, moist conditions. Adequate soil moisture is essential
- Dicotyledons controlled include chickweed, cleavers and speedwell
- Weed growth stopped rapidly though full effects may take 6-8 wk to develop
- Do not use on soils with more than 10% organic matter
- Do not apply during prolonged periods of cold weather when weeds fully dormant
- Various tank mixes effective against a wider range of dicotyledons are recommended. See label for details and for mixtures with other pesticides and sequential treatments

Crop Safety/Restrictions
- Maximum number of treatments 1 per crop

FOR FULL CONDITIONS OF USE ALWAYS READ THE PRODUCT LABEL

- Apply to brassicas from late autumn to late winter provided crop has at least 4 true leaves (spring cabbage, spring greens), 3-4 true leaves (seed crops, oilseed rape)
- Apply to established lucerne and sainfoin from mid-Oct to end Feb, to established red and white clover from Feb to mid-Mar
- After treatment do not sow brassicas or field beans for 2 wk, peas or runner beans for 8 wk, cereals or maize for 16 wk

Protective clothing/Label precautions
- 30, 54, 63, 67

Latest application/Harvest Interval(HI)
- HI 6 wk

84   carbofuran

A systemic carbamate insecticide and nematicide for soil treatment

Products					
1 Barclay Carbosect	Barclay	5% w/w	GR	05512
2 Rampart	Sipcam	5% w/w	GR	05166
3 Tripart Nex	Tripart	5% w/w	GR	05165
4 Yaltox	Bayer	5% w/w	GR	02371

Uses

Aphids, cabbage root fly, cabbage stem weevil, flea beetles, turnip root fly in SWEDES, TURNIPS [2-4]. Aphids, carrot fly, free-living nematodes in PARSNIPS [2-4]. Aphids, carrot fly, free-living nematodes in CARROTS [1-4]. Beet leaf miner, docking disorder vectors, flea beetles, free-living nematodes, millipedes, pygmy beetle, springtails, tortrix moths, wireworms in SUGAR BEET [1-4]. Beet leaf miner, docking disorder vectors, flea beetles, free-living nematodes, millipedes, pygmy beetle, springtails, wireworms in FODDER BEET, MANGELS [1-4]. Cabbage root fly, cabbage stem flea beetle, rape winter stem weevil in WINTER OILSEED RAPE [2-4]. Cabbage root fly, cabbage stem weevil, diamond-back moth, flea beetles in CHINESE CABBAGE, COLLARDS, KALE [2, 3]. Cabbage root fly, cabbage stem weevil, diamond-back moth, flea beetles in BROCCOLI, BRUSSELS SPROUTS, CABBAGES, CALABRESE, CAULIFLOWERS [1-4]. Cabbage root fly, soil pests in KOHLRABI *(off-label)* [4]. Flea beetles in RED BEET [2-4]. Frit fly in MAIZE, SWEETCORN [1-4]. Large narcissus fly in NARCISSI *(off-label)* [4]. Onion fly, stem nematodes in BULB ONIONS [1-4]. Potato cyst nematode in POTATOES [1-4]. Vine weevil in HARDY ORNAMENTAL NURSERY STOCK, POT PLANTS, STRAWBERRIES [2-4].

Notes

Efficacy
- Apply through a suitably calibrated granule applicator and incorporate into the soil
- Method of application, rate and timing vary with pest and crop. See label for details
- Performance is reduced in dry soil conditions
- Do not apply to potatoes in very wet or water-logged soils
- Controls only first generation carrot fly, use a follow-up application of a suitable specific insecticide for later attacks
- May be applied in mixture with granular fertilizers
- Do not apply to any site more than once in 2 yr to reduce the risk of enhanced biodegradation

Crop Safety/Restrictions
- Maximum number of treatments 1 per crop or yr
- When applied at drilling do not allow granules to come into contact with crop seed
- Use on carrots is limited to crops growing in mineral soils

- Use only on onions intended for harvest as mature bulbs
- Apply on ornamentals as surface treatment to moist soil after planting or potting up and follow immediately by thorough watering. Do not treat plants grown under cover, see label for details of species which have been treated safely
- On strawberries only apply after last harvest of year

Special precautions/Environmental safety
- This product contains an anticholinesterase carbamate compound. Do not use if under medical advice not to work with such compounds
- Keep in original container, tightly closed, in a safe place, under lock and key
- Harmful in contact with skin and if swallowed
- Dangerous to fish or other aquatic life. Do not contaminate surface waters or ditches with chemical or used container
- Dangerous to livestock. Bury or remove spillages
- Dangerous to game, wild birds and animals. Bury or remove spillages

Protective clothing/Label precautions
- A, B, C or D+E, H, K, M
- 2, 5a, 5c, 14, 16, 18, 22, 25, 28, 29, 36, 37, 40, 45, 52, 64, 67, 70, 78

Withholding period
- Keep all livestock out of treated areas for at least 2 wk

Latest application/Harvest Interval(HI)
- At drilling as band treatment on drilled winter oilseed rape or 6 wk pre-harvest as overall treatment on broadcast crops, at drilling for other drilled crops, at least 4 wk before release for sale or supply for ornamentals and pot plants.

Maximum Residue Levels (mg residue/kg food)
- hops 10; radishes 0.5; carrots, parsnips, garlic, onions, shallots 0.3; broccoli, cauliflowers, kohlrabi, tea 0.2; tree nuts (except hazelnuts), grapes, blackberries, dewberries, loganberries, raspberries, bilberries, cranberries, currants, gooseberries, wild berries, miscellaneous fruits, beetroot, horseradish, Jerusalem artichokes, parsley root, salsify, sweet potatoes, yams, spring onions, fruiting vegetables, leafy vegetables and herbs, peas (with and without pods), asparagus, cardoons, fennel, globe artichokes, rhubarb, fungi, lentils, peas, poppy seed, sesame seed, mustard seed, wheat, rye, barley, triticale, maize, animal products 0.1

---

85   carbon dioxide (commodity substance)
A gas for the control of trapped rodents and other vertebrates

Products	carbon dioxide	various	99.9%	GA

Uses   Birds, mice, rats in TRAPS.

Notes   **Efficacy**
- Use to control trapped rodent pests

FOR FULL CONDITIONS OF USE ALWAYS READ THE PRODUCT LABEL

- Use to control birds covered by general licences issued by the Agriculture and Environment Departments under Section 16(1) of the Wildlife and Countryside Act (1981) for the control of opportunistic bird species, where birds have been trapped or stupefied with alphachloralose/seconal

Special precautions/Environmental safety
- Operators must wear self-contained breathing apparatus when carbon dioxide levels are greater than 0.5% v/v
- Operators must be suitably trained and competent
- Unprotected persons and non-target animals must be excluded from the treatment enclosures and surrounding areas unless the carbon dioxide levels are below 0.5% v/v

Approval
- Only to be used where a licence has been issued in accordance with Section 16(1) of the Wildlife and Countryside Act 1981

---

86   carbosulfan

A systemic carbamate insecticide for control of soil pests

Products

1 Marshal 10G	RP Agric.	10% w/w	GR	06165
2 Marshal/suSCon	Fargro	10% w/w	CG	06978
3 Standon Carbosulfan 10G	Standon	10% w/w	GR	05671

Uses

Aphids, carrot fly, free-living nematodes in CARROTS, PARSNIPS [1, 3]. Aphids, flea beetles, free-living nematodes, mangold fly, millipedes, pygmy mangold beetle, springtails, symphylids, wireworms in FODDER BEET, MANGELS [3]. Aphids, flea beetles, free-living nematodes, mangold fly, millipedes, pygmy mangold beetle, springtails, symphylids, wireworms in SUGAR BEET [1, 3]. Cabbage root fly, cabbage stem weevil, flea beetles in COLLARDS [3]. Cabbage root fly, cabbage stem weevil, flea beetles in BROCCOLI, BRUSSELS SPROUTS, CABBAGES, CALABRESE, CAULIFLOWERS, SWEDES, TURNIPS [1, 3]. Large pine weevil in FORESTRY [2].

Notes

Efficacy
- At recommended rates seed is not damaged by contact with product
- Apply with suitable granule applicator feeding directly into seed furrow or immediately behind drill coulter (behind seed drill boot for brassicas) or use bow-wave technique
- For transplanted brassicas apply sub-surface with 'Leeds' coulter
- See label for details of suitable applicators and settings. Correct calibration is essential
- Where used in forestry, granules should be placed in the planting hole by hand or metered applicator before or after placing the tree [2]
- Where applied annually in intensive brassica growing areas enhanced biodegradation by soil organisms may lead to unsatisfactory control

Crop Safety/Restrictions
- Maximum number of treatments 1 per crop or tree
- Do not mix with compost for blockmaking or use in Hassy trays for brassicas
- Safe to Douglas Fir and Sitka Spruce. Other conifers may vary in their sensitivity [2]
- Forest trees may take 10-15 d to achieve full protection. An additional pre-planting insecticide dip or spray is advised [2]

Special precautions/Environmental safety
- This product contains an anticholinesterase carbamate compound. Do not use if under medical advice not to work with such compounds
- Harmful if swallowed
- Dangerous to game, wild birds and animals. Bury spillages
- Dangerous to fish or other aquatic life. Do not contaminate surface waters or ditches with chemical or used container
- Keep in original container, tightly closed, in a safe place, under lock and key

Protective clothing/Label precautions
- A, B, C, D, E, K, M [2]; B, C or D+E, H, K, M [1, 3]
- 2, 14, 16, 18, 25, 29, 36, 37, 45, 52 [1-3]; 5c, 22, 28, 64, 67, 70, 78 [1, 3]; 23, 63 [2]

Latest application/Harvest Interval(HI)
- HI leaf brassicas 60 d [1], 56 d [3]; sugar beet, fodder beet, mangels, swedes, turnips, carrots, parsnips 100 d

Maximum Residue Levels (mg residue/kg food)
- carrots, parsnips, tea 0.1; tree nuts, berries and small fruit, miscellaneous fruits, beetroot, celeriac, horseradish, Jerusalem artichokes, parsley root, radishes, salsify, sweet potatoes, yams, garlic, shallots, spring onions, tomatoes, peppers, aubergines, cucumbers, gherkins, courgettes, sweet corn, leaf vegetables and herbs, legume vegetables, asparagus, cardoons, fennel, globe artichokes, rhubarb, mushrooms, pulses, oilseeds (except sunflower seed, cotton seed), potatoes, cereals, animal products 0.05

---

87   carboxin

An anilide fungicide available only in mixtures

---

88   carboxin + gamma-HCH + thiram

A fungicide and insecticide dressing for rape seed

Products	Vitavax RS	Uniroyal	45:675:90 g/l	FS	06029

Uses   Canker, damping off, flea beetles in OILSEED RAPE.

Notes   **Efficacy**
- Prior to and during application product must be thoroughly agitated with drum agitator to ensure uniform mixing
- Keep at temperatures above 10°C prior to and during application

Crop Safety/Restrictions
- Maximum number of treatments 1 per batch
- Do not store treated seed for more than 3 mth

Special precautions/Environmental safety
- Harmful in contact with skin, by inhalation and if swallowed
- Irritating to eyes and skin

FOR FULL CONDITIONS OF USE ALWAYS READ THE PRODUCT LABEL

- Treated seed not to be used as food or feed. Do not re-use sack
- Harmful to fish or other aquatic life. Do not contaminate surface waters or ditches with chemical or used container

Protective clothing/Label precautions
- A, C
- 5a, 5b, 5c, 6a, 6b, 18, 22, 29, 36, 37, 53, 63, 65, 72, 73, 74, 75, 76, 77, 78

Maximum Residue Levels (mg residue/kg food)
- see gamma-HCH entry

89   carboxin + imazalil + thiabendazole
A fungicide seed dressing for barley and oats

Products   Vitaflo Extra   Uniroyal   300:20:25 g/l   FS   07048

Uses   Covered smut, leaf stripe, loose smut, net blotch, seedling blight and foot rot in SPRING BARLEY. Covered smut, leaf stripe, net blotch, seedling blight and foot rot in WINTER BARLEY. Loose smut, pyrenophora leaf spot, seedling blight and foot rot in OATS.

Notes   **Efficacy**
- Apply thropugh suitable liquid flowable seed treating equipment of the batch treatment or continuous flow type
- Preferably treat seed several days before sowing
- Stocks of treated seed carried over to following season should be tested for germination

Crop Safety/Restrictions
- Maximum number of treatments 1 per batch of seed
- Do not treat seed with moisture content above 16%. Store treated seed in a cool, dry, well ventilated place
- Do not apply to seed already treated with a fungicide
- Do not apply to cracked, split or sprouted seed

Special precautions/Environmental safety
- Treated seed not to be used as food or feed. Do not re-use sack. Treated seed must not be applied from the air
- Harmful to fish or aquatic life. Do not contaminate surface waters or ditches with chemical or used container

Protective clothing/Label precautions
- A, C, D, H, M
- 18, 22, 29, 36, 37, 53, 63, 65, 72, 73, 74, 76, 77

Latest application/Harvest Interval(HI)
- Before drilling

Maximum Residue Levels (mg residue/kg food)
- see imazalil and thiabendazole entries

90   carboxin + thiabendazole
A fungicide seed dressing for wheat and rye

Products   Vitaflo   Uniroyal   360:20 g/l   FS   06379

Uses   Bunt, loose smut, seedling blight and foot rot in WHEAT. Bunt, seedling blight and foot rot in RYE.

Notes   **Efficacy**
- Preferably apply through suitable liquid flowable seed treating equipment of the batch or continuous flow type
- Preferably treat seed several days before sowing
- Stocks of treated seed carried over to following season should be tested for germination

Crop Safety/Restrictions
- Maximum number of treatments 1 per batch of seed
- Do not treat seed with moisture content above 16%. Store treated seed in a cool, dry and well ventilated place
- Do not apply if seed already treated with a fungicide
- Treatment may lower germination capacity if seed not of good quality

Special precautions/Environmental safety
- Irritating to eyes and skin
- Treated seed not to be used as food or feed. Do not re-use sack
- Harmful to fish or aquatic life. Do not contaminate surface waters or ditches with chemical or used container

Protective clothing/Label precautions
- A, C, D, H, M
- 6a, 6b, 18, 22, 29, 36, 37, 53, 63, 67, 72, 73, 74, 76, 77

Latest application/Harvest Interval(HI)
- Before drilling

91   chloralose
A narcotic rodenticide used to kill mice

Products
1 Alphachloralose Concentrate	Rentokil	16.6% w/w	CB	01721
2 Alphachloralose Pure	Killgerm	100%	CB	00082

Uses   Mice in FARM BUILDINGS, GLASSHOUSES.

Notes   **Efficacy**
- Apply ready-mixed bait or concentrate mixed with suitable bait material in shallow trays where mouse droppings observed

FOR FULL CONDITIONS OF USE ALWAYS READ THE PRODUCT LABEL

- Lay bait at several points not more than 1.5 m apart
- Leave in position for several days until mouse activity ceases
- After treatment, clear up poison and bury residue

Special precautions/Environmental safety
- A chemical subject to the Poisons Rules 1982 and Poisons Act 1972
- Keep in original container, tighty closed, in a safe place, under lock and key
- Toxic if swallowed
- Do not use outside
- Prevent access by children and animals, particularly cats and dogs
- Should a domestic animal be affected keep the animal warm and quiet

Protective clothing/Label precautions
- 4c, 25, 29, 37, 64, 81, 82, 83, 84

---

92   chlordane

A persistent organochlorine earthworm killer approvals for which were revoked on 31 December 1992

---

93   chlorfenvinphos

A contact and ingested soil-applied organophosphorus insecticide

Products					
1 Birlane 24	Cyanamid	240 g/l	EC	07002
2 Birlane Granules	Cyanamid	10% w/w	GR	07003
3 Birlane Liquid Seed Treatment	RP Agric.	322 g/l	LS	06376
4 Sapecron 240 EC	Ciba Agric.	240 g/l	EC	01861

Uses

Cabbage root fly in RADISHES [1, 4]. Cabbage root fly in BROCCOLI, BRUSSELS SPROUTS, CABBAGES, CAULIFLOWERS, SWEDES, TURNIPS [1, 2, 4]. Cabbage root fly in KOHLRABI *(off-label)* [1, 2]. Carrot fly in PARSNIPS [1, 4]. Carrot fly in CARROTS [1, 2, 4]. Carrot fly in CELERIAC *(off-label)* [1, 2]. Colorado beetle in POTATOES *(off-label)* [1, 4]. Frit fly in MAIZE [4]. Insect control in MOOLI *(off-label)*, RADISHES *(off-label)*[2]. Mushroom flies in MUSHROOMS [1, 2, 4]. Wheat bulb fly, yellow cereal fly in WINTER WHEAT [1, 3, 4].

Notes

Efficacy
- Soil incorporation recommended for most treatments. Application method, timing, rate and frequency vary with pest, crop and soil type. See labels for details
- Overall pre-planting sprays are less effective than band treatment and should not be used in areas of heavy cabbage root fly infestation [4]
- Cabbage and cauliflower grown in peat blocks can be protected from cabbage root fly by incorporation into the peat used for blocking [2, 4]
- Efficacy reduced on highly organic or very dry soils unless crop irrigated
- Apply as seed treatment for wheat bulb fly control on highly organic soils [3]

Crop Safety/Restrictions
- Maximum number of treatments 1 per crop for most crops, but varies with dose rate and product - see label for details; 1 per batch of seed [3]

- On carrots the maximum total dose applied per crop must not exceed the equivalent of 3 (on mineral soils) or 4 (on organic soils) full dose applications
- Band sprays on sown brassicas may cause damage if applied post-emergence [1]
- Danger of scorch if applied to maize during very hot weather or if crop is under stress
- Do not apply in conjunction with seed treatments containing gamma-HCH
- Treated seed can be stored for 1 mth provided moisture content is 16% or less. Do not exceed drilling depth of 4 cm [3]

Special precautions/Environmental safety

- Chlorfenvinphos is subject to the Poisons Rules 1982 and the Poisons Act 1972. See notes in Section 1 [1, 3, 4]
- Keep in original container, tightly closed, in a safe place, under lock and key [2-4]
- This product contains an anticholinesterase organophosphorus compound. Do not use if under medical advice not to work with such compounds
- Toxic in contact with skin and if swallowed [1, 3, 4]. Irritating to eyes and skin [1, 4]
- Harmful if swallowed [2]
- Flammable [1, 3, 4]
- Harmful to game, wild birds and animals. Bury spillages [3]
- Dangerous to fish or other aquatic life. Do not contaminate surface waters or ditches with chemical or used container
- Treated seed not to be used as food or feed [3]

Protective clothing/Label precautions

- A, B, C, D, E, H, J, K, M [2]; A, C, H [1, 3, 4]
- 1, 4a, 4c, 16, 28, 79 [1, 3, 4]; 5c, 10a, 45, 78 [2]; 6a, 6b, 12c [1, 4]; 14, 18, 25, 29, 36, 37, 64, 70 [1-4]; 21 [2-4]; 23, 49, 54a, 65 [1]; 26, 27, 35, 66 [4]; 35 (3 wk), 51 [1, 2]; 52 [3, 4]; 67 [2, 3]; 72, 73, 74, 75, 76, 77, 80 [3]

Latest application/Harvest Interval(HI)

- Before end of Feb for winter wheat; pre- or at drilling [1, 4].
- HI brassicas, swedes, turnips, radish, kohlrabi, carrots, celery, celeriac, parsnips, maize, potatoes 3 wk

Approval

- Off-label approval to Jan 1997 for use on potatoes for Colorado beetle control as required by Statutory Notice (OLA 0628/92) [4]

Maximum Residue Levels (mg residue/kg food)

- citrus fruits 1; carrots, horseradish, parsnips, parsley root, salsify, swedes, turnips, garlic, onions, shallots, celery, rhubarb, potatoes 0.5; meat 0.2; tomatoes, peppers, aubergines, cucumbers, gherkins, courgettes, cauliflowers, Brussels sprouts, head cabbage, lettuce, beans, peas, leeks 0.1; pome fruits, apricots, peaches, nectarines, plums, grapes, strawberries, cane fruits, bilberries, cranberries, currants, gooseberries, bananas, mushrooms 0.05; milk 0.008

FOR FULL CONDITIONS OF USE ALWAYS READ THE PRODUCT LABEL

94 chloridazon
A residual pyridazinone herbicide for beet crops

Products	1 Barclay Champion	Barclay	430 g/l	SC	06903
	2 Better DF	Sipcam	65% w/w	SG	06250
	3 Better Flowable	Sipcam	430 g/l	SC	04924
	4 Gladiator DF	Tripart	65% w/w	SG	06342
	5 Luxan Chloridazon	Luxan	430 g/l	SC	06304
	6 Pyramin DF	BASF	65% w/w	SG	03438
	7 Starter Flowable	Truchem	430 g/l	SC	03421
	8 Stefes Chloridazon	Stefes	430 g/l	SC	07678
	9 Takron	BASF	430 g/l	SC	06237
	10 Tripart Gladiator	Tripart	430 g/l	SC	00986

Uses Annual dicotyledons, annual meadow grass in BULB ONIONS *(off-label)*, LEEKS *(off-label)*, SALAD ONIONS *(off-label)* [6]. Annual dicotyledons, annual meadow grass in SUGAR BEET [1-4, 6-10]. Annual dicotyledons, annual meadow grass in FODDER BEET, MANGELS [1-10].

Notes

Efficacy
- Absorbed by roots of germinating weeds and best results achieved pre-emergence of weeds or crop when soil moist and adequate rain falls after application
- Apply pre-emergence as soon as possible after drilling in mid-Mar to mid-Apr on fine, firm, clod-free seedbed
- Where crop drilled after mid-Apr or soil dry apply pre-drilling and incorporate to 2.5 cm immediately afterwards
- Application rate depends on soil type. See label for details
- Various tank mixes recommended on sugar beet for pre- and post-emergence use and as repeated low dose treatments. See label for details

Crop Safety/Restrictions
- Maximum number of treatments 1 per crop for fodder beet and mangels; 1 per crop [4-6], 3 per crop [8], 1 pre-emergence + 1 post-emergence [1, 2], 1 pre-emergence + 3 post-emergence [2, 6, 9] for sugar beet; 1 per crop for onions and leeks [6]
- Do not use on Coarse Sands, Sands or Fine Sands or where organic matter exceeds 5%
- Crop vigour may be reduced by treatment of crops growing under unfavourable conditions including poor tilth, drilling at incorrect depth, soil capping, physical damage, pest or disease damage, excess seed dressing, trace-element deficiency or a sudden rise in temperature after a cold spell
- In the event of crop failure only sugar or fodder beet, mangels or maize may be re-drilled on treated land after cultivation
- Winter cereals may be sown in autumn after ploughing. Any spring crop may follow treated beet crops harvested normally

Special precautions/Environmental safety
- Harmful if swallowed [7, 9]. Irritating to eyes, skin and respiratory system [10]
- Harmful to fish or other aquatic life. Do not contaminate surface waters or ditches with chemical or used container

Protective clothing/Label precautions
- A, C [8]; A [9, 10]
- 5c [7, 9]; 6a, 6b, 6c [10]; 10a, 70, 78 [9]; 18, 36, 37 [7-10]; 21, 29, 53, 63 [1-10]; 66 [1, 7-10]; 67 [2-6]

Latest application/Harvest Interval(HI)
- Pre-emergence for fodder beet and mangels; pre-emergence [4, 8], 8-true leaf stage [2-4], before leaves of crop meet in row [1, 5-7, 9] for sugar beet; up to and including second true leaf stage for onions and leeks [6]

Approval
- Off-label approval to Apr 1998 for use on salad and bulb onions and leeks (OLA 0349/93) [6]

95 chloridazon + chlorpropham + fenuron + propham

A residual pre-emergence herbicide for beet crops

Products					
Atlas Electrum	Atlas	200:30:20:120 g/l	SC	03548

Uses

Annual dicotyledons, annual grasses in FODDER BEET, MANGELS, SUGAR BEET.

Notes

Efficacy
- Best results achieved from application to firm, moist, clod-free seedbed when adequate rain falls afterwards
- Apply immediately after drilling pre-emergence of crop or weeds
- Do not disturb soil surface after application
- Dose rate depends on soil type - see label. Do not use on highly organic soils

Crop Safety/Restrictions
- Crops affected by poor growing conditions may be checked by treatment
- Excessive rainfall after application may check crop
- Do not incorporate chemical into soil
- In the event of crop failure only sugar or fodder beet or mangels should be re-drilled
- After lifting treated crops the land should be ploughed or cultivated repeatedly to 15 cm to dissipate residues

Special precautions/Environmental safety
- Irritating to skin, eyes and respiratory system
- Harmful to fish or other aquatic life. Do not contaminate surface waters or ditches with chemical or used container

Protective clothing/Label precautions
- A, C
- 6a, 6b, 6c, 18, 21, 28, 29, 36, 37, 53, 63, 66

FOR FULL CONDITIONS OF USE ALWAYS READ THE PRODUCT LABEL

96   chloridazon + ethofumesate
A residual pre-emergence herbicide for beet crops

Products

1 Magnum	BASF	275:170 g/l	SC	01237
2 Spectron	AgrEvo	211:200 g/l	SC	07284
3 Spectron	Schering	211:200 g/l	SC	03828

Uses   Annual dicotyledons, annual meadow grass, blackgrass in FODDER BEET, SUGAR BEET [1-3]. Annual dicotyledons, annual meadow grass, blackgrass in MANGELS [2, 3].

Notes   **Efficacy**
* Apply at or immediately after drilling pre-emergence of crop or weeds [2, 3], pre- or post-emergence up to cotyledon stage of weeds [1], on a fine, firm, clod-free seedbed
* Best results achieved when soil moist and adequate rain falls after spraying. Efficacy may be reduced if heavy rain falls just after incorporation
* A reduction in effectiveness may occur under conditions of low pH
* May be used on soil classes Loamy Sand - Silty Clay Loam [1]; may be used on all soils except Sands and those with more than 5% organic matter [2, 3]
* May be applied by conventional or repeat low dose method. See label for details
* Recommended for tank mixing with triallate on sugar beet incorporated pre-drilling and with various other beet herbicides early post-emergence [1]. See labels for details

Crop Safety/Restrictions
* Maximum number of treatments 1 per crop for sugar beet, fodder beet and mangels [2, 3]; 1 pre-emergence for fodder beet, 1 pre-emergence plus 3 post-emergence for sugar beet [1]
* Crop vigour may be reduced by treatment of crops growing under unfavourable conditions including poor tilth, drilling at incorrect depth, soil capping, physical damage, excess nitrogen, excess seed dressing, trace element deficiency or a sudden rise in temperature after a cold spell. Frost after pre-emergence treatment may check crop growth
* In the event of crop failure only sugar beet, fodder beet or mangels may be re-drilled
* Any crop may be sown 3 mth after spraying following ploughing to 15 cm

Special precautions/Environmental safety
* Harmful to fish or other aquatic life. Do not contaminate surface waters or ditches with chemical or used container

Protective clothing/Label precautions
* 21, 28, 29, 53, 60, 63, 66

Latest application/Harvest Interval(HI)
* Pre-emergence for sugar beet, fodder beet and mangels [2, 3], before crop leaves meet between rows for sugar beet [1]

Approval
* May be applied through CDA equipment. See notes in Section 1 [2, 3]

97   chloridazon + lenacil
A residual pre-emergence herbicide for beet crops

Products   Advizor   DuPont   200:133 g/l   SC   06571

Uses	Annual dicotyledons, annual meadow grass in FODDER BEET, MANGELS, SUGAR BEET.

Notes

Efficacy
* Apply at or immediately after drilling before emergence of crop or weeds
* Best results achieved from application to firm, moist, weed-free seedbed when adequate rain falls afterwards
* Application rate depends on soil type. See label for details
* May be used on Very Light, Light or Medium soils. Not on Sands, stony or gravelly soils, heavy soils or those with more than 10% organic matter
* Recommended for tank-mixing with Avadex BW incorporated pre-drilling on sugar beet and with various other herbicides pre- or early post-emergence. See label for details

Crop Safety/Restrictions
* Maximum number of treatments 1 per crop pre-drilling (sugar beet) or 1 pre-emergence (sugar beet, fodder beet, mangels) followed in either case by 1 post-emergence (or 3 post-emergence at low dose) per crop
* Crops should be drilled to at least 15 mm and the seed well covered
* Heavy rainfall after spraying may reduce crop stand especially when such rain is followed by very hot weather
* Only use in post-emergence mixtures on crops growing vigorously and not stressed by drought, pest attack, deficiency or other factors
* In the event of crop failure only beet crops may be re-drilled on treated land and no further application should be made for at least 4 mth. Any crop may be sown 4 mth after treatment following ploughing to 15 cm

Special precautions/Environmental safety
* Irritating to eyes, skin and respiratory system
* Harmful to fish or other aquatic life. Do not contaminate surface waters or ditches with chemical or used container

Protective clothing/Label precautions
* 6a, 6b, 6c, 18, 21, 28, 29, 36, 37, 53, 63, 65, 70

Latest application/Harvest Interval(HI)
* Before crop plants meet across rows

---

98   chloridazon + propachlor
A residual pre-emergence herbicide for use in onions and leeks

Products	Ashlade CP	Ashlade	86:400 g/l	SC	06481

Uses	Annual dicotyledons, annual grasses in BULB ONIONS, BULB ONIONS *(off-label, post-emergence)*, CHIVES *(off-label)*, LEEKS, LEEKS *(off-label, post-emergence)*, SALAD ONIONS, SALAD ONIONS *(off-label, post-emergence)*.

FOR FULL CONDITIONS OF USE ALWAYS READ THE PRODUCT LABEL

Notes

Efficacy
- Best results achieved from application to firm, moist, weed-free seedbed when adequate rain falls afterwards
- Apply pre-emergence of sown crops, preferably soon after drilling, before weeds emerge. Loose or fluffy seedbeds must be consolidated before application
- Apply to transplanted crops when soil has settled after planting
- Do not use on soils with more than 10% organic matter

Crop Safety/Restrictions
- Maximum number of treatments 1 per crop
- Ensure crops are drilled to 20 mm depth
- Crops stressed by nutrient deficiency, pests or diseases, poor growing conditions or pesticide damage may be checked by treatment, especially on sandy or gravelly soils
- In the event of crop failure only onions, leeks or maize should be planted
- Any crop can follow a treated onion or leek crop harvested normally as long as the ground is cultivated thoroughly before drilling

Special precautions/Environmental safety
- Irritating to eyes and skin
- Harmful to fish or other aquatic life. Do not contaminate surface waters or ditches with chemical or used container

Protective clothing/Label precautions
- A, C
- 6a, 6b, 14, 16, 18, 21, 25, 28, 29, 36, 37, 53, 63, 66, 78

Latest application/Harvest Interval(HI)
- Pre-emergence of crop; before 2 true leaf stage for onions and leeks (off-label).
- HI 12 wk for chives

Approval
- Off-label Approval to Jan 1997 for post-emergence use on onions, leeks (OLA 0081/92) and use on chives (OLA 0667/92)

99 chlormequat

A plant-growth regulator for reducing stem growth and lodging

Products					
1 Adjust	Mandops	620 g/l	SL	05589
2 Ashlade 460 CCC	Ashlade	460 g/l	SL	06474
3 Ashlade 700 CCC	Ashlade	700 g/l	SL	06473
4 Atlas 3C:645 Chlormequat	Atlas	645 g/l	SL	05710
5 Atlas Chlormequat 46	Atlas	460 g/l	SL	05660
6 Atlas Chlormequat 700	Atlas	700 g/l	SL	03402
7 Atlas Terbine	Atlas	730 g/l	SL	06523
8 Atlas Tricol	Atlas	670 g/l	SL	07190
9 Barclay Holdup	Barclay	700 g/l	SL	06799
10 Barleyquat B	Mandops	620 g/l	SL	06001
11 Barleyquat B	Mandops	620 g/l	SL	07051
12 Bettaquat B	Mandops	620 g/l	SL	06004
13 Bettaquat B	Mandops	620 g/l	SL	07050
14 CCC 700	FCC	700 g/l	SL	03366
15 Fargro Chlormequat	Fargro	460 g/l	SL	02600
16 Hyquat 70	Agrichem	700 g/l	SL	03364
17 Mandops Chlormequat 700	Mandops	700 g/l	SL	06002
18 Manipulator	Mandops	620 g/l	SL	05871
19 MSS Chlormequat 40	Mirfield	400 g/l	SL	01401
20 MSS Chlormequat 460	Mirfield	460 g/l	SL	03935
21 MSS Chlormequat 60	Mirfield	600 g/l	SL	03936
22 MSS Chlormequat 70	Mirfield	700 g/l	SL	03937
23 Quadrangle Chlormequat 700	Quadrangle	700 g/l	SL	03401
24 Standup 700	Vass	700 g/l	SL	03522
25 Stefes CCC	Stefes	460 g/l	SL	05959
26 Stefes CCC 640	Stefes	640 g/l	SL	06993
27 Stefes CCC 700	Stefes	700 g/l	SL	07116
28 Stefes CCC 720	Stefes	720 g/l	SL	05834
29 Tripart Brevis	Tripart	700 g/l	SL	03754
30 Uplift	United Phosphorus	700 g/l	SL	07527

Uses

Increasing yield, lodging control in WINTER BARLEY [1-11, 14, 17-22, 28-30]. Lodging control in TRITICALE [6]. Lodging control in SPRING WHEAT, WINTER WHEAT [1-9, 12-14, 16-29]. Lodging control in SPRING RYE [10, 11]. Lodging control in SPRING OATS, WINTER OATS [2-9, 12-14, 16, 17, 19-29]. Lodging control in WINTER RYE [2, 3, 10, 11, 29]. Lodging control in OATS, WHEAT [30]. Stem shortening in POT PLANTS [15]. Stem shortening in BEDDING PLANTS, CAMELLIAS, HIBISCUS TRIONUM, LILIES, PELARGONIUMS, POINSETTIAS [5, 6, 15]. Stem shortening in ORNAMENTALS [5, 6].

Notes

Efficacy
- Most effective results on cereals normally achieved from spring application, on wheat and rye from leaf sheath erect to first node detectable (GS 30-31), on oats at second node detectable (GS 32), on winter barley from mid-tillering to leaf sheath erect (GS 25-30). However, recommendations vary with product. See label for details
- In tank mixes with other pesticides optimum timing for herbicide action may differ from that for growth reduction. See label for details of tank mix recommendations

---

FOR FULL CONDITIONS OF USE ALWAYS READ THE PRODUCT LABEL

- Addition of approved non-ionic wetter recommended on oats
- At least 6 h, preferably 24 h, required before rain for maximum effectiveness. Do not apply to wet crops
- May be used on cereals undersown with grass or clovers

Crop Safety/Restrictions
- Maximum number of treatments 1 per crop for cereals (2 per crop for split application on winter wheat [5-7])
- Do not use on very late sown spring wheat or oats or on crops under stress
- Mixtures with liquid nitrogen fertilizers may cause scorch
- Do not use in tank mix with cyanazine [6]
- Do not use on spring barley

Special precautions/Environmental safety
- Harmful if swallowed [1, 2, 4, 7-18, 23-25, 29], in contact with skin [5, 6, 30], in contact with skin or if swallowed [3, 19-22, 27, 28]
- Harmful to fish or other aquatic life. Do not contaminate surface waters or ditches with chemical or used container [3]
- Wash equipment thoroughly with water and wetting agent immediately after use and spray out. Traces can cause harm to susceptible crops sprayed later [5-8]

Protective clothing/Label precautions
- A, C, H [22]; A, C [19, 21]; A [1-18, 20, 23-30]
- 5a [3, 19-22, 30]; 5c [1-29]; 14, 16 [22, 26, 28]; 18 [1-5, 9-30]; 19 [6-8]; 21, 29, 36, 37, 63, 70, 78 [1-30]; 25, 34, 44 [1, 2, 9-18, 23-25, 27, 29]; 28 [1-4, 6-20, 22-30]; 53 [3]; 54 [1, 2, 4-30]; 60 [1, 2, 9-21, 23-29]; 65 [1-3, 9-20, 23-25, 27, 29]; 66 [4-8, 21, 22, 26, 28, 30]

Latest application/Harvest Interval(HI)
- Varies with product. See label for details

Approval
- Approved for aerial application on wheat, oats [2-8, 14, 16, 17, 23, 29]; wheat [9]; winter barley [2, 3, 8, 10, 12, 14-16]; rye [2, 3]. See notes in Section 1

100   chlormequat + 2-chloroethylphosphonic acid
A plant growth regulator for use in cereals

Products					
Upgrade	RP Agric.	360:180 g/l	SL	06177

Uses   Lodging control in WINTER BARLEY, WINTER WHEAT.

Notes

Efficacy
- Apply before lodging has started. Best results obtained when crops growing vigorously
- Only crops growing under conditions of high fertility should be treated
- Recommended dose varies with growth stage. See labels for details and recommendations for use of sequential treatments
- Do not spray when crop wet or rain imminent

Crop Safety/Restrictions
- Maximum number of treatments 1 per crop

- Do not spray during cold weather or periods of night frost, when soil is very dry, when crop diseased or suffering pest damage, nutrient deficiency or herbicide stress
- If used on seed crops grown for certification inform seed merchant beforehand
- Do not use on spring barley

Special precautions/Environmental safety
- Harmful if swallowed and in contact with skin. Irritating to eyes
- Harmful to fish or other aquatic life. Do not contaminate surface waters or ditches with chemical or used container

Protective clothing/Label precautions
- A, C
- 5a, 5c, 6a, 18, 21, 25, 28, 29, 36, 37, 53, 60, 63, 66, 70, 78

Latest application/Harvest Interval(HI)
- Before flag leaf sheath opening (GS 47)

Maximum Residue Levels (mg residue/kg food)
- see 2-chloroethylphosphonic acid entry

101   chlormequat + choline chloride
A plant growth regulator for use in cereals and certain ornamentals

Products					
1 Ashlade 5C	Ashlade	460:320 g/l	SL	06227
2 Ashlade 700 5C	Ashlade	700:32 g/l	SL	07046
3 Atlas 460:46	Atlas	460:46 g/l	SL	06258
4 Atlas 5C Chlormequat	Atlas	460:320 g/l	SL	03084
5 Atlas Quintacel	Atlas	640:64 g/l	SL	06524
6 MSS Mircell	Mirfield	640:64 g/l	SL	06939
7 New 5C Cycocel	BASF	645:32 g/l	SL	01482
8 New 5C Cycocel	Hortichem	645:32 g/l	SL	01482

Uses   Increasing yield, lodging control in WINTER BARLEY [1-8]. Lodging control, stem shortening in LINSEED, WINTER OILSEED RAPE [7, 8]. Lodging control in SPRING OATS, SPRING WHEAT, WINTER OATS, WINTER WHEAT [1-8]. Lodging control in TRITICALE [3-8]. Lodging control in WINTER RYE [2, 7, 8]. Stem shortening in PELARGONIUMS, POINSETTIAS [4, 5, 7, 8]. Stem shortening in BEDDING PLANTS, CAMELLIAS, HIBISCUS TRIONUM, LILIES [4, 5]. Stem shortening in ORNAMENTALS [3-6].

Notes   **Efficacy**
- Influence on growth varies with crop and growth stage. Risk of lodging reduced by application at early stem extension. Root development and yield can be improved by earlier treatment
- Most effective results normally achieved from spring application. On winter barley an autumn treatment may also be useful. Timing of spray is critical and recommendations vary with product. See label for details
- Often used in tank-mixes with pesticides. Recommendations for mixtures and sequential treatments vary with product. See label for details

---

FOR FULL CONDITIONS OF USE ALWAYS READ THE PRODUCT LABEL

- Add authorised non-ionic wetter when spraying oats, oilseed rape or linseed
- At least 6 h required before rain for maximum effectiveness. Do not apply to wet crops
- May be used on cereals undersown with grass or clovers

Crop Safety/Restrictions
- Maximum number of treatments 1 per crop for spring wheat, oats, winter barley, rye, triticale, winter oilseed rape, linseed; 1 or 2 per crop for winter wheat (depending on dose); 1-3 per crop for ornamentals - see label for details
- Do not spray very late sown spring crops, crops on soils of low fertility, crops under stress from any cause or if frost expected
- Do not use on spring barley
- Mixtures with liquid nitrogen fertilizers may cause scorch

Special precautions/Environmental safety
- Harmful if swallowed [1, 3, 5-8], or in contact with skin [2, 4]
- Wash equipment thoroughly with water and wetting agent immediately after use and spray out. Traces can cause harm to susceptible crops spraayed later [4-6]

Protective clothing/Label precautions
- A [1-8]
- 5a [2, 4]; 5c [1-3, 5-8]; 18, 21, 28, 29, 36, 37, 54, 63, 70, 78 [1-8]; 60 [4]; 65 [2]; 66 [1, 3-8]

Latest application/Harvest Interval(HI)
- Before 3rd node detectable (GS 33) for oats; before 2nd (1st [2]) node detectable (GS 32) [1, 3-8] for winter wheat and rye; before 1st node detectable (GS 31) for spring wheat, winter barley, triticale; before first flower bud opens (GS 4,0) for winter oilseed rape; before first flower buds visible for linseed

Approval
- Approved for aerial application on wheat, oats, winter barley [1, 7, 8]; wheat, oats [1-6]; rye, triticale [7, 8]. See notes in Section 1

---

102   chlormequat + choline chloride + imazaquin
A plant growth regulator mixture for winter wheat

Products	Meteor	Cyanamid	368:28:0.8 g/l	SL	06505

Uses   Annual dicotyledons, perennial dicotyledons in AMENITY TURF, LAWNS, SPORTS TURF. Increasing yield, lodging control in WINTER WHEAT.

Notes

Efficacy
- Apply as single dose from leaf sheath lengthening up to and including 1st node detectable or as split dose, the first from tillers formed to leaf sheath lengthening, the second from leaf sheath erect up to and including 1st node detectable
- Apply to crops during good growing conditions or to those at risk from lodging
- On soils of low fertility, best results obtained where adequate nitrogen fertiliser used
- Do not apply when crop wet or rain imminent

Crop Safety/Restrictions
- Maximum number of applications 1 per crop (2 per crop at split dose)
- Do not treat durum wheat

* Do not apply to undersown crops

Special precautions/Environmental safety
* Harmful if swallowed. Irritating to eyes

Protective clothing/Label precautions
* A, C
* 5c, 6a, 18, 21, 25, 28, 29, 36, 37, 54, 63, 66, 70, 78

Latest application/Harvest Interval(HI)
* Before second node detectable (GS 31)

---

103   chlormequat with di-1-p-menthene

A plant-growth regulator for reducing stem growth and lodging

Products	Podquat	Mandops	470 g/l	SL	03003

Uses   Increasing yield, lodging control in BROAD BEANS, FIELD BEANS, OILSEED RAPE, PEAS.

Notes   **Efficacy**
* On winter oilseed rape and beans either apply as soon as possible after 3-leaf stage (GS 1, 3) until growth ceases followed by spring treatment in mid-Mar to early Apr or use a single spring spray. See label for details
* May be used at temperatures down to 1°C provided spray dries on leaves before rain, frost or snow occurs

Crop Safety/Restrictions
* Do not apply to plants covered by frost

Special precautions/Environmental safety
* Harmful if swallowed

Protective clothing/Label precautions
* A
* 5c, 18, 21, 28, 29, 36, 37, 54, 63, 70, 78

---

104   2-chloroethylphosphonic acid

A plant growth regulator for cereals and various horticultural crops
See also chlormequat + 2-chloroethylphosphonic acid

Products					
1 Cerone	RP Agric.	480 g/l	SL	06643
2 Ethrel C	Hortichem	480 g/l	SL	06995
3 Stefes Stance	Stefes	480 g/l	SL	06125

Uses   Basal bud stimulation in PROTECTED ROSES [2]. Fruit ripening in TOMATOES [2]. Increasing branching in PELARGONIUMS [2]. Inducing flowering in BROMELIADS [2]. Lodging control

---

FOR FULL CONDITIONS OF USE ALWAYS READ THE PRODUCT LABEL

in TRITICALE, WHEAT, WINTER BARLEY, WINTER RYE [1, 3]. Stem shortening in NARCISSI [2].

Notes

Efficacy
* Best results achieved on crops growing vigorously under conditions of high fertility
* May be used on winter barley from second node detectable to first awns visible (GS 32-49), on other cereals from flag leaf just visible to boots swollen (GS 37-45) preferably following chlormequat treatment. See label for details
* Do not spray crops when wet or if rain imminent
* Best results on horticultural crops when temperature does not fall below 10°C
* Use on tomatoes 17 d before planned pulling date [2]
* Apply as drench to daffodils when stems average 15 cm [2]
* Apply to glasshouse roses when new growth started after pruning [2]

Crop Safety/Restrictions
* Maximum number of treatments 1 per crop or yr
* Do not spray crops suffering from stress caused by any factor, during cold weather or period of night frost nor when soil very dry
* Do not apply to cereals within 10 d of herbicide or liquid fertilizer application
* Do not spray wheat or triticale where the leaf sheaths have split and the ear is visible
* Do not use on spring barley

Special precautions/Environmental safety
* Irritating to eyes and skin
* Harmful to fish or other aquatic life. Do not contaminate surface waters or ditches with chemical or used container
* Avoid accidental deposits on painted objects such as cars, trucks, aircraft
* Rinse aircraft windshields after each tank load [1]

Protective clothing/Label precautions
* A, C
* 6a, 6b, 18, 21, 25, 29, 36, 37, 53, 63, 66 [1-3]; 70 [2]

Latest application/Harvest Interval(HI)
* Before flag leaf sheath opening (GS 47) for winter and spring wheat, triticale; before first spikelet of inflorescence just visible (GS 51) for winter barley, winter rye.
* HI tomatoes 5 d

Approval
* Approved for aerial application on winter barley [1, 3]. See notes in Section 1
* Approval expiry: 31 Aug 95 [06643]

Maximum Residue Levels (mg residue/kg food)
* currants 5; pome fruits, cherries, tomatoes, peppers 3; rye, barley 0.5; wheat, triticale 0.2; tree nuts, tea, hops 0.1; apricots, peaches, plums, strawberries, blackberries, dewberries, loganberries, raspberries, bilberries, cranberries, gooseberries, wild berries, avocados, bananas, dates, kiwi fruit, kumquats, litchis, mangoes, passion fruit, pomegranates, root and tuber vegetables, garlic, shallots, spring onions, aubergines, cucumbers, gherkins, courgettes, melons, squashes, water melons, brassica vegetables, leaf vegetables and herbs, legume vegetables, stem vegetables, fungi, pulses, oilseeds, potatoes, oats, rice, animal products 0.05

105   2-chloroethylphosphonic acid + mepiquat chloride
A plant growth regulator for reducing lodging in cereals

Products

1	Stefes Mepiquat	Stefes	155:305 g/l	SL	06970
2	Terpal	BASF	155:305 g/l	SL	02103
3	Terpal	Clifton	155:305 g/l	SL	05026

Uses

Increasing yield in WINTER BARLEY *(low lodging situations)*. Lodging control in TRITICALE, WINTER BARLEY, WINTER OATS *(some cultivars only)*, WINTER RYE, WINTER WHEAT.

Notes

Efficacy
* Best results achieved on crops growing vigorously under conditions of high fertility
* Recommended dose and timing vary with crop, cultivar, growing conditions, previous treatment and desired degree of lodging control. See label for details
* Add an authorised non-ionic wetter to spray solution
* May be applied to crops undersown with grass or clovers
* Do not apply to crops if wet or rain expected

Crop Safety/Restrictions
* Maximum number of treatments 1 per crop (2 per crop at split dose [1])
* Recommendation for winter oats only applies to Pennal and Image
* Do not use on spring barley
* Do not spray crops damaged by herbicides or stressed by drought, waterlogging etc
* Do not treat crops on soils of low fertility unless adequately fertilized
* Do not use in a programme with any product containing 2-chloroethylphosphonic acid
* Late tillering may be increased with crops subject to moisture stress and may reduce quality of malting barley
* Do not apply to winter cultivars sown in spring or treat winter barley, triticale or winter rye on soils with more than 10% organic matter (winter wheat may be treated)
* Do not apply at temperatures above 21°C
* Do not use straw from treated cereals as a mulch or growing medium

Protective clothing/Label precautions
* 29, 54, 63, 66

Latest application/Harvest Interval(HI)
* Before ear visible (GS 49) for winter barley, winter wheat and triticale; boots swollen (GS 45) for winter oats; flag leaf just visible (GS 37) for winter rye

Maximum Residue Levels (mg residue/kg food)
* see 2-chloroethylphosphonic acid entry

FOR FULL CONDITIONS OF USE ALWAYS READ THE PRODUCT LABEL

106   chlorophacinone

An anticoagulant rodenticide

Products					
	1 Drat	RP Environ.	2.5 g/l	CB	05238
	2 Drat Bait	RP Environ.	0.006% w/w	RB	05239
	3 Endorats	American Products	0.005% w/w	RB	06503
	4 Karate Ready-to-Use Rat & Mouse Bait	Lever Industrial	0.006% w/w	RB	05321
	5 Karate Ready-to-Use Rodenticide Sachets	Lever Industrial	0.006% w/w	RB	05890
	6 Rat & Mouse Bait	B H & B	0.005% w/w	RB	00764
	7 Ruby Rat	Heatherington	0.005% w/w	RB	06059

Uses   Mice, rats, voles in FARM BUILDINGS.

Notes   **Efficacy**
- Chemical formulated with oil, thus improving weather resistance of bait
- Mix concentrate with any convenient bait, such as grain, apple, carrot or potato [1]
- Use in baiting programme. Lay small baits for mice, larger baits for rats
- Replenish baits every few days and remove unused bait when take ceases or after 7-10 d

Special precautions/Environmental safety
- Harmful in contact with skin and if swallowed [1]
- Prevent access to baits by children, domestic animals and birds; see label for other precautions required

Protective clothing/Label precautions
- A [1]
- 5a, 5c, 14, 16, 18, 22, 26, 27, 36, 37, 46, 54, 64, 65, 78, 85 [1]; 25, 29, 81, 82, 83, 84 [1-6]; 63, 67 [2-6]

107   chloropicrin

A highly toxic horticultural soil fumigant

Products					
	Chloropicrin Fumigant	Dewco-Lloyd	99.5%	LI	04216

Uses   Crown rot, nematodes, red core, verticillium wilt in STRAWBERRIES. Replant disease in HARDY ORNAMENTALS, TOP FRUIT.

Notes   **Efficacy**
- Treat pre- or post-planting
- Apply with specialised injection equipment
- For treating small areas or re-planting a single tree, a hand-operated injector may be used. Mark the area to be treated and inject to 22 cm at intervals of 22 cm
- Double roll within 1 h of treatment and leave undisturbed for at least 10 d

Crop Safety/Restrictions

- Polythene sheeting (150 gauge) should be progressively laid over soil as treatment proceeds. The margin of the sheeting around the treated area must be embedded or covered with treated soil. Remove progressively after at least 4 d provided good air movement conditions prevail
- Carry out a cress test before replanting treated soil

Special precautions/Environmental safety

- Very toxic by inhalation, in contact with skin and of swallowed
- Irritating to eyes, respiratory system and skin
- Before use, consult the code of practice for the fumigation of soil with chloropicrin. 2 fumigators must be present
- Avoid treatment or vapour release when persistent still air conditions prevail
- Remove contaminated gloves, boots or other clothing immediately and ventilate them in the open air until all odour is eliminated
- Dangerous to livestock
- Dangerous to game, wild birds and animals
- Dangerous to bees
- Dangerous to fish. Do not contaminate surface waters or ditches with chemical or used container

Protective clothing/Label precautions

- A, G, H, K, M
- 3a, 3b, 3c, 6a, 6b, 6c, 16, 18, 23, 25, 26, 27, 28, 29, 36, 37, 38 (until advised), 40 (until advised), 45, 48, 52, 54, 64, 67, 70, 79

Latest application/Harvest Interval(HI)

- Before planting

FOR FULL CONDITIONS OF USE ALWAYS READ THE PRODUCT LABEL

108   chlorothalonil

A protectant chlorophenyl fungicide for use in many crops

See also carbendazim + chlorothalonil
    carbendazim + chlorothalonil + maneb

Products					
1 Barclay Corrib 500	Barclay	500 g/l	SC	06392
2 Bombardier	Unicrop	500 g/l	SC	02675
3 Bravo 500	BASF	500 g/l	SC	05637
4 Bravo 500	ISK Biosciences	500 g/l	SC	05638
5 Bravo 720	ISK Biosciences	720 g/l	SC	05544
6 Contact 75	ISK Biosciences	75% w/w	WB	05563
7 Daconil Turf	ICI Professional	500 g/l	SC	03658
8 Daconil Turf	Zeneca Prof.	500 g/l	SC	06867
9 ISK 375	ISK Biosciences	75% w/w	WG	07455
10 Jupital	ISK Biosciences	500 g/l	SC	05554
11 Mainstay	Quadrangle	500 g/l	SC	05625
12 Miros DF	Sipcam	75% w/w	WG	04966
13 Sipcam UK Rover 500	Sipcam	500 g/l	SC	04165
14 Standon Chlorothalonil 50	Standon	500 g/l	SC	05922
15 Tripart Faber	Tripart	500 g/l	SC	05505
16 Tripart Faber	Tripart	500 g/l	SC	04549
17 Tripart Ultrafaber	Tripart	720 g/l	SC	05627

Uses

Alternaria, botrytis, damping off and wirestem, downy mildew, ring spot in LEAF BRASSICAS [1-5, 9-11, 13, 15-17]. Alternaria, downy mildew, ring spot in FLOWERHEAD BRASSICAS [2, 13, 16]. Anthracnose, brown patch, fusarium patch, grey snow mould, melting out, red thread, take-all patch in TURF [5, 7, 8]. Ascochyta, botrytis, mycosphaerella in PEAS [2-5, 9-17]. Blight, botrytis, leaf mould in TOMATOES [3, 4]. Blight in POTATOES [1-6, 9-17]. Botrytis, cane spot, powdery mildew in CANE FRUIT [3-5, 9-11, 15, 17]. Botrytis, downy mildew, ring spot in BRUSSELS SPROUTS [5, 11]. Botrytis, downy mildew in WINTER OILSEED RAPE [3-5, 9-11, 15, 17]. Botrytis, powdery mildew in CUCUMBERS [3]. Botrytis, powdery mildew in BLACKCURRANTS *(suppression)* [1]. Botrytis in STRAWBERRIES [3-5, 9-11, 15, 17]. Botrytis in PROTECTED ORNAMENTALS [3, 4]. Botrytis in PROTECTED CUCUMBERS [4]. Celery leaf spot, septoria leaf spot in CELERIAC *(off-label)* [4, 16, 17]. Celery leaf spot in CELERY [1, 3, 4, 9-11, 15]. Chocolate spot in FIELD BEANS [1-5, 9-17]. Currant leaf spot, leaf spot in GOOSEBERRIES [3-5, 9-11, 15, 17]. Currant leaf spot in REDCURRANTS [3-5, 9-11, 15, 17]. Currant leaf spot in BLACKCURRANTS [1-5, 9-11, 13, 15-17]. Downy mildew, powdery mildew in GRAPEVINES [5, 11]. Downy mildew in FLOWERHEAD BRASSICAS *(limited control)*, LEAF BRASSICAS *(limited control)* [2]. Downy mildew in HOPS [3-5, 9-11, 15, 17]. Downy mildew in BRASSICA SEED BEDS [4, 9, 10]. Dry bubble, wet bubble in MUSHROOMS [3-5, 9-11, 15, 17]. Glume blotch, leaf spot in SPRING WHEAT [1, 2, 12, 13, 16]. Glume blotch, leaf spot in WINTER WHEAT [1-5, 9-17]. Ink disease in IRISES [3, 4]. Leaf blotch in SPRING BARLEY, WINTER BARLEY [1]. Leaf rot, neck rot in ONIONS [1, 3-5, 9-11, 15, 17]. Powdery mildew in RASPBERRIES [4, 9-11].

Notes

Efficacy

- For some crops products differ in diseases listed as controlled. See label for details and for application rates, timing and number of sprays
- Apply as protective spray or as soon as disease appears and repeat at 7-21 d intervals
- Activity against Septoria may be reduced where serious mildew or rust present. In such conditions mix with suitable mildew or rust fungicide
- May be used at reduced rate on peas in tank mix with Ronilan FL [3]

- Do not mow or water turf for 24 h after treatment. Do not add surfactant or mix with liquid fertilizer
- For Botrytis control in strawberries important to start spraying early in flowering period and repeat at least 3 times at 10 d intervals

Crop Safety/Restrictions

- Maximum number of treatments 2 per crop for winter wheat, brassicas, oilseed rape, field beans, peas, onions, leeks or mushrooms (3 per crop for wheat [2, 12, 13, 16], 1 per crop for wheat, barley [1]); 4 per yr for celery, blackcurrants, cane fruit, redcurrants, gooseberries, strawberries; 5 per crop for potatoes [2, 13, 16]; 5 per yr for blackcurrants (see label) [2, 13, 16]; 6 per crop for onions [1, 3, 4]
- On strawberries some scorching of calyx may occur with protected crops

Special precautions/Environmental safety

- Irritating to eyes, skin and respiratory system [1-8, 10, 11, 13-17], to respiratory system [12]
- Risk of serious damage to eyes
- Dangerous to fish or other aquatic life. Do not contaminate surface waters or ditches with chemical or used container
- Do not allow direct spray from vehicle-mounted/drawn hydraulic sprayers to fall within 6 m, or from hand-held sprayers to within 2 m, of surface waters or ditches. Direct spray away from water [1, 2, 7, 8, 12-16]

Protective clothing/Label precautions

- A, C, D, H, M [9]; A, C [1-8, 10-17]
- 6a, 6b [1-8, 10, 11, 13-17]; 6c, 18, 22, 28, 29, 36, 37, 52, 63 [1-17]; 9 [9, 12]; 25, 26, 27, 54a, 70, 9 [1-8, 10-17]; 66 [6-8, 10-17]; 66, [1, 4]; 67 [9]; 66, [2, 3, 5]

Latest application/Harvest Interval(HI)

- Before ear emergence complete (GS 59) [1], before flowering (GS 60) [3-5, 10-12, 14-16], before grain watery ripe (GS 71) [2, 13] for wheat; before flowering for winter oilseed rape.
- HI potatoes zero (7 d [2, 13, 16]); tomatoes, cucumbers, peppers 12 h; mushrooms 24 h; currants, gooseberries, cane fruit, strawberries 3 d (currants 14 d [1], 28 d [2, 13, 16]); brassicas, celery 7 d; hops 10 d; field beans, peas, onions, leeks 14 d (field beans, onions 7 d [1])

Approval

- Approval for aerial spraying on wheat, barley, field beans, potatoes [1]; wheat, potatoes [2, 13, 16]; winter wheat, potatoes [3, 4, 10, 11, 15]; peas, field beans [3, 4, 10, 11]. See notes in Section 1
- Off-label approval to May 2000 for use on celeriac (OLA 0796/95)[4], (OLA 0797/95)[17]

Maximum Residue Levels (mg residue/kg food)

- cranberries, tomatoes, peppers, aubergines, peas (with pods) 2; grapes (table), cucumbers 1; garlic, onions, shallots, Brussels sprouts 0.5; tea, wheat, rye, barley, oats, triticale 0.1; maize, rice, animal products 0.01

FOR FULL CONDITIONS OF USE ALWAYS READ THE PRODUCT LABEL

109 chlorothalonil + cymoxanil

A systemic and protective fungicide for potato blight control

Products

	1 Guardian	ICI Agrochem.	56.7:6.3% w/w	WP	05663
	2 Guardian	Zeneca	56.7:6.3% w/w	WP	06676

Uses   Blight in EARLY POTATOES, MAINCROP POTATOES.

Notes   **Efficacy**
- Start spray treatments immediately after first blight warning, in mid-late Jun in blight prone areas, in early Jul elsewhere
- Repeat treatment at 10-14 d intervals depending on blight conditions

Special precautions/Environmental safety
- Irritating to eyes, skin and respiratory system
- Dangerous to fish or other aquatic life. Do not contaminate surface waters or ditches with chemical or used container

Protective clothing/Label precautions
- A, C
- 6a, 6b, 6c, 14, 18, 21, 28, 29, 36, 37, 52, 63, 65, 9

Latest application/Harvest Interval(HI)
- HI 10 d

Approval
- Approved for aerial application on potatoes. See notes in Section 1

Maximum Residue Levels (mg residue/kg food)
- see chlorothalonil entry

110 chlorothalonil + cyproconazole

A systemic protective and curative fungicide for cereals

Products

	1 Alto Elite	Sandoz	375:40 g/l	SC	05069
	2 Octolan	Sandoz	375:40 g/l	SC	06256

Uses   Brown rust, eyespot, net blotch, powdery mildew, rhynchosporium, yellow rust in WINTER BARLEY. Brown rust, eyespot, powdery mildew, septoria diseases, yellow rust in WINTER WHEAT. Brown rust, net blotch, powdery mildew, rhynchosporium, yellow rust in SPRING BARLEY.

Notes   **Efficacy**
- Apply at first signs of infection or as soon as disease becomes active
- A repeat application may be made if re-infection occurs
- For established mildew tank-mix with approved formulation of tridemorph
- When applied prior to third node detectable (GS 33) a useful reduction of eyespot will be obtained
- If infection of eyespot anticipated tank-mix with Sportak

Crop Safety/Restrictions
* Maximum number of treatments 2 per crop
* If applied to winter wheat in spring at GS 30-33 straw shortening may occur but yield is not reduced

Special precautions/Environmental safety
* Irritant. Risk of serious damage to eyes
* Do not apply at concentrations higher than recommended
* Dangerous to fish or other aquatic life. Do not contaminate surface waters or ditches with chemical or used container

Protective clothing/Label precautions
* A, C, H, M
* 6, 9, 18, 24, 28, 29, 36, 37, 52, 63, 66, 70

Latest application/Harvest Interval(HI)
* Up to and including emergence of ear complete (GS 59) for barley and winter wheat

Maximum Residue Levels (mg residue/kg food)
* see chlorothalonil entry

111   chlorothalonil + fenpropimorph
A systemic and protectant fungicide for use in winter wheat

Products					
1 BAS 438	BASF	250:187 g/l	SC	03451
2 Corbel CL	BASF	250:187 g/l	SC	04196

Uses   Brown rust, powdery mildew, septoria diseases, yellow rust in WINTER WHEAT.

Notes   **Efficacy**
* Apply before disease established from flag leaf visible to ear just completely emerged (GS 39-59)
* Addition of carbendazim recommended for additional sooty mould control and for later applications during ear emergence. See label for details and other recommended tank mixes

Crop Safety/Restrictions
* Maximum number of treatments 2 per crop (1 per crop at higher dose)
* Some crop scorch may occur if applied during high temperatures

Special precautions/Environmental safety
* Harmful by inhalation. Irritating to eyes, skin and respiratory system
* Dangerous to fish or other aquatic life. Do not contaminate surface waters or ditches with chemical or used container

Protective clothing/Label precautions
* A, C
* 5b, 6a, 6b, 6c, 9, 18, 21, 26, 27, 28, 29, 36, 37, 52, 60, 63, 66, 70, 78

FOR FULL CONDITIONS OF USE ALWAYS READ THE PRODUCT LABEL

Latest application/Harvest Interval(HI)
• HI 5 wk

Maximum Residue Levels (mg residue/kg food)
• see chlorothalonil entry

---

112   chlorothalonil + flutriafol
A systemic eradicant and protectant fungicide for winter wheat

Products

1 Halo	Zeneca	375:47 g/l	SC	06520
2 Impact Excel	ICI Agrochem.	300:47 g/l	SC	03758
3 Impact Excel	Zeneca	300:47 g/l	SC	06680
4 PP 375	Zeneca	375:47 g/l	SC	06681

Uses   Brown rust, glume blotch, leaf spot, powdery mildew, sooty moulds, yellow rust in WINTER WHEAT.

Notes   **Efficacy**
• Best results achieved by applying protective spray at flag leaf emergence (GS 37) and full ear emergence (GS 59). If disease already present spray before it reaches the top 2 leaves and ears
• Where mildew serious may be tank-mixed with fenpropidin

Crop Safety/Restrictions
• Maximum number of treatments 2 per crop
• On certain cultivars with erect leaves high transpiration can result in flag leaf tip scorch. This may be increased by treatment but does not affect yield

Special precautions/Environmental safety
• Irritating to eyes and skin. May cause sensitization by skin contact
• Dangerous to fish or other aquatic life. Do not contaminate surface waters or ditches with chemical or used container

Protective clothing/Label precautions
• A, C
• 6a, 6b, 10a, 14, 18, 28, 29, 36, 37, 52, 66, 70 [1-4]; 21, 9 [2, 3]; 22, 63 [1, 4]

Latest application/Harvest Interval(HI)
• Before beginning of flowering

Maximum Residue Levels (mg residue/kg food)
• see chlorothalonil entry

---

113   chlorothalonil + metalaxyl
A systemic and protectant fungicide for various field crops

Products   Folio 575 SC   Ciba Agric.   500:75 g/l   SC   05843

Uses   Downy mildew in BROAD BEANS, BULB ONIONS, FIELD BEANS, SALAD ONIONS, WINTER OILSEED RAPE. White blister in BRUSSELS SPROUTS. White tip in LEEKS.

Notes	**Efficacy**

Efficacy
- Apply at first signs of disease (oilseed rape, beans), at first signs of disease or when weather conditions favourable to disease (Brussels sprouts, onions, leeks)
- Repeat treatment at 14-21 d (14 d for oilseed rape and beans) intervals if necessary
- Oilseed rape crops most likely to benefit from treatment are infected crops between cotyledon and 3-leaf stage (GS 1,0-1,3)

Crop Safety/Restrictions
- Maximum number of treatments 2 per crop for oilseed rape and beans, 3 per crop for Brussels sprouts, onions and leeks

Special precautions/Environmental safety
- Irritating to skin and eyes
- Dangerous to fish or other aquatic life. Do not contaminate surface waters or ditches with chemical or used container

Protective clothing/Label precautions
- A, C
- 6a, 6b, 9, 18, 22, 29, 35, 36, 37, 52, 63, 65

Latest application/Harvest Interval(HI)
- HI field beans, broad beans, Brussels sprouts, onions, leeks 14 d

Maximum Residue Levels (mg residue/kg food)
- see chlorothalonil and metalaxyl entries

---

114   chlorothalonil + propamocarb hydrochloride

A contact and systemic fungicide mixture for blight control in potatoes

Products	Tattoo C	AgrEvo	375:375 g/l	SC	07623

Uses   Late blight in POTATOES.

Notes   **Efficacy**
- Commence treatment early in the season as soon as there is risk of infection
- In the absence of a blight warning treatment should start just before potatoes meet along the row
- Use only as a protectant. Stop use when blight readily visible (1% leaf area destroyed)
- Repeat sprays at 10-14 d intervals depending on blight infection risk. See label for details
- Complete blight spray programme after end Aug up to haulm destruction with protectant fungicides preferably fentin based

Crop Safety/Restrictions
- Maximum number of treatments 5 per crop
- Apply to dry foliage. Do not apply if rainfall or irrigation imminent

Special precautions/Environmental safety
- Irritant
- May cause serious damage to eyes

FOR FULL CONDITIONS OF USE ALWAYS READ THE PRODUCT LABEL

- May cause sensitisation by skin contact
- Dangerous to fish or other aquatic life. Do not contaminate surface waters or ditches with chemical or used container

Protective clothing/Label precautions
- A, C, H
- 6a, 9, 10a, 18, 21, 24, 28, 29, 36, 37, 52, 63, 66, 70

Latest application/Harvest Interval(HI)
- HI: 7 d

Maximum Residue Levels (mg residue/kg food)
- see chlorothalonil entry

---

115   chlorothalonil + propiconazole

A systemic and protectant fungicide for winter wheat and barley

Products	Sambarin 312.5 SC	Ciba Agric.	250:62.5 g/l	SC	05809

Uses   Brown rust, powdery mildew, septoria diseases, yellow rust in WINTER WHEAT. Mildew, net blotch, rhynchosporium in WINTER BARLEY.

Notes

Efficacy
- On wheat apply from start of flag leaf emergence up and including when ears just fully emerged (GS 37-59), on barley at any time to ears fully emerged (GS 59)
- Best results achieved from early treatment, especially if weather wet, or as soon as disease develops

Crop Safety/Restrictions
- Maximum number of treatments 2 per crop, 1 per crop if other propiconazole based fungicide used in programme

Special precautions/Environmental safety
- Irritating to eyes, skin and respiratory system
- Dangerous to fish or other aquatic life. Do not contaminate surface waters or ditches with chemical or used container

Protective clothing/Label precautions
- A, C
- 6a, 6b, 6c, 9, 14, 18, 22, 24, 28, 29, 35, 36, 37, 52, 63, 66, 70

Latest application/Harvest Interval(HI)
- Up to and including emergence of ear just complete (GS 59).
- HI 42 d

Maximum Residue Levels (mg residue/kg food)
- see chlorothalonil and propiconazole entries

---

116 chlorotoluron

A contact and residual urea herbicide for cereals

Products

1 Alpha Chlorotoluron 500	Makhteshim	500 g/l	SC	04848
2 Ashlade Tol-7	Ashlade	700 g/l	SC	06484
3 Atol	Ashlade	700 g/l	SC	07347
4 Stefes Toluron	Stefes	500 g/l	SC	05779
5 Tripart Ludorum	Tripart	500 g/l	SC	03059
6 Tripart Ludorum 700	Tripart	700 g/l	SC	03999

Uses

Annual dicotyledons, annual grasses, blackgrass, rough meadow grass, wild oats in TRITICALE [1-3, 5, 6]. Annual dicotyledons, annual grasses, blackgrass, rough meadow grass, wild oats in WINTER BARLEY, WINTER WHEAT [1-6]. Annual dicotyledons, annual grasses, blackgrass, rough meadow grass, wild oats in DURUM WHEAT [2, 5, 6].

Notes

Efficacy
* Best results achieved by application soon after drilling. Application in autumn controls most weeds germinating in early spring
* For wild oat control apply within 1 wk of drilling, not after 2-leaf stage. Blackgrass and meadow grasses controlled to 5 leaf, ryegrasses to 3 leaf stage
* Any trash or burnt straw should be buried and dispersed during seedbed preparation
* Do not use on soils with more than 10% organic matter
* Control may be reduced if prolonged dry conditions follow application
* Harrowing after treatment may reduce weed control

Crop Safety/Restrictions
* Maximum number of treatments 1 per crop
* Use only on listed crop varieties. See label. Ensure seed well covered at drilling
* Apply only as pre-emergence spray in durum wheat, pre- or post-emergence in wheat, barley or triticale
* Do not apply pre-emergence to crops sown after 30 Nov
* Do not apply to crops severely checked by waterlogging, pests, frost or other factors
* Do not use on undersown crops or those due to be undersown
* Do not apply post-emergence in mixture with liquid fertilizers
* Do not roll for 7 d before or after application to an emerged crop
* Crops on stony or gravelly soils may be damaged, especially after heavy rain

Special precautions/Environmental safety
* Harmful if swallowed [1]
* Irritating to skin and eyes
* Harmful to fish or other aquatic life. Do not contaminate surface waters or ditches with chemical or used container [3, 4]

Protective clothing/Label precautions
* A, C [1, 3]
* 5c [1]; 6a, 6b, 18, 21, 28, 29, 36, 37, 63, 66, 70 [1-6]; 53 [1, 3, 4]; 54 [2, 5, 6]; 60, 65 [2, 4-6]; 78 [1, 3]

FOR FULL CONDITIONS OF USE ALWAYS READ THE PRODUCT LABEL

Latest application/Harvest Interval(HI)
* Pre-emergence for durum wheat. Up to end of tillering (GS 30) for winter wheat, barley and triticale (but some variation with dose and volume of application - see label)
* Before 5 leaf stage (GS 25) and 31 Dec in year of sowing at highest dose or before end of tillering (GS 30) for reduced dose [3]

Approval
* May be applied through CDA equipment. See label for details. See notes in Section 1 [3-5]
* Approved for aerial application on winter wheat, winter barley, durum wheat, triticale [2, 5, 6]. See notes in Section 1

117   chlorotoluron + pendimethalin
A contact and residual herbicide mixture for winter cereals

Products	Totem	Cyanamid	300:200 g/l	SC	04670

Uses   Annual dicotyledons, annual grasses, blackgrass, wild oats in DURUM WHEAT, WINTER BARLEY, WINTER WHEAT.

Notes   **Efficacy**
* Best results achieved by application soon after drilling. Do not apply later than weed growth stage indicated on susceptibility list
* Apply to fine, firm seedbed. Any trash or straw should be dispersed during seedbed preparation
* Do not use on soils with more than 10% organic matter
* Best results obtained if rain falls within 7 d after application

Crop Safety/Restrictions
* Maximum number of treatments 1 per crop
* Use only on listed crop varieties. Do not apply to undersown crops
* May be applied pre- or post-emergence on winter wheat and barley, pre-emergence only on durum wheat
* Do not apply to crops drilled after 30 Nov
* Seed should be covered with at least 32 mm of settled soil. Do not apply pre-emergence if seed cover less than 32 mm
* On stony or gravelly soils there is risk of crop damage, especially if heavy rain falls soon after treatment
* Early sown crops may be prone to damage if application precedes or coincides with a period of rapid growth in autumn
* Do not apply to non-frost hardened crops if frost likely or during prolonged frost
* Avoid waterlogged soils and do not use on soils where surface water likely to accumulate
* Winter barley Gerbel may only be treated on heavy soils
* Do not roll emerged crops for 7 d before or after spraying
* In the event of crop failure in autumn land must be ploughed to at least 15 cm, after which spring wheat, spring barley, maize, potatoes, beans or peas may be grown
* After a dry season land must be well cultivated to at least 15 cm before drilling ryegrass

Special precautions/Environmental safety
* Dangerous to fish or other aquatic life. Do not contaminate surface waters or ditches with chemical or used container

Protective clothing/Label precautions
- 18, 21, 25, 28, 29, 36, 37, 52, 63, 66

Latest application/Harvest Interval(HI)
- Pre-emergence for durum wheat, before main shoot and 5 tillers (GS 25) for winter wheat and barley

---

118   chlorpropham

A residual carbamate herbicide and potato sprout suppressant

See also cetrimide + chlorpropham
                chloridazon + chlorpropham + fenuron + propham

Products

1 Atlas CIPC 40	Atlas	400 g/l	EC	03049
2 BL 500	Wheatley	500 g/l	HN	00279
3 MSS CIPC 40 EC	Mirfield	400 g/l	EC	01403
4 MSS CIPC 5 G	Mirfield	5% w/w	GR	01402
5 MSS CIPC 50 LF	Mirfield	500 g/l	EC	03285
6 MSS CIPC 50 M	Mirfield	500 g/l	EC	01404
7 MTM CIPC 40	MTM Agrochem.	400 g/l	EC	05895
8 Warefog 25	Mirfield	600 g/l	HN	02323

Uses

Annual dicotyledons, annual grasses, chickweed, polygonums in LETTUCE [3, 7]. Annual dicotyledons, annual grasses, chickweed, polygonums in FLOWER BULBS, LEEKS, ONIONS [1, 3]. Annual dicotyledons, annual grasses, chickweed, polygonums in CARROTS [1, 7]. Annual dicotyledons, annual grasses, chickweed, polygonums in ANNUAL FLOWERS, BLACKCURRANTS, CELERY, CHRYSANTHEMUMS, GOOSEBERRIES, LUCERNE, PARSLEY, STRAWBERRIES [7]. Sprout suppression in POTATOES [2, 4-6, 8]. Volunteer ryegrass in GRASS SEED CROPS *(off-label)* [1, 3, 7].

Notes

Efficacy
- Apply to freshly cultivated soil. Adequate rainfall must occur after spraying. Activity is greater in cold, wet than warm, dry conditions
- For sprout suppression apply with suitable fogging or rotary atomizer equipment or sprinkle granules over tubers before sprouting commences. Repeat applications may be needed. See label for details
- Cure potatoes according to label instructions before treatment and allow 3 wk between completion of loading into store and first treatment

Crop Safety/Restrictions
- Maximum number of treatments 1 per batch [4], 3 per batch [5], for potatoes; 1 per yr for grass seed crops [1, 3, 7]; 1 per crop for carrots, onions, leeks [1]; 2 per yr for flower bulbs [1]
- Apply to seeded crops pre-emergence of crop or weeds, to onions as soon as first crop seedlings visible, to planted crops a few days before planting, to bulbs immediately after planting, to fruit crops in late autumn-early winter. See label for further details
- Not to be used on grass seed crops if grass to be grazed or cut for fodder before 31 May following treatment [1, 3, 7]
- Excess rainfall after application may result in crop damage

---

FOR FULL CONDITIONS OF USE ALWAYS READ THE PRODUCT LABEL

- Do not use on sand, very light soils or soils low in organic matter
- Poor conditions at drilling or planting, soil compaction, surface capping, waterlogging or attack by pests may result in crop damage
- On crops under glass high temperatures and poor ventilation may cause crop damage
- Only clean, mature, disease-free potatoes should be treated for sprout suppression
- Do not use on potatoes for seed. Do not handle, dry or store seed potatoes or any other seed or bulbs in boxes or buildings in which potatoes are being or have been treated
- Do not remove potatoes for sale or processing for at least 21 d after application
- A minimum interval of 45 d must elapse between applications [6]

Special precautions/Environmental safety

- Harmful if swallowed [5] and in contact with skin [1, 6, 7]; harmful by inhalation or if swallowed [6]
- Irritating to eyes, skin [5] and respiratory system [1, 6]; irritating to eyes and respiratory system [8]
- Flammable [5, 6]; highly flammable [6]
- Harmful to fish or other aquatic life. Do not contaminate surface waters or ditches with chemical or used container [3, 4]
- Keep unprotected persons out of treated areas for at least 24 h after application [6]

Protective clothing/Label precautions

- A, C, D, E, H, J, M [6]; A, C, D, E [8]; A, C, D, G, H, J, M [5]; A, C [1-4, 7]
- 5a [1, 7]; 5b, 12b, 38, 66 [6]; 5c, 78 [1, 5-7]; 6a, 21 [1, 5, 6, 8]; 6b [1, 5, 6]; 6c [1, 6, 8]; 12c [5]; 18, 28, 36, 37, 63 [1, 3-8]; 22 [3, 4, 7]; 29, 70 [1, 3-7]; 30 [8]; 53 [3, 4]; 54 [1, 5-8]; 60 [7]; 65 [1, 3-5, 7, 8]

Latest application/Harvest Interval(HI)

- 3 wk before removal of potatoes from store for sale or processing [2, 4-6, 8]; 48 h before drilling carrots, onions (before 4 leaves on Silt soils) [1]; 14 d before planting onions, leeks [1]; before leaves unfurl for tulips, before other bulb crops 5 cm high [1]; before 28 Feb for grass seed crops

Approval

- Some products are formulated for application by thermal fogging. See label for details. See introductory notes [2, 8]
- Off-label approval to Jan 1997 for use on grass seed crops (OLA 0041/92) [1], (OLA 0040/92) [3]
- Approval expiry: 30 Sep 95 [02323]; replaced by MAFF 06776

---

119   chlorpropham with cetrimide

A soil-acting herbicide for lettuce under cold glass

Products					
Croptex Pewter	Hortichem	80:80 g/l	SC	02507

Uses   Annual dicotyledons, annual grasses, chickweed, polygonums in PROTECTED LETTUCE.

Notes   **Efficacy**
- Apply to drilled lettuce under cold glass within 24 h post-drilling, to transplanted crops pre-planting
- Adequate irrigation must be applied before or after treatment
- Best results achieved on firm soil of fine tilth, free from clods and weeds

Crop Safety/Restrictions
* Do not apply to crop foliage or use where seed has chitted
* Excess irrigation may cause temporary check to crop under certain circumstances
* Do not apply where tomatoes or brassicas are growing in the same house
* In the event of crop failure only lettuce should be grown within 2 mth

Special precautions/Environmental safety
* Harmful in contact with skin and if swallowed
* Irritating to eyes, skin and respiratory system
* Highly flammable
* Dangerous to fish or other aquatic life. Do not contaminate surface waters or ditches with chemical or used container

Protective clothing/Label precautions
* A, C
* 5a, 5c, 6a, 6b, 6c, 12b, 18, 21, 28, 29, 36, 37, 52, 60, 63, 65

Latest application/Harvest Interval(HI)
* Before crop emergence or pre-planting

---

120   chlorpropham + cresylic acid + fenuron
A residual herbicide for vegetables and ornamentals

Products	Atlas Red	Hortichem	200:-:50 g/l	EC	03091

Uses   Annual dicotyledons, annual grasses, chickweed in FLOWER BULBS, HARDY ORNAMENTAL NURSERY STOCK, LEEKS, ONIONS, PATHS AND DRIVES, RHUBARB, STANDING GROUND.

Notes   **Efficacy**
* Apply to soil freshly cultivated and free of established weed. Adequate rainfall must occur after spraying. Activity is greater in cold, wet than warm, dry conditions

Crop Safety/Restrictions
* Apply to seeded onions and leeks pre-emergence of crop and weeds or at post crook stage, to rhubarb and nursery stock in dormant season, avoiding foliage of conifers or evergreens
* Do not use on Sands, Very Light soils or soils low in organic matter
* Poor conditions at drilling or planting, soil compaction, surface capping, waterlogging or pest attack may result in crop damage
* In the event of crop failure only recommended crops should be planted in treated soil
* Plough or cultivate to 15 cm after harvest to dissipate any residues

Special precautions/Environmental safety
* Harmful if swallowed and in contact with skin
* Irritating to eyes, skin and respiratory system
* Flammable
* Dangerous to fish or other aquatic life. Do not contaminate surface waters or ditches with chemical or used container

---

FOR FULL CONDITIONS OF USE ALWAYS READ THE PRODUCT LABEL

Protective clothing/Label precautions
* A, C
* 5a, 5c, 6a, 6b, 6c, 12c, 18, 21, 25, 28, 29, 36, 37, 52, 60, 63, 67, 70

---

121 chlorpropham + diuron + propham

A residual herbicide for lettuce

Products					
	Atlas Pink C	Hortichem	25:6:100 g/l	EC	03095

Uses Annual dicotyledons, annual grasses in OUTDOOR LETTUCE.

Notes **Efficacy**
* Best results achieved from application to firm, moist, weed-free seedbed when adequate rain falls afterwards. Activity is greater in cold wet than warm, dry conditions
* Apply to soil after drilling or prior to transplanting
* Incorporation to 2.5-5 cm prior to sowing or planting can improve results under dry soil conditions or on organic soils

Crop Safety/Restrictions
* Maximum number of treatments 1 per crop
* Poor conditions at drilling or planting, soil compaction, surface capping, waterlogging and attack by pests may result in crop damage
* Treatment of lettuce under frames or tunnels may cause injury during sunny periods
* In the event of crop failure only lettuce should be replanted in treated soil
* Plough or cultivate to 10 cm after harvest to dissipate any residues

Special precautions/Environmental safety
* Harmful if swallowed and in contact with skin
* Irritating to eyes, skin and respiratory system
* Flammable

Protective clothing/Label precautions
* A
* 5a, 5c, 6a, 6b, 6c, 12c, 18, 21, 28, 29, 36, 37, 53, 63, 65, 70, 78

Latest application/Harvest Interval(HI)
* Post-drilling or prior to transplanting

---

122 chlorpropham + fenuron

A residual herbicide for vegetables and ornamentals

Products					
	Croptex Chrome	Hortichem	80:15 g/l	EC	02415

Uses Annual dicotyledons, annual grasses, chickweed in BROAD BEANS, FIELD BEANS, FLOWERS, LEEKS, MAIDEN STRAWBERRIES, ONIONS, PEAS, PROTECTED BULBS, SPINACH.

Notes

Efficacy
* Apply to soil freshly cultivated and free of established weeds. Adequate rainfall must occur after spraying. Activity is greater in cold, wet than warm, dry conditions

Crop Safety/Restrictions
* Apply to seeded crops pre-emergence of crop or weeds, to transplanted onions and leeks 10-14 d post-planting, to transplanted flowers and strawberries 5 d pre-transplanting
* Do not use on Sands, Very Light soils or soils low in organic matter
* Do not treat protected bulb or flower crops if other crops are being grown in same block of houses
* Poor conditions at drilling or planting, soil compaction, surface capping, waterlogging or attack by pests may result in crop damage
* In the event of crop failure only recommended crops should be replanted in treated soil
* Plough or cultivate to 15 cm after harvest to dissipate any residues

Special precautions/Environmental safety
* Harmful if swallowed and in contact with skin
* Irritating to eyes, skin and respiratory system
* Flammable
* Dangerous to fish or other aquatic life. Do not contaminate surface waters or ditches with chemical or used container

Protective clothing/Label precautions
* A, C
* 5a, 5c, 6a, 6b, 6c, 12c, 18, 21, 25, 28, 29, 36, 37, 52, 60, 63, 67, 70, 78

123   chlorpropham + fenuron + propham
A residual herbicide for beet crops

Products

1 Atlas Gold	Atlas	37.5:25:150 g/l	EC	03086
2 MSS Sugar Beet Herbicide	Mirfield	37.5:25:150 g/l	SC	02447
3 MTM Sugar Beet Herbicide	MTM Agrochem.	27.5:28.5:147 g/l	EC	05044
4 Sugar Beet Herbicide	United Phosphorus	27.5:28.5;147 g/l	EC	07525

Uses

Annual dicotyledons, annual grasses in FODDER BEET, MANGELS, RED BEET, SUGAR BEET.

Notes

Efficacy
* Apply to firm, moist, weed-free seedbed at drilling or up to 10 d after drilling in Mar, up to 7 d in early Apr, up to 3 d in mid-Apr pre-emergence of crop and weeds
* Activity is greater in cold, wet than warm, dry conditions
* Application rate varies with soil type and weather conditions. See label for details
* Avoid disturbance of soil surface after application
* Treatment recommended as particularly useful on organic/fen soils [1]

Crop Safety/Restrictions
* Maximum number of treatments 1 per crop
* Do not use on Very Light or Light soils

FOR FULL CONDITIONS OF USE ALWAYS READ THE PRODUCT LABEL

- When crop vigour is reduced by poor soil conditions, waterlogging, pest or disease attack, drilling too deep, rapid temperature change etc some crop damage may occur
- In case of crop failure only beet crops should be re-drilled within 12 wk of treatment

Special precautions/Environmental safety
- Harmful if swallowed and in contact with skin
- Irritating to skin, eyes and respiratory system
- Flammable [3, 4]
- Dangerous to fish or other aquatic life. Do not contaminate surface waters or ditches with chemical or used container [1, 2]

Protective clothing/Label precautions
- A, C [1, 2]; A [3, 4]
- 5a, 5c, 6a, 6b, 6c, 18, 21, 28, 29, 36, 37, 63, 70, 78 [1-4]; 12c, 60 [2-4]; 25, 54, 65 [1, 3, 4]; 52, 66 [2]

Latest application/Harvest Interval(HI)
- Immediately after drilling

---

124   chlorpropham + linuron
A residual and contact herbicide for use in bulb crops

Products					
Profalon	Hortichem	200:100 g/l	EC	01640

Uses   Annual dicotyledons, annual grasses in DAFFODILS, NARCISSI, TULIPS.

Notes

Efficacy
- Weeds controlled by combined residual action, requiring adequate soil moisture, and contact effect on young seedlings, requiring dry leaf surfaces for good control
- Do not spray during or immediately prior to rainfall
- Do not cultivate after spraying unless necessary

Crop Safety/Restrictions
- Maximum number of treatments 1 per crop
- Spray daffodils and narcissi pre-emergence or post-emergence before flower buds show. Spray tulips pre-emergence only
- Do not apply on Very Light soils or Silts low in humus or clay content

Special precautions/Environmental safety
- Irritating to skin, eyes and respiratory system
- Flammable
- Dangerous to fish or other aquatic life. Do not contaminate surface waters or ditches with chemical or used container
- Do not allow direct spray from vehicle mounted/drawn hydraulic sprayers to fall within 6 m of surface waters or ditches. Direct spray away from water
- Do not apply by hand-held sprayers

Protective clothing/Label precautions
- A, C, H
- 6a, 6b, 6c, 12c, 16, 18, 21, 28, 29, 36, 37, 52, 54a, 60, 63, 66, 70

Latest application/Harvest Interval(HI)
* Pre-emergence for tulips

Approval
* Approved for aerial application on bulbs and corms. See introductory notes

---

125   chlorpropham + pentanochlor
A contact and residual herbicide for horticultural crops

Products	Atlas Brown	Atlas	150:300 g/l	EC	03835

Uses   Annual dicotyledons, annual meadow grass in CARROTS, CELERIAC, CELERY, CHRYSANTHEMUMS, FENNEL, NARCISSI, OUTDOOR LEAF HERBS *(off-label)*, PARSLEY, PARSNIPS, TULIPS.

Notes   **Efficacy**
* Apply as pre- or post-weed emergence spray
* Best results by application to weeds up to 2-leaf stage on fine, firm, moist seedbed
* Greatest contact action achieved under warm, moist conditions, the short residual action greatest in earlier part of year

Crop Safety/Restrictions
* Maximum number of treatments 1 per yr for narcissi, tulips; 2 per yr for carrots, celeriac, celery, fennel, parsnips, chrysanthemums, outdoor leaf herbs
* Apply to carrots and related crops pre- or post-emergence after fully expanded cotyledon stage, to narcissi and tulips at any time before emergence
* Apply to chrysanthemums either pre-planting or after planting as carefully directed spray, avoiding foliage
* Any crop may be sown or planted after 4 wk following ploughing and cultivation

Special precautions/Environmental safety
* Irritating to eyes

Protective clothing/Label precautions
* C
* 6a, 18, 28, 29, 36, 37, 54, 63, 66

Latest application/Harvest Interval(HI)
* Pre-emergence for narcissi, tulips
* HI 28 d for carrots, parsnips [1]

Approval
* Off-label approval unlimited for use on outdoor leaf herbs (OLA 0249/93, 0602/93)

FOR FULL CONDITIONS OF USE ALWAYS READ THE PRODUCT LABEL

126   chlorpropham + propham
A plant growth regulator suppressing sprouting of potatoes

Products

1 Atlas Indigo	Atlas	220:30 g/l	HN	03087
2 Luxan Gro-Stop	Luxan	260:40 g/l	HN	06559
3 Pommetrol M	Fletcher	412:50 g/l	HN	01615

Uses

Sprout suppression in WARE POTATOES.

Notes

Efficacy
- Apply to clean tubers after curing but before any growth of shoots has commenced, normally 3-6 wk after entering storage. Treatment remains effective for 70-100 d
- Apply with thermal fog generator according to manufacturers recommendations
- Apply into vent ducts in well ventilated store where forced draft ventilation can be used
- Treatment is most effective at 7-9°C, above 10°C sprout suppression will not occur
- Do not use on outdoor clamps unless built with correct system of ducts [1]

Crop Safety/Restrictions
- Maximum number of treatments 3 per batch
- Maximum total dose must not exceed 60 ml/1000kg potatoes/batch
- Do not treat potatoes intended for seed
- Do not treat potatoes stored in wooden boxes [1, 3]
- Do not use on potatoes with a high level of skin spot infection
- Do not handle, store or dry seed potatoes, seed grain, bulbs or other seed in buildings where potatoes are being treated or have previously been treated

Special precautions/Environmental safety
- Harmful if swallowed and in contact with skin
- Irritating to eyes, skin and respiratory system
- Allow at least 3 wk between application and marketing or processing of treated potatoes

Protective clothing/Label precautions
- A, C, D, E, H, M [1]; A, C [2, 3]
- 5a, 5c, 6a, 6b, 6c, 18, 28, 36, 37, 54, 63, 65, 78 [1-3]; 21, 29, 60 [2, 3]; 76 [1]

Latest application/Harvest Interval(HI)
- 3 wk before removal from store for sale or processing

Approval
- Products formulated for ULV application by thermal fogging. See notes in Section 1
- Approval expiry: 30 Nov 96 [01615]

127 chlorpyrifos

A contact and ingested organophosphorus insecticide and acaricide

Products

1 Barclay Clinch	Barclay	480 g/l	EC	06148
2 Dursban 4	DowElanco	480 g/l	EC	05735
3 Lorsban T	Rigby Taylor	480 g/l	EC	05970
4 Spannit	PBI	480 g/l	EC	01992
5 Spannit Granules	PBI	6% w/w	GR	04048
6 Standon Chlorpyrifos	Standon	480 g/l	EC	07633
7 SuSCon Green Soil Insecticide	Fargro	10.34% w/w	GR	06312
8 Talon	FCC	480 g/l	EC	06017

Uses

Ambrosia beetle, elm bark beetle, larch shoot beetle, pine shoot beetle in CUT LOGS [2-4]. Aphids, apple blossom weevil, apple sucker, capsids, caterpillars, codling moth, red spider mites, sawflies, suckers, summer-fruit tortrix moth, tortrix moths, winter moth, woolly aphid in APPLES [1, 2, 4, 6]. Aphids, apple blossom weevil, capsids, caterpillars, codling moth, pear sucker, red spider mites, sawflies, suckers, tortrix moths, winter moth, woolly aphid in PEARS [1, 2, 4, 6]. Aphids, cabbage root fly, caterpillars, cutworms, leatherjackets in BRASSICAS [1, 2, 4, 5, 8]. Aphids, cabbage root fly, caterpillars, cutworms, leatherjackets in BROCCOLI, BRUSSELS SPROUTS, CABBAGES, CALABRESE, CAULIFLOWERS, CHINESE CABBAGE, COLLARDS, KALE, SWEDES, TURNIPS [6]. Aphids, capsids, caterpillars, red spider mites in GOOSEBERRIES [1, 2, 4, 6]. Aphids, capsids, caterpillars, red spider mites in CURRANTS [1, 2, 4]. Aphids, capsids, caterpillars, red spider mites in BLACKCURRANTS, REDCURRANTS, WHITECURRANTS [6]. Aphids, damson-hop aphid, red spider mites, tortrix moths, winter moth in PLUMS [1, 2, 4, 6]. Aphids, frit fly, leatherjackets, thrips, wheat bulb fly, wheat-blossom midges in BARLEY, OATS, WHEAT [1, 2, 4, 6, 8]. Aphids, raspberry beetle, raspberry cane midge, red spider mites in RASPBERRIES [1, 2, 4, 6]. Aphids, red spider mites, tortrix moths, vine weevil in STRAWBERRIES [1, 2, 4, 6]. Black pine beetle, pine weevil, vine weevil in FORESTRY TRANSPLANT LINES [2, 3]. Cabbage root fly in MOOLI *(off-label)*, RADISHES *(off-label)* [2]. Colorado beetle in POTATOES *(off-label)* [2]. Cutworms in LETTUCE *(outdoor only)*[2]. Cutworms in CARROTS, POTATOES [1, 2, 4, 6, 8]. Cutworms in BEETROOT, CELERY, LEEKS, ONIONS, PARSNIPS [1, 2, 6]. Frit fly, leatherjackets in GRASSLAND [1, 2, 4, 6, 8]. Frit fly, leatherjackets in AMENITY GRASS, GOLF COURSES [1, 3]. Frit fly in MAIZE, SWEETCORN [1, 2, 4, 6, 8]. Insect pests in PROTECTED VEGETABLES *(off-label)* [4]. Leatherjackets, pygmy mangold beetle in SUGAR BEET [1, 2, 4, 6]. Leatherjackets in PEAS [1, 2, 4, 6]. Rosy rustic moth in RHUBARB [2]. Vine weevil in CONIFERS [3]. Vine weevil in HARDY ORNAMENTAL NURSERY STOCK [7].

Notes

Efficacy

- Apply as a foliar or soil treatment for most uses, as granules or a drench for soil pests of brassicas, as a drench for vine weevil control, as a dip for forestry transplants
- Brassicas raised in plant-raising bed must be re-treated at transplanting
- Activity may be reduced when soil temperature below 5°C or on organic soils
- In dry conditions the effect of granules applied as a surface band may be reduced [5]
- For vine weevil control in hardy ornamental nursery stock incorporate in growing medium when plants first potted from rooted cutting stage [7]

FOR FULL CONDITIONS OF USE ALWAYS READ THE PRODUCT LABEL

Crop Safety/Restrictions
- Maximum number of treatments and timing vary with crop, product and pest. See label for details
- On carrots the maximum total dose applied per crop must not exceed the equivalent of 3 (on mineral soils) or 4 (on organic soils) full dose applications
- Do not apply to young lettuce plants [2] or treat potatoes under severe drought stress. The variety Desirée is particularly susceptible
- Do not apply to sugar beet under stress or within 4 d of applying a herbicide
- Do not mix with highly alkaline materials. Not compatible with zineb [2]
- In apples use pre-blossom up to pink/white bud and post-blossom after petal fall [2]
- Cuttings from treated grass should not be used as a mulch for at least 12 mth

Special precautions/Environmental safety
- Products contain an anticholinesterase organophosphorus compound. Do not use if under medical advice not to work with such compounds
- Harmful in contact with skin and if swallowed [1-4, 6, 8]
- Irritating to eyes and skin [4, 8], to eyes [1-3, 6, 8]
- Flammable [1-4, 6, 8]
- Dangerous to bees. Do not apply to crops in flower or to those in which bees are actively foraging. Do not apply when flowering weeds are present [1-4, 8]
- Dangerous to fish or other aquatic life (extremely dangerous [7]). Do not contaminate surface waters or ditches with chemical or used container [1-5, 7, 8]
- Do not allow direct spray from vehicle mounted sprayers to fall within 6 m of surface waters or ditches, from hand-held sprayers within 2 m. Direct spray away from water [1, 3]

Protective clothing/Label precautions
- A, C, H [2, 4, 6, 8]; A, C [1, 3]; A [5, 7]
- 1, 37, 52, 63 [1-8]; 2 [4, 5]; 5a, 5c, 12c, 18, 21, 28, 36, 66, 70, 78 [1-4, 6, 8]; 6a [1-4, 6-8]; 6b [4, 7]; 14, 29 [1-5, 7, 8]; 34, 68 [7]; 35 [1-5, 8]; 43 (14 d), 54a [6]; 48 [4, 6]; 48, [1-3, 8]; 67 [4, 5, 7]

Withholding period
- Keep livestock out of treated areas for at least 14 d after treatment [1, 2, 4-6]

Latest application/Harvest Interval(HI)
- End of Jul in year of harvest for sugar beet; 4 d after transplanting or at seedling emergence for brassicas (drench treatment) [6]
- HI cane fruit, strawberries 7 d [1, 2, 4, 6]; apples, pears, plums, currants, gooseberries, carrots, cereals 14 d; other crops 21 d [1, 2, 4, 6]; apples, pears, plums, currants, gooseberries, raspberries, strawberries, carrots, cereals, maize, grass, 14 d; peas, oilseed rape, brassicas, potatoes, swedes, turnips 21 d [5]; brassicas 6 wk [7]; brassicas, maize, sweetcorn, peas, beetroot, celery, leeks, onions, parsnips, potatoes 21 d [6]

Approval
- Off-label approval unlimited for aerial application to control Colorado beetle in potatoes as required by Statutory Notice (OLA 0796/91) [2]; to March 1988 for use on radish and mooli (OLA 0248/93) [2]; unlimited for use on a wide range of seedling brassicas and vegetables (OLA 0862/95)[4]

Maximum Residue Levels (mg residue/kg food)
• kiwi fruit 2; pome fruits, grapes, tomatoes, peppers, aubergines 0.5; citrus fruits 0.3; plums, strawberries 0.2; carrots, tea 0.1; tree nuts, apricots, bilberries, cranberries, currants, gooseberries, wild berries, avocados, dates, figs, kumquats, litchis, mangoes, passion fruit, pineapples, pomegranates, celeriac, horseradish, Jerusalem artichokes, parsley root, radishes, salsify, sweet potatoes, yams, garlic, shallots, spring onions, cucurbits, kohlrabi, cress, spinach, beet leaves, watercress, witloof, mushrooms, pulses, oilseeds, potatoes, cereals, poultry 0.05; milk, eggs 0.01

128   chlorpyrifos + dimethoate
A contact, systemic and fumigant insecticide for brassica crops

Products	Atlas Sheriff	Atlas	3.6:3.6% w/w	GR	04114

Uses   Aphids, cabbage root fly, cutworms, leatherjackets, wireworms in BROCCOLI, BRUSSELS SPROUTS, CABBAGES, CAULIFLOWERS, KALE, OILSEED RAPE, SWEDES, TURNIPS.

Notes   **Efficacy**
• Apply with suitable granule applicator (see label for details) as surface or sub-surface band treatment. Check calibration before applying
• May be used on drilled or transplanted crops and on nursery beds
• On nursery beds apply as overall treatment
• Apply by mid-Apr or at time of drilling or planting and repeat if necessary

Special precautions/Environmental safety
• Product contains organophosphorus compound. Do not use if under medical advice not to work with such compounds
• Dangerous to fish or other aquatic life. Do not contaminate surface waters or ditches with chemical or used container
• Harmful to game, wild birds and animals

Protective clothing/Label precautions
• A, C
• 1, 14, 18, 29, 35, 36, 37, 46, 52, 63, 67, 78

129   chlorpyrifos + disulfoton
A systemic and contact organophosphorus insecticide for brassicas

Products	Twinspan	PBI	4:6% w/w	GR	02255

Uses   Aphids, cabbage root fly in BROCCOLI, BRUSSELS SPROUTS, CABBAGES, CAULIFLOWERS.

FOR FULL CONDITIONS OF USE ALWAYS READ THE PRODUCT LABEL

Notes

Efficacy
- Apply with suitable band applicator as bow-wave treatment at drilling followed if necessary by surface band treatment 2 d after singling, or as sub-surface band treatment at transplanting. See label for details
- Can be used on all mineral and organic soils
- Crops treated in plant-raising bed must be treated again at transplanting
- Effect of surface band application may be reduced in dry weather

Crop Safety/Restrictions
- Do not treat mini-cauliflowers

Special precautions/Environmental safety
- This product contains an anticholinesterase organophosphorus compound. Do not use if under medical advice not to work with such compounds
- Harmful if swallowed and by inhalation
- Keep in original container, tightly closed, in a safe place, under lock and key
- Dangerous to game, wild birds and animals
- Dangerous to fish or other aquatic life. Do not contaminate surface waters or ditches with chemical or used container

Protective clothing/Label precautions
- A, B, C or D+E, H, K, M
- 1, 5b, 5c, 14, 16, 18, 22, 25, 26, 27, 28, 29, 35, 36, 37, 45, 52, 64, 67, 70, 78

Latest application/Harvest Interval(HI)
- HI 6 wk

Maximum Residue Levels (mg residue/kg food)
- see chlorpyrifos entry

---

130   chlorpyrifos-methyl

An organophosphorus insecticide and acaricide for grain store use

Products

1 Reldan 50	DowElanco	500 g/l	EC	05742
2 Smite	AgrEvo Environ.	300 g/l	EC	H5142

Uses

Flies in REFUSE TIPS [2]. Grain storage pests in STORED GRAIN [1]. Pre-harvest hygiene in GRAIN STORES [1]. Storage pests in STORED OILSEED RAPE [1].

Notes

Efficacy
- May be applied pre-harvest to surfaces of empty store and grain handling machinery and as admixture with grain [1]
- Apply to grain after drying to moisture content below 14%, cooling and cleaning [1]
- Insecticide may become depleted at grain surface if grain or rapeseed is being cooled by continuous extraction of air from the base leading to reduced mite control [1]
- Resistance to organophosphorus compounds sometimes occurs in insect and mite pests of stored products [1]
- Apply directly onto the surface of exposed refuse. Retreat as necessary [2]

Crop Safety/Restrictions
- Maximum number of treatments 1 per crop [1]

Special precautions/Environmental safety
- This product contains an anticholinesterase organophosphorus compound. Do not use if under medical advice not to work with such compounds
- Irritating to eyes and skin
- Flammable
- Dangerous to fish or other aquatic life. Do not contaminate surface waters or ditches with chemical or used container

Protective clothing/Label precautions
- A, C, H, M [1]; B, C, H [2]
- 1, 21, 25, 29, 36, 37, 63, 66, 70 [1]; 2, 14, 38, 39, 67 [2]; 6a, 6b, 12c, 18, 28, 52 [1, 2]

Latest application/Harvest Interval(HI)
- Treated barley must not be used for malting within 8 wk of application

Maximum Residue Levels (mg residue/kg food)
- pome fruits, strawberries, tomatoes, peppers, aubergines 0.5; grapes 0.2; tea, hops 0.1; grapefruit, limes, pomelos, tree nuts, apricots, cherries, plums, cane fruits, bilberries, cranberries, currants, gooseberries, wild berries, miscellaneous fruits, root and tuber vegetables, bulb vegetables, cucurbits, brassicas, leaf vegetables, legumes, stem vegetables, mushrooms, pulses, oilseeds, potatoes, rice, meat 0.05; milk, eggs 0.01

131   chlorsulfuron

A sulfonylurea herbicide formulations of which were voluntarily withdrawn in the UK in 1988

132   chlorthal-dimethyl

A residual benzoic acid herbicide for use in horticulture

Products					
1 Dacthal W-75	Hortichem	75% w/w	WP	05500
2 Dacthal W-75	ISK Biosciences	75% w/w	WP	05556

Uses   Annual dicotyledons in BLACKCURRANTS, BRASSICAS, GOOSEBERRIES, LEEKS, MUSTARD, OILSEED RAPE, ONIONS, ORNAMENTALS, RASPBERRIES, ROSES, RUNNER BEANS, SAGE, STRAWBERRIES, SWEDES, TURNIPS [1, 2]. Annual dicotyledons in OUTDOOR HERBS *(off-label)* [2]. Slender speedwell in TURF [1, 2].

Notes   **Efficacy**
- Best results on fine firm weed-free soil when adequate rain or irrigation follows
- Recommended alone on roses, runner beans, various ornamentals, strawberries and turf, in tank-mix with propachlor on brassicas, onions, leeks, sage, ornamentals, established strawberries and newly planted soft fruit
- Apply after drilling or planting prior to weed emergence. Rates and timing vary with crop and soil type. See label for details
- Do not use on organic soils

FOR FULL CONDITIONS OF USE ALWAYS READ THE PRODUCT LABEL

- For control of slender speedwell in turf apply when weeds growing actively. Do not mow for at least 3 d after treatment

Crop Safety/Restrictions
- Maximum number of treatments 1 per crop for brassicas, oilseed rape, mustard, leeks, onions, sage, newly planted bush and cane fruit; 1 per season for established strawberries
- Apply to brassicas pre-emergence or after 3-4 true leaf stage
- Do not apply mixture with propachlor to newly planted strawberries after rolling or application of other herbicides
- Do not use on strawberries between flowering and harvest
- Many types of ornamental have been treated successfully. See label for details. For species of unknown susceptibility treat a small number of plants first
- Do not use on turf where bent grasses form a major constituent of sward
- Do not plant lettuce within 6 mth of application, seeded turf within 2 mth, other crops within 3 mth. In the event of crop failure deep plough before re-drilling or planting
- Do not use on dwarf French beans

Protective clothing/Label precautions
- 29, 54, 63, 67

Latest application/Harvest Interval(HI)
- Before crop emergence for sown crops

Approval
- Off-label approval unlimited for use in outdoor herbs (see OLA notice for list of species) (OLA 1288/93)[2]

133   chlorthal-dimethyl + propachlor
A contact and residual herbicide for onions and brassicas

Products	Decimate	ISK Biosciences	225:216 g/l	SC	05626

Uses	Annual dicotyledons, annual grasses in BROCCOLI, BRUSSELS SPROUTS, CABBAGES, CAULIFLOWERS, MUSTARD, ONIONS.

Notes

Efficacy
- Best results achieved by application to fine, firm weed-free soil when adequate rain or irrigation follows
- Apply after drilling or planting prior to weed emergence
- Do not use on soils with more than 10% organic matter

Crop Safety/Restrictions
- Maximum number of treatments 1 per crop
- Apply to onions pre-emergence or from post-crook to young plant stage
- Apply to brassicas after drilling but pre-emergence, after 3-4 leaf stage or at any time after transplanting
- Pre-emergence treatment may check brassica growth but effect normally outgrown
- Young brassicas raised under glass should be hardened off before treatment. Do not use on protected crops
- Do not plant lettuce within 6 mth of application, seeded turf within 2 mth, other crops within 3 mth. In the event of crop failure deep plough before re-drilling or planting

Special precautions/Environmental safety
* Irritating to eyes and skin

Protective clothing/Label precautions
* A, C
* 6a, 6b, 14, 16, 18, 21, 25, 28, 29, 36, 37, 54, 63, 66

134 choline chloride

A plant growth regulator available only in mixtures

See also chlormequat + choline chloride
        chlormequat + choline chloride + imazaquin

135 clodinafop-propargyl

A contact acting herbicide for annual grass weed control in cereals

Products	Topik 240EC	Ciba Agric.	240 g/l	EC	07488

Uses   Blackgrass, rough meadow grass, wild oats in DURUM WHEAT, RYE, SPRING WHEAT, TRITICALE, WINTER WHEAT.

Notes   **Efficacy**
* Spray in autumn, winter or spring from 1 true leaf stage (GS11) to before second node detectable (GS 32)
* Spray when majority of weeds have germinated but before competition reduces yield
* Optimum control achieved when all grass weeds emerged. Wait for delayed germination on dry or cloddy seedbed
* A mineral oil additive is recommended to give more consistent control of very high blackgrass populations or for late season treatments. See label for details
* Weed control not affected by soil type, organic matter or straw residues
* Control may be reduced if rain falls within 1 h of treatment

Crop Safety/Restrictions
* Maximum total dose 0.25 l/ha/crop
* Do not treat crops under stress or suffering from water-logging, pest attack, disease or frost
* Do not treat crops undersown with grass mixtures
* An interval of 7 d must occur before or after treatment with products containing MCPA, mecoprop, 2,4-D, 2,4-DB or dichlorprop
* Only a broad leaved crop may be sown after failure of a treated crop. After normal harvest any broad leaved crop or wheat, durum wheat, rye, triticale or barley should be sown

Special precautions/Environmental safety
* Irritating to skin and eyes
* Flammable
* Harmful to fish or other aquatic life. Do not contaminate surface waters or ditches with chemical or used container

FOR FULL CONDITIONS OF USE ALWAYS READ THE PRODUCT LABEL

Protective clothing/Label precautions
* A, C, H, K
* 6a, 6b, 12c, 14, 18, 22, 29, 36, 53, 63, 67

Latest application/Harvest Interval(HI)
* Before second node detectable stage (GS 32)

---

136 clodinafop-propargyl + diflufenican

A contact and residual herbicide mixture for broad spectrum weed control in cereals

Products	Amazon TP	Ciba Agric.	240:500 g/l	KL	07681

Uses Annual dicotyledons, blackgrass, rough meadow grass, wild oats in DURUM WHEAT, TRITICALE, WINTER RYE, WINTER WHEAT.

Notes **Efficacy**
* Optimum weed control obtained when most grass weeds emerged and broad leaved weeds at susceptible stage before they compete with the crop
* Spray in autumn, winter or spring from first leaf unfolded stage (GS 11) to end Feb. Soil should be moist at time of treatment
* Delay treatment if dry or cloddy seedbeds favour late weed germination
* Grass weed control unaffected by seedbed condition but control of broad leaved weed sequires a fine, firm seedbed cleared of trash and straw
* Always use a recommended adjuvant/wetter. See label
* Speed of action depends on temperature and growing conditions and may appear slow in dry or cold weather

Crop Safety/Restrictions
* Maximum number of treatments 1 per crop
* Do not treat crops under stress or suffering from water-logging, pest attack, disease or frost
* Do not treat broadcast crops or those undersown with grass mixtures
* An interval of 7 d must occur before or after treatment with products containing MCPA, mecoprop, 2,4-D, 2,4-DB or dichlorprop
* Do not use on Sands, stony or gravelly soils or soil with over 10% organic matter
* Do not roll autumn treated crops until spring. Do not harrow at any time after treatment
* Only listed crops may be sown after failure of a treated crop - see label.
* After normal harvest cereals, field beans, oilseed rape, onions, leaf brassicas or sugar beet seed crops may be sown in the autumn. Land must be ploughed or cultivated to 15 cm except for cereals
* Successive treatments of any products containing diflufenican can lead to soil build up and inversion ploughing must precede sowing any non-cereal crop. Even where ploughing occurs some crops may be damaged - see label

Special precautions/Environmental safety
* Irritating to skin and eyes
* Harmful to fish or other aquatic life. Do not contaminate surface waters or ditches with chemical or used container

Protective clothing/Label precautions
* A, C, H, K

* 6a, 6b, 14, 18, 22, 25, 29, 36, 53, 63, 67

Latest application/Harvest Interval(HI)
* Before end of Feb in year of harvest

137 clodinafop-propargyl + trifluralin
A contact and residual herbicide for winter wheat

Products	Hawk	Ciba Agric.	240:480 g/l	KL	07614

Uses   Blackgrass, rough meadow grass, wild oats in WINTER WHEAT.

Notes

Efficacy
* Optimum weed control obtained when most grass weeds emerged and broad leaved weeds at susceptible stage before they compete with the crop
* Spray in autumn, winter or spring from first to third leaf unfolded stage (GS 11-13) but before fourth leaf unfolded (GS 14)
* Delay treatment if dry or cloddy seedbeds favour late weed germination
* Optimum weed control obtained in crops growing in a fine, firm seedbed cleared of trash and straw.
* Always use a recommended adjuvant/wetter. See label
* Speed of action depends on temperature and growing conditions and may appear slow in dry or cold weather

Crop Safety/Restrictions
* Maximum number of treatments 1 per crop
* Do not treat crops under stress or suffering from water-logging, pest attack, disease or frost
* Do not treat crops undersown with grass mixtures
* An interval of 7 d must occur before or after treatment with products containing MCPA, mecoprop, 2,4-D, 2,4-DB or dichlorprop
* Do not use on Sands, stony or gravelly soils or soils with over 10% organic matter
* After normal harvest only a broad leaved crop, wheat, durum wheat, rye or triticale should be sown. Before drilling or planting subsequent crops soil must be mouldboard ploughed to 15 cm
* After failure of a treated crop sow only a broad leaved crop (not sugar beet). 12 mth must elapse after treatment before sugar beet is drilled

Special precautions/Environmental safety
* Irritating to skin
* Flammable
* Harmful to fish or other aquatic life. Do not contaminate surface waters or ditches with chemical or used container

Protective clothing/Label precautions
* A, C, H, K
* 6b, 12c, 14, 18, 22, 25, 29, 36, 53, 63, 67

FOR FULL CONDITIONS OF USE ALWAYS READ THE PRODUCT LABEL

Latest application/Harvest Interval(HI)
* Before fourth crop leaf unfolded (GS 14)

138 clofentezine

A selective ovicidal tetrazine acaricide for use in top fruit

Products	Apollo 50 SC	Promark	500 g/l	SC	07242

Uses

Red spider mites in APPLES, BLACKCURRANTS *(off-label)*, CHERRIES, PEARS, PLUMS, RASPBERRIES *(off-label)*, STRAWBERRIES *(off-label)*.

Notes

Efficacy
* Acts on eggs and early motile stages of mites. For effective control total cover of plants is essential, particular care being needed to cover undersides of leaves
* For red spider mite control spray apples and pears between bud burst and pink bud, plums and cherries between white bud and first flower. Rust mite is also suppressed
* On established infestations apply in conjunction with an adult acaricide

Crop Safety/Restrictions
* Maximum number of treatments 1 per yr
* Product safe on predatory mites, bees and other predatory insects

Protective clothing/Label precautions
* 21, 29, 36, 37, 54, 60, 63, 67

Latest application/Harvest Interval(HI)
* 28 d for apple and pear, 8 wk for plum and cherry

Approval
* Off-label approval unlimited for use on blackcurrants, strawberries, raspberries (OLA 1250/95)[1]

139 clopyralid

A foliar, translocated picolinic herbicide for beets, brassicas etc
See also benazolin + clopyralid
*   bromoxynil + clopyralid*
*   bromoxynil + clopyralid + fluroxypyr*

Products	Dow Shield	DowElanco	200 g/l	SL	05578

Uses

Annual dicotyledons, corn marigold, creeping thistle, mayweeds, perennial dicotyledons in BROCCOLI, BRUSSELS SPROUTS, CABBAGES, CALABRESE, CAULIFLOWERS, CEREALS, ESTABLISHED GRASSLAND, FODDER BEET, FODDER RAPE, KALE, LINSEED, MAIZE, MANGELS, OILSEED RAPE, ONIONS, RED BEET, STRAWBERRIES, SUGAR BEET, SWEDES, SWEETCORN, TURNIPS, WOODY ORNAMENTALS. Annual dicotyledons, perennial dicotyledons in BROADLEAVED TREES *(off-label)*, CONIFERS *(off-label)*, GRASS SEED CROPS *(off-label)*, SAGE *(off-label)*. Annual dicotyledons in OUTDOOR HERBS *(off-label)*. Creeping thistle in ESTABLISHED GRASSLAND *(off-label, spot treatment)*. Mayweeds, thistles in HONESTY *(off-label)*.

Notes

Efficacy
* Best results achieved by application to young weed seedlings. Treat creeping thistle at rosette stage and repeat 3-4 wk later as directed
* High activity on weeds of Compositae family. For most crops recommended for use in tank mixes. See label for details
* Do not apply when crop damp or when rain expected within 6 h

Crop Safety/Restrictions
* Maximum total dose 2.0 l/ha/crop for winter oilseed rape; 1.5 l/ha/crop for beet crops, vegetable brassicas, fodder rape, swedes, turnips, onions, maize, sweetcorn, spring oilseed rape, strawberries; 1.0 l/ha/crop for established grassland, ornamental trees and shrubs; 0.5 l/ha/crop for linseed; 0.35 l/ha/crop for cereals
* Timing of application varies with weed problem, crop and other ingredient of tank mixes. See label for details
* Do not apply to cereals later than the second node detectable stage (GS 32)
* Do not use straw from treated cereals in compost or any other form for glasshouse crops. Straw may be used for strawing down strawberries
* Straw from treated grass seed crops or linseed should be baled and carted away. If incorporated do not plant winter beans in same year
* Do not use on onions at temperatures above 20°C
* Do not treat maiden strawberries or runner bed or apply to early leaf growth during blossom period or within 4 wk of picking. Aug or early Sep sprays may reduce yield
* Apply as directed spray in woody ornamentals, avoiding leaves, buds and green stems. Do not apply in root zone of families Compositae or Papilionaceae
* Do not plant susceptible autumn-sown crops in same year as treatment. Do not apply later than Jul where susceptible crops are to be planted in spring. See label for details

Special precautions/Environmental safety
* Do not contaminate surface waters or ditches with chemical or used container

Protective clothing/Label precautions
* A, C
* 21, 28, 29, 43 (7 d), 54, 60, 63, 66

Withholding period
* Keep livestock out of treated areas for at least 7 d and until foliage of any poisonous weeds such as ragwort has died and become unpalatable

Latest application/Harvest Interval(HI)
* Before 3rd node detectable (GS 33) for cereals; before flower buds visible from above for oilseed rape and linseed.
* HI grassland 7 d; strawberries 4 wk; maize, sweetcorn, onions, Brussels sprouts, broccoli, cabbage, cauliflowers, calabrese, kale, fodder rape, oilseed rape, swedes, turnips, sugar beet, red beet, fodder beet, mangels, sage, honesty 6 wk

Approval
* Off-label approval unlimited for use on established grassland (wiper application), grass seed crops (OLA 0662/92)[1], honesty (OLA 0480/93)[1]; to Jun 1997 for use on outdoor sage (OLA 0663/92)[1]; unlimited for use on coniferous and broadleaved trees in forestry (OLA 0757/92)[1]; to Mar 1999 for use on outdoor herbs (see OLA notice for list)(OLA 0514/95)[1]

FOR FULL CONDITIONS OF USE ALWAYS READ THE PRODUCT LABEL

140   clopyralid + diflufenican + MCPA
A selective herbicide for use in established turf

Products   Spearhead   RP Environ.   20:15:300 g/l   SL   07342

Uses   Annual dicotyledons, buttercups, corn marigold, creeping thistle, dandelions, mayweeds, perennial dicotyledons in TURF.

Notes   **Efficacy**
* Best results achieved by application when grass and weeds are actively growing
* Treatment during early part of the season is recommended, but not during drought

Crop Safety/Restrictions
* Maximum number of treatments 1 per yr
* Only use when sward is satisfactorily established and regular mowing has begun
* Turf sown in spring or early summer may be ready for treatment after 2 mth. Later sown turf should not be sprayed until growth is resumed in the following spring
* Avoid mowing within 3-4 d before or after treatment
* Do not use cuttings from treated area as a mulch for any crop
* Avoid drift. Small amounts of spray can cause serious injury to herbaceous plants, vegetables, fruit and glasshouse crops

Special precautions/Environmental safety
* Irritating to eyes and skin
* Harmful to fish or other aquatic life. Do not contaminate surface waters or ditches with chemical or used container

Protective clothing/Label precautions
* A, C
* 6a, 6b, 14, 18, 21, 28, 29, 36, 37, 43, 53, 63, 66

Withholding period
* Keep livestock out of treated areas

141   clopyralid + fluroxypyr + ioxynil
A contact and translocated herbicide mixture for use in cereals

Products   Hotspur   DowElanco   45:150:200 g/l   LI   05301

Uses   Annual dicotyledons, chickweed, cleavers, hemp-nettle, mayweeds, speedwells in BARLEY, DURUM WHEAT, OATS, RYE, TRITICALE, WHEAT.

Notes   **Efficacy**
* Best results achieved by application in good growing conditions, when weeds small and growing actively in a strongly competitive crop
* Do not spray if rain falling or imminent

Crop Safety/Restrictions
* Maximum number of treatments 1 per crop

- Apply to winter or spring cereals from 2-leaf stage to first node detectable (GS 12-31)
- Crops undersown with grass may be treated from tillering stage of grass. Do not spray crops undersown or about to be undersown with clover or other legumes
- Do not spray during drought, waterlogging, frost or extremes of temperature
- Do not use straw from treated crops in compost or any other form for glasshouse crops
- Do not apply with hand-held equipment or at concentrations higher than those recommended

Special precautions/Environmental safety
- Harmful if swallowed. Irritating to eyes and skin
- Do not apply by knapsack sprayer or at concentrations higher than recommended
- Flammable
- Dangerous to fish or other aquatic life. Do not contaminate surface waters or ditches with chemical or used container
- Harmful to bees. Do not apply to crops in flower or to those in which bees are actively foraging. Do not apply when flowering weeds are present

Protective clothing/Label precautions
- A, C
- 5c, 6a, 6b, 12c, 18, 21, 25, 28, 29, 36, 37, 43, 50, 52, 60, 63, 66, 70, 78

Withholding period
- Keep livestock out of treated areas until foliage of any poisonous weeds such as ragwort has died and become unpalatable

Latest application/Harvest Interval(HI)
- Before second node detectable (GS 32).
- HI 6 wk (animal consumption)

Approval
- Approval expiry: 28 Feb 95 [05301]

Maximum Residue Levels (mg residue/kg food)
- see ioxynil entry

---

142   clopyralid + fluroxypyr + triclopyr
A foliar acting herbicide mixture for grassland

Products	Pastor		DowElanco	50:108:139 g/l	EC	07440

Uses   Docks, stinging nettle, thistles in ESTABLISHED GRASSLAND.

Notes   **Efficacy**
- May be applied in spring or autumn depending on weeds present
- Treatment must be made when weeds and grass are actively growing
- Important to ensure sufficient leaf area for uptake especially on established docks and thistles

---

FOR FULL CONDITIONS OF USE ALWAYS READ THE PRODUCT LABEL

- On large well established docks and where there is a large soil seed reservoir further treatment in the following year may be needed
- To allow maximum translocation do not cut grass for 4 wk after treatment

Crop Safety/Restrictions
- Maximum number of treatments 1 per yr at full dose or 2 per yr at half dose
- Application during active growth ensures minimal check to grass
- Product may be used in established grassland which is under non-rotational setaside arrangements
- Do not spray in drought, very hot or very cold weather
- Do not treat grass less than 1 yr old or sports or amenity turf
- Product kills or severely checks clover and should not be used where clover is an important constituent of the sward
- Occasionally some yellowing of sward may occur after treatment which is quickly outgrown
- Do not roll or harrow 10 d before or 7 d after treatment
- Residues in incompletely decayed plant tissue may affect succeeding susceptible crops such as peas, beans and other legumes, carrots and related crops, potatoes, tomatoes, lettuce
- Do not plant susceptible autumn-sown crops in the same yr as treatment with product. Spring sown crops may follow if treatment was before end Jul in the previous yr
- Do not allow spray or drift to reach other crops, amenity plantings, gardens, ponds, lakes or water courses

Special precautions/Environmental safety
- Irritating to eyes and skin
- May cause sensitisation by skin contact
- Harmful to fish or other aquatic life. Do not contaminate surface waters or ditches with chemical or used container
- Wash out spray equipment thoroughly with water and detergent immediately after use. Traces of product could harm susceptible crops sprayed later

Protective clothing/Label precautions
- A, C
- 6a, 6b, 10a, 14, 18, 21, 28, 29, 34, 36, 37, 43 (7 d), 53, 63, 66, 70

Withholding period
- Keep livestock out of treated areas for at least 7 d following treatment and until foliage of poisonous weeds such as ragwort has died and become unpalatable

Latest application/Harvest Interval(HI)
- HI 7 d

---

143   clopyralid + propyzamide

A post-emergence herbicide for winter oilseed rape

Products					
1 Matrikerb	PBI	4.3:43% w/w	WP	01308
2 Matrikerb	Rohm & Haas	4.3:43% w/w	WP	02443

Uses

Annual dicotyledons, annual grasses, barren brome, mayweeds in WINTER OILSEED RAPE. Annual dicotyledons, annual grasses, mayweeds in EVENING PRIMROSE *(off-label)*.

Notes

Efficacy
- Apply from Oct to end Jan. Mayweed and groundsel may be controlled after emergence but before crop large enough to shield seedlings
- Best results achieved on fine, firm, moist soils when weeds germinating or small
- Do not use on soils with more than 10% organic matter
- Effectiveness reduced by surface organic debris, burnt straw or ash
- Do not apply if rainfall imminent or frost present on foliage

Crop Safety/Restrictions
- Maximum number of treatments 1 per crop
- Apply to crop as soon as possible after 3-true leaf stage (GS 1,3)
- Minimum period between spraying and drilling a following crop varies from 10 to 40 wk. See label for details

Protective clothing/Label precautions
- A, C
- 22, 28, 29, 35, 37, 54, 63, 67, 70

Latest application/Harvest Interval(HI)
- HI oilseed rape 6 wk; evening primrose 14 wk

Approval
- Off-label approval unlimited for use on evening primrose (OLA 0318/92)

144   clopyralid + triclopyr
A perennial and woody weed herbicide for use in grassland

Products   Grazon 90   DowElanco   60:240 g/l   EC   05456

Uses   Brambles, broom, docks, gorse, perennial dicotyledons, stinging nettle, thistles in ESTABLISHED GRASSLAND, GRASSY AREAS AROUND FARM. Perennial dicotyledons in ESTABLISHED GRASSLAND *(weed wiper - off-label)*.

Notes

Efficacy
- For good results must be applied to actively growing weeds
- Spray stinging nettle before flowering, docks in rosette stage in spring, creeping thistle before flower stems 15 cm high, brambles, broom and gorse in Jun-Aug
- Allow 2-3 wk regrowth after grazing or mowing before spraying perennial weeds
- Do not cut grass for 21 d before or 28 d after spraying

Crop Safety/Restrictions
- Maximum number of treatments 1 per yr
- Only use on permanent pasture or leys established for at least 1 yr
- Do not apply overall where clover is an important constituent of sward
- Do not roll or harrow within 7 d before or after spraying. Do not direct drill kale, swedes, turnips, grass or grass mixtures within 6 wk of spraying. Do not plant susceptible autumn-sown crops (eg winter beans) in same year as treatment. Do not spray after end July where susceptible crops to be planted next spring

FOR FULL CONDITIONS OF USE ALWAYS READ THE PRODUCT LABEL

• Do not allow drift onto other crops, amenity plantings or gardens

Special precautions/Environmental safety
• Irritating to eyes and skin
• Not to be used on food crops
• Dangerous to fish or other aquatic life. Do not contaminate surface waters or ditches with chemical or used container

Protective clothing/Label precautions
• A, C, H, M
• 6a, 6b, 14, 18, 21, 28, 29, 34, 36, 37, 43, 57, 63, 70

Withholding period
• Keep livestock out of treated areas for at least 7 d after spraying and until foliage of any poisonous weeds such as ragwort or buttercup has died down and become unpalatable

Latest application/Harvest Interval(HI)
• 7 d before grazing or harvest

Approval
• Off-label approval unlimited for use on established grassland via a tractor mounted/drawn weed wiper (OLA 0692/95)[1]

---

145   copper ammonium carbonate
A protectant copper fungicide

Products				
	Croptex Fungex	Hortichem	8.2% w/w (copper) SL	02888

Uses
Blight, leaf mould in OUTDOOR TOMATOES, TOMATOES. Cane spot in LOGANBERRIES, RASPBERRIES. Celery leaf spot in CELERY. Currant leaf spot in BLACKCURRANTS. Damping off in SEEDLINGS OF ORNAMENTALS. Leaf curl in PEACHES. Powdery mildew in CHRYSANTHEMUMS, CUCUMBERS.

Notes

Efficacy
• Apply spray to both sides of foliage
• With protected crops keep foliage dry before and after spraying

Crop Safety/Restrictions
• Maximum number of treatments 5 per crop for celery; 3 per yr for blackcurrants and loganberries; 2 per yr for peaches and raspberries
• Do not spray plants which are dry at the roots
• Ventilate glasshouse immediately after spraying

Special precautions/Environmental safety
• Harmful if swallowed. Risk of serious damage to eyes
• Harmful to fish or other aquatic life. Do not contaminate surface waters or ditches with chemical or used container
• Harmful to livestock

Protective clothing/Label precautions
• 5c, 9, 18, 29, 36, 37, 41, 53, 63, 65, 70, 78

Withholding period
• Keep all livestock out of treated areas for at least 3 wk. Bury or remove spillages

146   copper oxychloride
A protectant copper fungicide and bactericide

Products

1 Cuprokylt	Unicrop	50% w/w (copper)	WP	00604
2 Cuprokylt L	Unicrop	270 g/l (copper)	SC	02769
3 Cuprosana H	Unicrop	6% w/w (copper)	DP	00605
4 Headland Inorganic Liquid Copper	Headland	256 g/l	SC	07799

Uses

Bacterial canker in CHERRIES [1, 4]. Bacterial canker in PLUMS [1]. Blight in OUTDOOR TOMATOES, POTATOES [1, 4]. Buck-eye rot, damping off, foot rot in TOMATOES [1, 4]. Cane spot in CANE FRUIT [1, 4]. Canker in APPLES, PEARS [1, 2, 4]. Celery leaf spot in CELERY [1, 4]. Collar rot in APPLES *(off-label)* [2]. Downy mildew in HOPS [1, 3, 4]. Downy mildew in GRAPEVINES [1, 4]. Leaf curl in PEACHES [1]. Rust in BLACKCURRANTS [1, 4]. Spear rot in CALABRESE *(off-label)* [1].

Notes

Efficacy
• Spray crops when foliage dry. Do not spray if rain expected soon
• Spray interval commonly 10-14 d but varies with crop, see label for details
• If buck-eye rot occurs, spray soil surface and lower parts of tomato plants to protect unaffected fruit [1]
• To break cycle of bacterial canker infection apply drenching spray mid-Aug, mid-Sep and mid-Oct [1]
• Increase dose and volume with growth of hops. Complete cover is essential

Crop Safety/Restrictions
• Maximum number of treatments 3 per yr for apples, pears, cherries, blackberries (1 per yr for paintbrush or coarse spray on apples for collar rot control); 2 per crop for calabrese
• Some peach cultivars are sensitive to copper. Seek advice before spraying [1]
• Slight damage may occur to leaves of cherries and plums [1]

Special precautions/Environmental safety
• Harmful to livestock.
• Harmful to fish or other aquatic life. Do not contaminate surface waters or ditches with chemical or used container

Protective clothing/Label precautions
• 28 [3]; 29 [1, 3, 4]; 30 [2]; 41 [1, 2, 4]; 53, 63, 65 [1-4]

Withholding period
• Keep all livestock out of treated areas for at least 3 wk [1, 2, 4]

Latest application/Harvest Interval(HI)
• Before bud burst for apples and pears.
• HI calabrese 3 d [1]

FOR FULL CONDITIONS OF USE ALWAYS READ THE PRODUCT LABEL

Approval
- Approved for aerial application on potatoes. See notes in Section 1 [1]
- Off-label approval unlimited for use on apples (OLA 0328/92, 0329/92) [2]; to Apr 1997 for use on apples to control collar rot (OLA 0117/88) [2]; unlimited for use on calabrese (OLA 0993/92) [1]

147   copper oxychloride + maneb + sulfur
A protectant fungicide and yield stimulant for wheat and barley

Products					
1 Ashlade SMC Flowable	Ashlade	10:160:640 g/l	SC	06494
2 Tripart Senator Flowable	Tripart	10:160:640 g/l	SC	04561

Uses

Glume blotch, leaf spot, mildew in WHEAT [2]. Leaf blotch, mildew in BARLEY [2]. Leaf spot, powdery mildew in WINTER WHEAT [1].

Notes

Efficacy
- Use in a protectant programme covering period from first node detectable to beginning of ear emergence (GS 31-51), see label for details
- Treatment has little effect on established disease
- In addition to fungicidal effects treatment can also give nutritional benefits
- Do not apply when foliage is wet or rain imminent

Crop Safety/Restrictions
- Maximum number of treatments 3 per crop
- Do not apply to crops suffering stress from any cause
- Do not roll or harrow within 7 d of treatment
- Do not apply with any trace elements, liquid fertilizers or wetting agent

Special precautions/Environmental safety
- Irritating to skin and eyes
- Harmful to fish or other aquatic life. Do not contaminate surface waters or ditches with chemical or used container

Protective clothing/Label precautions
- A, C [1, 2]
- 5b, 5c, 28, 67, 70, 78 [2]; 6a, 6b, 18, 21, 29, 36, 37, 53, 63 [1, 2]; 66 [1]

Latest application/Harvest Interval(HI)
- Before ear fully emerged (GS 59)

Maximum Residue Levels (mg residue/kg food)
- see maneb entry

148   copper oxychloride + metalaxyl
A systemic and protectant fungicide mixture

Products					
Ridomil Plus 50 WP	Ciba Agric.	35:15% w/w	WP	01803

Uses	Collar rot in APPLES. Downy mildew in BROCCOLI *(off-label)*, BRUSSELS SPROUTS *(off-label)*, CABBAGES *(off-label)*, CAULIFLOWERS *(off-label)*, CUCUMBERS *(off-label)*, GRAPEVINES *(off-label)*, HOPS. Red core in STRAWBERRIES.

Notes	**Efficacy**
	• Drench soil at base of apples in Sep to Dec or Mar in first 2 yr after planting only
	• Apply drench to maiden strawberries immediately after planting and to established plants immediately new growth begins in autumn
	• Spray hops to run-off. Increase dose and volume with hop growth

Crop Safety/Restrictions
• Maximum number of treatments 3 per crop for brassicas; 1 per yr for strawberries; 6 per yr for grapevines; 8 per yr for hops, cucumbers
• Do not use on fruiting apple trees
• Use only on hops when well established

Special precautions/Environmental safety
• Irritating to eyes and skin
• Harmful to fish or other aquatic life. Do not contaminate surface waters or ditches with chemical or used container

Protective clothing/Label precautions
• A, C
• 6a, 6b, 18, 22, 28, 29, 35, 36, 37, 41, 53, 63, 67, 70

Withholding period
• Harmful to livestock. Keep all livestock out of treated areas for at least 3 wk. Bury or remove spillages

Latest application/Harvest Interval(HI)
• At planting for maiden strawberries, after harvest before 30 Nov for established strawberries.
• HI 14 d

Approval
• Off-label Approval to Feb 1997 for use in grapevines, Brussels sprouts, cabbages, broccoli, cauliflowers, protected cucumbers (OLA 0166/92)

Maximum Residue Levels (mg residue/kg food)
• see metalaxyl entry

---

149   copper sulfate
See Bordeaux Mixture

---

150   copper sulfate + sulfur
A contact fungicide for powdery and downy mildew control

Products	Top-Cop	Stoller	6:670 g/l	SC	04553

FOR FULL CONDITIONS OF USE ALWAYS READ THE PRODUCT LABEL

Uses   Blight in POTATOES, TOMATOES. Downy mildew, powdery mildew in HOPS. Mildew in GRAPEVINES, LEAF BRASSICAS, SWEDES, TURNIPS. Powdery mildew in SUGAR BEET.

Notes   **Efficacy**
* Apply at first blight warning or when signs of disease appear and repeat every 7-10 d as necessary
* Timing varies with crop and disease, see label for details

Crop Safety/Restrictions
* Do not apply to hops at or after the burr stage

Special precautions/Environmental safety
* Irritating to skin, eyes and respiratory system
* Dangerous to fish or other aquatic life. Do not contaminate surface waters or ditches with chemical or used container
* Harmful to livestock

Protective clothing/Label precautions
* 6a, 6b, 6c, 18, 21, 25, 26, 27, 29, 36, 37, 41, 52, 63, 65

Withholding period
* Keep livestock out of treated areas for at least 3 wk

151   coumatetralyl
An anticoagulant coumarin rodenticide

Products | | | | | |
|---|---|---|---|---|
| 1 Racumin Bait | Bayer | 0.0375% w/w | RB | 01679 |
| 2 Racumin Tracking Powder | Bayer | 0.75% w/w | TP | 01681 |

Uses   Rats in FARM BUILDINGS.

Notes   **Efficacy**
* Chemical is available as a bait and tracking powder and it is recommended that both formulations are used together
* Place bait in runs in sheltered positions near rat holes. Use at least 250 g per baiting point and examine at least every 2 d. Replenish as long as bait being eaten
* When no signs of activity seen for about 10 d remove unused bait
* Apply tracking powder in a 2.5-5 cm wide layer inside and across entrance to holes or blow into holes with dusting machine. Lay a 3 mm thick layer along runs in 30 cm long patches. Only use on exposed runs if well away from buildings

Special precautions/Environmental safety
* Cover bait or powder to prevent access by children, animals or birds

Protective clothing/Label precautions
* 28, 29, 63, 66, 67, 81, 82, 83, 84

152   cresylic acid

A contact fungicide, soil sterilant and insecticide

See also chlorpropham + cresylic acid + fenuron

Products

	1 Armillatox	Armillatox	30% v/v (phenol)	EC	06234
	2 Bray's Emulsion	Fargro	48% v/v	EC	00323

Uses

Algae, lichens, liverworts, mosses in PATHS [2]. Ants, slugs, woodlice in GLASSHOUSE STRUCTURES AND SURROUNDS [1, 2]. Canker, crown gall, honey fungus in TREES AND SHRUBS [1, 2]. Clubroot in BRASSICAS, WALLFLOWERS [1]. Mosses in TURF [1]. Overwintering pests in CANE FRUIT, CURRANTS, FRUIT TREES [2]. Rose replant disease in ROSES [1]. Soil-borne diseases in PROTECTED CROPS [2].

Notes

Efficacy
- To control honey fungus drench around the collar region of woody subjects. Avoid treating waterlogged or frozen soil. Apply a foliar and/or root feed in following season
- Plants should be re-treated annually to prevent reinvasion
- For canker control prune back branch to healthy wood, cut out diseased tissues and paint wound and surrounding area [2]
- For crown gall control loosen soil round collar and saturate area for 30 cm around plant [2]
- To control soil-borne diseases apply as drench in autumn as soon as crop removed
- Apply as winter wash in Dec or Jan to cover entire plant, with particular attention to cracks and crevices in bark [2]

Crop Safety/Restrictions
- Maximum number of treatments 1 per crop for brassicas
- Do not apply honey fungus control treatment to new plantings until established at least 12 mth. Minimize contact with feeding roots
- Do not replant sterilized areas around treated trees for 6-8 mth
- Do not allow spray to contact any green tissues
- Soil sterilized against damping off and foot rot may be planted after 7 wk
- Protect plants or grass under trees treated with winter wash to avoid scorch

Special precautions/Environmental safety
- Irritating to eyes, skin and respiratory system
- Dangerous to fish or other aquatic life. Do not contaminate surface waters or ditches with chemical or used container

Protective clothing/Label precautions
- A
- 6a, 6b, 6c, 18, 21, 28, 29, 36, 37, 52, 63, 65

Latest application/Harvest Interval(HI)
- 3 wk before planting for brassicas

FOR FULL CONDITIONS OF USE ALWAYS READ THE PRODUCT LABEL

153   cyanazine

A contact and residual triazine herbicide

See also bentazone + cyanazine + 2,4-DB

Products

1 Barclay Canter	Barclay	500 g/l	SC	07530
2 Fortrol	Cyanamid	500 g/l	SC	07009

Uses

Annual dicotyledons, annual grasses in ONIONS [1, 2]. Annual dicotyledons, annual grasses in BROAD BEANS *(Scotland only)*, CALABRESE *(off-label)*, COLLARDS *(off-label)*, FARM FORESTRY *(off-label)*, MAIZE, NARCISSI, POTATOES, SWEETCORN, TULIPS [2]. Annual dicotyledons, annual grasses in PEAS, SPRING FIELD BEANS, WINTER BARLEY, WINTER OILSEED RAPE, WINTER WHEAT [1, 2]. Annual dicotyledons, annual grasses in LINSEED, MAIZE/SWEETCORN, WARE POTATOES [1]. Annual dicotyledons in ONIONS *(organic fen soils)* [1]. Charlock in HONESTY *(off-label)* [2].

Notes

Efficacy

- Weeds controlled before emergence or at young seedling stage. Chemical must be carried into soil by rainfall
- Best results achieved when applied during mild, bright weather. Avoid applications in dull, cold or wet conditions
- Recommended for use alone, or in range of tank mixtures on cereals, rape, potatoes, linseed and peas. See label for details of tank mix partners and timings
- Residual activity normally persists for about 2 mth but on cereals all previous crop and weed residues must be removed or buried at least 15 cm deep
- Do not use as pre-emergence treatment on soils with more than 10% organic matter
- Some strains of blackgrass have developed resistance to many blackgrass herbicides which may lead to poor control

Crop Safety/Restrictions

- Maximum number of treatments 1 per crop (2 per crop for collards), unspecified for narcissi and tulips
- Apply pre- or post-emergence to winter cereals which must be drilled at least 25 mm deep. Do not spray within 7 d of rolling or harrowing
- Recommended for use on potatoes as tank mixture with pendimethalin, alone or as tank mixture on cereals, oilseed rape and peas (some varieties only). See label for details of timings, split applications and rates [2]
- Apply to oilseed rape after 1 Nov from 5-leaf stage when winter hardened. Do not apply after 31 Jan [2], or when the flower buds have become visible [1]
- Apply pre-emergence to potatoes as soon as possible after planting and at least 10 d before any shoots emerge
- Recommended pre- or post-emergence on some varieties of peas only. See label for details
- On onions only apply post-emergence to crops on fen soils with more than 10% organic matter after 2-true leaf stage
- On maize, sweetcorn, broad beans and field beans use as pre-emergence spray or as a pre-drilling incorporated treatment in maize or sweetcorn
- On flower bulbs treat pre- or early post-emergence [2]
- On linseed use on vigorously growing crops when 5-15 cm high and before the flower buds become visible
- Do not use on Sands, Very Light or stony soils
- Do not treat crops stressed by frost, waterlogging, drought, chemical damage etc. Heavy rain shortly after treatment may lead to damage, especially on lighter soils

Special precautions/Environmental safety
- Harmful if swallowed and in contact with skin
- Harmful to fish or other aquatic life. Do not contaminate surface waters or ditches with chemical or used container

Protective clothing/Label precautions
- A, C [1, 2]
- 5a, 5c, 18, 21, 28, 29, 36, 37, 53, 63, 70, 78 [1, 2]; 65 [2]; 66 [1]

Latest application/Harvest Interval(HI)
- Pre-emergence for potatoes, spring sown field beans, broad beans, maize and sweetcorn; before 31 Jan [2] or before flower buds visible [1] for winter oilseed rape; pre-emergence or before flower buds visible for peas; pre-emergence before end Oct or post-emergence before 2nd node detectable (GS 32) for cereals; before 5 true leaf stage for collards, before flower buds appear above the developing leaves for honesty.
- HI onions, leeks 8 wk, calabrese 11 wk

Approval
- Off-label Approval unlimited for use in farm forestry (OLA 0602/94) [2]

154 cyanazine + terbuthylazine
A foliar and soil acting herbicide for cereals

Products	Angle 567 SC	Ciba Agric.	306:261 g/l	SC	06254

Uses Annual dicotyledons, annual meadow grass in WINTER BARLEY, WINTER WHEAT.

Notes

Efficacy
- Best results obtained from application made early post-emergence of weeds
- Weeds germinating after application are controlled by root uptake, those present at application by leaf and root uptake
- Weed control may be reduced if heavy rain falls shortly after application or if there is prolonged dry weather
- Weeds germinating more than 2 mth after application may not be controlled
- Reduced weed control may result from presence of more than 10% organic matter or of surface ash and trash

Crop Safety/Restrictions
- Maximum number of treatments 1 per crop
- May be applied from 1 leaf unfolded on main shoot up to the 2 tiller stage (GS 11-22)
- Plant crop at normal depth of 25 mm; it is important that crop is well covered
- Do not apply to undersown crops or those due to be undersown
- Only spray healthy crops. Do not use on crops under stress or suffering from waterlogging, pest or disease attack, frost or when frost imminent
- Do not use on Sands or Very Light soils. There is a risk of crop damage on stony or gravelly soils, especially if heavy rain follows application
- Early crops (Sep drilled) may be prone to damage if spraying precedes or coincides with period of rapid growth in autumn

---

FOR FULL CONDITIONS OF USE ALWAYS READ THE PRODUCT LABEL

Special precautions/Environmental safety
- Harmful in contact with skin and if swallowed
- Irritating to skin and eyes
- May cause sensitization by skin contact
- Harmful to fish or other aquatic life. Do not contaminate surface waters or ditches with chemical or used container

Protective clothing/Label precautions
- A, C
- 5a, 5c, 6a, 6b, 10a, 21, 29, 53, 63, 66, 70

Latest application/Harvest Interval(HI)
- Before main shoot and 2-tiller stage (GS 22)

155   cycloxydim

A translocated post-emergence oxime herbicide for grass weed control

Products					
	1 Laser	BASF	200 g/l	EC	05251
	2 Stratos	BASF	200 g/l	EC	06891

Uses   Annual grasses, black bent, blackgrass, couch, creeping bent, volunteer cereals, wild oats in BRUSSELS SPROUTS, BULB ONIONS, CABBAGES, CARROTS, CAULIFLOWERS, DWARF BEANS, FIELD BEANS, FLOWER BULBS, FODDER BEET, LINSEED, MANGELS, PARSNIPS, PEAS, POTATOES, SALAD ONIONS, SPRING OILSEED RAPE, STRAWBERRIES, SUGAR BEET, SWEDES, WINTER OILSEED RAPE. Green cover in SETASIDE.

Notes   **Efficacy**
- Best results achieved when weeds small and have not begun to compete with crop. Effectiveness reduced by drought, cool conditions or stress. Weeds emerging after application are not controlled
- Foliage death usually complete after 3-4 wk but longer under cool conditions, especially late treatments to winter oilseed rape
- Perennial grasses should have sufficient foliage to absorb spray and should not be cultivated for at least 14 d after treatment
- On established couch pre-planting cultivation recommended to fragment rhizomes and encourage uniform emergence
- Split applications to volunteer wheat and barley at GS 12-14 will often give adequate control in winter oilseed rape. See label for details
- Apply to dry foliage when rain not expected for at least 2 h
- Must be used with Actipron adjuvant

Crop Safety/Restrictions
- Maximum number of treatments 1 per crop for spring oilseed rape; 2 per crop (the second at reduced dose) for other crops
- Maximum total dose for winter oilseed rape 2.25 l/ha per crop
- Recommended time of application varies with crop. See label for details
- On peas a crystal violet wax test should be done if leaf wax likely to have been affected by weather conditions or other chemical treatment. The wax test is essential if other products are to be sprayed after treatment

- May be used on ornamental bulbs when crop 5-10 cm tall. Product has been used on tulips, narcissi, hyacinths and irises but some subjects may be more sensitive and growers advised to check tolerance on small number of plants before treating the rest of the crop
- May be applied to setaside where the green cover is made up predominantly of tolerant crops listed on label. Use on industrial crops of linseed and oilseed rape on setaside also permitted.
- Do not apply to crops damaged or stressed by adverse weather, pest or disease attack or other pesticide treatment
- Prevent drift onto other crops, especially cereals and grass
- Guideline intervals for sowing succeeding crops after failed treated crop: field beans, peas, sugar beet, rape, kale, swedes, radish, white clover, lucerne 1 wk; onions, dwarf French beans 4 wk; wheat, barley, maize 6 wk; oats 8 wk

Special precautions/Environmental safety
- Irritating to skin and eyes
- Green cover treated on setaside must not be grazed by livestock or harvested for human or animal consumption

Protective clothing/Label precautions
- A, C
- 6a, 6b, 18, 21, 29, 36, 37, 54, 63, 66

Latest application/Harvest Interval(HI)
- Before canopy prevents adequate spray penetration.
- HI cabbage, cauliflower, salad onions 4 wk; peas, dwarf French beans 5 wk; bulb onions, carrots, parsnips, strawberries 6 wk; sugar and fodder beet, mangels, potatoes, field beans, swedes, Brussels sprouts 8 wk; winter and spring oilseed rape, linseed 12 wk

---

156   cyhexatin

An organotin acaricide approvals for use of which were revoked in November 1987

---

157   cymoxanil

A urea fungicide available only in mixtures
See also chlorothalonil + cymoxanil

---

158   cymoxanil + mancozeb

A protectant and systemic fungicide for potato blight control

Products					
1 Ashlade Solace	Ashlade	4.5:68% w/w	WP	06472
2 Curzate M	DuPont	4.5:68% w/w	WP	04343
3 Fytospore	Zeneca	5.25:71.6% w/w	WP	06517
4 Standon Cymoxanil Extra	Standon	4.5:68% w/w	WP	06807
5 Stefes Blight Spray	Stefes	4.5:68% w/w	WP	05811
6 Systol M	Quadrangle	4.5:68% w/w	WP	07098

FOR FULL CONDITIONS OF USE ALWAYS READ THE PRODUCT LABEL

Uses Blight in POTATOES [1-6]. Blight in POTATOES UNDER PROTECTION *(off-label)* [3].

Notes **Efficacy**
* Apply immediately after blight warning or as soon as local conditions dictate and repeat at 10-14 d intervals until haulm dies down or is burnt off
* Spray interval should not be more than 10 d in irrigated crops

Crop Safety/Restrictions
* Maximum number of treatments 6 per crop
* At least 10 days must elapse between treatments

Special precautions/Environmental safety
* Irritating to eyes, skin and respiratory system
* Harmful to fish or other aquatic life. Do not contaminate surface waters or ditches with chemical or used container

Protective clothing/Label precautions
* A, C [3]; A [1, 2, 4-6]
* 6a, 6b, 6c, 18, 21, 28, 29, 36, 37, 53, 63 [1-6]; 35 [6]; 65 [1]; 67 [2-6]

Latest application/Harvest Interval(HI)
* HI zero [2]; 7 d [5]

Approval
* Off-label Approval unlimited for blight control on potatoes grown under protection for mini and microtuber production (OLA 1560/93)[3]
* Approved for aerial application on potatoes. See notes in Section 1 [1-6]

Maximum Residue Levels (mg residue/kg food)
* see mancozeb entry

159   cymoxanil + mancozeb + oxadixyl
A systemic and contact protective fungicide for potatoes

Products

1 Ripost Pepite	Sandoz	3.2:56:8% w/w	WG	06485
2 Trustan	DuPont	3.2:56:8% w/w	WP	05022
3 Trustan WDG	DuPont	3.2:56:8% w/w	WG	05050

Uses Blight in POTATOES.

Notes **Efficacy**
* Apply first spray as soon as risk of blight infection or official warning. In absence of warning, spray just before crop meets within row
* Repeat spray every 10-14 d according to blight risk
* Do not treat crops already showing blight infection
* Do not apply within 2-3 h of rainfall or irrigation and only apply to dry foliage
* Use fentin based fungicide for final spray in blight control programme

Crop Safety/Restrictions
* Maximum number of treatments 5 per crop. No more than 5 sprays of any phenylamide based product should be applied in one season

Special precautions/Environmental safety
* Irritating to eyes, skin and respiratory system
* May cause sensitization by skin contact
* Harmful to fish or other aquatic life. Do not contaminate surface waters or ditches with chemical or used container

Protective clothing/Label precautions
* A, C
* 6a, 6b, 6c, 10a, 18, 23, 24, 28, 29, 36, 37, 53, 63, 67

Latest application/Harvest Interval(HI)
* Before end of Aug.
* HI zero

Approval
* Approved for aerial application on potatoes. See notes in Section 1

Maximum Residue Levels (mg residue/kg food)
* see mancozeb entry

160 cypermethrin
A contact and stomach acting pyrethroid insecticide

Products					
1 Ambush C	Zeneca	100 g/l		EC	06632
2 Ashlade Cypermethrin 10 EC	Ashlade	100 g/l		EC	06229
3 Barclay Cypersect XL	Barclay	100 g/l		EC	06509
4 Cyper 10	Quadrangle	100 g/l		EC	03242
5 Cyperkill 10	Chiltern	100 g/l		EC	04119
6 Cyperkill 5	Mitchell Cotts	50 g/l		EC	00625
7 Cypertox	FCC	100 g/l		EC	05122
8 Luxan Cypermethrin 10	Luxan	100 g/l		EC	06283
9 Ripcord	Cyanamid	100 g/l		EC	07014
10 Siege II	Chemsearch	0.1% w/w		ME	H5251
11 Standon Cypermethrin	Standon	100 g/l		EC	06818
12 Stefes Cypermethrin	Stefes	100 g/l		EC	05635
13 Stefes Cypermethrin 2	Stefes	100 g/l		EC	05719
14 Toppel 10	United Phosphorus	100 g/l		EC	06516
15 Vassgro Cypermethrin Insecticide	Vass	100 g/l		EC	03240

Uses Aphids, apple sucker, capsids, caterpillars, codling moth, sawflies, tortrix moths, winter moth in APPLES [1, 3, 5, 6, 8, 11-14]. Aphids, barley yellow dwarf virus vectors, yellow cereal fly in WINTER BARLEY, WINTER WHEAT [3, 6, 8]. Aphids, barley yellow dwarf virus vectors, yellow cereal fly in WINTER CEREALS [1, 2, 4, 5, 7-15]. Aphids, cabbage stem flea beetle, caterpillars, cutworms, whitefly in SWEDES [1, 3, 8, 11, 13, 14]. Aphids, cabbage stem flea beetle, caterpillars, mealy aphid, whitefly in CAULIFLOWERS [1, 3, 6, 7, 11]. Aphids, cabbage stem flea beetle, caterpillars, whitefly in BRUSSELS SPROUTS, CABBAGES [1, 3, 6, 7, 11, 14]. Aphids, cabbage stem flea beetle, caterpillars in KALE [6, 7, 12]. Aphids, cabbage stem flea beetle, caterpillars in BRASSICAS [2, 4, 5, 8]. Aphids, capsids, caterpillars, cutworms, leaf

FOR FULL CONDITIONS OF USE ALWAYS READ THE PRODUCT LABEL

miners, thrips, whitefly in ORNAMENTALS, PROTECTED ORNAMENTALS [1, 3, 11]. Aphids, capsids, caterpillars, pear sucker, suckers, tortrix moths in PEARS [1-3, 5-8, 11-15]. Aphids, caterpillars, cutworms in CELERY [1, 3, 11]. Aphids, caterpillars, cutworms in LETTUCE [1, 11]. Aphids, caterpillars, tortrix moths, winter moth in PLUMS [1, 3, 8, 12-15]. Aphids, caterpillars, tortrix moths, winter moth in CHERRIES [1, 3, 8, 13-15]. Aphids, pea and bean weevils, pea moth in PEAS [1-9, 11-15]. Aphids in DURUM WHEAT, RYE, TRITICALE, WINTER OATS [3]. Aphids in HOPS [1, 3, 7, 8, 11-14]. Asparagus beetle in ASPARAGUS *(off-label)* [1]. Cabbage stem flea beetle, rape winter stem weevil in WINTER OILSEED RAPE [3]. Cabbage stem flea beetle, rape winter stem weevil in OILSEED RAPE [1, 2, 4-9, 11-15]. Capsids, caterpillars, strawberry tortrix in STRAWBERRIES [1, 3, 5-8, 11, 14]. Capsids, gooseberry sawfly, sawflies in GOOSEBERRIES [1-3, 5-9, 11, 14]. Capsids, sawflies in BLACKCURRANTS, REDCURRANTS [1-3, 5, 7-9, 11, 14]. Caterpillars, mealy aphid, whitefly in BROCCOLI [1, 3, 11]. Caterpillars, mealy aphid, whitefly in CALABRESE [1]. Caterpillars, whitefly in FRENCH BEANS [1]. Cockroaches, flour beetles, grain beetles, moths in FOOD STORAGE AREAS [10]. Cutworms, leaf miners, whitefly in CUCUMBERS, PROTECTED CELERY, PROTECTED LETTUCE [1]. Cutworms in ROOT CROPS [5, 14]. Cutworms in CARROTS, PARSNIPS, POTATOES, SUGAR BEET [1, 3, 8, 11, 13]. Frit fly in GRASS RE-SEEDS [1, 3, 8, 11, 14]. Insect pests in ORNAMENTALS *(off-label)*, PROTECTED VEGETABLES *(off-label)* [6]. Leaf miners in ARTICHOKES *(protected crops, off-label)*, ASPARAGUS *(protected crops, off-label)*, AUBERGINES *(protected crops, off-label)*, BEANS *(protected crops, off-label)*, CELERY *(protected crops, off-label)*, CHICORY *(protected crops, off-label)*, CHINESE CABBAGE *(protected crops, off-label)*, COURGETTES *(protected crops, off-label)*, CUCUMBERS *(protected crops, off-label)*, ENDIVES *(protected crops, off-label)*, FENNEL *(protected crops, off-label)*, GARLIC *(protected crops, off-label)*, GHERKINS *(protected crops, off-label)*, LEEKS *(protected crops, off-label)*, LETTUCE *(protected crops, off-label)*, MARROWS *(protected crops, off-label)*, ONIONS *(protected crops, off-label)*, ORNAMENTALS *(protected crops, off-label)*, POTATOES *(protected crops, off-label)*, PROTECTED MELONS *(protected crops, off-label)*, PROTECTED SQUASHES *(protected crops, off-label)*, PUMPKINS *(protected crops, off-label)*, RHUBARB *(protected crops, off-label)*, SHALLOTS *(protected crops, off-label)*, SPINACH BEET *(protected crops, off-label)*, SPINACH *(protected crops, off-label)* [7]. Pea and bean weevils in BEANS [5, 8, 13, 14]. Pea and bean weevils in FIELD BEANS [3, 11]. Pod midge, pollen beetles, seed weevil in OILSEED RAPE *(restricted area)* [14].

Notes

Efficacy
- Products combine rapid action, good persistence, and high activity on Lepidoptera. Most organophosphorus and organochlorine-resistant pests are susceptible
- As effect is mainly via contact good coverage is essential for effective action. Spray volume should be increased on dense crops
- A repeat spray after 10-14 d is needed for some pests of outdoor crops, several sprays at shorter intervals for whitefly and other glasshouse pests
- Rates and timing of sprays vary with crop and pest. See label for details
- Add a non-ionic wetter to improve results on leaf brassicas. In Brussels sprouts use of a drop-leg sprayer may be beneficial

Crop Safety/Restrictions
- Maximum number of treatments varies with crop and product. See label or approval notice for details
- Test spray sample of new or unusual ornamentals before committing whole batches

Special precautions/Environmental safety
- Harmful in contact with skin or if swallowed [1-5, 7-9, 11-15]
- Irritating to eyes and skin [1-9, 11-15]; may cause sensitisation by skin contact [5]
- Flammable

- Dangerous to bees (Extremely dangerous to bees [5]). Do not apply to crops in flower or to those in which bees are actively foraging except as directed on peas. Do not apply when flowering weeds are present
- Extremely dangerous to fish or other aquatic life. Do not contaminate surface waters or ditches with chemical or used container
- Do not allow direct spray from ground based sprayer vehicles to fall within 6 m, or from hand-held sprayers within 2 m of surface waters or ditches. Direct spray away from water

Protective clothing/Label precautions
- A, C
- 5a [2, 4, 7, 9, 11, 14]; 5c, 10a, 16, 48a [5]; 6a, 6b, 12c, 18, 28, 48, 66 [1-9, 11-14]; 14, 21, 24, 51, 70, 78 [1, 2, 4-9, 11-14]; 23 [3, 5]; 29, 36, 37, 63 [1-14]; 50, 54a [3]; 5c [2, 4, 6, 7, 9, 11, 14]

Latest application/Harvest Interval(HI)
- Before end of Mar for winter cereals and oilseed rape (end of flowering for oilseed rape [5]), white bud stage for cherries and plums [11]

Approval
- Off-label Approval unlimited for use on protected celery, fennel, lettuce, endive, chicory, cucumbers, cucurbits (OLA 0591/92)[6]; to Mar 1999 for use on outdoor asparagus (OLA 0498/94)[1]

Maximum Residue Levels (mg residue/kg food)
- hops 30; citrus fruits, apricots, peaches, nectarines, wild berries, lettuces, herbs 2; Chinese cabbage, kale, wild mushrooms 1; grapes, cane fruits, tomatoes, peppers, aubergines, broccoli, cauliflowers, spinach, beet leaves, beans (with pods), leeks 0.5; cucumbers, gherkins, courgettes, kohlrabi, poppy seed, sesame seed, sunflower seed, rape seed, cotton seed, meat (except poultry) 0.2; garlic, onions, shallots 0.1; tree nuts, cranberries, miscellaneous fruits, root and tuber vegetables (except radishes, swedes, turnips), spring onions, sweetcorn, watercress, asparagus, cardoons, rhubarb, cultivated mushrooms, beans, peanuts, soya beans, mustard seed, potatoes, triticale, maize, rice, poultry, eggs 0.05; milk 0.02

161   cyproconazole
A contact and systemic conazole fungicide for cereals

See also carbendazim + cyproconazole
chlorothalonil + cyproconazole

Products					
1 Alto 100 SL	Sandoz	100 g/l	SL	05065
2 Aplan	Sandoz	100 g/l	SL	06121
3 Barclay Shandon	Barclay	100 g/l	SL	06464
4 Landgold Cyproconazole 100	Landgold	100 g/l	SL	06463
5 Standon Cyproconazole	Standon	100 g/l	SL	07751

Uses Brown rust, eyespot, glume blotch, powdery mildew, septoria diseases, septoria leaf spot, yellow rust in WINTER WHEAT. Brown rust, eyespot, net blotch, powdery mildew, rhynchosporium, yellow rust in SPRING BARLEY, WINTER BARLEY.

FOR FULL CONDITIONS OF USE ALWAYS READ THE PRODUCT LABEL

Notes

Efficacy
- Apply at start of disease development or as preventive treatment and repeat as necessary
- Most effective time of treatment varies with disease and use of tank mixes may be desirable. See label for details
- When applied early between the start of stem elongation and the fourth node detectable stage (GS 30-34) for control of other diseases, useful reduction of eyespot may be obtained

Crop Safety/Restrictions
- Maximum number of treatments 3 per crop
- Application to winter wheat in spring between the start of stem elongation and the third node detectable stage (GS 30-33) may cause straw shortening, but does not cause loss of yield

Special precautions/Environmental safety
- Harmful to fish or other aquatic life. Do not contaminate surface waters or ditches with chemical or used container

Protective clothing/Label precautions
- A
- 53, 63 [1-5]; 60 [1-3, 5]; 66 [4, 5]; 66; [1-3]

Latest application/Harvest Interval(HI)
- Up to and including beginning of anthesis (GS 61) for wheat; up to and including emergence of ear complete (GS 59) for barley

162   cyproconazole + prochloraz
A broad spectrum protective and curative fungicide for cereals

Products

1 Sportak Delta 460	AgrEvo	48:320 g/l	EC	07224
2 Sportak Delta 460	Schering	48:320 g/l	EC	05107

Uses

Brown rust, eyespot, glume blotch, leaf spot, powdery mildew, yellow rust in SPRING WHEAT, WINTER WHEAT. Brown rust, eyespot, leaf spot, powdery mildew, rhynchosporium, yellow rust in WINTER RYE. Brown rust, eyespot, net blotch, powdery mildew, rhynchosporium, yellow rust in SPRING BARLEY, WINTER BARLEY.

Notes

Efficacy
- To control foliar diseases apply before infection starts spreading to younger leaves and repeat if necessary
- For established mildew use in tank mixture with an approved morpholine fungicide eg tridemorph
- For eyespot control apply in spring from when leaf sheaths start to erect up to and including 3rd node detectable stage (GS 30-33). Effective against eyespot and *Septoria tritici* resistant to MBC fungicides
- See label for details of timing for foliar disease control
- When applied to control eyespot may give some control of sharp eyespot and Fusarium if present
- Good spray cover of stem bases and leaves is essential
- A period of at least 3 h without rain should follow spraying for full effectiveness

Crop Safety/Restrictions
- Maximum number of treatments 2 per crop

- Application to winter wheat in spring at GS 30-33 may cause straw shortening but does not cause loss of yield

Special precautions/Environmental safety
- Irritating to eyes and skin
- Dangerous to fish or other aquatic life. Do not contaminate surface waters or ditches with chemical or used container
- Flammable

Protective clothing/Label precautions
- A, C
- 6a, 6b, 18, 22, 29, 36, 37, 52, 63, 66

Latest application/Harvest Interval(HI)
- Up to and including emergence of ear complete (GS 59)

---

163 cyproconazole + tridemorph

A broad spectrum contact and systemic fungicide for cereals

Products					
1 Alto Major	Sandoz	80:350 g/l	EC	06979
2 Moot	Sandoz	80:350 g/l	EC	06990

Uses

Brown rust, eyespot, powdery mildew, septoria diseases, sooty moulds, yellow rust in WINTER WHEAT. Brown rust, powdery mildew, yellow rust in BARLEY. Eyespot, net blotch, rhynchosporium in WINTER BARLEY. Net blotch, rhynchosporium in SPRING BARLEY.

Notes

Efficacy
- Apply when disease is first active
- Most effective time of treatment varies with disease. Tank mixtures and/or repeat treatments may be desirable. See label for details
- When applied between GS 30-34 for control of other diseases, useful reduction of eyespot is achieved

Crop Safety/Restrictions
- Maximum number of treatments 2 per crop in the spring/summer (winter wheat, spring barley), 1 in autumn and 2 in spring/summer (winter barley)
- Application to winter wheat in spring at GS 30-33 may cause straw shortening but does not cause loss of yield

Special precautions/Environmental safety
- Harmful if swallowed. Irritating to eyes and skin
- Dangerous to fish or other aquatic life. Do not contaminate surface waters or ditches with chemical or used container

Protective clothing/Label precautions
- A, C [1, 2]
- 5c, 6a, 6b, 12c, 14, 18, 22, 29, 36, 37, 52, 63, 66, 70, 78 [1, 2]; 16, 41 [1]; 43 (14 d) [2]

FOR FULL CONDITIONS OF USE ALWAYS READ THE PRODUCT LABEL

Withholding period
* Keep livestock out of treated areas for at least 14 d following treatment

Latest application/Harvest Interval(HI)
* Up to and including emergence of ear complete (GS 59) (winter and spring barley), up to and including beginning of anthesis (GS 60) (winter wheat)

164   cyromazine
A triazine insect larvicide

Products   Neporex 2SG   Ciba Agric.   2% w/w   SG   06985

Uses   Flies in LIVESTOCK HOUSES, MANURE HEAPS.

Notes   **Efficacy**
* Granules can be scattered dry or applied in water
* Commence treatment when houseflies begin to breed
* Ensure that all breeding sites are treated
* Repeat after 2-3 wk for heavy populations
* Use an adulticide if large numbers of adult flies present

Crop Safety/Restrictions
* Do not use continuously in intensive or controlled environment animal units
* Do not feed treated poultry manure to stock
* Spread treated manure only onto grassland or arable land prior to cultivation
* Allow at least 4 wk between spreading and grazing or cropping
* Do not use on manure intended for production of mushroom compost

Special precautions/Environmental safety
* Harmful to fish. Do not contaminate surface waters or ditches with chemical or used container

Protective clothing/Label precautions
* A, B, C, H, M
* 18, 29, 36, 37, 53, 63, 65

165   2,4-D

A translocated phenoxy herbicide for use in cereals and grass

See also amitrole + 2,4-D + diuron

Products

1 Agricorn D	FCC	500 g/l	SL	00056
2 Atlas 2,4-D	Atlas	470 g/l	SL	03052
3 Barclay Haybob	Barclay	500 g/l	SL	07597
4 Campbell's Dioweed 50	MTM Agrochem.	500 g/l	SL	00401
5 Dicotox Extra	RP Environ.	400 g/l	EC	05330
6 Dormone	RP Environ.	465 g/l	SL	05412
7 Forester	Vitax	500 g/l	EC	00914
8 Headland Staff	Headland	470 g/l	SL	07189
9 HY-D	Agrichem	470 g/l	SL	06278
10 MSS 2,4-D Amine	Mirfield	500 g/l	SL	01391
11 MSS 2,4-D Ester	Mirfield	500 g/l	EC	01393
12 Syford	Vitax	500 g/l	SL	02062

Uses
Annual dicotyledons, heather, perennial dicotyledons, willows, woody weeds in CONIFER PLANTATIONS, FORESTRY [5, 11]. Annual dicotyledons, perennial dicotyledons in SPRING RYE [1, 4, 10]. Annual dicotyledons, perennial dicotyledons in SPORTS TURF [1, 4, 6, 7, 10, 12]. Annual dicotyledons, perennial dicotyledons in WINTER RYE [1, 2, 4, 8-10]. Annual dicotyledons, perennial dicotyledons in ESTABLISHED GRASSLAND [1-4, 7-9, 11, 12]. Annual dicotyledons, perennial dicotyledons in WHEAT [1-4, 6, 8-12]. Annual dicotyledons, perennial dicotyledons in BARLEY [1-4, 8-12]. Annual dicotyledons, perennial dicotyledons in AMENITY TURF [1, 3-7, 10-12]. Annual dicotyledons, perennial dicotyledons in GRASSLAND [10]. Annual dicotyledons, perennial dicotyledons in GRASS SEED CROPS [12]. Annual dicotyledons, perennial dicotyledons in WINTER OATS [2, 8-11]. Annual dicotyledons, perennial dicotyledons in SPRING OATS [9]. Annual dicotyledons, perennial dicotyledons in RYE [3, 11]. Aquatic weeds, perennial dicotyledons in WATER OR WATERSIDE AREAS [2, 6, 9]. Aquatic weeds in AQUATIC SITUATIONS [10].

Notes   **Efficacy**
* Best results achieved by spraying weeds in seedling to young plant stage when growing actively in a strongly competing crop
* Most effective stage for spraying perennials varies with species. See label for details
* Spray aquatic weeds when in active growth between May and Sep [2, 6]
* Do not spray if rain falling or imminent
* Do not cut grass or graze for at least 7 d after spraying

Crop Safety/Restrictions
* Maximum number of treatments 1 per crop; 1 per yr in forestry, 2 per yr (3 [11]) on amenity turf
* Spray winter cereals in spring when leaf-sheath erect but before first node detectable (GS 31), spring cereals from 5-leaf stage to before first node detectable (GS 15-31)
* Do not use on newly sown leys containing clover
* Do not spray grass seed crops after ear emergence
* Do not spray within 6 mth of laying turf or sowing fine grass

---

FOR FULL CONDITIONS OF USE ALWAYS READ THE PRODUCT LABEL

- Selective treatment of resistant conifers can be made in Aug when growth ceased and plants hardened off, spray must be directed if applied earlier. See label for details
- Do not plant conifers until at least 1 mth after treatment
- Do not spray crops stressed by cold weather or drought or if frost expected
- Do not use shortly before or after sowing any crop
- Do not direct drill brassicas or grass/clover mixtures within 3 wk of application
- Do not roll or harrow within 7 d before or after spraying

Special precautions/Environmental safety

- Harmful in contact with skin and if swallowed
- Irritating to eyes [2-4, 6, 7, 12]
- Harmful to fish or other aquatic life. Do not contaminate surface waters or ditches with chemical or used container
- May be used to control aquatic weed in presence of fish if used in strict accordance with directions for waterweed control and precautions needed for aquatic use [2, 6, 9, 10]
- Water containing the herbicide must not be used for irrigation purposes within 3 wk of treatment or until the concentration in water is below 0.05 ppm [2, 6, 9, 10]

Protective clothing/Label precautions

- A, C, D, H, M [1, 2, 4, 6-9, 12]; A, C, H, M [3, 5, 10, 11]
- 5a, 5c, 18, 21, 29, 36, 37, 53, 63, 66, 70, 78 [1-12]; 6a [2-7, 12]; 6b, 10a [5]; 43 (2 wk) [3, 5, 10, 11]; 43, 65 [1, 2, 4, 6-9, 12]; 55, 56 [2, 6, 9]

Withholding period

- Keep livestock out of treated areas for at least 2 wk and until foliage of any any poisonous weeds such as ragwort has died and become unpalatable

Latest application/Harvest Interval(HI)

- Before first node detectable stage (GS 31] in cereals

Approval

- May be applied through CDA equipment. See label for details. See notes in Section 1 on ULV application [12]
- Approved for aquatic weed control. See notes in Section 1 on use of herbicides in or near water [2, 6, 9]

---

166 2,4-D + dicamba

A translocated herbicide for use on turf

Products					
	1 Green-Up Weedfree Spot Weedkiller for Lawns	Vitax		AE	06321
	2 Keychem SWK 333	Keychem	200:35 g/l	EC	05045
	3 New Estermone	Vitax	200:35 g/l	EC	06336

Uses   Annual dicotyledons, perennial dicotyledons in AMENITY TURF, LAWNS.

Notes   **Efficacy**
- Best results achieved by application when weeds growing actively from Apr to Sep [1, 2], in spring or early summer (later with irrigation and feeding) [3]
- More resistant weeds may need repeat treatment after 3 wk
- Use aerosol formulation as a spot treatment [1]
- Do not use during drought conditions or mow for 3 d before or after treatment

Crop Safety/Restrictions
* Maximum number of treatments 2 per yr [2], 3 per yr [3]
* Do not treat turf established for less than 6 wk [1], newly sown or turfed areas [2-4]
* Avoid spray drift onto cultivated crops or ornamentals
* Do not re-seed for 6 wk after application [3]

Special precautions/Environmental safety
* Harmful to fish or other aquatic life. Do not contaminate surface waters or ditches with chemical or used container

Protective clothing/Label precautions
* A, C, H, M
* 21, 29, 53 [1-3]; 37 [1]; 43, 63, 65 [2, 3]

Withholding period
* Keep livestock out of treated areas for at least 2 wk and until foliage of any poisonous weeds such as ragwort has died and become unpalatable

167   2,4-D + dicamba + ferrous sulfate
A herbicide/fertilizer combination for moss and weed control in turf

Products				
1 Longlife Renovator	Zeneca Prof.	0.36:0.05:3.0% w/wMG	07697
2 Renovator	Zeneca Prof.	0.36:0.05:3.0% w/wMG	06860

Uses

Annual dicotyledons, mosses, perennial dicotyledons in AMENITY TURF.

Notes   **Efficacy**
* Apply from mid-Apr to mid-Aug when weeds are growing
* Apply with a suitable calibrated fertilizer distributor
* Where regrowth of moss or weeds occurs a repeat treatment may be made after 4-6 wk
* Do not cut grass at least 3 d before and at least 4 d after treatment
* Avoid treatment to wet grass or during drought. If no rain falls within 48 h water in thoroughly
* Do not apply during freezing conditions or when rain imminent

Crop Safety/Restrictions
* Do not apply to fine turf established less than 12 mth
* The first 4 mowings after treatment should not be used to mulch cultivated plants unless composted at least 6 mth
* Avoid walking on treated areas until it has rained or they have been watered

Special precautions/Environmental safety
* Harmful to fish and other aquatic life. Do not contaminate surface waters or ditches with chemical or used container
* Dangerous to aquatic higher plants

Protective clothing/Label precautions
* A, C, H, M

FOR FULL CONDITIONS OF USE ALWAYS READ THE PRODUCT LABEL

* 22, 29, 43 (2 wk), 53, 54, 63, 67

Withholding period
* Keep livestock out of treated areas for at least 2 wk following treatment and until foliage of any poisonous weeds such as ragwort has died and become unpalatable

---

168   2,4-D + dicamba + mecoprop
A translocated herbicide for perennial and woody weed control

Products   Weed and Brushkiller (New   Vitax   144:32:144 g/l   EC   07072
                Formulation)

Uses   Annual dicotyledons, brambles, perennial dicotyledons, stinging nettle, woody weeds in AMENITY GRASS, NON-CROP AREAS, ROUGH GRAZING.

Notes   **Efficacy**
* Mix with water for application as foliar spray. Apply Mar-Jul on perennials when in active growth, Jun-Sep on woody plants, May-Jul on gorse
* Do not spray in drought conditions
* Mix with paraffin or other light oil for application as basal bark treatment on established trees in Jul-Sep or as stump treatment in Jan-Mar

Crop Safety/Restrictions
* Avoid spray drift onto crops or ornamentals. Do not use in hot or windy conditions
* Do not cultivate or replant land for at least 6 wk after treatment
* Do not cut or graze for at least 7 d after treatment

Special precautions/Environmental safety
* Irritating to eyes
* Harmful to fish or other aquatic life. Do not contaminate surface waters or ditches with chemical or used container

Protective clothing/Label precautions
* A, C, D, H, M
* 6a, 18, 21, 29, 36, 37, 43, 53, 63, 65

Withholding period
* Keep livestock out of treated areas for at least 2 wk and until foliage of any poisonous weeds such as ragwort has died and become unpalatable

---

169   2,4-D + dicamba + triclopyr
A translocated herbicide for perennial and woody weed control

Products   Broadshot   Cyanamid   200:85:65 g/l   EC   07141

Uses   Annual dicotyledons, brambles, docks, gorse, japanese knotweed, perennial dicotyledons, rhododendrons, stinging nettle, thistles, woody weeds in ESTABLISHED GRASSLAND, FORESTRY, NON-CROP AREAS.

Notes	**Efficacy**
	• Apply as foliar spray to herbaceous or woody weeds. Timing and growth stage for best results vary with species. See label for details
	• Dilute with water for stump treatment, with paraffin for basal bark treatment
	• May be applied at 1/3 or 1/8 dilution with ropewick applicators
	• Do not roll or harrow grassland for 7 d before or after spraying

Crop Safety/Restrictions
- Do not use on pasture established less than 1 yr or on grass grown for seed
- Where clover a valued constituent of sward only use as a spot treatment
- Do not graze for 7 d or mow for 14 d after treatment
- Do not direct drill grass, clover or brassicas for at least 6 wk after treatment
- Sprays may be applied in pines, spruce and fir providing drift is avoided. Optimum time is mid-autumn when tree growth ceased but weeds not yet senescent
- Do not plant trees for 1-3 mth after spraying depending on dose applied. See label
- Avoid spray drift into greenhouses or onto crops or ornamentals. Vapour drift may occur in hot conditions

Special precautions/Environmental safety
- Harmful if swallowed. Irritating to eyes and skin
- Dangerous to fish or other aquatic life. Do not contaminate surface waters or ditches with chemical or used container

Protective clothing/Label precautions
- A, C, D, H, M
- 5c, 6a, 6b, 18, 21, 29, 36, 37, 43, 52, 63, 66, 70, 78

Withholding period
- Keep livestock out of treated area for at least 2 wk or until foliage of any poisonous weeds such as ragwort or buttercup has died and become unpalatable

---

170   2,4-D + dichlorprop + MCPA + mecoprop
A translocated herbicide for use in apple and pear orchards

Products	1 Campbell's New Camppex	MTM Agrochem.	34:133:53:164 g/l	SL	00414
	2 Camppex	United Phosphorus	34:133:53:164 g/l	SL	06809

Uses	Annual dicotyledons, chickweed, cleavers, perennial dicotyledons in APPLES, PEARS.

Notes	**Efficacy**
	• Use on emerged weeds in established apple and pear orchards (from 1 yr after planting) as directed application
	• Spray when weeds in active growth and at growth stage recommended on label
	• Effectiveness may be reduced by rain within 12 h
	• Do not roll, harrow or cut grass crops on orchard floor within at least 3 d before or after spraying

---

FOR FULL CONDITIONS OF USE ALWAYS READ THE PRODUCT LABEL

Crop Safety/Restrictions
* Applications must be made around, and not directly to, trees. Do not allow drift onto trees
* Do not spray during blossom period. Do not spray to run-off
* Avoid spray drift onto neighbouring crops

Special precautions/Environmental safety
* Harmful in contact with skin and if swallowed. Irritating to eyes
* Harmful to fish or other aquatic life. Do not contaminate surface waters or ditches with chemical or used container

Protective clothing/Label precautions
* A, C, D, H, M
* 5a, 5c, 6a, 18, 21, 29, 36, 37, 43, 53, 63, 66, 70, 78

Withholding period
* Keep livestock out of treated areas for at least 2 wk and until foliage of any poisonous weeds such as ragwort has died and become unpalatable

Maximum Residue Levels (mg residue/kg food)
* see dichlorprop entry

---

171   2,4-D + mecoprop
A translocated herbicide for use in turf and grassland

Products					
	1 BH CMPP/2,4-D	RP Environ.	116:250 g/l	SL	05393
	2 Cleanrun	Zeneca Prof.	0.26:0.55% w/w	MG	06857
	3 Longlife Cleanrun	Zeneca Prof.	0.26:0.55% w/w	MG	06914
	4 Mascot Selective Weedkiller	Rigby Taylor	78.5:200 g/l	SL	03423
	5 Nomix Turf Selective Herbicide	Nomix-Chipman	6.7:13.4% w/w	UL	06777
	6 Selective Weedkiller	Yule Catto	12.75:26.95% w/w	SL	06579
	7 Supertox 30	RP Environ.	90:190 g/l	SL	05340
	8 Sydex	Vitax	125:250 g/l	SL	06412
	9 Vitax Weed 'N' Feed Extra	Vitax	0.19:0.371% w/w	SA	06506

Uses

Annual dicotyledons, perennial dicotyledons in AMENITY TURF, LAWNS, SPORTS TURF [1, 4-8]. Annual dicotyledons, perennial dicotyledons in ESTABLISHED GRASSLAND, GRASS SEED CROPS [8]. Annual dicotyledons, perennial dicotyledons in TURF [2, 3].

Notes

Efficacy
* May be applied from Apr to Sep, best results in May-Jun when weeds in active growth
* Some species may need repeat treatment after 4-8 wk
* Do not spray during drought conditions or when rain imminent
* Do not mow for 2-4 d before or 1 d after treatment (3 d after on some labels)
* For best results apply fertiliser 1-2 wk before treatment
* Water thoroughly if no rain falls within 48 h of treatment [2, 3]

Crop Safety/Restrictions
* Do not spray ULV formulation at temperatures above 25-26°C [5]
* Do not use first 4 mowings for mulching unless composted for at least 6 mth (do not use first 2 mowings for composting on some labels)
* Do not resow treated areas for at least 6 wk (8 wk on some labels)

- Avoid spray drift onto nearby trees, shrubs, vegetables or flowers
- Avoid walking on treated areas until it has rained or they have been watered [2, 3]

Special precautions/Environmental safety
- Harmful in contact with skin [1], in contact with skin and if swallowed [4, 6, 8]
- Irritating to eyes [6-8]; Irritating to eyes and skin [4]; irritating to skin, eyes and respiratory system [5]
- Harmful to fish or other aquatic life. Do not contaminate surface waters or ditches with chemical or used container [1-3, 5, 6, 8]
- Dangerous to aquatic higher plants [2, 3]

Protective clothing/Label precautions
- A, C, D, H, M [1, 4, 6-9]; A, C, H, M [2, 3, 5]
- 5a, 78 [1, 4, 6, 8]; 5c [4, 6, 8]; 6a [4, 6-8]; 6b, 54 [4]; 14, 28 [5]; 18, 21, 36, 37 [1, 4-8]; 22 [2, 3]; 29, 63 [1-8]; 43 (2 wk), 67 [2, 3, 5]; 43, 65, 70 [1, 4, 6-8]; 53 [1-3, 5, 6, 8]

Withholding period
- Keep livestock out of treated areas for at least 2 wk and until foliage of any poisonous weeds such as ragwort has died and become unpalatable

Approval
- May be applied through CDA equipment. See label for details. See notes in Section 1 on ULV application [5]

172   2,4-D + picloram

A persistent translocated herbicide for non-crop land

Products					
Atladox HI		Nomix-Chipman	240:65 g/l	SL	05559

Uses

Annual dicotyledons, brambles, creeping thistle, docks, japanese knotweed, perennial dicotyledons, ragwort, scrub, woody weeds in AMENITY GRASS, NON-CROP GRASS, ROAD VERGES.

Notes

Efficacy
- Apply as overall foliar spray during period of active growth when foliage well developed

Crop Safety/Restrictions
- Maximum number of treatments 1 per yr
- Do not apply around desirable trees or shrubs where roots may absorb chemical
- Prevent leaching into areas where desirable plants are present
- Avoid drift of spray onto desirable plants
- Do not use cuttings from treated grass for mulching or composting

Special precautions/Environmental safety
- Irritating to eyes
- Flammable
- Harmful to fish or other aquatic life. Do not contaminate surface waters or ditches with chemical or used container

---

FOR FULL CONDITIONS OF USE ALWAYS READ THE PRODUCT LABEL

Protective clothing/Label precautions
* A, C, D, H, M
* 6a, 12c, 21, 29, 43, 53, 63, 66

Withholding period
* Keep livestock out of treated areas for at least 2 wk and until foliage of any poisonous weeds such as ragwort has died and become unpalatable

173   dalapon
A translocated chloroalkanoic acid grass herbicide no longer marketed alone but still approved for use

174   dalapon + dichlobenil
A persistent herbicide for use in woody plants and non-crop areas

Products	Fydulan G		Nomix-Chipman	10:6.75% w/w	GR	00958

Uses   Annual weeds, bracken, perennial grasses, rushes in FORESTRY, HEDGES. Annual weeds, perennial grasses in ROSES, WOODY ORNAMENTALS. Total vegetation control in NON-CROP AREAS.

Notes

Efficacy
* Apply evenly with suitable granule applicator. Best results achieved by application in late winter/early spring to slightly moist soil, especially if rain falls soon after
* May be used on all soil types. Do not disturb soil after application
* For forestry use apply as a 1 m wide band
* In non-crop situations apply before end of spring

Crop Safety/Restrictions
* Use in ornamentals before onset of spring growth, usually Feb-early Mar. See label for list of tolerant species
* Use in roses established for at least 2 yr before bud growth starts, usually early Feb
* Use in forestry before end of winter. Allow at least 2 mth before planting. Apply after planting as soon as soil settled
* Do not apply to areas underplanted with bulbs or herbaceous stock
* Do not apply near hops or glasshouses, nor if nearby crop foliage wet
* Store well away from bulbs, corms, tubers or seed

Special precautions/Environmental safety
* Irritating to eyes, skin and respiratory system

Protective clothing/Label precautions
* 6a, 6b, 6c, 22, 29, 36, 37, 63, 67

Approval
* Cleared for aquatic weed control. See introductory notes on use of herbicides in or near water

175   daminozide

A hydrazide plant growth regulator for use in certain ornamentals

Products

1 B-Nine	Hortichem	85% w/w	SP	04468
2 Dazide	Fine	85% w/w	SP	02691

Uses   Internode reduction in POT PLANTS [2]. Internode reduction in AZALEAS, BEDDING PLANTS, CHRYSANTHEMUMS, HYDRANGEAS [1, 2]. Internode reduction in POINSETTIAS [1].

Notes

Efficacy
* Best results obtained by application in late afternoon when glasshouse has cooled down
* Spray when foliage dry
* A tank mix of reduced rates of B-Nine and New 5C Cycocel is recommended for use on poinsettias - see label for details [1]

Crop Safety/Restrictions
* Maximum number of treatments 2 per crop
* Apply only to turgid, well watered plants. Do not water for 24 h after spraying
* Do not use on chrysanthemum Fandango
* Do not mix with other spray chemicals except as recommended above

Special precautions/Environmental safety
* Irritating to eyes [2]
* Harmful to fish or other aquatic life. Do not contaminate surface waters or ditches with chemical or used container

Protective clothing/Label precautions
* A, C, H
* 6a, 18, 22, 28, 29, 36, 37, 53, 63, 65

Approval
* Sales of daminozide for use on food crops were halted worldwide by the manufacturer in October 1989. Sales for use on flower crops were not affected

Maximum Residue Levels (mg residue/kg food)
* tea, hops 0.1; tree nuts, oilseeds, animal products 0.05; citrus fruits, pome fruits, stone fruits, berries and small fruit, miscellaneous fruits, vegetables, pulses, potatoes, cereals 0.02

176   dazomet

A methyl isothiocyanate releasing soil fumigant

Products

1 Basamid	BASF	98-99% w/w	GR	00192
2 Basamid	Hortichem	98-99% w/w	GR	07204

FOR FULL CONDITIONS OF USE ALWAYS READ THE PRODUCT LABEL

Uses Nematodes, soil insects, soil-borne diseases, weed seeds in FIELD CROPS, PROTECTED CROPS, SOIL LOAM OR TURF FOR COMPOST, VEGETABLES.

Notes

Efficacy
* Product acts by releasing methyl isothiocyanate in contact with moist soil
* Soil sterilization is carried out after harvesting one crop and before planting the next
* The soil must be of fine tilth, free of clods and evenly moist to the depth of sterilization
* Soil moisture must not be less than 50% of water-holding capacity or oversaturated. If too dry, water at least 7-14 d before treatment
* Do not treat ground where water table may rise into treated layer
* In order to obtain short treatment times it is recommended to treat soils when soil temperature is above 7°C. Treatment should be used outdoors before winter rains make soil too wet to cultivate - usually early Nov
* For club root control treat only in summer when soil temperature above 10°C
* Where onion white rot a problem unlikely to give effective control where inoculum level high or crop under stress
* Apply granules with suitable applicators, mix into soil immediately to desired depth and seal surface with polythene sheeting, by flooding or heavy rolling. See label for suitable application and incorporation machinery
* With 'planting through' technique polythene seal is left in place to form mulch into which new crop can be planted

Crop Safety/Restrictions
* Maximum number of treatments 1 per crop or batch of soil
* With 'planting through' technique no gas release cultivations are made and safety test with cress is particularly important. Conduct cress test on soil samples from centre as well as edges of bed. Observe minimum of 30 d from application to cress test in soils at 10°C or above, at least 50 d in soils below 10°C. See label for details
* In all other situations 14-28 d after treatment cultivate lightly to allow gas to disperse and conduct cress test after a further 14-28 d (timing depends on soil type and temperature). Do not treat structures containing live plants or any ground within 1 m of live plants

Special precautions/Environmental safety
* Harmful if swallowed

Protective clothing/Label precautions
* A, M
* 5c, 16, 18, 22, 28, 29, 36, 37, 54, 63, 67, 70, 78. See label for additional precautions.

Latest application/Harvest Interval(HI)
* Pre-planting

177   2,4-DB

A translocated phenoxy herbicide for use in lucerne

See also bentazone + cyanazine + 2,4-DB

Products

1 Campbell's DB Straight	MTM Agrochem.	300 g/l	SL	00394
2 DB Straight	United Phosphorus	300 g/l	SL	07523

Uses Annual dicotyledons, thistles in CEREALS UNDERSOWN LUCERNE, LUCERNE.

Notes

Efficacy
- Best results achieved on young seedling weeds under good growing conditions. Treatment less effective in cold weather and dry soil conditions
- Rain within 12 h may reduce effectiveness

Crop Safety/Restrictions
- Apply after first trifoliate leaf stage of lucerne. Optimum time 3-4 trifoliate leaves
- Spray lucerne undersown in cereals as soon as possible after first trifoliate leaf stage, while barley or oats from 2-leaf stage (GS 12), wheat from 3-leaf stage (GS 13), to before start of jointing (GS 30)
- Do not allow spray drift onto neighbouring crops

Special precautions/Environmental safety
- Harmful to fish or other aquatic life. Do not contaminate surface waters or ditches with chemical or used container

Protective clothing/Label precautions
- 21, 29, 43, 53, 63, 66, 70

Withholding period
- Keep livestock out of treated areas until foliage of any poisonous weeds such as ragwort has died and become unpalatable

Approval
- Approval expiry: 31 Dec 96 [00394]; replaced by MAFF 06814

---

178 2,4-DB + linuron + MCPA
A translocated herbicide for undersown cereals and grass

Products	Alistell	Zeneca	220:30:30 g/l	EC	06515

Uses Annual dicotyledons in CLOVERS, SEEDLING GRASSLAND, UNDERSOWN CEREALS.

Notes

Efficacy
- Best results achieved on young seedling weeds growing actively in warm, moist weather
- May be applied at any time of year provided crop at correct stage and weather suitable
- Avoid spraying if rain falling or imminent

Crop Safety/Restrictions
- Maximum number of treatments 1 per crop or yr
- Spray winter cereals when fully tillered but before first node detectable (GS 29-30)
- Spray spring wheat from 5-fully expanded leaf stage (GS 15), barley and oats from 2-fully expanded leaves (GS 12)
- Do not spray cereals undersown with lucerne, peas or beans
- Apply to clovers after 1-trifoliate leaf, to grasses after 2-fully expanded leaf stage. Do not spray established clover seed crops
- Do not spray in conditions of drought, waterlogging or extremes of temperature
- In frosty weather clover leaf scorch may occur but damage normally outgrown

---

FOR FULL CONDITIONS OF USE ALWAYS READ THE PRODUCT LABEL

- Do not use on sand or soils with more than 10% organic matter
- Do not roll or harrow within 7 d before or after spraying
- Avoid drift of spray or vapour onto susceptible crops

Special precautions/Environmental safety
- Harmful in contact with skin and if swallowed. Irritating to eyes and skin
- Dangerous to fish or other aquatic life. Do not contaminate surface waters or ditches with chemical or used container
- Do not allow direct spray from vehicle mounted/drawn hydraulic sprayers to fall within 6 m of surface waters or ditches. Direct spray away from water
- Do not apply by hand-held sprayers

Protective clothing/Label precautions
- A, C, H
- 5a, 5c, 6a, 6b, 18, 21, 25, 28, 29, 36, 37, 43, 52, 54a, 63, 66, 70, 78

Withholding period
- Keep livestock out of treated areas for at least 2 wk and until foliage of any poisonous weeds such as ragwort has died and become unpalatable

Latest application/Harvest Interval(HI)
- Before first node detectable (GS 31) for undersown cereals; 2 wk before grazing for grassland

179 2,4-DB + MCPA
A translocated herbicide for cereals, clovers and leys

Products

1 Agrichem DB Plus	Agrichem	243:40 g/l	SL	00044
2 Campbell's Redlegor	MTM Agrochem.	244:44 g/l	SL	00421
3 MSS 2,4-DB + MCPA	Mirfield	280 g/l (total)	SL	01392
4 Redlegor	United Phosphorus	244:44 g/l	SL	07519

Uses

Annual dicotyledons, perennial dicotyledons, polygonums in UNDERSOWN CEREALS [1-4]. Annual dicotyledons, perennial dicotyledons, polygonums in CEREALS [1, 3]. Annual dicotyledons, perennial dicotyledons, polygonums in SEEDLING LEYS [2-4]. Annual dicotyledons, perennial dicotyledons, polygonums in CLOVERS [3]. Annual dicotyledons, perennial dicotyledons in LEYS [1].

Notes

Efficacy
- Best results achieved on young seedling weeds under good growing conditions
- Spray thistles and other perennials when 10-20 cm high provided clover at correct stage
- Effectiveness may be reduced by rain within 12 h, by very cold conditions or drought

Crop Safety/Restrictions
- Apply in spring to winter cereals from leaf sheath erect stage, to spring barley or oats from 2-leaf stage (GS 12), to spring wheat from 5-leaf stage (GS 15)
- Spray clovers as soon as possible after first trifoliate leaf, grasses after 2-3 leaf stage
- Red clover may suffer temporary distortion after treatment
- Do not spray established clover crops or lucerne
- Do not roll or harrow within 7 d before or after spraying
- Do not spray immediately before or after sowing any crop
- Avoid drift onto neighbouring sensitive crops

Special precautions/Environmental safety
• Harmful to fish or other aquatic life. Do not contaminate surface waters or ditches with chemical or used container

Protective clothing/Label precautions
• 18, 37, 60, 63, 70 [1, 3]; 21, 29, 43, 53, 66 [1-4]

Withholding period
• Keep livestock out of treated areas until foliage of any poisonous weeds such as ragwort has died and become unpalatable

Latest application/Harvest Interval(HI)
• Before first node detectable (GS 31) for cereals

180   deltamethrin

A pyrethroid insecticide with contact and residual activity

Products					
1 Crackdown	Roussel Uclaf	10 g/l	SC	H5097
2 Decis	AgrEvo	25 g/l	EC	07172
3 Decis	Hoechst	25 g/l	EC	00657
4 Decis	Hoechst	25 g/l	EC	06311
5 Landgold Deltaland	Landgold	25 g/l	EC	07480
6 Standon Deltamethrin	Standon	25 g/l	EC	07053
7 Thripstick	Aquaspersions	0.125 g/l	AL	02134

Uses   Alder flea beetle, pollen beetles in EVENING PRIMROSE *(off-label)* [3]. American serpentine leaf miner, non-indigenous leaf miners, western flower thrips in PROTECTED VEGETABLES *(off-label)* [3]. American serpentine leaf miner, western flower thrips in PROTECTED ORNAMENTALS *(off-label)*[3]. Aphids, barley yellow dwarf virus vectors, yellow cereal fly in CEREALS [2-6]. Aphids, capsids, caterpillars, codling moth, sawflies, suckers, tortrix moths in APPLES [2-6]. Aphids, caterpillars, damson-hop aphid, plum fruit moth, sawflies in PLUMS [2-6]. Aphids, caterpillars, mealy bugs, scale insects, whitefly in PEPPERS, PROTECTED CUCUMBERS, PROTECTED ORNAMENTALS, PROTECTED POT PLANTS, PROTECTED TOMATOES [2-4]. Aphids, caterpillars in CELERIAC *(off-label)*, CHINESE CABBAGE *(off-label)*, OUTDOOR LETTUCE *(off-label)* [3, 4]. Aphids, caterpillars in CELERY *(off-label)*, FENNEL *(off-label)*, RADICCHIO *(off-label)* [3]. Aphids, damson-hop aphid in HOPS [2-6]. Aphids, pea and bean weevils, pea moth in PEAS [2-6]. Aphids, pollen beetles, virus yellows vectors in OILSEED RAPE [2-5]. Aphids in GRASS SEED CROPS *(off-label)* [3]. Aphids in ORNAMENTALS [2-4]. Beet virus yellows vectors, cabbage stem flea beetle, rape winter stem weevil in WINTER OILSEED RAPE [1-6]. Brassica pod midge, cabbage seed weevil, cabbage stem weevil, pollen beetles in MUSTARD [2-5]. Brassica pod midge, cabbage seed weevil, cabbage stem weevil, pollen beetles in SPRING OILSEED RAPE [2-4]. Capsids, scale insects, thrips in FLOWERS, NURSERY STOCK, WOODY ORNAMENTALS [2-4, 6]. Caterpillars, cutworms in OUTDOOR LETTUCE [2-6]. Caterpillars, leaf miners in MUSHROOM HOUSES *(off-label)*, TOMATO HOUSES *(off-label)* [7]. Caterpillars in BROCCOLI, BRUSSELS SPROUTS, CABBAGES, CAULIFLOWERS, KALE, SWEDES, TURNIPS [5, 6]. Colorado beetle in POTATOES *(off-label)* [3]. Flea beetles in SORREL *(off-label)* [3]. Flea beetles in SUGAR BEET [2-6]. Flea beetles in LEAF BRASSICAS, ROOT BRASSICAS [2-4]. Flour beetles, grain beetles, grain moths, grain storage pests in GRAIN STORES [1]. Insect pests in GARLIC *(off-label)*, LEEKS *(off-label)*, ONIONS *(off-label)* [3]. Leafhoppers in MARJORAM

FOR FULL CONDITIONS OF USE ALWAYS READ THE PRODUCT LABEL

(off-label)[3]. Non-indigenous leaf miners, western flower thrips in PROTECTED HERBS (off-label) [3]. Pea and bean weevils in FIELD BEANS [2-6]. Pea and bean weevils in BROAD BEANS [2-5]. Raspberry beetle in RASPBERRIES [2-6]. Suckers in PEARS [2-6]. Thrips in PROTECTED GARLIC (off-label), PROTECTED LEEKS (off-label), PROTECTED ONIONS (off-label)[3]. Weevils in LUCERNE (off-label) [3].

Notes

Efficacy
- A contact and stomach poison with 3-4 wk persistence, particularly effective on caterpillars and sucking insects
- Normally applied at first signs of damage with follow-up treatments where necessary at 10-14 d intervals. Rates, timing and recommended combinations with other pesticides vary with crop and pest. See label for details [2-4, 6]
- Spray is rainfast within 1 h. May be applied in frosty weather provided foliage not covered in ice [2-4, 6]
- Temperatures above 35°C may reduce effectiveness or persistence
- Spray Thripstick without dilution onto floors and floor covering of crop houses. Use within 2 mth of purchase [7]

Crop Safety/Restrictions
- Maximum number of treatments varies with crop and pest, 4 per crop for wheat and barley, only 1 application between 1 Apr and 31 Aug; see label or off-label approval notice for other crops
- Do not apply more than 1 aphicide treatment to cereals in summer
- Do not spray crops suffering from drought or other physical stress
- Test spray small numbers of ornamentals before committing whole batches [2-4]
- Consult processer before treating crops for processing

Special precautions/Environmental safety
- Harmful in contact with skin or if swallowed. Irritating to eyes and skin
- Flammable [2-4, 6, 7]
- Approved respirator must be worn for application by fogging. See OLA for details [3]
- Dangerous to bees. Do not apply to crops in flower or to those in which bees are actively foraging except as directed (see labels). Do not apply when flowering weeds are present
- Extremely dangerous to fish or other aquatic life. Do not contaminate surface waters or ditches with chemical or used container
- Do not apply to a cereal crop if any product containing a pyrethroid insecticide or dimethoate has been applied to that crop after the start of ear emergence (GS 51)
- Do not allow direct spray from ground based spray vehicles to fall within 6 m, or from hand-held sprayers within 2 m, of surface waters or ditches. Direct spray away from water
- Do not spray cereals after 31 Mar in the year of harvest within 6 m of the outside edge of the crop

Protective clothing/Label precautions
- A, C, H
- 5a, 5c, 6a, 6b, 12c, 16, 18, 21, 28, 29, 36, 37, 51, 63, 66, 70, 78 [1-7]; 48, 60, 61 [1-4, 6, 7]; 49, 54a [5]

Latest application/Harvest Interval(HI)
- Before grain watery ripe (GS 71) for wheat and barley [6]; before soft dough stage (GS 85) for wheat, barley, oats [2-4]; end of flowering (GS 5,0) for oilseed rape. See label or off-label approval notice for other crops

Approval
- Off-label approval unlimited for use on outdoor lettuce and Chinese cabbage (OLA 0699, 0700/93) [3, 4]; to Jul 1998 for use on outdoor celeriac (OLA 0776, 0777/93)[3, 4]; unlimited for control of Colorado beetle in potatoes as required by Statutory Notice (OLA 0038/92); to May 1997 for use in salad and bulb onions, leeks, garlic (OLA 1326/93)[3]; unlimited for use in tomato houses (OLA 0180/92)[7], to Feb 1996 for use in mushroom houses (OLA 0181/92)[7]; unlimited for application by fogging in non-edible glasshouse ornamentals and glasshouse vegetables (OLA 0190/92)[3]; to Feb 1996 for application by fogging in glasshouse vegetables and glasshouse herbs (OLA 0191/92)[3]; to Jun 1997 for use on protected onions, leeks, garlic, sorrel, fennel, radicchio, marjoram, lettuce, Chinese cabbage and celery and to outdoor fennel, celery and lucerne (OLA 0622/92)[3]; unlimited for use on grass seed crops, evening primrose and outdoor sorrel, radicchio and marjoram (OLA 0623/92)[3]

Maximum Residue Levels (mg residue/kg food)
- tea, hops 5; pulses, cereals 1; Chinese cabbage, kale, lettuces, spinach, beet leaves, herbs 0.5; currants, gooseberries, tomatoes, peppers, aubergines, beans (with pods) 0.2; pome fruits, stone fruits, grapes, garlic, onions, shallots, cucumbers, gherkins, courgettes, broccoli, cauliflowers, Brussels sprouts, head cabbages, peas (with pods), rape seed 0.1; citrus fruits, tree nuts, strawberries, bilberries, cranberries, wild berries, miscellaneous fruit (except kiwi fruit, olives), root and tuber vegetables, melons, squashes, watermelons, kohlrabi, watercress, peas (without pods), stem vegetables (except globe artichokes), mushrooms, oilseeds (except rape seed), early potatoes, poultry, eggs 0.01

181   deltamethrin + heptenophos

A systemic aphicide with knockdown, vapour and anti-feeding activity
See also heptenophos

Products					
1 Decisquick	AgrEvo	25:400 g/l	EC	07312
2 Decisquick	Hoechst	25:400 g/l	EC	03117

Uses

Aphids, beet mild yellowing virus vectors, beet virus yellows vectors in SUGAR BEET. Aphids, leaf roll virus vectors, potato mosaic virus vectors in POTATOES. Aphids, pea enation virus vectors, pea moth in PEAS. Aphids in OUTDOOR LETTUCE *(off-label)*, PROTECTED LETTUCE *(off-label)*. Insect pests in BROCCOLI *(off-label)*, BRUSSELS SPROUTS *(off-label)*, CABBAGES *(off-label)*, CALABRESE *(off-label)*, CAULIFLOWERS *(off-label)*, FENNEL *(off-label)*, KOHLRABI *(off-label)*, RADICCHIO *(off-label)*, ROSCOFF CAULIFLOWER *(off-label)*.

Notes

Efficacy
- Works by combination of contact, systemic, vapour and anti-feeding activity and persists for 10-14 d, longer in cool and shorter in hot weather
- Spray peas as soon as aphids seen and repeat once if necessary
- Spray potatoes as soon as first aphids found or before aphid build-up depending on variety (see label for details) and repeat if necessary. Spray seed crops at 80% emergence, whether aphids present or not, with up to 3 further sprays at 7-14 d intervals
- Spray sugar beet when aphids first seen or warning issued and repeat once if necessary
- Product rainfast within 2 h. Do not spray wet crops

FOR FULL CONDITIONS OF USE ALWAYS READ THE PRODUCT LABEL

Crop Safety/Restrictions
* Maximum number of treatments 2 per crop for peas and sugar beet; 4 per crop for potatoes
* Do not spray crops suffering from severe moisture stress

Special precautions/Environmental safety
* This product contains an anticholinesterase organophosphorus compound. Do not use if under medical advice not to work with such compounds
* Toxic if swallowed. Harmful in contact with skin. Irritating to eyes and skin
* Flammable
* Dangerous to bees. Do not apply to crops in flower or to those in which bees are actively foraging except as directed on peas. Do not apply when flowering weeds are present
* Extremely dangerous to fish or other aquatic life. Do not contaminate surface waters or ditches with chemical or used container
* Do not allow direct spray from vehicle mounted/drawn hydraulic sprayers to fall within 6 m, or from hand-held sprayers to fall within 2 m, of surface waters or ditches. Direct spray away from water

Protective clothing/Label precautions
* A, C, H [1, 2]
* 1, 4c, 5a, 6a, 6b, 12c, 14, 16, 18, 21, 28, 29, 35, 36, 37, 49, 51, 63, 66, 70, 79 [1, 2]; 54a [1]

Latest application/Harvest Interval(HI)
* HI 24 h

Approval
* Off-label approval unlimited for use on leaf and flowerhead brassicas (OLA 0760/93); to Jul 1998 for use on kohlrabi (OLA 0761/93); unlimited for use on outdoor radicchio (OLA 1027/92); to Oct 1997 for use on protected radicchio, outdoor and protected fennel (OLA 1028/92); unlimited for use on outdoor lettuce and protected lettuce seedlings (OLA 0316/93)

182   demeton-S-methyl

A systemic and contact organophosphorus insecticide and acaricide

Products   Campbell's DSM   MTM Agrochem.   580 g/l   EC   00405

Uses   Aphids, bryobia mites, leafhoppers, pear sucker, red spider mites, sawflies, woolly aphid in APPLES, PEARS. Aphids, mangold fly in BEET CROPS. Aphids, pea midge in PEAS. Aphids, red spider mites, sawflies in CHERRIES, PLUMS. Aphids, red spider mites in CHRYSANTHEMUMS, CUCUMBERS, CURRANTS, DAMSONS, GOOSEBERRIES, ORNAMENTALS, PROTECTED CARNATIONS, RASPBERRIES, STRAWBERRIES, TOMATOES. Aphids, thrips in MANGE-TOUT PEAS *(off-label)*. Aphids in BRASSICAS, BROAD BEANS, CARROTS, CELERY, CEREALS, DWARF BEANS, FIELD BEANS, GRASS SEED CROPS, LETTUCE, PARSNIPS, POTATOES, RUNNER BEANS.

Notes   **Efficacy**
* For several crops different products vary in the range of pests listed as controlled by treatment. See label for details
* Apply as spray when pest appears and repeat as necessary. Dose, number and timing of sprays vary with pest and crop. See label for details
* May also be applied by soil watering on many ornamentals. Do not apply to dry soil

- Do not spray waxy-leaved plants such as brassicas or sugar beet under hot, dry conditions. Delay until late evening or early morning. Do not spray wilting plants
- For bryobia mite control spray at temperatures above 13°C
- Strains of red spider mite, peach-potato aphid in sugar beet and some other pests resistant to organophosphorus compound have developed in certain areas

Crop Safety/Restrictions
- Maximum number of treatments 1 per crop for mangels, fodder beet, red beet, cereals, cucumbers, lettuce, gooseberries, ornamentals; 2 per crop for peas, apples, pears, plums, cherries, peaches, apricots, currants, raspberries, tomatoes, chrysanthemums, carnations; 3 per crop for sugar beet; 4 per crop for potatoes
- On carrots the maximum total dose applied per crop must not exceed the equivalent of 3 (on mineral soils) or 4 (on organic soils) full dose applications
- Do not treat brassicas after the end of Oct. Do not spray cauliflower curds
- Do not spray carnations in bloom as residues may cause unpleasant odours
- Do not mix with fentin compound or alkaline sprays

Special precautions/Environmental safety
- Demeton-S-methyl is subject to the Poisons Rules 1982 and the Poisons Act 1972. See notes in Section 1
- Toxic in contact with skin, by inhalation and if swallowed
- Flammable
- This product contains an anticholinesterase organophosphorus compound. Do not use if under medical advice not to work with such compounds
- Do not apply with hand-held ULV equipment
- Harmful to livestock
- Harmful to game, birds and animals
- Harmful to bees. Do not apply to crops in flower or to those in which bees are actively foraging. Do not apply when flowering weeds are present
- Harmful to fish or other aquatic life. Do not contaminate surface waters or ditches with chemical or used container

Protective clothing/Label precautions
- A, C, H, M
- 1, 4a, 4b, 4c, 12c, 14, 16, 18, 21, 23, 24, 25, 26, 27, 28, 29, 36, 37, 41, 46, 50, 53, 64, 66, 70, 79

Withholding period
- Keep all livestock out of treated areas for at least 2 wk

Latest application/Harvest Interval(HI)
- Before flowers open for peas.
- HI mangels, fodder beet, red beet 10 d (not to be used as fodder for 21 d); other crops 21 d

Approval
- Approved for aerial application on cereals, peas, potatoes, sugar beet, carrots, brassicas, field beans. See notes in Section 1
- Off-label approval unlimited for use on mange-tout peas (OLA 0398/92)

FOR FULL CONDITIONS OF USE ALWAYS READ THE PRODUCT LABEL

183 desmetryn

A contact triazine herbicide for use on brassica crops

Products   Semeron 25WP   Ciba Agric.   25% w/w   WP   01916

Uses   Annual dicotyledons, fat-hen in BROCCOLI *(off-label)*, BRUSSELS SPROUTS, CABBAGES, CALABRESE *(off-label)*, CAULIFLOWERS *(off-label)*, COLLARDS, FODDER RAPE, KALE, KALE *(off-label)*.

Notes

Efficacy
* May be applied from Mar to mid-Nov to weeds up to 5-100 mm high (fat-hen to 350 mm)
* A higher dose required in drier eastern than western areas. See label for details
* Do not apply when rain imminent. Rain within 24 h reduces effectiveness
* May be tank-mixed with aziprotryne or clopyralid. See label for details

Crop Safety/Restrictions
* Maximum number of treatments 1 per crop
* Apply to drilled crops after plants have 3 true leaves and are at least 125 mm high, to transplants at least 2 wk after planting and after reaching above stage
* Do not spray poor, thin crops or seed crops intended for producing Maris Kestrel kale as one of the parent lines is susceptible
* Do not spray in frosty weather or if crops affected by frost or cabbage root fly
* Do not spray for at least 7 d after applying insecticide
* Leaf scorch may occur after spraying but crop recovers quickly without effect on yield

Special precautions/Environmental safety
* Do not contaminate surface waters or ditches with chemical or used container

Protective clothing/Label precautions
* 22, 29, 35, 54, 63, 67

Latest application/Harvest Interval(HI)
* HI 4 wk (5 wk for calabrese)

Approval
* Off-label approval unlimited for use on broccoli, kale, cauliflowers (OLA 344/93), calabrese (OLA 0985/93)

184 diazinon

A contact organophosphorus insecticide for soil and foliar treatment

Products   Darlingtons Diazinon Granules   Darmycel   5% w/w   GR   05674

Uses   Mushroom flies in MUSHROOMS.

Notes

Efficacy
* Incorporate in mushroom compost at spawning for mushroom fly control, it is essential to mix in the granules thoroughly

Crop Safety/Restrictions
* Maximum number of treatments 1 per spawning

Special precautions/Environmental safety
* This product contains an anticholinesterase organophosphorus compound. Do not use if under medical advice not to work with such compounds
* Harmful in contact with skin
* Harmful to game, wild birds and animals
* Harmful to fish or other aquatic life. Do not contaminate surface waters or ditches with chemical or used container

Protective clothing/Label precautions
* A
* 1, 22, 29, 35, 36, 37, 46, 53, 63, 67, 70

Latest application/Harvest Interval(HI)
* HI 2 wk

Maximum Residue Levels (mg residue/kg food)
* meat 0.7; citrus fruit, pome fruits, apricots, peaches, nectarines, plums, grapes, cane fruits, bilberries, cranberries, currants, gooseberries, bananas, carrots, horseradish, parsnips, parsley root, salsify, swedes, turnips, garlic, onions, shallots, tomatoes, peppers, aubergines, cucumbers, gherkins, courgettes, cauliflowers, head cabbages, lettuce, beans (with pods), peas (with pods), celery, leeks, rhubarb, cultivated mushrooms, potatoes 0.5; cereals 0.05; milk 0.02

185   dicamba

A translocated benzoic herbicide for control of bracken and perennial weeds

See also 2,4-D + dicamba
2,4-D + dicamba + ferrous sulfate
2,4-D + dicamba + mecoprop
2,4-D + dicamba + triclopyr
bifenox + dicamba

Products	1 Tracker	Cyanamid	480 g/l	SL	07149
	2 Tracker	PBI	480 g/l	SL	03847

Uses   Bracken in AMENITY GRASS, FORESTRY, NON-CROP AREAS. Perennial dicotyledons in AMENITY GRASS *(wiper application)*, GRASSLAND *(wiper application)*.

Notes   **Efficacy**
* For bracken control apply to soil over litter as a concentrated solution which must be washed down to roots by rainfall
* Apply in Mar, Apr or early May in season of planting or 1 yr after planting as 2-5 cm band midway between tree rows
* Where trees planted in furrows delay application until second year
* Frond production affected up to 1 m from band of application

---

FOR FULL CONDITIONS OF USE ALWAYS READ THE PRODUCT LABEL

- Control may be reduced where rhizomes broken up by ploughing
- For dock and perennial weed control in grassland use in diluted form with a ropewick applicator when weeds growing vigorously but before senescence of flower heads

Crop Safety/Restrictions
- Maximum number of treatments 1 per yr
- Avoid contact with tree foliage or application in rooting zone of trees
- Do not use where run-off likely to occur after heavy rain
- Avoid drift onto susceptible crops. Tomatoes may be affected by vapour drift at a considerable distance

Protective clothing/Label precautions
- 21, 29, 43, 54, 63, 65

Withholding period
- Keep livestock out of treated areas for at least 2 wk and until foliage of any poisonous weeds such as ragwort and buttercups has died and become unpalatable

---

186   dicamba + dichlorprop + MCPA
A translocated herbicide mixture for turf

Products	Intrepid	Miracle Prof.	20.8:333:166.5 g/l SL	07819

Uses   Annual dicotyledons, perennial dicotyledons in TURF.

Notes   **Efficacy**
- Apply between Apr and Sep when weeds growing actively
- Do not mow for at least 3 d before and after treatment so that there is sufficient leaf growth for spray uptake and sufficient time for translocation
- If re-growth occurs or new weeds germinate re-treatment recommended

Crop Safety/Restrictions
- Do not use during drought unless irrigation is carried out before and after treatment
- Do not apply during freezing conditions or when heavy rain is imminent
- Do not treat new turf until established for about 6 mth after seeding or turfing
- The first 4 mowings after treatment should not be used to mulch cultivated plants unless composted for at least 6 mth
- Do not re-seed turf within 8 wk of last treatment

Special precautions/Environmental safety
- Harmful if swallowed
- Irritating to skin
- Risk of serious damage to eyes
- May cause sensitisation by skin contact
- Harmful to fish. Do not contaminate surface waters or ditches with chemical or used container

Protective clothing/Label precautions
- A, C
- 5c, 6b, 9, 10a, 18, 21, 28, 29, 36, 37, 43 (2 wk), 53, 63, 66, 70, 78

Withholding period
* Keep livestock out of treated areas for at least 2 wk if poisonous weeds such as ragwort are present

Maximum Residue Levels (mg residue/kg food)
* see dichlorprop entry

187 dicamba + maleic hydrazide + MCPA
A herbicide/plant growth regulator mixture for amenity grass

Products					
Mazide Selective	Vitax	6:200:75 g/l	SL	05753

Uses Annual dicotyledons, growth retardation, perennial dicotyledons in AMENITY GRASS, ROAD VERGES.

Notes

Efficacy
* Best results achieved by application in Apr-May when grass and weeds growing actively but before weeds have started to flower
* May be used either as one annual spray or as spring spray repeated after 8-10 wk

Crop Safety/Restrictions
* Maximum number of treatments 2 per yr
* Do not use on fine turf

Special precautions/Environmental safety
* Harmful to fish or other aquatic life. Do not contaminate surface waters or ditches with chemical or used container

Protective clothing/Label precautions
* 21, 29, 43, 53, 63, 66

Withholding period
* Keep livestock out of treated areas until foliage of any poisonous weeds such as ragwort has died and become unpalatable

Maximum Residue Levels (mg residue/kg food)
* see maleic hydrazide entry

FOR FULL CONDITIONS OF USE ALWAYS READ THE PRODUCT LABEL

188   dicamba + MCPA + mecoprop
A translocated herbicide for cereals and grassland

Products

1 Barclay Hat-Trick	Barclay	25:200:400 g/l	SL	07579
2 Campbell's Field Marshal	MTM Agrochem.	18:360:160 g/l	LI	00406
3 Campbell's Grassland Herbicide	MTM Agrochem.	25:200:400 g/l	SL	06157
4 Fisons Tritox	Levington	14:155:91 g/l	SL	03013
5 Headland Relay	Headland	25:200:400 g/l	LI	07684
6 Herrisol	Bayer	18:252:84 g/l	SL	01048
7 Hyprone	Agrichem	18.7:106:194 g/l	SL	01093
8 Hysward	Agrichem	1.6:9.3:17.0% w/w	SL	01096
9 Pasturol	FCC	25:200:400 g/l	SL	01545
10 Quad-Ban	Quadrangle	19.5:245:86.5 g/l	SL	03114
11 Springcorn Extra	FCC	18:360:160 g/l	SL	02004
12 Stefes Banlene Plus	Stefes	18:252:84 g/l	SL	07688
13 Tribute	Nomix-Chipman	18:252:84 g/l	SL	06921
14 Tritox	Levington	14:155:91 g/l	SL	07502

Uses

Annual dicotyledons, chickweed, cleavers, docks, mayweeds, perennial dicotyledons, polygonums in BARLEY, OATS, WHEAT [2, 6, 7, 10-12]. Annual dicotyledons, chickweed, cleavers, docks, mayweeds, perennial dicotyledons, polygonums in GRASS SEED CROPS [6, 7, 10, 12]. Annual dicotyledons, chickweed, cleavers, docks, mayweeds, perennial dicotyledons, polygonums in RYE [6, 10, 12]. Annual dicotyledons, chickweed, cleavers, docks, mayweeds, perennial dicotyledons, polygonums in LEYS, PERMANENT PASTURE [1, 3, 5, 6, 9, 10, 12]. Annual dicotyledons, chickweed, docks, mayweeds, perennial dicotyledons in ESTABLISHED GRASSLAND [8, 12]. Annual dicotyledons, docks, perennial dicotyledons in APPLES, PEARS [6, 12]. Annual dicotyledons, perennial dicotyledons in AMENITY TURF, SPORTS TURF [4, 13, 14].

Notes

Efficacy
* Best results achieved on young, actively growing weeds up to 15 cm high in a strongly competitive crop, perennials when well developed but before flowering. See label for details of susceptibility
* Spray grassland and turf from early spring to Oct when grasses growing actively
* Do not spray in rain, when rain imminent or in drought

Crop Safety/Restrictions
* Maximum number of treatments 1 per crop [6, 10]
* Spray winter cereals from main leaf sheath erect and at least 5 cm high but before first node detectable (GS 31), spring cereals from 5-leaf stage to before first node detectable (GS 15-31) (some formulations only recommend treatment in spring)
* Do not spray cereals undersown with clovers or legumes, to be undersown with grass or legumes or grassland where clovers or other legumes are important
* Spray newly sown grass or undersown crops when grass seedlings have 2-3 leaves
* Do not spray grass seed crops later than 5 wk before emergence of seed heads
* In orchards avoid drift of spray onto tree foliage. Do not spray during blossom period
* Do not roll, harrow, cut, graze for 3-7 d before or after treatment (products vary)
* Do not use first 4 mowings after treatment for mulch unless composted for at least 6 mth. See label for other precautions [4, 13]
* Do not use on newly sown turf in the year of establishment. Allow 6-8 wk after treatment before seeding bare patches [4]

Special precautions/Environmental safety
* Irritating to eyes
* Harmful if swallowed [2, 11]. Harmful in contact with skin and if swallowed [1, 3, 5, 9]
* Harmful to fish or other aquatic life. Do not contaminate surface waters or ditches with chemical or used container

Protective clothing/Label precautions
* A, C
* 5a, 5c, 78 [1-3, 5-11, 13]; 6a, 18, 21, 29, 36, 37, 53, 63, 66 [1-14]; 6b [2-11, 13, 14]; 43 (2 wk) [12]; 43, 70 [1-11, 13, 14]

Withholding period
* Keep livestock out of treated areas for at least 2 wk and until foliage of any poisonous weeds such as ragwort has died and become unpalatable

Latest application/Harvest Interval(HI)
* Before first node detectable (GS 31) for cereals
* HI 7 d before cutting or grazing grass

---

189   dicamba + MCPA + mecoprop-P
A translocated herbicide for cereals, grassland and amenity grass

Products					
1	Mascot Super Selective-P	Rigby Taylor	15:100:100 g/l	SL	06106
2	MSS Mircam Plus	Mirfield	19.5:245:43.3 g/l	SL	01416
3	Tritox	Levington	12:139:41 g/l	SL	07764

Uses   Annual dicotyledons, chickweed, cleavers, mayweeds, perennial dicotyledons, polygonums in BARLEY, GRASS SEED, GRASSLAND, OATS, TURF, WHEAT [2]. Annual dicotyledons, perennial dicotyledons in AMENITY TURF, SPORTS TURF [1, 3].

Notes

Efficacy
* Treatment should be made from early spring to Oct when weeds are growing actively
* For best results apply in fine warm weather, preferably when soil is moist. Do not spray before or during rain or in drought
* Application of fertilizer 1-2 wk before spraying aids weed control in turf

Crop Safety/Restrictions
* Maximum number of treatments 1 per yr or per crop [2]; 2 per yr on amenity grass [1]
* Apply to winter cereals from the leaf sheath erect stage (GS 30), and to spring cereals from the five expanded leaf stage (GS 15) [2]
* Do not apply to cereals after the first node is detectable (GS 31) [2]
* Spray grass seed crops 4-6 wk before flower heads begin to emerge (Timothy 6 wk) [2]
* Do not use on turf in year of establishment [1, 3]. Allow 6-8 wk after treatment before seeding bare patches
* Do not use first 4 mowings after treatment as mulch for at least 6 mth [3]
* Do not roll or harrow within 7 d before or after treatment
* Do not treat rye or cereals undersown or to be undersown with anything except grass [2]

FOR FULL CONDITIONS OF USE ALWAYS READ THE PRODUCT LABEL

- Do not close mow turf for 3-4 d before or after treatment, however, on fine closely mown turf do not mow for 24 h before or after treatment
- Do not treat newly sown turf and do not reseed for 6-8 wk after treatment [2]
- Grass cuttings should not be used for mulching but may be composted and used not less than 6 mth later [1]

Special precautions/Environmental safety
- Irritating to eyes and skin. Harmful if swallowed and in contact with skin [2]. Irritating to eyes and skin [3]
- Harmful to fish or other aquatic life. Do not contaminate surface waters or ditches with chemical or used container
- Avoid drift especially near susceptible vegetable crops and glasshouses

Protective clothing/Label precautions
- A, C
- 5a, 5c, 78 [2]; 6a, 6b, 18, 21, 29, 36, 37, 43, 53, 63, 70 [1-3]; 28, 67 [1]; 66 [2, 3]

Withholding period
- Keep livestock out of treated areas for at least 2 wk and until foliage of any poisonous weeds such as ragwort has died and become unpalatable

190 dicamba + mecoprop

A translocated post-emergence herbicide for cereals and grassland

Products					
1	Condox	Zeneca	112:265 g/l	SL	06519
2	Hyban	Agrichem	18.7:300 g/l	SL	01084
3	Hygrass	Agrichem	18.7:300 g/l	SL	01090

Uses   Annual dicotyledons, chickweed, cleavers, mayweeds, perennial dicotyledons, polygonums in BARLEY, OATS, WHEAT [2]. Annual dicotyledons, docks, perennial dicotyledons, thistles in LEYS, PERMANENT PASTURE [1, 3].

Notes   ### Efficacy
- Best results by application to young weed seedlings under good growing conditions
- Do not spray in cold, frosty or windy weather, when rain expected or in drought
- Spray perennial weeds in grassland just before flowering, docks in rosette stage in late spring with repeat spray in Aug-Oct if necessary
- Allow 10-14 d regrowth of docks after grazing

Crop Safety/Restrictions
- Apply to winter or spring cereals in spring from 5-fully expanded leaf stage to first node detectable (GS 15-31)
- Do not spray cereals undersown or to be undersown
- Do not apply to leys established for less than 2 yr [1, 2]
- Do not apply to permanent pastures where clovers are an essential part of the sward
- Do not roll, harrow, cut or graze within 7 d before or after spraying

Special precautions/Environmental safety
- Harmful in contact with skin [1]
- Irritating to eyes [1-3]
- Harmful to fish or other aquatic life. Do not contaminate surface waters or ditches with chemical or used container

Protective clothing/Label precautions
* A, C
* 5a [1]; 6a, 18, 21, 29, 36, 37, 43, 53, 63, 65, 70, 78 [1-3]

Withholding period
* Keep livestock out of treated areas until foliage of any poisonous weeds such as ragwort has died and become unpalatable

---

191 dicamba + mecoprop + triclopyr
A translocated herbicide for perennial and woody weed in grassland

Products	Fettel	Zeneca	78:130:72 g/l	EC	06399

Uses   Annual dicotyledons, brambles, broom, docks, gorse, perennial dicotyledons, stinging nettle, thistles, woody weeds in ESTABLISHED LEYS, NON-CROP AREAS, PERMANENT PASTURE.

Notes

Efficacy
* Apply to actively growing perennials before flowering shoots appear, usually late spring or summer. Later application can be made to brambles and regrowth after cutting
* With well established weeds retreatment may be needed in following season
* Do not spray in drought, frost, extremes of temperature or if rain falling or imminent

Crop Safety/Restrictions
* Maximum number of treatments 1 per yr
* Do not use on leys established for less than 2 yr
* Do not apply overall where clovers are an essential part of sward
* Do not roll or harrow within 7 d before or after treatment
* Avoid drift onto nearby susceptible crops. Drift of vapour may occur in hot conditions
* Do not apply near desirable plants where chemical may move and contact roots
* Do not sow grass, clovers or brassicas by direct drilling or minimum cultivation for at least 6 wk after spraying

Special precautions/Environmental safety
* Irritating to eyes and skin
* Dangerous to fish or other aquatic life. Do not contaminate surface waters or ditches with chemical or used container
* Not to be used on food crops

Protective clothing/Label precautions
* A, C
* 6a, 6b, 18, 21, 28, 29, 34, 36, 37, 43, 52, 63, 65

Withholding period
* Keep livestock out of treated areas for at least 2 wk and until foliage of poisonous weeds such as ragwort has died and become unpalatable

---

FOR FULL CONDITIONS OF USE ALWAYS READ THE PRODUCT LABEL

192   dicamba + mecoprop-P
A translocated post-emergence herbicide for cereals

Products	MSS Mircam	Mirfield	18.7:150 g/l	SL	01415

Uses   Annual dicotyledons, chickweed, cleavers, mayweeds, perennial dicotyledons, polygonums in BARLEY, OATS, WHEAT.

Notes

Efficacy
* Best results by application in warm, moist weather when weeds are actively growing
* Do not spray in cold or frosty conditions
* Do not spray if rain expected within 6 h
* Ensure any lime is washed off crop and weeds before spraying

Crop Safety/Restrictions
* Apply to winter sown crops from leaf sheath erect stage but before first node detectable (GS 30-31)
* Apply to spring sown cereals from 5 expanded leaf stage but before first node is detectable (GS 15-31)
* Transient crop prostration may occur after spraying but recovery is rapid
* Do not treat undersown grass until tillering begins
* Do not spray cereals undersown with clover or legume mixtures
* Do not roll, harrow within 7 d before or after spraying
* Do not treat crops suffering from stress from any cause
* Avoid treatment when drift may damage neighbouring susceptible crops

Special precautions/Environmental safety
* Harmful if swallowed and in contact with skin
* Irritating to eyes and skin
* Harmful to fish or other aquatic life. Do not contaminate surface waters or ditches with chemical or used container

Protective clothing/Label precautions
* A, C
* 5a, 5c, 6a, 6b, 18, 21, 29, 36, 37, 43 (2 wk), 53, 63, 66, 70, 78

Withholding period
* Keep livestock out of treated areas for at least 2 wk and until foliage of poisonous weeds such as ragwort has died and become unpalatable

193   dicamba + paclobutrazol
A plant growth regulator/herbicide mixture for amenity grass

Products					
1 Holdfast D	ICI Professional	25:250 g/l	SC	05056
2 Holdfast D	Zeneca Prof.	25:250 g/l	SC	06858

Uses   Annual dicotyledons, growth retardation, perennial dicotyledons in AMENITY GRASS.

PESTICIDE PROFILES

Notes

Efficacy
- Uptake mainly via roots and aided by rain after application
- Apply 2 wk before grass growth starts in spring (mid-Mar to mid-May). Can also be applied up to Jul if grass growing rapidly and in Aug/Sep to reduce autumn flush
- Grass growth retarded for 3 mth or more
- Do not apply during period of drought or frost
- Where coarse grasses (eg cocksfoot) dominant addition of maleic hydrazide or mefluidide recommended

Crop Safety/Restrictions
- Maximum number of treatments 1 per yr
- Use only on permanent grassland, not on grass within 1 yr after sowing
- Use only on areas of restricted or limited public access
- Do not use where food crops or ornamentals to be grown or grass resown within 5 yr
- Fine-leaved grasses can be retarded more than others leading to an uneven appearance
- Do not use cuttings from treated grass for composting or mulch

Special precautions/Environmental safety
- Not to be used on food crops
- Harmful to fish or other aquatic life. Do not contaminate surface waters or ditches with chemical or used container

Protective clothing/Label precautions
- 21, 29, 34, 43, 53, 63, 66

Withholding period
- Keep livestock out of treated areas for at least 2 yr after treatment

194   dichlobenil

A residual benzonitrile herbicide for woody crops and non-crop areas
See also dalapon + dichlobenil

Products

1 Casoron G	ICI Agrochem.	6.75% w/w	GR	00448
2 Casoron G	Zeneca	6.75% w/w	GR	00448
3 Casoron G	ICI Professional	6.75% w/w	GR	00448
4 Casoron G	Zeneca Prof.	6.75% w/w	GR	06854
5 Casoron G-SR	ICI Professional	20% w/w	GR	00451
6 Casoron G-SR	Zeneca Prof.	20% w/w	GR	06856
7 Casoron G4	ICI Professional	4% w/w	GR	02406
8 Casoron G4	Zeneca Prof.	4% w/w	GR	06855
9 Prefix D	Cyanamid	6.75% w/w	GR	07013

Uses   Annual weeds, perennial dicotyledons, perennial grasses in ESTABLISHED WOODY ORNAMENTALS, ROSES [1-4, 7-9]. Annual weeds, perennial dicotyledons, perennial grasses in APPLES, BLACKBERRIES, BLACKCURRANTS, GOOSEBERRIES, LOGANBERRIES, PEARS, RASPBERRIES, REDCURRANTS [1-3, 9]. Annual weeds, perennial dicotyledons, perennial grasses in RHUBARB *(off-label)* [1-3]. Aquatic weeds in AQUATIC SITUATIONS [4-6]. Total vegetation control in NON-CROP AREAS [1-4, 9]. Volunteer potatoes in POTATO DUMPS *(blight prevention)* [1-3].

FOR FULL CONDITIONS OF USE ALWAYS READ THE PRODUCT LABEL

Notes

Efficacy
- Best results achieved by application in winter to moist soil during cool weather, particularly if rain follows soon after. Do not disturb treated soil by hoeing
- Lower rates control annuals, higher rates perennials, see label for details
- Residual activity lasts 3-6 mth with selective, up to 12 mth with non-selective rates
- Apply to potato dump sites before emergence of potatoes for blight prevention
- For control of emergent, floating and submerged aquatics apply to water surface in early spring. Intended for use in still or sluggish flowing water [4-6]
- Do not use on fen peat or moss soils

Crop Safety/Restrictions
- Maximum number of treatments 1 per yr
- Apply to crops in dormant period. See label for details of timing and rates
- Do not treat crops established for less than 2 yr. See label for lists of resistant and sensitive species. Do not treat stone fruit or Norway spruce (Christmas trees)
- Do not apply in or near glasshouses, near hops or areas underplanted with bulbs, annuals or herbaceous stock
- Do not apply to frozen, snow-covered or waterlogged ground or when crop foliage wet
- Do not apply to sites less than 18 mth before replanting or sowing
- Store well away from corms, bulbs, tubers and seed

Special precautions/Environmental safety
- Keep poultry out of treated areas for 7 d [9]
- Do not use treated water for irrigation purposes within 2 wk of treatment [1-3, 8]
- Harmful to fish or other aquatic life. Do not contaminate surface waters or ditches with chemical or used container

Protective clothing/Label precautions
- 22, 36, 37 [1]; 29 [1, 9]; 30 [2-8]; 42, 58 [9]; 53, 70 [2-4, 7-9]; 55 [4-6]; 56 (14 d) [4]; 56 [5, 6]; 63, 67 [1-9]

Latest application/Harvest Interval(HI)
- Early Mar for shrubs and trees; early Apr for apples, pears, currants, gooseberries; before bud movement for cane fruit, roses; before growth starts for potato dumps

Approval
- Cleared for aquatic weed control. See notes in Section 1 on use of herbicides in or near water [4-6]
- Off label approval unlimited for use on established rhubarb in dormant period (late autumn) (OLA 0371/93)[1-3]

195   dichlofluanid

A protectant sulfamide fungicide for horticultural crops

Products	Elvaron WG	Bayer	50% w/w	WG	04855

Uses   Black spot, mildew in ROSES. Botrytis, leaf mould in TOMATOES. Botrytis in AUBERGINES *(off-label)*, CANE FRUIT, CURRANTS, GOOSEBERRIES, GRAPEVINES, NON-EDIBLE ORNAMENTALS *(off-label)*, PEPPERS *(off-label)*, STRAWBERRIES. Cane blight, cane spot, mildew in RASPBERRIES. Downy mildew in PROTECTED BRASSICA SEEDLINGS. Fire in TULIPS.

Notes

Efficacy
- Apply as a protective programme of sprays. See label for recommended timings
- Treatment also gives reduction of mildew and spur blight on raspberries and useful control of mildew on strawberries
- Apply to brassica seedlings under cover as soon as first seedlings appear and repeat after 3, 5, 7 and 10 d using a wetting agent. Thereafter apply 3 sprays at weekly intervals

Crop Safety/Restrictions
- Maximum number of treatments 8 per crop for brassicas under cover, tomatoes, peppers, aubergines, non-edible ornamentals under glass and outdoor grapes for wine-making; 6 per crop for currants; 4 per crop for gooseberries. For other fruit crops number depends on dose applied and restriction on total for season
- Do not use on strawberries under glass or polythene. Plants which have been under cover recently may still be susceptible to leaf scorch
- On raspberries up to 7 applications may be made, up to 6 on currants, up to 8 on grapes for winemaking

Special precautions/Environmental safety
- Harmful to fish or other aquatic life. Do not contaminate surface waters or ditches with chemical or used container

Protective clothing/Label precautions
- 22, 28, 29, 53, 63, 67

Latest application/Harvest Interval(HI)
- Before 2 true leaf stage for brassicas; before 8 expanded leaves for planted out cabbage, cauliflower, broccoli.
- HI raspberries 1 wk; strawberries 2 wk; broccoli, cabbage 3 wk; currants, gooseberries, loganberries, blackberries, grapes (fresh consumption) 3 wk; grapes (winemaking) 5 wk; protected tomatoes, peppers, aubergines, non-edible ornamentals 3 d

Approval
- Off-label approval unlimited on protected peppers, aubergines and non-edible ornamentals (OLA 0167/93)

196   dichlorophen

A chlorophenol moss-killer, fungicide, bactericide and algicide

Products

1	50/50 Liquid Mosskiller	Vitax	360 g/l	SL	07191
2	Enforcer	Zeneca Prof.	360 g/l	SL	07079
3	Fungo	Dax		SL	H4768
4	Fungo Exterior	Dax	340 g/l	SL	H4768
5	Halophen RE 49	McMillan	101.4 g/l	SL	04636
6	Mascot Mosskiller	Rigby Taylor	340 g/l	SL	02439
7	Nomix-Chipman Mosskiller	Nomix-Chipman	340 g/l	SL	06271
8	Panacide M	Coalite	360 g/l	SL	05611
9	Panacide M21	Coalite	165 g/l	SL	H4923
10	Panacide TS	Coalite	480 g/l	SL	05612
11	Panaclean 736	Coalite	45 g/l	SL	H5075
12	Super Mosstox	RP Environ.	340 g/l	SL	05339

FOR FULL CONDITIONS OF USE ALWAYS READ THE PRODUCT LABEL

Uses   Algae, fungus diseases, mosses, red thread in TURF [1-4, 6-8, 10, 12]. Algae, fungus diseases, mosses in HARD SURFACES [1-4, 6, 9, 11, 12]. Algae, fungus diseases, mosses in PATHS AND DRIVES [1, 3, 4, 6, 12]. Black leg, gangrene, silver scurf, soft rot in POTATOES *(in store)* [5]. Fungus diseases in MUSHROOM BOXES [3, 4].

Notes   **Efficacy**
* For moss control in turf spray at any time when moss is growing. Supplement spraying with other measures to improve fertility and drainage
* Rake out dead moss 2-3 wk after treatment
* To control mushroom diseases use on trays, floors etc as part of an environmental hygiene programme [3, 4]
* Apply within 10 d of storing potatoes [5]

Crop Safety/Restrictions
* Maximum number of treatments 2 per batch of seed [5]
* Must not be applied to growing mushroom crops [3, 4]

Special precautions/Environmental safety
* Harmful if swallowed [3, 4, 7, 8, 10, 11]. Irritating to eyes and skin [1, 3-5, 7-10, 12]
* Irritating to respiratory system [5]
* May cause sensitisation by skin contact and by inhalation [5]
* Only to be used by professional operators, keep unprotected persons and animals away while treatment in progress [1, 5-8]
* Operate fogging equipment outside the store [5]
* Prevent surface run-off from entering storm drains [1, 8-11]
* Harmful [6, 12], dangerous [1, 7-11], to fish or other aquatic life. Do not contaminate surface waters or ditches with chemical or used container
* Store product between 5°C and 30 °C out of direct sunlight [5]

Protective clothing/Label precautions
* A, C, H, M [1-4, 7, 8, 10, 12]; A, C, H [5, 6]; A, E, H [11]; A, H [9]
* 5a, 26 [7]; 5c, 6a, 22, 29, 37, 53, 63, 65 [3, 4]; 6a, 6b, 29 [1, 5, 7-11]; 6c, 10a, 10b, 14a, 38 (24 h) [5]; 14, 39 [1, 8-11]; 16 [1, 5, 7, 8, 10]; 18, 36, 37 [1, 5, 7-12]; 21 [1, 6, 12]; 22 [5, 8-11]; 27 [1, 7]; 28, 67 [9, 11]; 30 [12]; 38 (48 h) [1, 2, 9]; 38 [8, 10, 11]; 52 [1, 7-11]; 53 [5, 6, 12]; 55a [1, 8, 9, 11]; 63 [1, 5, 6, 8-12]; 65 [1, 5, 6, 8, 10, 12]; 78 [8, 10]

Withholding period
* Keep unprotected persons out of treated store for at least 24 h [5]
* Keep unprotected persons out of treated areas for 48 h or until surfaces are dry [1]

Latest application/Harvest Interval(HI)
* 3 wk before removal from store for sale or processing [5]

Approval
* Product formulated for use undiluted through fogging machine. See notes on ULV application in Section 1 [5]

197   dichlorophen + ferrous sulfate
A mosskiller/fertilizer mixture for use on turf

Products	1 Aitken's Lawn Sand Plus	Aitken	0.3:10% w/w	GR	04542
	2 SHL Lawn Sand Plus	Sinclair	0.3:10% w/w	GR	04439

Uses   Mosses in AMENITY TURF.

Notes   **Efficacy**
* Apply to established turf from late spring to early autumn when soil moist but not when grass wet or damp with dew
* Heavy infestations may need a repeat treatment

Crop Safety/Restrictions
* Maximum number of treatments 2 per yr
* Do not treat newly sown grass or freshly laid turf for the first year
* Do not apply during drought or freezing conditions or when rain imminent
* Avoid walking on treated areas until it has rained or turf has been watered
* Do not mow for 3-4 d before or after application

Special precautions/Environmental safety
* Harmful to fish or other aquatic life. Do not contaminate surface waters or ditches with chemical or used container

Protective clothing/Label precautions
* 22, 29, 53, 63

198   1,3-dichloropropene
A halogenated hydrocarbon soil nematicide

Products   Telone II   DowElanco   94% w/w   VP   05749

Uses   Free-living nematodes, potato cyst nematode in POTATOES. Nematodes, stem nematodes, virus vectors in STRAWBERRIES. Nematodes, virus vectors in HOPS, RASPBERRIES. Stem and bulb nematodes in NARCISSI.

Notes   **Efficacy**
* Before treatment soil should be in friable seed bed condition above 5°C, with adequate moisture and all crop remains decomposed. Tilling to 30 cm improves results
* Do not use on heavy clays or soils with many large stones
* Apply with sub-surface injector combined with soil-sealing roller or, in hops, with a hollow tined injector. See label for details of suitable machinery
* Leave soil undisturbed for 21 d after treatment. Wet or cold soils need longer exposure
* For control of potato-cyst nematode contact firm's representative, check nematode level and use in integrated control programme
* For use in hops apply in May or Jun following the yr of grubbing and leave as long as possible before replanting
* Do not use water to clean out apparatus nor use aluminium or magnesium alloy containers which may corrode

Crop Safety/Restrictions
* Maximum number of treatments 2 per yr for hops, 1 per yr for other crops
* Do not drill or plant until odour of fumigant is eliminated

FOR FULL CONDITIONS OF USE ALWAYS READ THE PRODUCT LABEL

• Do not use fertilizer containing ammonium salts after treatment. Use only nitrate nitrogen fertilizers until crop well established and soil temperature above 18°C
• Allow at least 6 wk after treatment before planting raspberries or strawberries

Special precautions/Environmental safety
• Toxic if swallowed. Harmful in contact with skin and by inhalation
• Irritating to eyes, skin and respiratory system
• Flammable
• Do not contaminate surface waters or ditches with chemical or used container

Protective clothing/Label precautions
• A, C, M
• 4c, 5a, 5b, 6a, 6b, 6c, 12c, 14, 16, 18, 22, 26, 27, 28, 29, 36, 37, 54, 64, 65, 70, 79

199 dichlorprop

A translocated phenoxy herbicide for use in cereals

See also 2,4-D + dichlorprop + MCPA + mecoprop
dicamba + dichlorprop + MCPA

Products	MSS 2,4-DP	Mirfield	500 g/l	SL	01394

Uses   Annual dicotyledons, black bindweed, perennial dicotyledons, redshank in BARLEY, OATS, RYE, UNDERSOWN CEREALS, WHEAT.

Notes   **Efficacy**
• Best results achieved by application to young seedling weeds in good growing conditions in a strongly competing crop
• Effectiveness may be reduced by rain within 6 h of spraying, by very cold weather or by drought conditions

Crop Safety/Restrictions
• Spray winter crops in spring from fully tillered stage to before first node detectable (GS 29-30), spring crops from 1-leaf unfolded to before first node detectable (GS 11-30)
• Spray crops undersown with grass as soon as possible after grass begins to tiller
• Do not spray crops undersown with legumes, lucerne, peas or beans
• Do not roll or harrow within 7 d before or after treatment
• Avoid spray drifting onto nearby susceptible crops

Special precautions/Environmental safety
• Harmful in contact with skin and if swallowed

Protective clothing/Label precautions
• 5a, 5c, 18, 21, 29, 36, 37, 43, 54, 63, 66, 70, 78

Withholding period
• Keep livestock out of treated areas until foliage of any poisonous weeds such as ragwort has died and become unpalatable

Maximum Residue Levels (mg residue/kg food)
• tea, hops 0.1; all other products (except cereals and animal products) 0.05

200 dichlorprop + ferrous sulfate + MCPA
A herbicide/fertilizer combination for moss and weed control in turf

Products SHL Turf Feed and Weed + Sinclair 0.474:20:0.366% DP 04438
Mosskiller w/w

Uses Annual dicotyledons, mosses, perennial dicotyledons in AMENITY TURF.

Notes **Efficacy**
* Apply from late spring to early autumn when grass in active growth and soil moist
* A repeat treatment may be needed after 4-6 wk to control perennial weeds or if moss regrows

Crop Safety/Restrictions
* Do not treat newly sown grass or freshly laid turf for the first year
* Do not apply during drought or freezing conditions or when rain imminent
* Avoid walking on treated areas until it has rained or turf has been watered
* Do not mow for 3-4 d before or after application
* Do not use first 4 mowings after treatment for composting

Special precautions/Environmental safety
* Harmful to fish or other aquatic life. Do not contaminate surface waters or ditches with chemical or used container

Protective clothing/Label precautions
* 22, 29, 53, 63

Approval
* Approval expiry: 20 Jun 95

Maximum Residue Levels (mg residue/kg food)
* see dichlorprop entry

201 dichlorprop + MCPA
A translocated herbicide for use in cereals and turf

See also dichlorprop

Products

1 Campbell's Redipon Extra	MTM Agrochem.	350:150 g/l	SL	00420
2 MSS 2,4-DP+MCPA	Mirfield	500 g/l (total)	SL	01396
3 Redipon Extra	United Phosphorus	350:150 g/l	SL	07518
4 SHL Turf Feed and Weed	Sinclair	0.474:0.366 g/l	DP	04437

Uses Annual dicotyledons, black bindweed, hemp-nettle, perennial dicotyledons, redshank in BARLEY, OATS, WHEAT [1-3]. Annual dicotyledons, buttercups, clovers, daisies, dandelions, perennial dicotyledons in AMENITY GRASS [4].

FOR FULL CONDITIONS OF USE ALWAYS READ THE PRODUCT LABEL

Notes

Efficacy
- Best results achieved by application to young seedling weeds in active growth
- Do not spray during cold weather, if rain or frost expected, if crop wet or in drought
- Use on turf from Apr-Sep. Avoid close mowing for 3-4 d before or after treatment [4]

Crop Safety/Restrictions
- Apply to winter cereals in spring from fully tillered, leaf-sheath erect stage but before first node detectable (GS 31), to spring barley when crop has 5 leaves unfolded but before first node detectable (GS 15-31), to spring oats when 2 leaves unfolded but before first node detectable (GS 12-31) [1]
- Apply to winter and spring cereals from fully tillered stage before 2nd node detectable (GS 32) [2]
- Do not use on undersown crops
- Do not spray crops suffering from herbicide damage or stress
- Do not roll or harrow for 7 d before or after spraying
- Avoid drift of spray onto nearby susceptible crops
- Do not use on turf in year of sowing [4]

Special precautions/Environmental safety
- Harmful if swallowed and in contact with skin
- Harmful to fish or other aquatic life. Do not contaminate surface waters or ditches with chemical or used container [2]

Protective clothing/Label precautions
- A, C [1, 3]; P [2]
- 5a, 5c, 18, 21, 28, 30, 36, 37, 43, 63, 66, 70, 78 [1-4]; 53 [2]; 54 [1, 3, 4]

Withholding period
- Keep livestock out of treated areas until foliage of any poisonous weeds such as ragwort has died and become unpalatable

Maximum Residue Levels (mg residue/kg food)
- see dichlorprop entry

---

202   dichlorvos

A contact and fumigant organophosphorus insecticide

Products

1 Darlingtons Dichlorvos	Darmycel	500 g/l	EC	05699
2 Nuvan 500 EC	Ciba Agric.	500 g/l	EC	03861

Uses

Beetles, flies, mosquitoes, poultry ectoparasites in POULTRY HOUSES [2]. Mushroom flies in MUSHROOM HOUSES [1]. Non-indigenous leaf miners, western flower thrips in PROTECTED BRASSICA SEEDLINGS *(off-label)*, PROTECTED HERBS *(off-label)*, PROTECTED ORNAMENTALS *(off-label)*, PROTECTED VEGETABLES *(off-label)* [2].

Notes

Efficacy
- Spray walls, roof, floor and ventilators of mushroom houses. Avoid spraying mushroom beds. Use as an aerosol spray or wet spray [1]
- For use in poultry houses apply as surface spray, cold fog or mist using suitable applicator. Direct away from poultry. See label for details [2]

Crop Safety/Restrictions
* Do not use on chrysanthemums in flower or on roses. Phytotoxicity may occur on cucumbers and other cucurbits

Special precautions/Environmental safety
* Dichlorvos is subject to the Poisons Rules 1982 and the Poisons Act 1972. See notes in Section 1
* This product contains an anticholinesterase organophosphorus compound. Do not use if under medical advice not to work with such compounds
* Toxic in contact with skin or if swallowed
* Very toxic by inhalation
* May cause sensitisation by skin contact
* Flammable
* Keep in original container, tightly closed, in a safe place, under lock and key
* Keep unprotected persons out of treated areas for at least 12 h
* Harmful to game, wild birds and animals
* Extremely dangerous to bees. Do not apply to crops in flower or to those in which bees are actively foraging. Do not apply when flowering weeds are present
* Extremely dangerous to fish or other aquatic life. Do not contaminate surface waters or ditches with chemical or used container

Protective clothing/Label precautions
* A, C, H
* 1, 4a, 4b, 4c, 12c, 14, 16, 18, 21, 25, 28, 29, 35, 36, 37, 38, 46, 48a, 51, 64, 66, 70, 79

Latest application/Harvest Interval(HI)
* HI Brussels sprouts, cabbage 14 wk; cauliflower 9 wk; broccoli, calabrese 7 wk; kohlrabi 1 wk; cucurbits 48 h; other edible crops 24 h

Approval
* May be applied as a ULV spray or as a fog. See label for details and for protective clothing requirements for fogging treatment. See notes in Section 1 on ULV application [1, 2]
* Off-label Approval unlimited for high volume application to protected vegetables for control of western flower thrips (OLA 1590/94) [2], (OLA 0377/95)[1]; unlimited for high volume application to protected non-edible ornamentals and protected vegetables for control of western flower thrips (OLA 0882/92) [2]; to Nov 1997 (OLA 1388/93 [2]) or unlimited (OLA 1387/93 [2]) for use as a fog in a range of protected non-edible ornamentals and protected edible crops for control of western flower thrips and non-indigenous leaf miners. See approval notices for details of crops which may be treated but note that all off-label uses on protected crops of asparagus, celery, chicory, Chinese cabbage, fennel, kale, leeks, marrows, melons and squashes were revoked in July 1994

Maximum Residue Levels (mg residue/kg food)
* lettuce 1; carrots, horseradish, parsnips, parsley root, salsify, swedes, turnips, garlic, onions, shallots, tomatoes, peppers, aubergines, cucumbers, gherkins, courgettes, cauliflowers, Brussels sprouts, head cabbages, beans (with pods), peas (with pods), celery, leeks, rhubarb, cultivated mushrooms, potatoes 0.5; citrus fruits, pome fruits, apricots, peaches, nectarines, plums, grapes, cane fruits, bilberries, cranberries, currants, gooseberries, bananas 0.1; meat, eggs 0.05; milk 0.02

FOR FULL CONDITIONS OF USE ALWAYS READ THE PRODUCT LABEL

203 diclofop-methyl

A translocated phenoxypropionic herbicide for grass weed control

Products					
1 Hoegrass	AgrEvo	378 g/l	EC	07323
2 Hoegrass	Hoechst	378 g/l	EC	01063

Uses

Annual grasses, awned canary grass, blackgrass, rough meadow grass, ryegrass, wild oats, yorkshire fog in BARLEY, BROAD BEANS, BROCCOLI, BRUSSELS SPROUTS, CABBAGES, CARROTS, CELERY, DURUM WHEAT, DWARF BEANS, FIELD BEANS, LETTUCE, LINSEED, LUCERNE, LUPINS, MUSTARD, OILSEED RAPE, ONIONS, PARSNIPS, PEAS, POTATOES, RED BEET, SUGAR BEET, TRITICALE, WHEAT, WINTER RYE.

Notes

Efficacy

* Best results when applied in autumn, winter or spring to seedling grasses up to 3-4 leaf stage. See label for details of rates and timing. Annual meadow-grass not controlled
* Do not spray if foliage wet or covered in ice. Effectiveness reduced by dry conditions
* Spray is rainfast within 1 h of application
* Do not spray after a sudden drop in temperature or a period of warm days/cold nights
* Recommended for tank mixture with a range of herbicides, growth regulators and other pesticides. See label for details
* Do not mix with hormone herbicides

Crop Safety/Restrictions

* Maximum number of treatments 2 per crop (up to a maximum of 4.5 l/ha) for cereals; 1 per crop for broad-leaved crops
* Apply to wheat, rye and triticale at any time after crop emergence
* Apply to winter barley from emergence to no later than leaf sheath erect (GS 07-30), to spring barley from emergence to 4-fully expanded leaves and 2-tillers (GS 07-22)
* Apply to broad-leaved crops from 100% emergence (cotyledons fully expanded), to onions from 1-true leaf onward (not at crook stage)
* In Scotland and Northern England (north of North Yorkshire) do not treat spring cereals; winter cereals may only be treated until the end of Feb and only at the low rate
* Do not use on cereal crops undersown with grasses
* Do not roll or harrow cereals within 7 d of spraying
* Do not spray crops under stress, suffering from drought or waterlogging or if soil compacted

Special precautions/Environmental safety

* Flammable
* Dangerous to fish or other aquatic life. Do not contaminate surface waters or ditches with chemical or used container

Protective clothing/Label precautions

* A, C, H
* 12c, 21, 29, 35, 52, 63, 66

Withholding period

* Keep livestock out of treated areas for at least 7 d

Latest application/Harvest Interval(HI)

* Before first node detectable (GS 31) for winter barley; before 5 expanded leaves and 3 tillers (GS 15, 23) for spring barley.

• HI wheat, rye, triticale, broad-leaved crops 6 wk

204 diclofop-methyl + fenoxaprop-P-ethyl
A foliar acting herbicide for grass control in barley

Products	1 Tigress	AgrEvo	313:14 g/l	EC	07337
	2 Tigress	Hoechst	313:14 g/l	EC	05976

Uses Blackgrass, ryegrass, wild oats in SPRING BARLEY, WINTER BARLEY.

Notes

Efficacy
• Apply from emergence of crop up to and including flag leaf fully emerged (GS 39)
• Wild oats and blackgrass controlled from 2 fully expanded leaves to end of tillering but before 1st node detectable, ryegrass from 2 fully expanded leaves to mid-tillering
• Spray is rainfast from 1 h after spraying. Do not apply to wet or icy foliage
• Can be sprayed in frosty weather provided crop hardened off
• Any conditions resulting in moisture stress may reduce effectiveness

Crop Safety/Restrictions
• Maximum number of treatments 2 per crop (up to a maximum of 4.5 l/ha)
• Do not apply to undersown crops or those due to be undersown
• Broadcast crops should be sprayed post-emergence after root system well established
• Do not roll or harrow within 1 wk
• Do not spray crops under stress from drought, waterlogging etc
• Do not spray immediately before or after a sudden drop in temperature, a period of warm days/cold nights or when extremely low temperatures forecast

Special precautions/Environmental safety
• Irritant. May cause sensitization by skin contact
• Dangerous to fish or other aquatic life. Do not contaminate surface waters or ditches with chemical or used containers

Protective clothing/Label precautions
• A, C, H
• 10a, 18, 29, 35, 36, 37, 52, 63, 66

Withholding period
• Keep livestock out of treated areas for at least 7 d

Latest application/Harvest Interval(HI)
• Before flag leaf sheath erect (GS 41).
• HI 6 wk

FOR FULL CONDITIONS OF USE ALWAYS READ THE PRODUCT LABEL

205   dicloran

A protectant nitroaniline fungicide used as a glasshouse fumigant

Products   Fumite Dicloran Smoke   Hortichem   14 g   FU   00930

Uses   Botrytis, rhizoctonia in CUCUMBERS, PROTECTED LETTUCE, PROTECTED ORNAMENTALS, TOMATOES.

Notes   **Efficacy**
* Treat at first sign of disease and at 14 d intervals if necessary
* For best results fumigate in late afternoon or evening when temperature is at least 16°C. Keep house closed overnight or for at least 4 h. Do not fumigate in bright sunshine, windy conditions or when temperature too high
* Water plants, paths and straw used as mulches several hours before fumigating. Ensure that foliage is dry before treatment

Crop Safety/Restrictions
* Do not fumigate young seedlings or plants being hardened off
* Treat only crops showing strong growth. Do not treat crops which are dry at the roots
* Cut all open carnation or chrysanthemum blooms before fumigating
* Test treat small numbers of new or unusual plants before committing whole batches

Special precautions/Environmental safety
* Irritating to eyes and respiratory system
* Ventilate glasshouse thoroughly before re-entering
* Harmful to fish or other aquatic life. Do not contaminate surface waters or ditches with chemical or used container

Protective clothing/Label precautions
* 6a, 6c, 18, 28, 29, 35, 36, 53, 63, 67

Latest application/Harvest Interval(HI)
* HI cucumbers, tomatoes 2 d; lettuce 14 d

206   dicofol

A non-systemic organochlorine acaricide for horticultural use

Products
1 Fumite Dicofol Smoke	Hortichem	5 or 10 g	FU	00931
2 Kelthane	Rohm & Haas	18.5% w/w	EC	01131

Uses   Red spider mites, tarsonemid mites in STRAWBERRIES [2]. Red spider mites in TOMATOES [1, 2]. Red spider mites in LETTUCE, ORNAMENTALS, PROTECTED CUCUMBERS [1]. Red spider mites in APPLES, CUCUMBERS, HOPS [2].

Notes   **Efficacy**
* Spray apples when winter eggs hatched and summer eggs laid and repeat 3 wk later if necessary. Spray other crops at any time before stated harvest interval
* Strawberries may be sprayed after picking where necessary
* Apply fumigant treatment at first sign of infestation and repeat every 14 d

- Do not fumigate in bright sunshine, when foliage wet or roots dry
- Resistance has developed in some areas, in which case sprays are unlikely to give satisfactory control, especially after second or subsequent application

Crop Safety/Restrictions
- Maximum number of treatments 1 per crop for protected tomato, cucumber, lettuce and ornamentals [1]; 2 per yr for apples and hops, 2 per crop for strawberries, protected cucumbers and tomatoes [2]
- Do not fumigate young seedlings or plants which are being hardened off
- Consult supplier before treating crops other than those listed on label [1]
- Do not spray apples within 4 wk of petal fall
- Do not spray cucumber or tomato seedlings before mid-May or in bright sunshine

Special precautions/Environmental safety
- Harmful in contact with skin and if swallowed [2]
- Irritating to eyes and respiratory system [1]. May cause sensitization by skin contact [2]
- Flammable [2]
- Keep unprotected persons out of treated areas for at least 4.5 h after treatment [1]

Protective clothing/Label precautions
- A, C [2]
- 5a, 5c, 10a, 12c, 14, 21, 25, 54, 63, 65, 70, 78 [2]; 6a, 6c, 67 [1]; 18, 28, 29, 35, 36, 37 [1, 2]

Latest application/Harvest Interval(HI)
- HI edible crops 2 d [1]; tomatoes, cucumbers 2 d [2]; strawberries 7 d [2]; apples 2 wk [2]; hops 4 wk [2]

Approval
- Approval expiry: 31 Oct 96 [00931]

Maximum Residue Levels (mg residue/kg food)
- citrus fruits, pome fruits, apricots, peaches, nectarines, plums, grapes, cane fruits, bilberries, cranberries, currants, gooseberries, bananas, carrots, horseradish, parsnips, parsley root, salsify, swedes, turnips, garlic, onions, shallots, cauliflowers, Brussels sprouts, head cabbages, lettuce, beans (with pods), peas (with pods), celery, leeks, rhubarb, cultivated mushrooms, potatoes 5; cucumbers, gherkins, courgettes 2; tomatoes, peppers, aubergines 1

207   dicofol + tetradifon
A contact acaricide mixture for fruit and glasshouse crops

Products	Childion	Hortichem	170:62.5 g/l	EC	03821

Uses   Broad mite, red spider mites, tarsonemid mites in PROTECTED CUCUMBERS, PROTECTED ORNAMENTALS, PROTECTED TOMATOES. Bryobia mites, leaf and bud mite, red spider mites in APPLES, PEARS. Red spider mites, tarsonemid mites in STRAWBERRIES. Red spider mites in BLACKCURRANTS, HOPS.

FOR FULL CONDITIONS OF USE ALWAYS READ THE PRODUCT LABEL

Notes

Efficacy
* Apply to hops and glasshouse crops as soon as mites appear. Other crops should be sprayed post blossom, strawberries and blackcurrants being treated only if mites appear. Repeat as necessary in all cases

Crop Safety/Restrictions
* Maximum number of treatments 2 per yr for apples, pears; 2 per yr pre-harvest, 1 per yr post-harvest for strawberries, blackcurrants
* A minimum of 3 wk must elapse between treatments for tomatoes, cucumbers, ornamentals
* Do not apply to apples or pears until 3 wk after petal fall, to strawberries during blossoming or to young plants or cucumbers in bright sunshine or before mid-May
* Treat small numbers of new glasshouse subjects before committing whole batches
* Do not apply to cissus, dahlias, ficus, gloxinias, impatiens, kalanchoes, primulas or stephanotis

Special precautions/Environmental safety
* Harmful in contact with skin or if swallowed. Irritating to eyes and skin
* May cause sensitization by skin contact
* Flammable

Protective clothing/Label precautions
* A, C
* 5a, 5c, 6a, 6b, 10a, 12c, 18, 21, 29, 36, 37, 54, 63, 65, 70, 78

Latest application/Harvest Interval(HI)
* HI apples, pears 14 d; hops 28 d; strawberry, blackcurrent 7 d; glasshouse tomatoes and cucumbers 2 d

Maximum Residue Levels (mg residue/kg food)
* see dicofol entry

208 dieldrin

A persistent organochlorine insecticide approvals for which were revoked in May 1989

209 dienochlor

An organochlorine acaricide for use on ornamentals

Products	Pentac Aquaflow	DowElanco	480 g/l	EC	05741

Uses

Glasshouse red spider mite in PROTECTED ORNAMENTALS.

Notes

Efficacy
* Spray roses when pest first seen and repeat 7 d later
* Spray ornamentals when pest first seen. One application normally sufficient but when problem severe repeat 7 d later
* Add wetter/spreader and spray surfaces to point of run-off

Crop Safety/Restrictions
* Maximum number of treatments 2 per yr

- Do not use on rare or unusual plants without first testing on small scale

Special precautions/Environmental safety
- Irritating to eyes and skin
- Dangerous to fish or other aquatic life. Do not contaminate surface waters or ditches with chemical or used container
- Do not use on edible crops

Protective clothing/Label precautions
- A, C
- 6a, 6b, 18, 21, 28, 29, 34, 36, 37, 52, 63, 66, 70

210   difenacoum

An anticoagulant coumarin rodenticide

See also calciferol + difenacoum

Products					
1 Deosan Rataway	Deosan	0.005% w/w	RB	05560
2 Deosan Rataway Bait Bags	Deosan	0.005% w/w	RB	05562
3 Killgerm Rat Rods	Killgerm	0.005% w/w	RB	05154
4 Killgerm Wax Bait	Killgerm	0.005% w/w	RB	04096
5 Neokil	Sorex	0.005% w/w	RB	05564
6 Neosorexa Ratpacks	Sorex	0.005% w/w	RB	04653
7 Neosorexa Ready to Use Rat and Mouse Bait	Sorex	0.005% w/w	RB	01474
8 Ratak	ICI Professional	0.005% w/w	RB	02586
9 Ratak	Zeneca Prof.	0.005% w/w	RB	06828
10 Ratak Wax Blocks	ICI Professional	0.005% w/w	RB	04217

Uses   Mice, rats in FARM BUILDINGS, FARMYARDS.

Notes   **Efficacy**
- Chemical is a chronic poison and rodents need to feed several times before accumulating a lethal dose. Effective against rodents resistant to other commonly used anticoagulants
- Apply ready-to-use baits in baiting programme
- Lay small baits about 1 m apart throughout infested areas for mice, larger baits for rats near holes and along runs
- Cover baits by placing in bait boxes, drain pipes or under boards
- Inspect bait sites frequently and top up as long as there is evidence of feeding

Special precautions/Environmental safety
- Only for use by farmers, horticulturists and other professional users
- Cover bait to prevent access by children, animals or birds

Protective clothing/Label precautions
- 25, 29, 63, 67, 81, 82, 83, 84

FOR FULL CONDITIONS OF USE ALWAYS READ THE PRODUCT LABEL

211   difenoconazole

A diphenyl-ether triazole protectant and curative fungicide for wheat

Products   Plover   Ciba Agric.   250 g/l   EC   07232

Uses   Brown rust, septoria, yellow rust in WINTER WHEAT.

Notes   **Efficacy**
* For most effective control of Septoria, apply as part of a programme of sprays which includes a suitable flag leaf treatment
* Adequate control of yellow rust may require an earlier appropriate treatment
* Product is fully rainfast 2 h after application

Crop Safety/Restrictions
* Maximum number of treatments 1 per crop
* Apply at any time from ear fully emerged stage but before early milk-ripe stage (GS 59-73)

Special precautions/Environmental safety
* Irritating to eyes
* May cause sensitization by skin contact
* Harmful to fish or other aquatic life. Do not contaminate surface waters or ditches with chemical or used container
* May have adverse effect on ladybirds and parasites of insects

Protective clothing/Label precautions
* A, C
* 6a, 10a, 18, 22, 29, 36, 37, 53, 63, 65

Latest application/Harvest Interval(HI)
* Before grain early milk-ripe stage (GS 73)

212   difenzoquat

A post-emergence quaternary ammonium herbicide for wild oat control

Products   1 Avenge 2   Cyanamid   150 g/l   SL   03241
               2 Match   Cyanamid   64% w/w   SG   07186

Uses   Powdery mildew in WINTER BARLEY *(reduction)*, WINTER WHEAT *(reduction)* [2]. Wild oats in BARLEY, DURUM WHEAT, MAIZE, RYE, RYEGRASS SEED CROPS, TRITICALE, WHEAT [1].

Notes   **Efficacy**
* Autumn and winter spraying controls wild oats from 2-leaf stage to mid-tillering, spring treatment to end of tillering [1]
* Dose rate varies with season and growth-stage of weed. See label for details [1]
* Spring treatment on barley provides control of powdery mildew as well as wild oats [1]
* Do not apply if rain expected within 6 h
* Recommended for tank mixture with a range of herbicides, growth regulators and other pesticides. See label for details. Other products can be applied after 24 h

Crop Safety/Restrictions
* Maximum number of treatments 2 per crop for winter cereals, durum wheat and ryegrass seed crops; 1 per crop for spring cereals and maize [1]
* Maximum total dose 0.94 kg/ha per crop [2]
* Apply to all varieties of winter wheat and winter barley when mildew is present on lower leaves and before it spreads to new growth [2]
* Apply to crops from 2-leaf stage to 50% of plants with 3 nodes detectable (GS 12-33) [1]
* Only apply to named cultivars of wheat, durum wheat and triticale. See label for details [1]
* Apply to crops undersown with ryegrass and clover either before emergence or after grass has reached 3-leaf stage [1]
* Spray ryegrass seed crops after 3-leaf stage, maize as soon as weeds at susceptible stage [1]
* Do not spray crops suffering stress from waterlogging, drought or other factors [1]
* Temporary yellowing may follow application, especially under extremes of temperature, but is normally outgrown rapidly

Special precautions/Environmental safety
* Harmful if swallowed and in contact with skin
* Irritating to skin [1]
* Risk of serious damage to eyes
* Harmful to fish or other aquatic life. Do not contaminate surface waters or ditches with chemical or used container
* Harmful to livestock

Protective clothing/Label precautions
* A, C [1, 2]
* 5a, 6b, 41, 65 [1]; 5c, 9, 18, 21, 26, 27, 28, 29, 36, 37, 53, 63, 70, 78 [1, 2]; 43 (6 wk), 67 [2]

Withholding period
* Keep all livestock out of treated areas for at least 6 wk

Latest application/Harvest Interval(HI)
* Before flag leaf just visible (GS 39) for cereals and maize; 6 wk before grazing for ryegrass seed crops [1]
* Before early milk stage (GS 73) in winter wheat; before medium milk stage (GS 75) in winter barley

213   diflubenzuron
A selective, persistent, contact and stomach acting insecticide

Products	Dimilin WP	Zeneca	25% w/w	WP	06656

Uses	Browntail moth, caterpillars, pine looper, winter moth in FORESTRY. Cabbage moth, cabbage white butterfly, diamond-back moth in BRASSICAS. Caterpillars, codling moth, earwigs, leaf miners, pear sucker, rust mite, sawflies, tortrix moths, winter moth in PEARS. Caterpillars, codling moth, earwigs, leaf miners, rust mite, sawflies, tortrix moths, winter moth in APPLES. Caterpillars in AMENITY TREES AND SHRUBS, ORNAMENTALS, PEPPERS, PROTECTED

FOR FULL CONDITIONS OF USE ALWAYS READ THE PRODUCT LABEL

CROPS, TOMATOES. Plum fruit moth, tortrix moths, winter moth in PLUMS. Sawflies, winter moth in BLACKCURRANTS. Sciarid flies in MUSHROOMS.

Notes

Efficacy
- Acts by disrupting chitin synthesis and preventing hatching of eggs. Most active on young caterpillars and most effective control achieved by spraying as eggs start to hatch
- Dose and timing of spray treatments vary with pest and crop. See label for details
- Addition of wetter recommended for use on brassicas and pears
- Negligible effect on Phytoseiulus or Encarsia used for biological control or on bees
- Apply as case mixing treatment on mushrooms or as post-casing drench

Crop Safety/Restrictions
- Maximum number of treatments 3 per yr for apples, pears, plums, blackcurrants; 2 per crop for brassicas; 4 per crop for tomatoes, peppers; 1 per crop for mushrooms
- Consult supplier regarding range of ornamentals which can be treated and check for varietal differences on a small sample in first instance
- Do not use as a compost drench or incorporated treatment on ornamental crops

Protective clothing/Label precautions
- 29, 35, 54, 63, 67, 70

Latest application/Harvest Interval(HI)
- HI tomatoes, peppers 24 h; apples, pears, plums, blackcurrants, brassicas 14 d

Maximum Residue Levels (mg residue/kg food)
- citrus fruits, pome fruits, plums, tomatoes, peppers, aubergines, Brussels sprouts, head cabbages 1; cultivated mushrooms 0.1; meat, milk, eggs 0.05

---

214 diflufenican

A shoot absorbed anilide herbicide available only in mixtures

See also clodinafop-propargyl + diflufenican
clopyralid + diflufenican + MCPA

---

215 diflufenican + isoproturon

A contact and residual herbicide for use in winter cereals

Products	Panther	RP Agric.	50:500 g/l	SC	06491

Uses

Annual dicotyledons, annual grasses, blackgrass, wild oats in TRITICALE, WINTER BARLEY, WINTER RYE, WINTER WHEAT.

Notes

Efficacy
- May be applied in autumn or spring (but only post-emergence on wheat or barley). Best control normally achieved by early post-emergence treatment
- Best results by application to fine, firm seedbed moist at or after application
- Weeds controlled from before emergence to 6 true leaf stage
- Apply to moist, but not waterlogged, soils
- Any trash or ash should be buried during seedbed preparation

Crop Safety/Restrictions
* Maximum number of treatments 1 per crop. Maximum total dose of isoproturon 2.5 kg a.i./ha per yr
* Spray winter wheat and barley post-emergence before crop reaches second node detectable stage (GS 32)
* On triticale and winter rye only treat named varieties and apply as pre-emergence spray
* Drill crop to normal depth (25 mm) and ensure seed well covered
* Do not use on other cereals, broadcast or undersown crops or crops to be undersown
* Do not use on Sands or Very Light soils, or those that are very stony or gravelly or on soils with more than 10% organic matter
* Do not spray when heavy rain is forecast or on crops suffering from stress, frost, deficiency, pest or disease attack
* Do not harrow after application nor roll autumn-treated crops until spring
* See label for details of time and land preparation needed before sowing other crops (12 wk for most crops). In the event of crop failure winter wheat can be redrilled without ploughing, other crops only after ploughing - see label for details
* Successive treatments of any products containing diflufenican can lead to soil build up and inversion ploughing must precede sowing any non-cereal crop. Even where ploughing occurs some crops may be damaged

Special precautions/Environmental safety
* Harmful to fish or other aquatic life. Do not contaminate surface waters or ditches with chemical or used container
* Do not spray where soils are cracked, to avoid run-off through drains
* Avoid direct or indirect contamination of water by use of buffer strips

Protective clothing/Label precautions
* 30, 53, 60, 63, 66

Latest application/Harvest Interval(HI)
* Before second node detectable (GS 32) for wheat and barley; pre-emergence of crop for triticale and rye

---

216   diflufenican + terbuthylazine
A foliar and residual acting herbicide

Products	Bolero	Ciba Agric.	200:400 g/l	SC	07436

Uses	Annual dicotyledons, annual meadow grass in WINTER BARLEY, WINTER WHEAT.

Notes   **Efficacy**
* Apply to healthy crops from when crop has 1 leaf unfolded (GS 11) up to the end of tillering. Best results obtained from applications made early post-emergence of weeds
* Product is foliar and soil acting. Weeds germinating after treatment are controlled by root uptake
* Prolonged dry weather or heavy rain after treatment may reduce control

---

FOR FULL CONDITIONS OF USE ALWAYS READ THE PRODUCT LABEL

- Weed control may be reduced by presence of high levels of organic matter (over 10%) or surface trash

Crop Safety/Restrictions
- Maximum number of treatments 1 per crop
- Ensure seed covered by 2.5 cm consolidated soil before spraying
- Do not use on crops under stress or suffering from water-logging, pest attack, disease or frost. Do not spray when frost imminent
- Do not apply to undersown crops or those to be undersown
- Sep drilled crops may be damaged if treatment precedes or coincides with a period of rapid growth
- Do not use on Sand or Very Light soils. On stony or gravelly soils crop damage may occur if heavy rain follows treatment
- After harvest treated soils should be inversion ploughed to 15 cm before drilling field beans, leaf brassicas, winter oilseed rape, winter onions or sugar beet seed crops. Autumn sown cereals can be drilled as normal
- Restrictions apply to crops that may be sown after failure of a treated crop. See label for details of crops, cultivations and intervals
- Successive treatments of any products containing diflufenican can lead to soil build up and inversion ploughing must precede sowing any non-cereal crop. Even where ploughing occurs some crops may be damaged - see label

Special precautions/Environmental safety
- Harmful in contact with skin and if swallowed
- Harmful to fish or other aquatic life. Do not contaminate surface waters or ditches with chemical or used container

Protective clothing/Label precautions
- A
- 5a, 5c, 21, 29, 53, 60, 63, 66, 70

Latest application/Harvest Interval(HI)
- Before ear at 1 cm stage (GS 30)

---

217   diflufenican + trifluralin

A contact and residual herbicide for use in winter cereals

Products	Ardent		RP Agric.	40:400 g/l	SC	06203

Uses   Annual dicotyledons, annual meadow grass in WINTER BARLEY, WINTER WHEAT.

Notes   **Efficacy**
- Apply pre- or early post-emergence of crop up to maximum recommended weed size
- Weeds controlled pre-emergence or up to 4 leaf stage (2 leaf for annual meadow grass)
- Best results achieved on fine, firm, moist seedbeds
- Do not use on soils with more than 10% organic matter nor on waterlogged soil
- Any residues from straw burning should be buried or dispersed prior to drilling

Crop Safety/Restrictions
- Maximum number of applications 1 per crop
- Do not use on other cereals, nor on broadcast or undersown crops

- Ensure crop is evenly drilled and seed well covered
- Do not harrow after application
- Do not use on Sands or Very Light soils, or those that are very stony or gravelly
- Do not direct drill broad-leaved crops in autumn following treatment
- Do not broadcast oilseed rape or other brassicas after treatment of previous crop
- See label for details of time and land preparation needed before sowing other crops and for crops which may be planted in the event of crop failure
- Successive treatments of any products containing diflufenican can lead to soil build up and inversion ploughing must precede sowing any non-cereal crop. Even where ploughing occurs some crops may be damaged

Special precautions/Environmental safety
- Irritating to eyes
- Harmful to fish or other aquatic life. Do not contaminate surface waters or ditches with chemical or used container

Protective clothing/Label precautions
- A, C
- 6a, 18, 21, 25, 29, 36, 37, 53, 60, 63, 66, 70

Latest application/Harvest Interval(HI)
- Before second tiller (GS 22) or end of Nov, whichever is sooner

218   dikegulac

A growth regulator for use on hedges and ornamentals

Products	Atrinal	RP Environ.	200 g/l	SL	05389

Uses   Growth retardation in HEDGES, WOODY ORNAMENTALS. Increasing branching in AZALEAS, BEGONIAS, FUCHSIAS, KALANCHOES, ORNAMENTALS, VERBENA.

Notes   **Efficacy**
- For controlling hedge growth apply when in full leaf and growing actively, normally May-Jun. Cut just before spraying but leave a good leaf coverage
- Spray on a mild, calm day when no rain expected
- Recommended concentration varies with species. See label for details
- For promoting side branching in ornamentals mainly applied from Mar to Sep when plants actively growing. Details of treatment recommended vary with species. See label

Crop Safety/Restrictions
- Only spray healthy hedges at least 3 yr old
- Do not spray more than once in a season
- Slight, temporary yellowing of shoot tips may occur a few weeks after treatment

Protective clothing/Label precautions
- 29, 54, 63, 65

FOR FULL CONDITIONS OF USE ALWAYS READ THE PRODUCT LABEL

219   di-1-p-menthene

An antitranspirant and coating agent available only in mixtures

See also chlormequat + di-1-p-menthene

220   dimethoate

A contact and systemic organophosphorus insecticide and acaricide

See also chlorpyrifos + dimethoate

Products

1 Ashlade Dimethoate	Ashlade	400 g/l	EC	04814
2 Atlas Dimethoate 40	Atlas	400 g/l	EC	03044
3 BASF Dimethoate 40	BASF	400 g/l	EC	00199
4 Danadim Dimethoate 40	Cheminova	400 g/l	EC	07351
5 MTM Dimethoate 40	MTM Agrochem.	400 g/l	EC	05693
6 Turbair Systemic Insecticide	Fargro	10 g/l	UL	02250

Uses

Aphids, bryobia mites, capsids, red spider mites, sawflies, suckers in APPLES, PEARS [1-3, 5, 6]. Aphids, capsids, leafhoppers, red spider mites in CANE FRUIT [5, 6]. Aphids, capsids, red spider mites in GOOSEBERRIES [3, 6]. Aphids, cherry fruit moth, red spider mites in CHERRIES [5, 6]. Aphids, leaf miners, red spider mites in FLOWERS, ORNAMENTALS, WOODY ORNAMENTALS [1, 3, 5, 6]. Aphids, leaf miners in PROTECTED CARNATIONS, ROSES [5]. Aphids, mangold fly in FODDER BEET [4]. Aphids, pea midge, thrips in PEAS [1-6]. Aphids, red spider mites, sawflies in PLUMS [1-3, 5, 6]. Aphids, red spider mites in TOMATOES [3, 5]. Aphids, red spider mites in HOPS *(Fuggles only)* [3]. Aphids, red spider mites in BLACKCURRANTS [2, 3, 6]. Aphids, red spider mites in STRAWBERRIES [1-3, 5, 6]. Aphids, red spider mites in CURRANTS [5]. Aphids, wheat bulb fly in WHEAT [1-6]. Aphids in WATERCRESS *(off-label)* [3, 6]. Aphids in LETTUCE, PROTECTED LETTUCE [3, 5, 6]. Aphids in MANGEL SEED CROPS *(excluding Myzus persicae)*, SUGAR BEET SEED CROPS *(excluding Myzus persicae)* [3, 5]. Aphids in BROAD BEANS, FRENCH BEANS, RUNNER BEANS [3, 4, 6]. Aphids in CEREALS [3-6]. Aphids in RASPBERRIES [3]. Aphids in CELERY, PARSNIPS [6]. Aphids in BEET CROPS *(excluding Myzus persicae)* [1-3, 5, 6]. Aphids in POTATOES *(excluding Myzus persicae)* [1-6]. Aphids in GRASS SEED CROPS [1-3]. Aphids in CARROTS [1, 2, 4-6]. Aphids in BARLEY [1]. Aphids in FIELD BEANS [5, 6]. Aphids in MANGELS *(excluding Myzus persicae)*, RED BEET *(excluding Myzus persicae)*, SUGAR BEET *(excluding Myzus persicae)* [4]. Cabbage aphid in BRASSICAS [1-6]. Insect pests in BULB ONIONS *(off-label)*, CELERY *(off-label)*, GARLIC *(off-label)*, GRAPEVINES *(off-label)*, LEEKS *(off-label)*, MARROWS *(off-label)*, SALAD ONIONS *(off-label)*, SWEDES *(off-label)*, SWEETCORN *(off-label)*, TURNIPS *(off-label)* [3]. Mangold fly in BEET CROPS [1-3, 5, 6]. Mangold fly in MANGELS, RED BEET, SUGAR BEET [4].

Notes

Efficacy

- Chemical has quick knock-down effect and systemic activity lasts for up to 14 d
- With some crops, products differ in range of pests listed as controlled. Uses section above provides summary. See labels for details
- For most pests apply when pest first seen and repeat 2-3 wk later or as necessary. Timing and number of sprays varies with crop and pest. See label for details
- Best results achieved when crop growing vigorously. Systemic activity reduced when crops suffering from drought or other stress
- In hot weather apply in early morning or late evening
- Do not tank mix with alkaline materials. See label for recommended tank-mixes

- Populations of aphids and mites resistant to organophosphorus compound have developed in some areas, especially in beet crops; in such cases products are unlikely to give satisfactory control. Must not be used for control of peach-potato aphid (*Myzus persicae*)in potatoes or beet crops
- Consult processor before spraying crops grown for processing

Crop Safety/Restrictions
- Maximum number of treatments 8 per crop for hops, tomatoes; 7 per crop for seed potatoes; 6 per crop for brassicas, peas, outdoor lettuce, strawberries, swedes, turnips; 4 per crop for cereals, grass seed crops, carrots, apples, pears, plums, cherries, gooseberries, cane fruit; 3 per crop for blackcurrants, 2 per crop for beans, beet crops (1 per crop [4]), celery; 1 per crop for protected lettuce between Oct and Feb
- On carrots the maximum total dose applied per crop must not exceed the equivalent of 3 (on mineral soils) or 4 (on organic soils) full dose applications
- Do not treat chrysanthemums or ornamental prunus [2, 3, 5]. Do not treat peaches or plants in flower [2, 5, 6]
- Test for varietal susceptibility on all unusual plants or new cultivars

Special precautions/Environmental safety
- This product contains an anticholinesterase organophosphorus compound. Do not use if under medical advice not to work with such compounds
- Harmful in contact with skin/in contact with skin or if swallowed (products vary see label for details)
- May cause sensitisation by skin contact
- Flammable
- Harmful to game, wild birds and animals. Bury all spillages
- Harmful to livestock
- Likely to cause adverse effects on beneficial arthropods
- Dangerous to bees. Do not apply to crops in flower, except as directed on peas, or to those in which bees are actively foraging. Do not apply when flowering weeds are present
- Surface residues may also cause bee mortality following spraying
- Dangerous to fish or other aquatic life. Do not contaminate surface waters or ditches with chemical or used container
- Do not treat cereals after 1 Apr within 6 m of edge of crop
- Must not be applied to cereals if any product containing a pyrethroid insecticide or dimethoate has been sprayed after the start of ear emergence (GS 51)

Protective clothing/Label precautions
- A, C, H, J, M
- 1, 5c, 10a, 12c, 14, 28, 29, 46, 70, 78 [1-6]; 5a, 18, 36, 37 [1-3, 5, 6]; 6a, 6b, 21 [4]; 16, 41, 49, 66 [1, 3-6]; 22, 35 (7 d), 41 (7 d), 48, 65 [2]; 23, 25, 35, 52, 60, 64 [1, 3, 5, 6]; 53, 63 [2, 4]

Withholding period
- Keep all livestock out of treated areas for at least 7 d

Latest application/Harvest Interval(HI)
- Before 30 Jun in yr of harvest for potatoes and beet crops; before 31 Mar in yr of harvest for aerial application to cereals

FOR FULL CONDITIONS OF USE ALWAYS READ THE PRODUCT LABEL

• HI apples, pears 35 d; protected lettuce, blackcurrants 28 d; field beans, plums, cherries, raspberries, gooseberries, strawberries 21 d; cereals, grass seed crops, carrots, peas, beans, hops 14 d; watercress 10 d; brassicas, outdoor lettuce, tomatoes, celery 7 d

Approval
• Approved for aerial application on sugar beet, peas, potatoes, brassicas, carrots [2]; cereals, peas, ware potatoes, sugar beet [3, 4]; wheat, barley, oats, peas, potatoes, fodder beet, sugar beet, mangels [5]; wheat, peas, ware potatoes, sugar beet, brassicas, carrots [6]. See notes in Section 1
• Off label approval to Jan 1997 for use in celery, leeks, marrows, salad onions, sweetcorn, grapevines (OLA 0390/94)[3]; unlimited for use on compost for propagation of protected and outdoor watercress (OLA 0394/94)[3]; to Aug 1996 for use on watercress beds (OLA 0388/94)[3]; to Dec 1996 for use on watercress beds OLA 0392/94)[6]; to Dec 1998 for use on watercress beds (OLA 0392/94)[6]; to Jul 2000 for use on watercress beds (OLA 0993/95)[3]; unlimited for use on outdoor and protected vegetables (see approval notice for details) (OLA 0389/94)[3]; to Oct 1998 for use on protected spring onions, leeks, celery, and on bulb onions, garlic, swedes, turnips (OLA 0391/94)[3]

Maximum Residue Levels (mg residue/kg food)
• citrus fruits, apricots, peaches, nectarines, plums, bilberries, cranberries, currants, gooseberries, cucumbers, gherkins, courgettes, cauliflowers, Brussels sprouts, head cabbages, lettuce, beans (with pods) 2; pome fruits, grapes, cane fruits, bananas, carrots, horseradish, parsnips, parsley root, salsify, swedes, turnips, garlic, onions, shallots, tomatoes, peppers, aubergines, peas (with pods), celery, leeks, rhubarb, cultivated mushrooms 1; potatoes 0.05

---

221   dimethomorph
A cinnamic acid fungicide with translaminar activity available only in mixtures

---

222   dimethomorph + mancozeb
A systemic and protectant fungicide for potato blight control

Products	Invader	Cyanamid	7.5:66.7% w/w	WG	06989

Uses   Blight in POTATOES.

Notes   **Efficacy**
• Commence treatment as soon as there is a risk of blight infection
• In the absence of a warning treatment should start before the crop meets along the rows
• Repeat treatments every 10-14 d depending in the degree of infection risk
• Irrigated crops should be regarded as at high risk and treated every 10 d
• For best results good spray coverage of the foliage is essential

Crop Safety/Restrictions
• Maximum number of treatments 8 per crop

Special precautions/Environmental safety
• Irritant. May cause sensitization by skin contact
• Oxidising agent. Contact with combustible material may cause fire

- Dangerous to fish or other aquatic life. Do not contaminate surface waters or ditches with chemical or used container
- Do not allow direct spray from vehicle-mounted/drawn sprayers to fall within 6 m of surface waters or ditches. Direct spray away from water

Protective clothing/Label precautions
- A
- 10a, 12d, 14, 16, 18, 21, 25, 28, 29, 36, 37, 52, 54a, 63

Latest application/Harvest Interval(HI)
- HI 7 d before harvest

Maximum Residue Levels (mg residue/kg food)
- see mancozeb entry

---

223   dinocap

A protectant dinitrophenyl fungicide for powdery mildew control

Products	Karathane Liquid	Rohm & Haas	350 g/l	EC	01126

Uses   Powdery mildew in APPLES, CHRYSANTHEMUMS, ROSES, STRAWBERRIES.

Notes

Efficacy
- Spray at 7-14 d intervals to maintain protective film
- Product must be applied to roses before disease becomes established
- Regular use on apples suppresses red spider mites and rust mites

Crop Safety/Restrictions
- Maximum number of treatments 10 per crop
- When applied to apples during blossom period may cause spotting but no adverse effect on pollination or fruit set. Do not apply to Golden Delicious during blossom period
- Do not apply when temperature is above 24°C. Do not apply with white oils
- Certain chrysanthemum cultivars may be susceptible
- May cause petal spotting on white roses

Special precautions/Environmental safety
- Harmful by inhalation, if swallowed and in contact with skin
- Irritating to eyes and skin. May cause sensitization by skin contact
- Flammable
- Dangerous to fish or other aquatic life. Do not contaminate surface waters or ditches with chemical or used container

Protective clothing/Label precautions
- A, C, J, M
- 6a, 6b, 10a, 12c, 16, 18, 23, 25, 26, 27, 28, 29, 35, 36, 37, 52, 63, 65, 78

Latest application/Harvest Interval(HI)
- HI apples 14 d, strawberries 7 d

FOR FULL CONDITIONS OF USE ALWAYS READ THE PRODUCT LABEL

224   dinoseb
A dinitrophenol herbicide and dessicant approvals for use and storage of which were revoked in 1988

225   diphacinone
An anticoagulant rodenticide

Products					
	1 Tomcat Rat and Mouse Bait	Antec	0.005%w/w	RB	07171
	2 Tomcat Rat and Mouse Blox	Antec	0.005% w/w	BB	07230

Uses

Mice, rats in AGRICULTURAL PREMISES.

Notes

Efficacy
* Products formulated using human food grade ingredients, flavour enhancers and paraffin
* Rodents must consume bait for 3-5 d to produce mortality after 6-15 d
* Maintain uninterrupted supply of fresh bait for 10-15 d or until signs of rodent activity cease
* Remove and replace stale, damp or mouldy bait
* Bait product [1] available loose or in sachets

Special precautions/Environmental safety
* Prevent access to the baits by children, birds and other animals, particularly dogs, cats and pigs

Protective clothing/Label precautions
* 25, 29, 54, 63, 67, 78, 81, 83, 84

226   diquat
A non-residual bipyridyl contact herbicide and crop desiccant

Products					
	1 Barclay Desiquat	Barclay	200 g/l	SL	04969
	2 Landgold Diquat	Landgold	200 g/l	SL	05974
	3 Midstream	ICI Professional	100 g/l	PC	01348
	4 Midstream	Zeneca Prof.	100 g/l	PC	06824
	5 Reglone	ICI Agrochem. Profes	200 g/l	SL	04444
	6 Reglone	Zeneca	200 g/l	SL	06703
	7 Reglone	Zeneca Prof.	200 g/l	SL	06703
	8 Standon Diquat	Standon	200 g/l	SL	05587
	9 Stefes Diquat	Stefes	200 g/l	SL	05493

Uses

Annual dicotyledons, pre-harvest desiccation in NAVY BEANS *(off-label)*[5]. Annual dicotyledons in ASPARAGUS *(off-label)*, SAGE *(off-label)* [5]. Annual dicotyledons in ROW CROPS [6]. Aquatic weeds in AREAS OF WATER [3-7]. Chemical stripping in HOPS [6]. Pre-harvest desiccation in HONESTY *(off-label)* [5, 6]. Pre-harvest desiccation in BORAGE *(off-label)*, MUSTARD *(off-label)*, SOYA BEANS *(off-label)* [5]. Pre-harvest desiccation in COMBINING PEAS [1, 2, 8, 9]. Pre-harvest desiccation in FIELD BEANS, LAID BARLEY AND OATS, OILSEED RAPE, POTATOES [1, 2, 6, 8, 9]. Pre-harvest desiccation in LINSEED

[1, 2, 6, 8]. Pre-harvest desiccation in CLOVERS [6, 9]. Pre-harvest desiccation in EVENING PRIMROSE, LUPINS, PEAS [6].

Notes

Efficacy
- Acts rapidly on green parts of plants and rainfast in 15 min
- Best results achieved by spraying in bright light and low humidity conditions
- Add Agral wetter for improved weed control, not on aquatic weeds [6]
- Apply to potatoes when tubers the desired size, to other crops when mature or approaching maturity. See label for details of timing and of period to be left before harvesting potatoes and for timing of pea and bean desiccation sprays
- Spray linseed when seed matured evenly over whole field; direct combining can normally begin 10-20 d after spraying
- Apply as hop stripping treatment when shoots have reached top wire
- Apply to floating and submerged aquatic weeds in still or slow moving water [3-6]
- Do not spray in muddy water [3-6]

Crop Safety/Restrictions
- Maximum number of treatments 1 per crop for potatoes, oilseed rape, peas and beans; 2 per crop for potatoes and evening primrose, 1 per crop for other recommended crops [6]
- Do not apply haulm destruction treatment when soil dry. Tubers may be damaged if spray applied during or shortly after dry periods. See label for details of maximum allowable soil-moisture deficit and varietal drought resistance scores
- Apply to laid barley and oats to be used only for stock feed, only on peas for harvesting dry, on field beans for pigeon and animal feed only
- Do not add wetters to desiccant sprays for potatoes or processing peas. Agral wetter may be added to spray for use on peas to be used as animal fodder [6]
- For weed control in crops apply as overall spray before crop emergence or as interrow spray in row crops. Keep off crop foliage
- Do not apply through mist-blower or in windy conditions

Special precautions/Environmental safety
- Harmful if swallowed. Irritating to eyes and skin
- Harmful to livestock
- Do not use treated straw or haulm as animal feed or bedding within 4 d of spraying
- Do not use treated water for human consumption within 24 h or for overhead irrigation within 10 d of treatment [3, 5, 6]
- Do not use on barley, oats and field beans intended for human consumption

Protective clothing/Label precautions
- A, C [1-9]
- 5c, 6a, 6b, 17, 18, 21, 28, 29, 36, 37, 41, 44, 54, 55, 56, 63, 65, 70, 78 [1-9]

Withholding period
- Keep all livestock out of treated areas for at least 24 h [1, 2, 5, 6, 9]

Latest application/Harvest Interval(HI)
- Pre-emergence for sugar beet, other vegetables and bulbs; before 1 Jul for hops [6]

Approval
- Approved for aquatic weed control. See notes in Section 1 on use of herbicides in or near water [3-6]

FOR FULL CONDITIONS OF USE ALWAYS READ THE PRODUCT LABEL

• Off-label Approval to Jan 1997 for use on navy beans (OLA 0025/92) [5]; to June 1997 for use on soya beans (OLA 0627/92) [5]; unlimited for use on mustard, sage, asparagus (pre-emergence) (OLA 0626/92)[5], borage (OLA 0490/92)[5], honesty (OLA 0587)[5], (0588/93) [6]

227   diquat + paraquat
A non-residual bipyridyl contact herbicide and crop desiccant

Products					
1 Parable	Zeneca	100:100 g/l	SL	06692
2 PDQ	Zeneca	80:120 g/l	SL	06518

Uses   Annual dicotyledons, annual grasses, chemical stripping, perennial non-rhizomatous grasses, volunteer cereals in HOPS [1, 2]. Annual dicotyledons, annual grasses, perennial non-rhizomatous grasses, volunteer cereals in FIELD CROPS *(autumn stubble, pre-planting/sowing, sward destruction)*, FORESTRY, FRUIT CROPS, NON-CROP AREAS, POTATOES, ROW CROPS, STRAWBERRIES [1, 2]. Annual dicotyledons, annual grasses, perennial non-rhizomatous grasses in FLOWER BULBS, HARDY ORNAMENTALS [1].

Notes   **Efficacy**
• Apply to young emerged weeds less than 15 cm high, annual grasses must have at least 2 leaves when sprayed. Best results when foliage dry and weather dull and humid
• Addition of approved non-ionic wetter recommended for control of certain species and with low dose rates. See label for details
• Allow at least 4 d between spraying and cultivation, as long a period as possible after spraying creeping perennial grasses
• For chemical stripping apply in July or after hops have reached top wire. Do not use on hops under drought conditions
• Chemical rapidly inactivated in moist soil, activity reduced in dirty or muddy water
• Spray is rainfast in 10 min

Crop Safety/Restrictions
• Maximum number of treatments 2 per yr for grassland destruction; 1 per yr or crop in other crops
• Apply up to just before sown crops emerge or just before planting
• On sandy or immature peat soils and on forest nursery seedbeds allow 3 d between spraying and planting
• Where trash or dying weeds are left on surface allow at least 3 d before planting
• In potatoes spray earlies up to 10% emergence, maincrop to 40% emergence, provided plants are less than 15 cm high. Do not use post-emergence on potatoes from diseased or small tubers or under very hot, dry conditions
• Use guarded no-drift sprayers to kill interrow weed and strawberry runners
• Apply to fruit crops as a directed spray, preferably in dormant season
• If spraying bulbs at end of season ensure all crop foliage is detached from bulbs. Do not use on very sandy soils

Special precautions/Environmental safety
• Paraquat is subject to the Poisons Rules 1982 and the Poisons Act 1972. See notes in Section 1
• Toxic if swallowed. Paraquat can kill if swallowed
• Harmful in contact with skin. Irritating to eyes and skin
• Keep in original container, tightly closed, in a safe place, under lock and key

- Do not put in a food or drinks container. Keep out of reach of children
- Not to be used by amateur gardeners
- Harmful to animals and livestock
- Paraquat can be harmful to hares, where possible spray stubbles early in the day

Protective clothing/Label precautions
- A, C
- 4c, 5a, 6a, 6b, 17, 18, 21, 23, 24, 28, 29, 36, 37, 41, 47, 54, 64, 65, 70, 79

Withholding period
- Keep all livestock out of treated areas for at least 24 h

Maximum Residue Levels (mg residue/kg food)
- see paraquat entry

228   disulfoton

A systemic organophosphorus aphicide and insecticide

See also chlorpyrifos + disulfoton

Products					
1 Campbell's Disulfoton FE 10	MTM Agrochem.	10% w/w		GR	00402
2 Disyston P 10	Bayer	10% w/w		GR	00715
3 MTM Disulfoton P 10	MTM Agrochem.	10% w/w		GR	05094
4 Prosper	United Phosphorus	10% w/w		GR	07532

Uses

Aphids, carrot fly in CARROTS, PARSNIPS [1-4]. Aphids, mangold fly in SUGAR BEET [2-4]. Aphids, mangold fly in FODDER BEET, MANGELS [3, 4]. Aphids, pea midge, thrips in PEAS [1]. Aphids in BRUSSELS SPROUTS, CABBAGES, CAULIFLOWERS, POTATOES [1-4]. Aphids in BROCCOLI [1, 3, 4]. Aphids in BROAD BEANS, FIELD BEANS [2-4]. Aphids in FRENCH BEANS, MARROWS, PARSLEY, RUNNER BEANS, STRAWBERRIES [2]. Aphids in SUGAR BEET STECKLINGS [4]. Carrot fly in CELERY [2].

Notes

Efficacy
- Granules based on fullers earth (FE) are applied at drilling or planting, those based on pumice (P) can also be used as foliar treatments
- For effective results granules applied at drilling or planting should be incorporated in soil. See label for details of suitable application machinery
- One application is normally sufficient, a split application may be used on carrots
- Persistence may be reduced in fen peat soils

Crop Safety/Restrictions
- Maximum number of treatments 1 per crop
- On carrots the maximum total dose applied per crop must not exceed the equivalent of 3 (on mineral soils) or 4 (on organic soils) full dose applications
- On potatoes treat only second early and maincrop varieties
- Consult processor before treating crops grown for processing

FOR FULL CONDITIONS OF USE ALWAYS READ THE PRODUCT LABEL

Special precautions/Environmental safety
* This product contains an anticholinesterase organophosphorus compound. Do not use if under medical advice not to work with such compounds
* Toxic in contact with skin, by inhalation or if swallowed
* Keep in original container, tightly closed, in a safe place, under lock and key
* Dangerous to livestock
* Dangerous to game, wild birds and animals
* Applied correctly treatment will not harm bees
* Dangerous to fish or other aquatic life. Do not contaminate surface waters or ditches with chemical or used container

Protective clothing/Label precautions
* A, B, C, C or D + E, H, J, K, M
* 1, 4a, 4b, 4c, 16, 18, 22, 23, 25, 28, 29, 35, 36, 37, 45, 52, 64, 67, 70, 79

Withholding period
* Keep all livestock out of treated areas for at least 6 wk. Bury or remove spillages

Latest application/Harvest Interval(HI)
* Before drilling or planting for protected marrow [714].
* HI 6 wk

Approval
* Approved for aerial application on brassicas, beans, sugar beet, carrots [2]. See notes in Section 1

---

229   disulfoton + quinalphos
A systemic and contact organophosphorus insecticide for brassicas

Products	Knave	Hortichem	5:2.5% w/w	GR	02534

Uses Aphids, cabbage root fly in BROCCOLI, BRUSSELS SPROUTS, CABBAGES, CALABRESE, CAULIFLOWERS.

Notes

Efficacy
* Apply as band treatment with suitable granule applicator at drilling using 'bow-wave' technique. See label for details
* Apply as a sub-surface band to crops transplanted after mid-Apr using a 'Leeds' coulter set so that roots are planted below treated layer of soil
* Product can be used on all mineral and organic soils

Crop Safety/Restrictions
* Maximum number of treatments 1 per crop
* Consult processor before using on crops for processing

Special precautions/Environmental safety
* Disulfoton and quinalphos are subject to the Poisons Rules 1982 and Poisons Act 1972. See notes in Section 1
* This product contains two anticholinesterase organophosphorus compounds. Do not use if under medical advice not to work with such compounds
* Toxic in contact with skin, by inhalation or if swallowed
* Keep in original container, tightly closed, in a safe place, under lock and key

- Dangerous to game, wild birds and animals, bury spillages
- Extremely dangerous to fish or other aquatic life. Do not contaminate surface waters or ditches with chemical or used container

Protective clothing/Label precautions
- A, B, C, H or L, M
- 1, 4a, 4b, 4c, 14, 16, 18, 22, 25, 28, 29, 35, 36, 37, 45, 51, 64, 67, 70, 79

Latest application/Harvest Interval(HI)
- HI 6 wk

230 dithianon

A protectant and eradicant quinone fungicide for scab control

Products					
1 Barclay Cluster	Barclay	750 g/l	SC	07467
2 Dithianon Flowable	Cyanamid	750 g/l	SC	07007

Uses   Scab in APPLES, PEARS.

Notes   **Efficacy**
- Apply at bud-burst and repeat every 10-14 d until danger of scab infection ceases
- Application at high rate within 48 h of a Mills period prevents new infection
- Spray programme also reduces summer infection with apple canker

Crop Safety/Restrictions
- Maximum number of treatments 8 per crop
- Do not use on Golden Delicious apples after green cluster
- Do not mix with lime sulfur or highly alkaline products

Special precautions/Environmental safety
- Harmful if swallowed.
- Irritating to eyes and skin [2]

Protective clothing/Label precautions
- 5c, 18, 21, 28, 29, 36, 37, 54, 63, 66, 70, 78 [1, 2]; 6a, 6b, 60 [2]

Latest application/Harvest Interval(HI)
- HI 4 wk

231 dithianon + penconazole

A systemic and protective fungicide for use in apples

Products				
Topas D 275 SC	Ciba Agric.	250:25 g/l	SC	05855

Uses   Powdery mildew, scab in APPLES.

FOR FULL CONDITIONS OF USE ALWAYS READ THE PRODUCT LABEL

Notes

Efficacy
- Treatment controls both leaf and fruit scab
- High anti-sporulant activity reduces development of powdery mildew from primary infections and controls spread of secondary mildew
- Apply as protective foliar spray from bud-burst and continue every 10-14 d (7 d with high infection pressure) until extension growth ceases

Crop Safety/Restrictions
- Maximum number of treatments 10 per crop. Do not use more than 10 sprays of any penconazole product per season
- May be used on all apple varieties except Golden Delicious, which should not be sprayed after green cluster

Special precautions/Environmental safety
- Harmful if swallowed. Irritating to eyes and skin
- Harmful to fish or other aquatic life. Do not contaminate surface waters or ditches with chemical or used container

Protective clothing/Label precautions
- A, C
- 5c, 6a, 6b, 14, 18, 21, 28, 29, 35, 36, 37, 53, 63, 66, 70, 78

Latest application/Harvest Interval(HI)
- HI 28 d

---

232   diuron

A residual urea herbicide for non-crop areas and woody crops

See also amitrole + 2,4-D + diuron
          amitrole + bromacil + diuron
          bromacil + diuron
          chlorpropham + diuron + propham

Products

1 Chipko Diuron 80	Nomix-Chipman	80% w/w	WP	00497
2 Chipman Diuron Flowable	Nomix-Chipman	500 g/l	SC	05701
3 Diuron 80 WP	RP Environ.	80% w/w	WP	05199
4 Freeway	RP Environ.	500 g/l	SC	06047
5 Karmex	DuPont	80% w/w	WP	01128
6 MSS Diuron 50 FL	Mirfield	500 g/l	SC	04786
7 MSS Diuron 50 FL	Mirfield	500 g/l	SC	07160
8 MTM Diuron 80	MTM Agrochem.	80% w/w	WP	04958
9 Unicrop Flowable Diuron	Unicrop	500 g/l	SC	02270
10 UPL Diuron 80	United Phosphorus	80% w/w	WP	07619

Uses

Annual dicotyledons, annual grasses in ESTABLISHED WOODY ORNAMENTALS, WOODY NURSERY STOCK [1, 3, 4]. Annual dicotyledons, annual grasses in APPLES, PEARS [9]. Annual dicotyledons, annual grasses in AMENITY TREES AND SHRUBS [4, 9]. Annual weeds, perennial weeds in NON-CROP AREAS [1-10].

Notes

Efficacy
- Best results when applied to moist soil and rain falls soon afterwards
- Length of residual activity may be reduced on heavy or highly organic soils
- Selective rates must be applied to weed-free soil and activity persists for 2-3 mth

- Around trees and shrubs spray in late winter or early spring [4]
- Application for total vegetation control may be at any time of year, best results obtained in late winter to early spring
- Application to frozen ground not recommended

Crop Safety/Restrictions
- Maximum number of treatments 1 per yr on land not intended for cropping (high rate), 2 per yr around trees and shrubs (low rate) [3, 4]
- Apply to weed-free soil in apple and pear orchards established for at least 1 yr during Feb-Mar
- Apply to weed-free soil in blackcurrants and gooseberries established for at least 1 yr in early spring, or as split application in Oct-Nov and before bud-burst
- Apply to established asparagus before crop emergence
- Do not treat trees and shrubs less than 5 cm tall or established less than 12 mth. See label for list of sensitive species [4]
- Do not use on Sands or Very Light soils or those that are gravelly or where less than 1% organic matter
- Do not apply to areas intended for replanting during next 12 mth (2 yr for vegetables)
- Do not apply non-selective rates on or near desirable plants where chemical may be washed into contact with roots
- Application for amenity use should only take place between the beginning of Feb and end of Apr (May in Scotland)

Special precautions/Environmental safety
- Irritating to eyes, skin and respiratory system (to eyes and skin [3, 4])
- Harmful to fish or other aquatic life. Do not contaminate surface waters or ditches with chemical or used container
- Do not apply over drains or in drainage channels or gullies
- Avoid run-off when using on paved and similar surfaces

Protective clothing/Label precautions
- A, C [2, 5]; D [8, 10]
- 6a, 6b [1-7, 9]; 6c, 67 [1-6, 8-10]; 18, 28, 29, 36, 37, 53, 63 [1-10]; 21 [7, 8, 10]; 22 [1-6, 9]; 66 [7]; 70 [8, 10]

Latest application/Harvest Interval(HI)
- HI apples, pears, blackcurrants, gooseberries, asparagus 60 d

Approval
- Approval expiry: 30 Apr 95 [04786]

---

233   diuron + glyphosate
A non-selective residual herbicide mixture for non-crop and amenity use

Products Touché Nomix-Chipman 217.6:145.3 g/l UL 06797

Uses Annual dicotyledons, annual grasses, perennial dicotyledons, perennial grasses in AMENITY TREES AND SHRUBS, NON-CROP AREAS.

FOR FULL CONDITIONS OF USE ALWAYS READ THE PRODUCT LABEL

Notes

Efficacy
* May be applied at any time of year when weeds are green and actively growing
* Perennial grasses are susceptible when tillering and making new rhizome growth, normally when plants have 4-5 new leaves
* Perennial dicotyledons most susceptible if treated at or near flowering but will be severely checked if treated at other times when growing actively
* Most species of germinating weeds are controlled. See label for more resistant species
* Weed control may be reduced on heavy or highly organic soils
* At least 6 and preferably 24 h rain-free must follow spraying. Rain soon after treatment may reduce initial weed control
* Apply with Nomix System equipment

Crop Safety/Restrictions
* Maximum number of treatments 1 per yr
* Ornamental trees and shrubs should be established for at least 12 mth and be at least 50 mm tall
* Among ornamental trees and shrubs take care to avoid foliage or the stems of young plants
* Do not allow spray to contact desired plants or crops
* Do not use on tree and shrub nurseries or before planting
* Do not treat ornamental trees and shrubs on Sands or Very Light soils or those that are gravelly or with less than 1% organic matter
* Do not decant, connect directly to Nomix applicator

Special precautions/Environmental safety
* Harmful to fish or other aquatic life. Do not contaminate surface waters or ditches with chemical or used container

Protective clothing/Label precautions
* H, M
* 14, 22, 28, 29, 53, 63, 67

Approval
* Approved for use through ULV applicators. See notes in Section 1

Maximum Residue Levels (mg residue/kg food)
* see glyphosate entry

234   diuron + paraquat
A total herbicide with contact and residual activity

Products   Dexuron   Nomix-Chipman   300:100 g/l   SC   07169

Uses   Annual dicotyledons, annual grasses, perennial dicotyledons, perennial grasses in ESTABLISHED WOODY ORNAMENTALS, WOODY NURSERY STOCK. Total vegetation control in NON-CROP AREAS, PATHS.

Notes

Efficacy
* Apply to emerged weeds at any time of year. Best results achieved by application in spring or early summer
* Effectiveness not reduced by rain soon after treatment

Crop Safety/Restrictions
* Maximum number of treatments 1 per yr
* Avoid contact of spray with green bark, buds or foliage of desirable trees or shrubs
* In nurseries use low dose rate as inter-row spray, not more than once per year

Special precautions/Environmental safety
* Paraquat is subject to the Poisons Rules 1982 and the Poisons Act 1972. See notes in Section 1
* Keep in original container, tightly closed, in a safe place, under lock and key
* Toxic if swallowed
* Harmful in contact with skin. Irritating to eyes and skin
* Harmful to fish or other aquatic life. Do not contaminate surface waters or ditches with chemical or used container

Protective clothing/Label precautions
* A, C
* 4c, 5a, 6a, 6b, 14, 16, 18, 21, 28, 29, 36, 43, 53, 64, 66, 70, 79

Withholding period
* Keep livestock out of treated areas for at least 2 wk and until foliage of any poisonous weeds such as ragwort has died and become unpalatable

Maximum Residue Levels (mg residue/kg food)
* see paraquat entry

---

235 DNOC
A dinitrophenyl insecticide all approvals for which were revoked in December 1989

---

236 dodemorph
A systemic morpholine fungicide for powdery mildew control

Products	F238	BASF	400 g/l	EC	00206

Uses Powdery mildew in ROSES.

Notes **Efficacy**
* Spray roses every 10-14 d during mildew period or every 7 d and at increased dose if cleaning up established infection or if disease pressure high
* Add Citowett when treating rose varieties which are difficult to wet
* Product has negligible effect on Phytoseiulus spp being used to control red spider mites

Crop Safety/Restrictions
* Do not use on seedling roses
* Do not apply to roses under hot, sunny conditions, particularly under glass, but spray early in the morning or during the evening. Increase the humidity some hours before spraying

---

FOR FULL CONDITIONS OF USE ALWAYS READ THE PRODUCT LABEL

• Check tolerance of new varieties before treating rest of crop

Special precautions/Environmental safety
• Irritating to skin. Risk of serious damage to eyes
• Flammable
• Harmful to fish or other aquatic life. Do not contaminate surface waters or ditches with chemical or used container

Protective clothing/Label precautions
• A, C
• 6b, 9, 12c, 16, 18, 22, 26, 27, 28, 29, 36, 37, 53, 60, 63, 66

---

237   dodine
A protectant and eradicant guanidine fungicide

Products

1 Barclay Dodex	Barclay	450 g/l	SC	05655
2 Radspor FL	Truchem	450 g/l	SC	01685

Uses

Black spot in ROSES [1]. Currant leaf spot in BLACKCURRANTS, GOOSEBERRIES [1, 2]. Scab in APPLES, PEARS [1, 2].

Notes

Efficacy
• Apply protective spray on apples and pears at bud-burst and at 10-14 d intervals until late Jun to early Jul
• Apply post-infection spray within 36 h of rain responsible for initiating infection. Where scab already present spray prevents production of spores
• Apply to blackcurrants immediately after flowering, repeat every 10-14 d to within 1 mth of harvest and once or twice post-harvest

Crop Safety/Restrictions
• Do not apply in very cold weather (under 5°C) or under slow drying conditions to pears or dessert apples during bloom or immediately after petal fall
• Do not mix with lime sulfur or tetradifon

Special precautions/Environmental safety
• Harmful if swallowed and by skin contact.
• Irritating to eyes and skin
• May cause sensitisation by skin contact
• Harmful to fish or other aquatic life. Do not contaminate surface waters or ditches with chemical or used container

Protective clothing/Label precautions
• A, C [1, 2]
• 5a, 10a, 21, 28 [2]; 5c, 6a, 6b, 18, 29, 36, 37, 53, 63, 65, 70, 78 [1, 2]; 6c, 22 [1]

Latest application/Harvest Interval(HI)
• Early Jul for apples and pears [1]
• Early Jul for dessert apples; pre-blossom for culinary apples and pears [2]
• HI blackcurrants 1 mth

---

238 endosulfan

A contact and ingested organochlorine insecticide and acaricide

Products

1 Thiodan 20	Promark	200 g/l	EC	02122
2 Thiodan 20	Promark	200 g/l	EC	07335
3 Thiodan 35 EC	Promark	350 g/l	EC	02123
4 Thiodan 35 EC	Promark	352 g/l	EC	07336

Uses

Big-bud mite in BLACKCURRANTS [1, 2]. Blackberry mite in BLACKBERRIES [1, 2]. Bulb scale mite in NARCISSI [1, 2]. Colorado beetle in POTATOES *(off-label)* [3]. Damson-hop aphid in HOPS [3, 4]. Insect pests in HARDY ORNAMENTALS *(off-label)*, PROTECTED ORNAMENTALS *(off-label)* [3]. Pod midge, pollen beetles, seed weevil in MUSTARD, OILSEED RAPE [3, 4]. Tarsonemid mites in STRAWBERRIES [1, 2].

Notes

Efficacy
* Adjust spray volume to achieve total cover. See label for minimum dilutions
* On blackcurrants apply 3 sprays, at first flower, end of flowering and fruit set [1]
* On blackberries apply 3 sprays at 14 d intervals before flowering [1]
* On strawberries apply immediately after whole crop picked, where a second crop to be picked in autumn within 1 wk of mowing old foliage [1]
* Apply as drench to boxed narcissi a few days after bringing into greenhouse [1]
* On hops apply up to 6 sprays at 10-14 d intervals [2]
* On oilseed rape and mustard apply 1 or 2 sprays. See label for timing [2]

Special precautions/Environmental safety
* Endosulfan is subject to the Poisons Rules 1982 and the Poisons Act 1972. See notes in Section 1
* Toxic if swallowed. Harmful in contact with skin
* Irritating to the respiratory system
* Flammable
* Keep in original container, tightly closed, in a safe place, under lock and key
* Keep unprotected persons out of treated areas for at least 1 d
* Hops must not be handled for at least 4 d after spraying unless wearing rubber gloves
* Disbudding must not take place until at least 24 h after treatment
* Dangerous to livestock
* Harmful to bees. Do not apply at flowering stage except as directed on hops, oilseed rape and mustard. Keep down flowering weeds. When using on oilseed rape and mustard remove hives and warn local beekeepers
* Extremely dangerous to fish or aquatic life. Do not contaminate surface waters or ditches with chemical or used container

Protective clothing/Label precautions
* A, C, H, J, K, L, M
* 4c, 5a, 6c, 12c, 14, 16, 18, 21, 25, 28, 29, 35, 36, 37, 38, 40, 50, 51, 64, 66, 70, 79

Withholding period
* Keep all livestock out of treated areas for at least 3 wk

FOR FULL CONDITIONS OF USE ALWAYS READ THE PRODUCT LABEL

Latest application/Harvest Interval(HI)
* Before end of flowering for oilseed rape, mustard
*
* HI blackcurrants, blackberries, strawberries 6 wk; hops, potatoes 3 wk; ornamental crops 24 h before intended sale

Approval
* Off-label approval unlimited for use on outdoor and protected ornamentals (OLA 0656/92) [2]; to Jan 1997 for use to control Colorado beetle on potatoes as required by Statutory Notice (OLA 0065/92) [2]

Maximum Residue Levels (mg residue/kg food)
* tea 30; citrus fruits, pome fruits, apricots, peaches, nectarines, plums, grapes, cane fruits, bilberries, cranberries, currants, gooseberries, bananas, swedes, turnips, tomatoes, peppers, aubergines, cucumbers, gherkins, courgettes, cauliflowers, Brussels sprouts, head cabbages, lettuce, beans (with pods), peas (with pods), celery, leeks, rhubarb 2; garlic, onions, shallots 1; carrots, horseradish, parsnips, parsley root, salsify, potatoes 0.2; cereals 0.1

---

239   epoxiconazole

A systemic, protectant and curative triazole fungicide for use in cereals

Products					
	1 Epic	BASF	125 g/l	SC	07237
	2 Opus	BASF	125 g/l	SC	07236

Uses Brown rust, net blotch, powdery mildew, rhynchosporium, yellow rust in BARLEY. Brown rust, powdery mildew, septoria, yellow rust in WINTER WHEAT. Eyespot, fusarium ear blight, sooty moulds in WINTER WHEAT *(reduction)*. Eyespot in WINTER BARLEY *(reduction)*.

Notes **Efficacy**
* Apply at the start of foliar disease attack
* Optimum effect against eyespot achieved by spraying between leaf-sheath erect and second node detectable stages (GS 30-32)
* Best control of ear diseases of wheat obtained by treatment during ear emergence
* For Septoria spray after third node detectable stage (GS 33) when weather favouring disease development has occurred
* Mildew control improved by use of tank mixtures. See label for details

Crop Safety/Restrictions
* Maximum number of treatments 2 per crop
* Product may cause damage to broad-leaved plant species
* Avoid spray drift onto neighbouring crops

Special precautions/Environmental safety
* Dangerous to fish or other aquatic life. Do not contaminate surface waters or ditches with chemical or used container
* Do not allow direct spray from vehicle mounted/drawn sprayers to fall within 6 m of surface waters or ditches. Direct spray away from water

Protective clothing/Label precautions
* A
* 18, 36, 37, 52, 54a, 63, 66

Latest application/Harvest Interval(HI)
* Before crop commences flowering (GS 60) in winter wheat; up to and including emergence of ear just complete (GS 59) in barley

240   epoxiconazole + fenpropimorph

A systemic, protectant and curative fungicide mixture for cereals

Products					
1 Eclipse	BASF	84:250 g/l	SE	07362
2 Opus Team	BASF	84:250 g/l	SE	07361

Uses   Brown rust, net blotch, powdery mildew, rhynchosporium, yellow rust in BARLEY. Brown rust, powdery mildew, septoria, yellow rust in WINTER WHEAT. Eyespot, fusarium ear blight, sooty moulds in WINTER WHEAT *(reduction)*. Eyespot in WINTER BARLEY *(reduction)*.

Notes   **Efficacy**
* Apply at the start of foliar disease attack
* Optimum effect against eyespot achieved by spraying between leaf-sheath erect and second node detectable stages (GS 30-32)
* Best control of ear diseaes of wheat obtained by treatment during ear emergence
* For Septoria spray after third node detectable stage (GS 33) when weather favouring disease development has occurred

Crop Safety/Restrictions
* Maximum number of treatments 2 per crop
* Product may cause damage to broad-leaved plant species
* Avoid spray drift onto neighbouring crops

Special precautions/Environmental safety
* Irritating to skin
* Harmful to fish or other aquatic life. Do not contaminate surface waters or ditches with chemical or used container
* Do not allow direct spray from vehicle mounted/drawn sprayers to fall within 6 m of surface waters or ditches. Direct spray away from water

Protective clothing/Label precautions
* A
* 6b, 18, 36, 37, 53, 54a, 63, 66

Latest application/Harvest Interval(HI)
* Before crop commences flowering (GS 60) in winter wheat; up to and including emergence of ear just complete (GS 59) in barley

241   epoxiconazole + tridemorph

A systemic, protectant and curative fungicide mixture for cereals

Products				
Opus Plus	BASF	125:375 g/l	SE	07363

FOR FULL CONDITIONS OF USE ALWAYS READ THE PRODUCT LABEL

Uses   Brown rust, net blotch, powdery mildew, rhynchosporium, yellow rust in BARLEY. Brown rust, powdery mildew, septoria, yellow rust in WINTER WHEAT. Eyespot, fusarium ear blight, sooty moulds in WINTER WHEAT *(reduction)*. Eyespot in WINTER BARLEY *(reduction)*.

Notes   **Efficacy**
* Apply at the start of foliar disease attack
* Optimum effect against eyespot achieved by spraying between leaf-sheath erect and second node detectable stages (GS 30-32)
* Best control of ear diseases of wheat obtained by treatment during ear emergence
* For Septoria spray after third node detectable stage (GS 33) when weather favouring disease development has occurred

Crop Safety/Restrictions
* Maximum number of treatments 2 per crop
* Product may cause damage to broad-leaved plant species
* Avoid spray drift onto neighbouring crops

Special precautions/Environmental safety
* Irritating to skin
* Harmful to fish or other aquatic life. Do not contaminate surface waters or ditches with chemical or used container
* Do not allow direct spray from vehicle mounted/drawn sprayers to fall within 6 m of surface waters or ditches. Direct spray away from water

Protective clothing/Label precautions
* A
* 6b, 18, 36, 37, 43 (14 d), 53, 54a, 63, 66

Withholding period
* Keep livestock out of treated areas for at least 14 d

Latest application/Harvest Interval(HI)
* Before crop commences flowering (GS 60) in winter wheat; up to and including emergence of ear just complete (GS 59) in barley

---

242   esfenvalerate
A contact and ingested pyrethroid insecticide

Products   Sumi-Alpha   Cyanamid   25 g/l   EC   07207

Uses   Aphids in WINTER BARLEY, WINTER WHEAT. Caterpillars in BROCCOLI, BRUSSELS SPROUTS, CABBAGES, CAULIFLOWERS, CHINESE CABBAGE, COLLARDS, KALE. Flea beetles in SPRING OILSEED RAPE. Pea and bean weevils, pea moth in PEAS. Pea and bean weevils in FIELD BEANS. Pollen beetles in OILSEED RAPE.

Notes   **Efficacy**
* Apply when pests first seen or at first sign of attack and repeat as necessary
* For pea moth apply according to PGRO or ADAS warning or as indicated by pheromone traps

Crop Safety/Restrictions
* Maximum number of treatments 2 per crop for winter wheat, winter barley, field beans; 3 per crop for oilseed rape, peas; 6 per crop for broccoli, Brussels sprouts, cauliflower, Chinese cabbage, collards, kale; 8 per crop for cabbage
* Consult processor before using on crops for processing

Special precautions/Environmental safety
* Harmful if swallowed, irritating to skin and eyes
* Extremely dangerous to bees. Do not apply at flowering stage except as directed on oilseed rape and peas. Keep down flowering weeds
* Extremely dangerous to fish or other aquatic life. Do not contaminate surface waters or ditches with chemical or used container
* Do not allow spray from vehicle sprayers to fall within 6 m, or from hand held sprayers to within 2 m of surface waters or ditches. Direct spray away from water

Protective clothing/Label precautions
* A, C, H
* 5c, 6a, 6b, 16, 18, 21, 28, 29, 36, 37, 48a, 51, 63, 66, 70, 78

Latest application/Harvest Interval(HI)
* 31 Mar for winter wheat and barley; before first flowers open for field beans and oilseed rape (GS 3,7); up to and including pods set on 4th truss for peas (GS 204).
* HI brassicas 7 d

---

243   ethirimol
A pyrimidinol fungicide available only in mixtures

---

244   ethirimol + flutriafol + thiabendazole
A systemic non-mercurial fungicide fungicide for seed treatment in barley

Products					
1 Ferrax	Bayer	400:30:10 g/l	FS	05284
2 Ferrax	Zeneca	400:30:10 g/l	FS	06662

Uses   Brown rust, covered smut, leaf blotch, leaf stripe, loose smut, net blotch, powdery mildew, rhynchosporium, seedling blight and foot rot, yellow rust in SPRING BARLEY, WINTER BARLEY.

Notes   **Efficacy**
* Apply with suitable liquid seed treatment machinery. See label for details
* Controls seed and soil-borne diseases and early attacks of foliar diseases. Additional treatment may be required later
* Recalibrate drill for treated seed
* Do not use treated seed on soils with more than 20% organic matter
* Disease control may be reduced under dry conditions
* May be mixed with Gammasan 30 provided both seed dressings applied together. Do not apply to seed already treated with another seed treatment

---

FOR FULL CONDITIONS OF USE ALWAYS READ THE PRODUCT LABEL

PESTICIDE PROFILES

Crop Safety/Restrictions
* Maximum number of treatments 1 per batch
* Do not apply to seed with more than 16% moisture
* Emergence of seed may be delayed, especially under poor germination conditions
* Treated seed should be stored in cool, well ventilated conditions and drilled as soon as possible. Test germination if stored until next season

Special precautions/Environmental safety
* Irritating to eyes and skin
* Seed must only be treated by means which incorporate engineering controls for workers' protection and a means of accurately dispensing the dose
* Treated seed not to be used as food or feed. Do not re-use sack
* Harmful to game, wild birds and animals. Bury spillages
* Harmful to fish or other aquatic life. Do not contaminate surface waters or ditches with chemical or used container

Protective clothing/Label precautions
* A, B, C, D, H, K, M
* 6a, 6b, 14, 18, 21, 29, 36, 37, 53, 63, 65, 70, 72, 73, 74, 75, 76, 77, 78

Latest application/Harvest Interval(HI)
* Before drilling

245 ethofumesate

A benzofuran herbicide for grass weed control in various crops

See also bromoxynil + ethofumesate + ioxynil
chloridazon + ethofumesate
desmedipham + ethofumesate

Products

1 Atlas Thor	Atlas	200 g/l	EC	06966
2 Barclay Keeper 200	Barclay	200 g/l	EC	05266
3 Nortron	AgrEvo	200 g/l	EC	07266
4 Nortron	Schering	200 g/l	EC	03853
5 Salute	United Phosphorus	200 g/l	EC	07660
6 Standon Ethofumesate 200	Standon	200 g/l	EC	05668
7 Stefes Fumat	Stefes	200 g/l	EC	05525

Uses

Annual dicotyledons, annual meadow grass, blackgrass in FODDER BEET, MANGELS, SUGAR BEET [1-7]. Annual dicotyledons, annual meadow grass, blackgrass in RED BEET [1-6]. Annual grasses, blackgrass, chickweed, cleavers, volunteer cereals in LEYS [2-5]. Annual grasses, blackgrass, chickweed, cleavers, volunteer cereals in ESTABLISHED GRASSLAND, GRASS SEED CROPS [3-5]. Annual grasses, blackgrass, chickweed, cleavers, volunteer cereals in AMENITY TURF [3, 4]. Annual grasses, chickweed, cleavers in HORSERADISH *(off-label)* [4]. Annual grasses, chickweed in GARLIC *(off-label)*, ONIONS *(off-label)* [4].

Notes

Efficacy
* May be applied pre- or post-emergence of crop or weeds
* Apply in beet crops in tank mixes with other pre- or post-emergence herbicides. Recommendations vary for different mixtures. See label for details
* In grass crops apply to moist soil as soon as possible after sowing or post-emergence when crop in active growth, normally mid-Oct to mid-Dec. See label for details

- Volunteer cereals not well controlled pre-emergence, weed grasses should be sprayed before fully tillered
- Do not use on soils with more than 10% organic matter
- Grass crops may be sprayed during rain or when wet. Not recommended in very dry conditions or prolonged frost
- Do not graze or cut grass for 14 d after, or roll less than 7 d before or after spraying

Crop Safety/Restrictions
- Maximum number of treatments 1 pre- plus 1 post-emergence (2 at reduced dose) per crop or yr for beet crops and grassland; 3 per crop for onions and garlic
- Safe timing on beet crops varies with other ingredient of tank mix. See label for details
- May be used in Italian, hybrid and perennial ryegrass, timothy, cocksfoot, meadow fescue and tall fescue. Apply pre-emergence to autumn-sown leys, post-emergence after 2 to 3-leaf stage. See label for details
- Do not use on swards reseeded without ploughing
- Clovers will be killed or severely checked
- Any crop may be sown 3 mth after application of mixtures in beet crops following ploughing, 5 mth after application in grass crops

Special precautions/Environmental safety
- Flammable
- Harmful to fish or other aquatic life. Do not contaminate surface waters or ditches with chemical or used container

Protective clothing/Label precautions
- 12c, 21, 28, 29, 53, 63, 66 [1-7]; 60 [2-7]

Latest application/Harvest Interval(HI)
- When crops meet across rows for beet crops and mangels; 14 d before cutting or grazing for grass leys; pre-emergence for horseradish.
- HI onions, garlic 3 mth

Approval
- Off-label approval unlimited for use on horseradish (OLA 0247/93) [2]; unlimited for use on onions and garlic (OLA 0704/91) [2]

---

246   ethofumesate + phenmedipham
A contact and residual herbicide for use in sugar beet

Products					
1	Betosip Combi	Sipcam	100:80 g/l	EC	07235
2	Stefes Medimat 2	Stefes	100:80 g/l	EC	07577

Uses   Annual dicotyledons in SUGAR BEET.

Notes   **Efficacy**
- Best results achieved by repeat applications to cotyledon stage weeds. Larger susceptible weeds not killed by first treatment usually checked and controlled by second application
- Apply on all soil types at 7-10 d intervals

FOR FULL CONDITIONS OF USE ALWAYS READ THE PRODUCT LABEL

- Do not spray wet foliage or if rain imminent
- Spray must be applied low volume. See label for details
- Interval between mixing spray and completion of spraying should not exceed 2 h to avoid crystallization

Crop Safety/Restrictions
- Maximum number of treatments 2 per crop
- Apply reduced dose from when the crop has fully expanded cotyledons or full dose from 2 fully expanded true leaf stage
- Spray in evening if daytime temperatures above 21°C expected
- Avoid or delay treatment if frost expected within 7 d
- Avoid or delay treating crops under stress from wind damage, manganese or lime deficiency, pest or disease attack etc
- Check from which recovery may not be complete may occur if treatment made during conditions of sharp diurnal temperature fluctuation

Special precautions/Environmental safety
- Irritating to eyes [1]. Irritating to skin and respiratory system [2]
- Harmful in contact with skin [2]
- Flammable [2]
- May cause sensitisation by skin contact [1]
- Harmful to fish or aquatic life. Do not contaminate surface waters or ditches with chemical or used container
- Extra care necessary to avoid drift because product is recommended for use as a fine spray.
- Spray volumes must not exceed those recommended
- Product may cause non-reinforced PVC pipes and hoses to soften and swell. Wherever possible use reinforced PVC or synthetic rubber

Protective clothing/Label precautions
- A, C [1, 2]
- 5a, 6b, 6c, 12c, 29, 70 [2]; 6a, 10a [1]; 18, 21, 28, 36, 37, 53, 63, 66 [1, 2]

Latest application/Harvest Interval(HI)
- Before crop foliage meets in the rows

---

247   ethoprophos
An organophosphorus nematicide and insecticide

Products	Mocap 10G	RP Agric.	10% w/w	GR	06166

Uses   Potato cyst nematode, wireworms in POTATOES.

Notes   ### Efficacy
- Broadcast shortly before or during final soil preparation with suitable fertilizer spreader and incorporate immediately to 10-15 cm. See label for details
- Treatment can be applied on all soil types. Control of pests reduced on organic soils
- Effectiveness dependent on soil moisture. Drought after application may reduce control

Crop Safety/Restrictions
- Maximum number of treatments 1 per crop

Special precautions/Environmental safety
* This product contains an anticholinesterase organophosphorous compound. Do not use if under medical advice not to work with such compounds
* Harmful by inhalation and in contact with skin
* Dangerous to game, wild birds and animals
* Dangerous to fish or other aquatic life. Do not contaminate surface waters or ditches with chemical or used container

Protective clothing/Label precautions
* A, B, C or D + E, H, K, M
* 1, 5a, 5b, 14, 16, 18, 21, 25, 28, 29, 36, 37, 45, 52, 63, 67, 70, 78

Latest application/Harvest Interval(HI)
* Pre-planting of crop

---

248 etridiazole
A protective thiadiazole fungicide for soil or compost incorporation

Products

1 Aaterra WP		ICI Agrochem.	35% w/w	WP	03795
2 Aaterra WP		Zeneca	35% w/w	WP	06625

Uses

Damping off in CUCUMBERS, MUSTARD AND CRESS, PEPPERS, TOMATOES, VEGETABLE SEEDLINGS, VEGETABLE TRANSPLANTS [1, 2]. Flour moths, grain beetles in STORED GRAIN [1]. Insect pests in CELERIAC [1]. Phytophthora, pythium in TULIPS [1, 2]. Phytophthora, pythium in WATERCRESS *(off-label)* [1]. Phytophthora in CONTAINER-GROWN STOCK, HARDY ORNAMENTAL NURSERY STOCK [1, 2]. Red core in STRAWBERRIES *(Scotland only)* [1, 2]. Root diseases in ROCKWOOL TOMATOES *(off-label)* [1, 2].

Notes

Efficacy
* Best results obtained when incorporated thoroughly into soil or compost. Drench application also recommended, especially for strawberries
* Do not apply to wet soil
* Treat compost as soon as possible before use

Crop Safety/Restrictions
* Maximum number of treatments 1 per crop
* After drenching wash spray residue from crop foliage
* Do not use on escallonia, pyracantha, gloxinia spp., pansies or lettuces
* Germination of lettuce in previously treated soil may be impaired
* Do not drench seedlings until well established
* When treating compost for blocking reduce dose by 50%
* Test on small numbers of plants in advance when treating subjects of unknown susceptibility or using compost with more than 20% inert material
* Apply only to spring planted strawberries established for at least 3 mth

Special precautions/Environmental safety
* Irritating to eyes and skin

FOR FULL CONDITIONS OF USE ALWAYS READ THE PRODUCT LABEL

Protective clothing/Label precautions
* A, C
* 6a, 6b, 14, 16, 18, 22, 28, 29, 36, 37, 54, 63, 67, 70

Latest application/Harvest Interval(HI)
* 24 h after seeding watercress.
* HI tomatoes, peppers, cucumbers, mustard and cress 3 d; strawberries 6 mth

Approval
* Off-label Approval unlimited for use on tomatoes grown on rock wool (OLA 0600/94)[2], 0601/94)[1]; unlimited for use on watercress propagation beds (OLA 0131/92)[1]

249 etrimfos

A contact organophosphorus insecticide for stored grain crops

Products

1	Satisfar	Nickerson Seeds	525 g/l	EC	04180
2	Satisfar Dust	Nickerson Seeds	2% w/w	DP	04085

Uses

Grain beetles, grain storage mites, grain storage pests, grain weevils in GRAIN STORES, STORED GRAIN, STORED OILSEED RAPE.

Notes

Efficacy
* Spray internal surfaces of clean store with knapsack or motorized sprayer
* Allow sufficient time for pests to emerge from hiding places and make contact with chemical before grain is stored
* Apply as admixture treatment to grain as it enters store using a suitable stored grain sprayer or automatic seed treater if using dust
* Controls malathion-resistant beetles and gamma-HCH-resistant mites

Crop Safety/Restrictions
* Maximum number of treatments 1 per batch
* Cool grain to below 15°C before treatment and storage. Moisture content of grain should not exceed 16%

Special precautions/Environmental safety
* This product contains an anticholinesterase organophosphorus compound. Do not use if under medical advice not to work with such compounds
* Harmful to game, wild birds and animals
* Dangerous to fish or other aquatic life. Do not contaminate surface waters or ditches with chemical or used container

Protective clothing/Label precautions
* A, C
* 22, 28, 29, 46, 52, 63, 67

Latest application/Harvest Interval(HI)
* On entry into store [2]; on removal from store [1]

Maximum Residue Levels (mg residue/kg food)
* cereals (except rice) 5

250   fatty acids
A soap concentrate insecticide and acaricide

Products   Savona   Koppert   49% w/w   SL   06057

Uses   Aphids, mealy bugs, red spider mites, scale insects, whitefly in BEANS, BRUSSELS SPROUTS, CABBAGES, CUCUMBERS, FRUIT TREES, LETTUCE, PEAS, PEPPERS, PUMPKINS, TOMATOES, WOODY ORNAMENTALS. Aphids in WATERCRESS *(off-label)*.

Notes   **Efficacy**
* Use only soft or rain water for diluting spray
* For glasshouse use apply when insects first seen and repeat as necessary
* To control whitefly spray when required and use biological control after 12 h

Crop Safety/Restrictions
* Do not use on new transplants, newly rooted cuttings or plants under stress
* Do not use on specified susceptible shrubs. See label for details

Special precautions/Environmental safety
* Harmful to fish or other aquatic life. Do not contaminate surface waters or ditches with chemical or used container

Protective clothing/Label precautions
* 12c, 30, 53, 60, 63, 65

Latest application/Harvest Interval(HI)
* HI zero

Approval
* Off-label approval to Apr 1999 for use on watercress grown outdoors (OLA 0735/94)

251   fenarimol
A systemic curative and protective pyrimidine fungicide

Products   1 Rimidin   Rigby Taylor   120 g/l   SC   05907
                 2 Rubigan   DowElanco   120 g/l   SC   05489

Uses   American gooseberry mildew in BLACKCURRANTS, GOOSEBERRIES [2]. Dollar spot, fusarium patch, red thread in AMENITY TURF [1]. Powdery mildew, scab in APPLES [2]. Powdery mildew in CUCUMBERS, GRAPEVINES, ORNAMENTALS, PROTECTED PEPPERS *(off-label)*, PROTECTED TOMATOES *(off-label)*, RASPBERRIES, ROSES, STRAWBERRIES [2].

Notes   **Efficacy**
* Recommended spray interval varies from 7-14 d depending on crop and climatic conditions. See label for timing details [2]

FOR FULL CONDITIONS OF USE ALWAYS READ THE PRODUCT LABEL

- Efficient coverage and short spray intervals essential, especially for scab control [2]
- Mildew spray programme also controls low to moderate levels of blackcurrant leaf spot [2]
- Spray strawberry runner beds and vine nurseries regularly throughout season [2]
- May be used on cucumbers where Phytoseiulus and Encarsia are being used for biological control [2]
- For turf disease control spray as preventive treatment and repeat as necessary. See label for details [1]
- Do not mow within 24 h after treatment [1]

Crop Safety/Restrictions
- Maximum number of treatments 15 per crop for apples; 3 per yr for raspberries; 3 per crop for protected tomatoes
- Do not use on trees that are under stress from drought, severe pest damage, mineral deficiency or poor soil conditions
- Test treat small numbers of ornamentals of unknown susceptibility in advance [2]

Special precautions/Environmental safety
- Dangerous to fish or other aquatic life. Do not contaminate surface waters or ditches with chemical or used container

Protective clothing/Label precautions
- A, C
- 21, 25, 28, 29, 35, 37, 52, 63, 66

Latest application/Harvest Interval(HI)
- HI tomatoes, cucumbers 2 d; other crops 14 d

Approval
- Off-label approval unlimited for use on protected tomatoes, protected peppers (OLA 0645/94) [2]

Maximum Residue Levels (mg residue/kg food)
- hops 5; currants, goosberries 1; pome fruits, grapes 0.3; tea 0.05; citrus fruits, tree nuts, blackberries, dewberries, loganberries, bilberries, cranberries, wild berries, miscellaneous fruits, root and tuber vegetables, bulb vegetables, sweet corn, brassica vegetables, leaf vegetables and herbs, beans (with and without pods), stem vegetables (except globe artichokes), fungi, pulses, oilseeds, potatoes, rye, oats, triticale, maize, rice, animal products (except liver and kidney) 0.02

252   fenbuconazole

A systemic protectant and curative triazole fungicide for top fruit

Products	Indar 5EW	AgrEvo	50 g/l	EW	07580

Uses   Scab in APPLES, PEARS.

Notes   **Efficacy**
- Most effective when used as part of a routine preventative programme from bud burst to onset of petal fall
- After petal fall, tank mix with other protectant fungicides to enhance scab control
- Safe to use on all main commercial varieties of apples and pears in UK

- See label for recommended spray intervals. In period of rapid growth or high disease pressure, a 7 d interval should be used

Crop Safety/Restrictions
- Maximum total dose 14 l/ha per yr
- Consult processors before using on pears for processing

Special precautions/Environmental safety
- Irritant. May cause serious damage to eyes
- Harmful to fish or other aquatic life. Do not contaminate surface waters or ditches with chemical or used container
- Do not harvest crops for human or animal consumption for at least 4 wk after last application
- Effect on parasites and predators used in IPM systems not fully established and safety cannot be assumed

Protective clothing/Label precautions
- A, C
- 6, 9, 18, 21, 29, 35 (4 wk), 36, 37, 53, 67

Latest application/Harvest Interval(HI)
- 28 d before harvest

---

253   fenbuconazole + fenpropimorph
A broad spectrum fungicide mixture for cereals

Products	Myriad	RP Agric.	37.5:281 g/l	EC	07584

Uses   Brown rust, glume blotch, leaf spot, powdery mildew, yellow rust in WINTER WHEAT. Brown rust, powdery mildew, rhynchosporium, yellow rust in SPRING BARLEY, WINTER BARLEY.

Notes   **Efficacy**
- Product has eradicant and protectant activity but control is generally less effective if treatment is delayed until disease established
- First treatment may from the tillering stage (GS 21). If disease pressure persists further treatment may be necessary up to and including ear just completely emerged stage (GS 59)
- Crops treated at flag leaf stage may need a second spray to protect the ear, especially against glume blotch

Crop Safety/Restrictions
- Maximum number of treatments 2 per crop

Special precautions/Environmental safety
- Irritating to eyes and skin
- Dangerous to fish or other aquatic life. Do not contaminate surface waters or ditches with chemical or used container
- Do not allow direct spray from ground-based vehicle mounted/drawn sprayers to fall within 6 m, or from hand-held sprayers to within 2 m, of surface waters or ditches. Direct spray away from water

---

FOR FULL CONDITIONS OF USE ALWAYS READ THE PRODUCT LABEL

Protective clothing/Label precautions
* A
* 6a, 6b, 18, 21, 28, 29, 36, 37, 52, 54a, 60, 63, 66

Latest application/Harvest Interval(HI)
* Before flowering stage (GS 59)

254 fenbutatin oxide

A selective contact and ingested organotin acaricide

Products	Torque	Zeneca	50% w/w	WP	03303

Uses   Red spider mites, straw mites in PROTECTED CUCUMBERS, PROTECTED ORNAMENTALS, PROTECTED TOMATOES, TUNNEL GROWN STRAWBERRIES.

Notes   **Efficacy**
* Active on larvae and adult mites. Spray may take 7-10 d to effect complete kill but mites cease feeding and crop damage stops almost immediately
* Apply as soon as mites first appear and repeat as necessary
* On tunnel-grown strawberries apply when mites first appear, usually before flowering starts, and repeat 10-14 d later. A post-harvest spray is also recommended
* Not harmful to bees or to Encarsia or Phytoseiulus being used for biological control

Crop Safety/Restrictions
* Allow at least 10 d between spray applications and do not apply within 10 d of a previous spray
* Do not add wetters or mix with anything other than water
* Do not use white petroleum oil and Torque within 28 d of each other or any other pesticide and Torque within 7 d of each other. Do not apply to crops which are under stress for any reason
* On subjects of unknown susceptibility test treat on a small number of plants in advance

Special precautions/Environmental safety
* Irritating to eyes, skin and respiratory system
* Extremely dangerous to fish or other aquatic life. Do not contaminate surface waters or ditches with chemical or used container

Protective clothing/Label precautions
* A, C, D, H, M
* 6a, 6b, 6c, 18, 21, 28, 29, 36, 37, 51, 63, 70, 78

Latest application/Harvest Interval(HI)
* HI glasshouse cucumbers, tomatoes 3 d; tunnel-grown strawberries 7 d

255   fenitrothion
A broad spectrum, contact organophosphorus insecticide

Products

1 Antec Durakil	Antec	41.6% w/w	WP	05147
2 Dicofen	PBI	500 g/l	EC	00693
3 EC-Kill	Antec	500 g/l	EC	H5468
4 Micromite	Grampian	500 g/l	EC	H4480
5 Unicrop Fenitrothion 50	Unicrop	500 g/l	EC	02267

Uses

Ants, bedbugs, beetles, cockroaches, crickets, earwigs, fleas, flies, mites, moths, silverfish in REFUSE TIPS [1, 3]. Ants, bedbugs, beetles, cockroaches, crickets, earwigs, fleas, flies, mites, moths, silverfish in FOOD STORAGE AREAS, LIVESTOCK HOUSES [3]. Aphids, apple blossom weevil, capsids, codling moth, sawflies, suckers, tortrix moths, winter moth in APPLES [2, 5]. Aphids, caterpillars in PEARS, PLUMS [2]. Aphids, pea and bean weevils, pea midge, pea moth, thrips in PEAS [2, 5]. Capsids, sawflies in BLACKCURRANTS, GOOSEBERRIES [2, 5]. Flies in MANURE HEAPS [1]. Flour beetles, grain beetles, grain weevils in GRAIN STORES [1, 2, 5]. Frit fly in MAIZE, SWEETCORN [2, 5]. Leatherjackets, saddle gall midge, thrips, wheat-blossom midges in WHEAT [2, 5]. Leatherjackets, saddle gall midge, thrips in BARLEY [2, 5]. Leatherjackets, saddle gall midge in RYE, TRITICALE [2, 5]. Leatherjackets, saddle gall midge in DURUM WHEAT [5]. Poultry house pests in POULTRY HOUSES [1, 3, 4]. Raspberry beetle, raspberry cane midge in RASPBERRIES [2, 5]. Thrips in LEEKS *(off-label)*, SALAD ONIONS *(off-label)*, SWEETCORN *(off-label)* [2]. Tortrix moths in STRAWBERRIES [2, 5].

Notes

Efficacy
* Number and timing of sprays vary with disease and crop. See label for details
* Apply as bran bait for leatherjacket control in cereals
* For cane midge add suitable spreader and apply to lower 60 cm of young canes
* For control of grain store pests apply to all surfaces of empty stores before filling with grain. Remove dust and debris before applying
* For use in poultry houses apply to breeding sites etc after depopulation [1, 4]

Crop Safety/Restrictions
* Maximum number of treatments 1 per crop for cereals, maize, sweetcorn; 2 per crop or yr for peas, plums, blackcurrants, gooseberries, strawberries, grain stores; 3 per crop for apples, pears; 4 per crop for raspberries
* Do not mix with magnesium sulfate or highly alkaline materials
* On raspberries may cause slight yellowing of leaves of some varieties which closely resembles that caused by certain viruses

Special precautions/Environmental safety
* This product contains an anticholinesterase organophosphorus compound. Do not use if under medical advice not to work with such compounds
* Harmful in contact with skin or if swallowed [2-5]. Irritating to eyes and skin [2, 5]
* Flammable
* Harmful to livestock
* Harmful to game, wild birds and animals
* Harmful to fish or other aquatic life. Do not contaminate surface waters or ditches with chemical or used container

FOR FULL CONDITIONS OF USE ALWAYS READ THE PRODUCT LABEL

- Dangerous to bees. Do not apply to crops in flower or to those in which bees are actively foraging. Do not apply when flowering weeds are present.

Protective clothing/Label precautions
- A, C [1, 2, 4, 5]; A, H [3]
- 1, 12c, 18, 28, 29, 36, 37, 53, 63, 78 [1-5]; 5a, 5c [2, 3, 5]; 6a, 6b, 41, [2, 5]; 14, 21, 25, 26, 27, 35, 46, 48, 65, 70 [1, 2, 4, 5]; 22, 38 (48 h), 67 [3]; 41 [1, 4]

Withholding period
- Keep all livestock out of treated areas for at least 7 d [2, 5]
- Keep unprotected persons and animals out of treated areas for 48 h or until surfaces are dry [3]

Latest application/Harvest Interval(HI)
- HI raspberries 7 d; other crops 2 wk

Approval
- Approved for aerial application on cereals, peas. See notes in Section 1[2, 5]
- Off-label approval to Feb 1997 for use on leeks, salad onions, sweetcorn (OLA 0869/92) [2]

Maximum Residue Levels (mg residue/kg food)
- cereals 5; citrus fruits 2; pome fruits, apricots, peaches, nectarines, plums, grapes, cane fruits, bilberries, cranberries, currants, gooseberries, bananas, carrots, horseradish, parsnips, parsley root, salsify, swedes, turnips, garlic, onions, shallots, tomatoes, peppers, aubergines, cucumbers, gherkins, courgettes, cauliflowers, Brussels sprouts, head cabbages, lettuce, beans (with pods), peas (with pods), celery, leeks, rhubarb, cultivated mushrooms 0.5; potatoes 0.05

---

256   fenitrothion + permethrin + resmethrin

An organophosphate/pyrethroid insecticide mixture for grain stores

| **Products** | Turbair Grain Store Insecticide | Graincare | 10:20:2 g/l | UL | 02238 |

Uses   Flour beetles, grain beetles, grain moths, grain weevils in GRAIN STORES.

Notes

Efficacy
- Apply as a surface spray in empty stores using a suitable fan-assisted ULV sprayer
- Clean store thoroughly before applying
- Combines knock-down effect with up to 5 mth residual activity
- Should not be mixed with other sprays

Special precautions/Environmental safety
- This product contains an anticholinesterase organophosphorus compound. Do not use if under medical advice not to work with such compounds
- Irritating to eyes, skin and respiratory system
- Highly flammable
- Extremely dangerous to fish or other aquatic life. Do not contaminate surface waters or ditches with chemical or used container

Protective clothing/Label precautions
- A, C
- 1, 6a, 6b, 6c, 12b, 18, 21, 28, 29, 36, 37, 51, 63, 65, 70

Approval
• Product formulated for application as a ULV spray. See label for details. See notes in Section 1 on ULV application

Maximum Residue Levels (mg residue/kg food)
• see fenitrothion and permethrin entries

---

257   fenoxaprop-ethyl
A phenoxypropionic acid herbicide for grass weed control in wheat

Products					
	Landgold Fenoxaprop	Landgold	60 g/l	EW	06352

Uses   Blackgrass, wild oats in WINTER WHEAT.

Notes   **Efficacy**
• Apply post-weed emergence from 2-leaf to flag leaf ligule visible stage of weeds (GS 12-39)
• Treat awned canary grass from 2-leaf stage up to end of tillering but before first node detectable (GS 31)
• Susceptible weeds stop growing within 3 d of spraying and die usually within 2-4 wk
• Spray is rainfast 1 h after application
• Do not spray onto wet foliage or leaves covered with ice
• Dry conditions resulting in moisture stress may reduce effectiveness

Crop Safety/Restrictions
• Maximum number of treatments 2 per crop
• Do not apply to barley, durum wheat, undersown crops or crops to be undersown
• Treat from crop emergence up to and including flag leaf ligule just visible (GS 39)
• Do not mix with products containing bifenox or hormone herbicides
• Do not roll or harrow within 1 wk of spraying
• Do not spray crops under stress, crops suffering from drought, waterlogging or nutrient deficiency or those which have been grazed or if soil compacted

Special precautions/Environmental safety
• Dangerous to fish or other aquatic life. Do not contaminate surface waters or ditches with chemical or used container

Protective clothing/Label precautions
• A, C
• 21, 29, 52, 63, 66

Latest application/Harvest Interval(HI)
• Before flag leaf sheath extending (GS 41)

FOR FULL CONDITIONS OF USE ALWAYS READ THE PRODUCT LABEL

258   fenoxaprop-P-ethyl

A phenoxypropionic acid herbicide for use in wheat

See also diclofop-methyl + fenoxaprop-P-ethyl

Products					
	1 Cheetah Super	AgrEvo	55 g/l	EC	07309
	2 Cheetah Super	Hoechst	55 g/l	EC	05825

Uses   Blackgrass, canary grass, loose silky bent, rough meadow grass, timothy, wild oats in SPRING WHEAT, WINTER WHEAT.

Notes

Efficacy
* Treat weeds from crop 2-leaf stage to flag leaf just visible (GS 12-37). For awned canary-grass from 2 leaves to the end of tillering
* A second application may be made in spring where susceptible weeds emerge after an autumn application
* Spray is rainfast 1 h after application
* Product may be sprayed in frosty weather provided crop hardened off but do not spray wet foliage or leaves covered with ice
* Dry conditions resulting in moisture stress may reduce effectiveness

Crop Safety/Restrictions
* Maximum number of treatments 2 per crop
* Treat from crop emergence up to and including flag leaf ligule just visible (GS 39)
* Do not apply to barley, durum wheat, undersown crops or crops to be undersown
* Do not roll or harrow within 1 wk of spraying
* Do not spray crops under stress, suffering from drought, waterlogging or nutrient deficiency or those grazed or if soil compacted
* Broadcast crops should be sprayed post-emergence after plants have developed well-established root system
* Avoid spraying immediately before or after a sudden drop in temperature or a period of warm days/cold nights
* Do not mix with products containing bifenox or hormones

Special precautions/Environmental safety
* Irritating to eyes and skin
* Dangerous to fish or other aquatic life. Do not contaminate surface waters or ditches with chemical or used container

Protective clothing/Label precautions
* A, C
* 6a, 6b, 18, 21, 29, 36, 37, 52, 63, 66

Latest application/Harvest Interval(HI)
* Before flag leaf sheath extending (GS 41)

259   fenpropathrin

A contact and ingested pyrethroid acaricide and insecticide

Products					
	Meothrin	Cyanamid	100 g/l	EC	07206

Uses	Caterpillars, damson-hop aphid, red spider mites, two-spotted spider mite in HOPS. Caterpillars, red spider mites in APPLES, BLACKCURRANTS. Two-spotted spider mite in ROSES, STRAWBERRIES.

Notes	**Efficacy**
	• Acts on motile stages of mites and gives rapid kill
	• Apply on apples as post-blossom spray and repeat 3-4 wk later if necessary
	• Apply on hops when hatch of spring eggs complete in May and repeat if necessary
	• Apply on roses when pest first seen and repeat as necessary
	• Apply on strawberries as a pre-flowering spray and repeat after harvest if necessary

Crop Safety/Restrictions
• Maximum number of treatments 2 per crop for apples and hops; 2 per yr, 1 pre-flowering and 1 post-harvest for strawberries; 3 per yr for blackcurrants

Special precautions/Environmental safety
• Toxic in contact with skin or if swallowed. Irritating to eyes and skin
• Flammable
• Extremely dangerous to bees. Do not apply to crops in flower or to those in which bees are actively foraging except as directed in blackcurrants. Do not apply when flowering weeds are present
• Extremely dangerous to fish or other aquatic life. Do not contaminate surface waters or ditches with chemical or used container
• Do not operate air-assisted sprayers within 18 m of surface water or ditches, other wheeled sprayers within 6 m, hand-held sprayers within 2 m. Direct spray away from water

Protective clothing/Label precautions
• A, C, H, J
• 4a, 4c, 6a, 6b, 12c, 14, 16, 18, 21, 28, 29, 35, 36, 37, 48a, 51, 63, 65, 70, 79

Latest application/Harvest Interval(HI)
• Pre-flowering for strawberries; 14 d after end of flowering for blackcurrants.
• HI apples and hops 7 d

260   fenpropidin

A systemic, curative and protective piperidine (morpholine) fungicide

Products	1 Mallard 750 EC	Ciba Agric.	750 g/l	EC	06079
	2 Patrol	Zeneca	750 g/l	EC	06693
	3 Tern 750 EC	Ciba Agric.	750 g/l	EC	05643

Uses	Brown rust, glume blotch, leaf spot, powdery mildew, yellow rust in WHEAT. Brown rust, leaf blotch, powdery mildew, yellow rust in BARLEY.

Notes	**Efficacy**
	• Best results obtained when applied at early stage of disease development. See label for details of recommended timing alone and in mixtures

FOR FULL CONDITIONS OF USE ALWAYS READ THE PRODUCT LABEL

- Disease control enhanced by vapour-phase activity. Control can persist for 4-6 wk
- Alternate with triazole fungicides to discourage build-up of resistance

Crop Safety/Restrictions
- Maximum number of treatments 3 per crop (up to 2 in yr of harvest) for winter crops; 2 per crop for spring crops

Special precautions/Environmental safety
- Harmful in contact with skin and if swallowed. Irritating to skin
- Risk of serious damage to eyes
- Dangerous to fish or other aquatic life. Do not contaminate surface waters or ditches with chemical or used container

Protective clothing/Label precautions
- A, C
- 5a, 5c, 6b, 9, 14, 16, 18, 21, 29, 35, 36, 37, 52, 63, 66, 70, 78

Latest application/Harvest Interval(HI)
- Up to and including ear emergence complete (GS 59).
- HI 5 wk

---

261   fenpropidin + prochloraz
A contact and systemic fungicide mixture for cereals

Products	Sponsor	AgrEvo	250:250 g/l	EC	07136

Uses   Brown rust, eyespot, glume blotch, leaf spot, powdery mildew, yellow rust in WHEAT. Brown rust, eyespot, leaf blotch, net blotch, powdery mildew, yellow rust in BARLEY.

Notes   **Efficacy**
- For eyespot apply in spring from when leaf sheaths erect up to and including third node detectable stage (GS 33)
- Product effective against strains of eyespot resistant to benzimidazole fungicides
- Protection of winter cereals througout the season usually requires at least 2 treatments. See label for details of timing
- For foliar diseases treat in spring or early summer at first sign of infection on new growth
- Best protection against ear diseases only achieved by spraying at full ear emergence

Crop Safety/Restrictions
- Maximum number of treatments 2 per crop

Special precautions/Environmental safety
- Harmful if swallowed
- Harmful to fish or other aquatic life. Do not contaminate surface waters or ditches with chemical or used container
- Do not allow direct spray from vehicle mounted/drawn hydraulic sprayers to fall within 6 m, or from hand-held sprayers to fall within 2 m, of surface waters or ditches
- Do not harvest crops for human or animal consumption for at least 5 wk after last application

Protective clothing/Label precautions
- A, C, H

• 5c, 14, 16, 18, 22, 29, 35 (5 wk), 36, 37, 53, 54a, 63, 66, 70, 78

Latest application/Harvest Interval(HI)
• Up to and including ear emergence complete (GS 59)

262 fenpropidin + propiconazole
A systemic curative and protective fungicide for cereals

Products

1 Legend	ICI Agrochem.	450:100 g/l	EC	04979
2 Legend	Zeneca	450:100 g/l	EC	06682
3 Sheen 550 EC	Ciba Agric.	450:100 g/l	EC	05980
4 Zulu 550 EC	Ciba Agric.	450:100 g/l	EC	06937

Uses

Brown rust, leaf blotch, net blotch, powdery mildew, yellow rust in SPRING BARLEY, WINTER BARLEY. Brown rust, powdery mildew, septoria diseases, yellow rust in SPRING WHEAT, WINTER WHEAT.

Notes

Efficacy
• Best results obtained when applied at early stage of disease development. See label for details of recommended timing

Crop Safety/Restrictions
• Maximum number of treatments (including any products containing propiconazole) 3 per crop (2 in yr of harvest) for winter crops; 2 per crop for spring crops
• Transient crop scorch can result on some wheat varieties, particularly those with erect flag leaves. Avoid spraying crops under stress

Special precautions/Environmental safety
• Harmful in contact with skin and if swallowed. Irritating to skin
• May cause sensitization by skin contact
• Risk of serious damage to eyes
• Dangerous to fish or other aquatic life. Do not contaminate surface waters or ditches with chemical or used container

Protective clothing/Label precautions
• A, C
• 5a, 5c, 6b, 9, 10a, 14, 16, 18, 23, 29, 35 (5 wk), 36, 37, 52, 63, 66, 70, 78

Latest application/Harvest Interval(HI)
• Up to and including grain watery ripe (GS 71) for wheat; up to and including emergence of ear just complete (GS 59) for barley.
• HI 5 wk

Maximum Residue Levels (mg residue/kg food)
• see propiconazole entry

FOR FULL CONDITIONS OF USE ALWAYS READ THE PRODUCT LABEL

263   fenpropidin + tebuconazole
A broad spectrum fungicide mixture for cereals

Products	Monicle	Bayer	300:200 g/l	EC	07375

Uses   Botrytis, brown rust, fusarium ear blight, glume blotch, powdery mildew, septoria leaf spot, sooty moulds, yellow rust in WHEAT. Brown rust, net blotch, powdery mildew, rhynchosporium, yellow rust in BARLEY.

Notes

Efficacy
* Best disease control and yield benefit obtained when applied at early stage of disease development before infection spreads to new growth
* To protect the flag leaf and ear from Septoria diseases apply from flag leaf emergence to ear fully emerged (GS 37-59)
* Applications once foliar symptoms of *Septoria tritici* are already present on upper leaves will be less effective

Crop Safety/Restrictions
* Maximum total dose 2.5 l/ha per crop
* Occasional slight temporary leaf speckling may occur on wheat but this has not been shown to reduce yield response or disease control
* Do not treat durum wheat

Special precautions/Environmental safety
* Harmful if swallowed and in contact with skin
* Irritating to skin
* Risk of serious damage to eyes
* Extremely dangerous to fish or other aquatic life. Do not contaminate surface waters or ditches with chemical or used container
* Do not allow direct spray from ground based/vehicle drawn sprayers to fall within 6 m, or from hand-held sprayers to fall within 2 m, of surface waters or ditches. Direct spray away from water

Protective clothing/Label precautions
* A, C, H
* 5a, 5c, 6b, 9, 14, 16, 18, 24, 29, 35 (5 wk), 36, 37, 51, 54a, 63, 66, 70, 78

Latest application/Harvest Interval(HI)
* Before milky ripe stage (GS 73) for winter wheat; up to and including ear emergence complete (GS 59) for barley and spring wheat

264 fenpropimorph

A contact and systemic morpholine fungicide

See also chlorothalonil + fenpropimorph
epoxiconazole + fenpropimorph
fenbuconazole + fenpropimorph

Products

1	Aura 750 EC	Ciba Agric.	750 g/l	EC	05705
2	BAS 421F	BASF	750 g/l	EC	06127
3	Corbel	BASF	750 g/l	EC	00578
4	Keetak	BASF	750 g/l	EC	06950
5	Mistral	RP Agric.	750 g/l	EC	06199
6	Standon Fenpropimorph 750	Standon	750 g/l	EC	05654
7	Widgeon 750 EC	Ciba Agric.	750 g/l	EC	06101

Uses

Alternaria, light leaf spot, ring spot in BRUSSELS SPROUTS, CABBAGES [2]. Brown rust, powdery mildew, rhynchosporium, yellow rust in BARLEY [1-7]. Brown rust, powdery mildew, yellow rust in WHEAT [1-7]. Brown rust, yellow rust in TRITICALE [1-3, 5, 7]. Crown rot in CARROTS *(off-label)*, HORSERADISH *(off-label)*, PARSLEY ROOT *(off-label)*, PARSNIPS *(off-label)*, SALSIFY *(off-label)* [3]. Powdery mildew in SOFT FRUIT *(off-label)* [3]. Powdery mildew in RYE [1-3, 5, 7]. Powdery mildew in OATS [1-7]. Rust in LEEKS [1-3, 7]. Rust in FIELD BEANS [1-4, 6, 7].

Notes

Efficacy
* On cereals spray at start of disease attack. See label for recommended tank mixes
* On field beans a second application may be needed after 2-3 wk
* On Brussels sprouts and cabbages apply with Bavistin FL at start of disease attack [2]
* Product rainfast after 2 h [2]

Crop Safety/Restrictions
* Maximum number of treatments 2 per crop for spring cereals and field beans; 3 per crop (up to 2 in yr of harvest with minimum interval of 3 mth between autumn and spring/summer treatments) for winter cereals; 5 per crop for Brussels sprouts and cabbage; 6 per crop for leeks
* Scorch may occur if applied during frosty weather or in high temperatures [2]
* Consult processors before using on crops for processing [2]

Special precautions/Environmental safety
* Harmful by inhalation. Irritating to eyes, skin and respiratory system
* Dangerous to fish or other aquatic life. Do not contaminate surface waters or ditches with chemical or used container

Protective clothing/Label precautions
* A, C
* 5b, 6a, 6b, 6c, 18, 21, 26, 27, 28, 29, 36, 37, 52, 63, 66, 70, 78

Latest application/Harvest Interval(HI)
* HI Brussels sprouts, cabbage 2 wk; leeks 3 wk; cereals, field beans 5 wk

FOR FULL CONDITIONS OF USE ALWAYS READ THE PRODUCT LABEL

Approval
- Approved for aerial application on wheat, barley, oats, rye, triticale and field beans. See notes in Section 1[3]
- Off-label approval to Jul 1998 for use on outdoor carrot, parsnip, parsley root, salsify, horseradish (OLA 1058/94)[3]; unlimited for use on cane and soft fruit (OLA 0787/95)[3]

265   fenpropimorph + flusilazole

A broad-spectrum eradicant and protectant fungicide mixture for cereals

Products					
Colstar	DuPont	375:160 g/l	EC	06783

Uses   Brown rust, net blotch, powdery mildew, rhynchosporium, yellow rust in BARLEY. Brown rust, powdery mildew, septoria diseases, yellow rust in WINTER WHEAT.

Notes

Efficacy
- Disease control is more effective if treatment made at an early stage of disease development
- Treat winter cereals in spring or early summer before diseases spread to new growth
- Spring barley should be treated when diseases are first evident
- Treatment may be repeated after 3-4 wk if necessary

Crop Safety/Restrictions
- Maximum number of treatments (including other products containing flusilazole) 3 per crop (winter wheat), 2 per crop (winter or spring barley)
- Do not apply to crops under stress
- Do not apply during frosty weather

Special precautions/Environmental safety
- Irritating to eyes
- Flammable
- Dangerous to fish or other aquatic life. Do not contaminate surface waters or ditches with chemical or used container

Protective clothing/Label precautions
- A, C
- 6a, 12c, 18, 24, 28, 29, 36, 37, 52, 63, 66

Latest application/Harvest Interval(HI)
- Before early milk stage (GS 73)

266   fenpropimorph + flusilazole + tridemorph

A systemic eradicant and protectant fungicide mixture for cereals

Products					
Duk 51	DuPont	275:160:100 g/l	EC	06764

Uses   Brown rust, net blotch, powdery mildew, rhynchosporium, yellow rust in SPRING BARLEY. Brown rust, powdery mildew, rhynchosporium, yellow rust in WINTER BARLEY. Brown rust, powdery mildew, septoria diseases, yellow rust in WINTER WHEAT. Net blotch in WINTER BARLEY *(reduction)*.

Notes

Efficacy
- Apply at the start of disease attack. Treatment is most effective when disease levels are low at application
- Effectiveness is reduced by rain within 2 h of treatment

Crop Safety/Restrictions
- Maximum number of treatments (including other products containing flusilazole) 3 per crop for winter wheat; 2 per crop for barley
- Do not apply to crops under stress
- Application during frosty weather, high temperatures or drought may cause some crop scorch

Special precautions/Environmental safety
- Harmful if swallowed
- Irritating to skin
- Risk of serious damage to eyes
- Harmful to fish or other aquatic life. Do not contaminate surface waters or ditches with chemical or used container

Protective clothing/Label precautions
- A, C
- 5c, 6b, 9, 18, 24, 28, 36, 37, 43 (14 d), 53, 63, 66, 70, 78

Withholding period
- Keep livestock out of treated areas for at least 14 d after treatment

Latest application/Harvest Interval(HI)
- Before early milk stage (GS73)

---

267   fenpropimorph + gamma-HCH + thiram
A combined fungicide and insecticide seed treatment

| **Products** | Lindex-Plus FS | DowElanco | 43:545:73 g/l | FS | 03934 |

Uses   Alternaria, canker, damping off, flea beetles in OILSEED RAPE.

Notes

Efficacy
- Do not treat seed with a moisture content exceeding 9%
- Keep treated seed in a cool, dry place, sow as soon as possible after treatment and do not store for more than 6 mth

Crop Safety/Restrictions
- Maximum number of treatments 1 per batch
- Do not mix with other treatments
- Control may be reduced if strains of fungi tolerant to either fungicide develop
- Treatment may reduce germination, especially if seed is grown, harvested or stored under adverse conditions

---

FOR FULL CONDITIONS OF USE ALWAYS READ THE PRODUCT LABEL

Special precautions/Environmental safety
- Product only to be used by agricultural contractors
- Toxic if swallowed. Irritating to skin, eyes and respiratory system
- Dangerous to fish or other aquatic life. Do not contaminate surface waters or ditches with chemical or used container

Protective clothing/Label precautions
- A, C
- 4c, 6a, 6b, 6c, 18, 21, 29, 36, 37, 52, 60, 61, 63, 65, 70, 73, 74, 75, 77, 79

Latest application/Harvest Interval(HI)
- Before drilling

Maximum Residue Levels (mg residue/kg food)
- see gamma-HCH entry

---

268   fenpropimorph + prochloraz
A fungicide mixture for late season disease control in cereals

Products					
	1 Sprint	AgrEvo	375:225 g/l	EC	07291
	2 Sprint	Schering	375:225 g/l	EC	03986

Uses
Brown rust, eyespot, net blotch, powdery mildew, rhynchosporium, yellow rust in BARLEY. Brown rust, eyespot, powdery mildew, rhynchosporium, septoria, yellow rust in WINTER RYE. Brown rust, eyespot, powdery mildew, septoria, yellow rust in SPRING WHEAT, WINTER WHEAT.

Notes

Efficacy
- In wheat, if disease present, spray as soon as ligule of flag leaf visible (GS 39). If no disease, delay until first signs appear or use as protectant treatment before flowering
- In barley spray when disease appears on new growth or as protective spray when flag leaf ligule just visible (GS 39)

Crop Safety/Restrictions
- Maximum number of treatments 2 per crop
- Do not apply to crops suffering from stress

Special precautions/Environmental safety
- Harmful in contact with skin
- Flammable
- Dangerous to fish or other aquatic life. Do not contaminate surface waters or ditches with chemical or used container

Protective clothing/Label precautions
- A, C
- 5a, 12c, 18, 21, 29, 35, 36, 37, 52, 60, 63, 66, 70, 78

Latest application/Harvest Interval(HI)
- Up to and including ear emergence just complete (GS 59).
- HI 5 wk

---

269   fenpropimorph + propiconazole
A fungicide mixture for control of leaf diseases of cereals

Products

1 Decade 500 EC	Ciba Agric.	375:125 g/l	EC	05757
2 Glint 500 EC	Ciba Agric.	375:125 g/l	EC	04126
3 Mantle 425 EC	Ciba Agric.	300:125 g/l	EC	05715

Uses   Brown rust, mildew, net blotch, rhynchosporium, yellow rust in WINTER BARLEY. Brown rust, mildew, septoria, sooty moulds, yellow rust in WINTER WHEAT. Mildew, net blotch, rhynchosporium in SPRING BARLEY.

Notes   **Efficacy**
* Apply at start of disease attack
* Spray winter wheat between GS 30-32 against early Septoria, mildew and rusts, at flag leaf against late foliar and ear disease, and repeat 28 d later if necessary [1, 2]
* Spray spring barley immediately disease appears in crop
* Spray winter barley at risk from foliar disease at GS 30-31, and repeat at flag leaf if necessary, or spray at awn emergence (GS 49) [1, 2]. See label for details

Crop Safety/Restrictions
* Maximum number of treatments 2 per crop [1, 2]; 2 per crop for spring wheat and barley, 3 per crop for winter crops (1 in autumn at least 3 mth before 2 in spring/summer) [3]
* If used in programme with other products containing fenpropimorph observe maximum total doses per season permitted for wheat and barley

Special precautions/Environmental safety
* Harmful by inhalation. Irritating to eyes, skin and respiratory system
* Dangerous to fish or other aquatic life. Do not contaminate surface waters or ditches with chemical or used container

Protective clothing/Label precautions
* A, C
* 5b, 6a, 6b, 6c, 18, 21, 26, 27, 28, 29, 36, 37, 52, 63, 66, 70, 78

Latest application/Harvest Interval(HI)
* HI 5 wk

Maximum Residue Levels (mg residue/kg food)
* see propiconazole entry

270   fenpropimorph + tridemorph
A systemic mixture of morpholine fungicides

Products

1 BAS 46402F	BASF	500:250 g/l	EC	03313
2 Gemini	BASF	500:250 g/l	EC	05684

Uses   Powdery mildew, rhynchosporium in BARLEY. Powdery mildew in WINTER WHEAT.

FOR FULL CONDITIONS OF USE ALWAYS READ THE PRODUCT LABEL

Notes

Efficacy
- Spray when disease first starts to build up
- Control of Rhynchosporium can be improved by mixture with carbendazim

Crop Safety/Restrictions
- Maximum number of treatments 2 per crop in spring barley, 3 per crop (with maximum of 2 between Jan 1 and harvest) on winter wheat and winter barley
- Crop scorch may follow treatment, especially on winter wheat, particularly if applied in high temperatures or high light intensity or during frosty weather
- Avoid drift on to neighbouring crops

Special precautions/Environmental safety
- Harmful if swallowed. Irritating to skin
- Risk of serious damage to eyes
- Dangerous to fish or other aquatic life. Do not contaminate surface waters or ditches with chemical or used container

Protective clothing/Label precautions
- A, C
- 5c, 6b, 9, 18, 21, 26, 27, 28, 29, 36, 37, 43 (14 d), 52, 63, 66, 70, 78

Withholding period
- Keep livestock out of treated areas for at least 14 d

Latest application/Harvest Interval(HI)
- HI 5 wk

271   fenpyroximate

A mitochondrial electron transport inhibitor acaricide for apples

Products	Sequel	Promark	51.3 g/l	SC	07624

Uses   Fruit tree red spider mite in APPLES.

Notes

Efficacy
- Kills motile stages of fruit tree red spider mite. Best results achieved if applied in warm weather
- Total spray cover of trees essential. Use higher volumes for large trees
- Apply when majority of winter eggs have hatched

Crop Safety/Restrictions
- Maximum number of treatments 1 per yr
- Other mitochondrial electron transport inhibitor (METI) acaricides should not be applied to the same crop in the same year
- Consult processor before use on crops for processing

Special precautions/Environmental safety
- Irritant. Risk of serious damage to eyes
- Dangerous to fish or other aquatic life. Do not contaminate surface waters or ditches with chemical or used container
- Do not allow direct spray from air assisted sprayers to fall within 38 m of surface waters or ditches. Direct spray away from water

Protective clothing/Label precautions
* A, C, H
* 6, 9, 18, 21, 26, 29, 36, 37, 52, 63, 66,

Latest application/Harvest Interval(HI)
* HI 2 wk

272 fentin acetate
An organotin fungicide available only in mixtures

273 fentin acetate + maneb
A curative and protectant fungicide for use in potatoes

Products					
1 Brestan 60 SP	AgrEvo	54:16% w/w	WB	07305
2 Brestan 60 SP	Hoechst	54:16% w/w	WB	06042

Uses Blight in POTATOES.

Notes

Efficacy
* Commence spraying before infection occurs, before haulm meets across the rows or at first blight warning, whichever occurs first
* Repeat at 7-14 d intervals until growth ceases or haulm burnt off
* Do not spray if rain imminent

Crop Safety/Restrictions
* Maximum number of treatments 6 per crop

Special precautions/Environmental safety
* Fentin acetate is subject to the Poisons Rules 1982 and the Poisons Act 1972. See notes in Section 1
* Harmful if swallowed. Irritating to eyes, skin and respiratory system
* Keep in original container, tightly closed, in a safe place, under lock and key
* Harmful to livestock
* Harmful to fish or other aquatic life. Do not contaminate surface waters or ditches with chemical or used container

Protective clothing/Label precautions
* A, C, H, J, M
* 5c, 6a, 6b, 6c, 14, 16, 18, 22, 28, 29, 35, 36, 37, 41, 53, 64, 67, 70, 78

Withholding period
* Keep all livestock out of treated areas for at least 1 wk

Latest application/Harvest Interval(HI)
* HI 7 d

FOR FULL CONDITIONS OF USE ALWAYS READ THE PRODUCT LABEL

Maximum Residue Levels (mg residue/kg food)
• see maneb entry

274   fentin hydroxide
A curative and protectant organotin fungicide

Products

1 Ashlade Flotin	Ashlade	625 g/l	SC	06223
2 MSS Flotin 480	Mirfield	480 g/l	SC	07616
3 Super-Tin 4L	Chiltern	480 g/l	SC	02995

Uses   Blight in POTATOES.

Notes   **Efficacy**
• Apply to potatoes before haulm meets across rows or on receipt of blight warning and repeat at 7-14 d intervals throughout season. Late sprays protect against tuber blight
• Best results achieved by using for 2 final sprays of blight control programme. Allow at least 14d after complete death of haulm before lifting unless for immediate sale

Crop Safety/Restrictions
• Maximum number of treatments 6 per crop [3]; 2 per crop [1]
• Drought stress [1] or mixtures with emulsifiable concentrates [2, 3] may cause localised leaf spotting
• A minimum interval of 7d (10 d [1]) between applications must be observed [2]

Special precautions/Environmental safety
• Fentin hydroxide is subject to the Poisons Rules 1982 and the Poisons Act 1972. See notes in Section 1
• Harmful if swallowed [2, 3] and in contact with skin [1]
• Irritating to eyes and skin [1], to eyes, skin and respiratory system [2, 3]
• Keep in original container, tightly closed, in a safe place, under lock and key [2, 3]
• Harmful (dangerous [1])to livestock
• Harmful (dangerous [3]) to fish or other aquatic life. Do not contaminate surface waters or ditches with chemical or used container

Protective clothing/Label precautions
• A, C, H
• 5a, 40 (7 d), 63, 67 [1]; 5c, 6a, 16, 18, 21, 28, 29, 36, 37, 53, 70, 78 [1-3]; 6b [1, 2]; 6c, 14, 25, 26, 41 (7 d), 60, 64, 66 [2, 3]

Withholding period
• Keep all livestock out of treated areas for at least 1 wk

Latest application/Harvest Interval(HI)
• HI potatoes 1 wk [1]

Approval
• Approved for aerial application on potatoes. See notes in Section 1 [1]

275   fenuron
A urea herbicide available only in mixtures

See also chloridazon + chlorpropham + fenuron + propham
    chlorpropham + cresylic acid + fenuron
    chlorpropham + fenuron
    chlorpropham + fenuron + propham

276   ferbam + maneb + zineb
A protectant dithiocarbamate complex fungicide

Products					
Trimanzone WP	Intracrop	10:65:10% w/w	WP	05860

Uses   Blight in POTATOES. Fungus diseases in LEEKS *(off-label)*.

Notes

Efficacy
* For blight control apply before haulm meets across the rows or as soon as blight warning received, whichever occurs first. Repeat every 10-14 d

Crop Safety/Restrictions
* Observe minimum period of 10 d between applications

Special precautions/Environmental safety
* Irritating to skin, eyes and respiratory system

Protective clothing/Label precautions
* A, C
* 6a, 6b, 6c, 18, 21, 29, 35, 36, 37, 54, 63, 67

Latest application/Harvest Interval(HI)
* HI potatoes, leeks 7 d

Approval
* Off-label approval unlimited for use on leeks (OLA 0599/92)[1]

Maximum Residue Levels (mg residue/kg food)
* see maneb entry

FOR FULL CONDITIONS OF USE ALWAYS READ THE PRODUCT LABEL

277 ferrous sulfate

A herbicide/fertilizer combination for moss control in turf

See also 2,4-D + dicamba + ferrous sulfate
dichlorophen + ferrous sulfate
dichlorprop + ferrous sulfate + MCPA

Products					
1 Aitken's Lawn Sand	Aitken	10% w/w	SA	05253
2 Elliott's Lawn Sand	Elliott	17% w/w	SA	04860
3 Elliott's Mosskiller	Elliott	45% w/w	GR	04909
4 Fisons Greenmaster Autumn	Levington	6% w/w	GR	03211
5 Fisons Greenmaster Mosskiller	Levington	8.9% w/w	GR	00881
6 Greenmaster Autumn	Levington	6% w/w	GR	07508
7 Greenmaster Mosskiller	Levington	8.9% w/w	GR	07509
8 Maxicrop Moss Killer & Conditioner	Maxicrop	16.4% w/w	SL	04635
9 SHL Lawn Sand	Sinclair	10% w/w	SA	05254
10 Taylors Lawn Sand	Rigby Taylor	7.35% w/w	SA	04451
11 Vitax Lawn Sand	Vitax	8.6% w/w	SA	04352
12 Vitax Microgran 2	Vitax	20% w/w	MG	04541
13 Vitax Turf Tonic	Vitax	15% w/w	SA	04354
14 Walkover Mosskiller	Allen	16.4% w/w	SL	04662

Uses

Mosses in TURF.

Notes

Efficacy
* Apply autumn treatment from Sep onward but not when heavy rain expected or in frosty weather [4]
* Apply from Mar to Sep except during drought or when soil frozen [5]
* For best results apply when light showers expected, mow 3 d before treatment, do not mow or walk on treated area until well watered and water after 2 d if no rain
* Rake out dead moss thoroughly 7-14 d after treatment
* Fertilizer component of autumn treatment encourages root growth, that of mosskiller formulations promotes tillering

Crop Safety/Restrictions
* If spilt on paving, concrete, clothes etc brush off immediately to avoid discolouration

Special precautions/Environmental safety
* Harmful to fish or other aquatic life. Do not contaminate surface waters or ditches with chemical or used container [4, 5]

Protective clothing/Label precautions
* 29, 63, 67 [1-14]; 53 [4-7]; 54 [1-3, 11-14]

278 flamprop-M-isopropyl

A translocated, post-emergence arylalanine wild-oat herbicide

Products					
1 Commando	Cyanamid	200 g/l	EC	07005
2 Stefes Flamprop	Stefes	200 g/l	EC	05789

Uses	Wild oats in BARLEY, DURUM WHEAT, RYE, TRITICALE, WHEAT.

Notes

Efficacy
- Must be applied when crop and weeds growing actively under conditions of warm, moist days and warm nights
- Timing determined mainly by crop growth stage. Best control at later stages of wild oats but not after weeds visible above crop
- Best results achieved before 3rd node detectable stage (GS 33) in barley, or 4th node detectable (GS 34) in wheat
- Good spray coverage and retention essential. Spray rainfast after 2 h
- Addition of Swirl adjuvant oil recommended for use on winter wheat and winter barley
- Do not treat thin, open crops
- See label for recommended tank mixes and sequential treatments

Crop Safety/Restrictions
- Maximum number of treatments 1 per crop
- May be used on crops undersown with ryegrass and clover
- Do not use on crops under stress or during periods of high temperature
- Do not roll or harrow within 7 d of spraying
- Do not mix with any herbicide other than fluroxypyr

Special precautions/Environmental safety
- Irritating to eyes and skin
- Flammable
- Do not use straw from barley treated after stage GS 33 or wheat treated after stage GS 34 as feed or bedding for animals
- Harmful to fish or other aquatic life. Do not contaminate surface waters or ditches with chemical or used container

Protective clothing/Label precautions
- A, C
- 6a, 6b, 12c, 14, 16, 18, 21, 28, 29, 36, 37, 53, 60, 63, 66, 70

Latest application/Harvest Interval(HI)
- Before flag leaf sheath opening (GS 47) for wheat; before first awns visible (GS 49) for barley; before 3rd node detectable (GS 33) for rye and triticale; before 4th node detectable (GS 34) for durum wheat

Approval
- Approved for aerial application on wheat, durum wheat, barley, rye, triticale. See notes in Section 1

FOR FULL CONDITIONS OF USE ALWAYS READ THE PRODUCT LABEL

279 fluazifop-P-butyl

A phenoxypropionic acid grass herbicide for broadleaved crops

Products

1 Barclay Winner	Barclay	125 g/l	EC	06986
2 Citadel	Zeneca	125 g/l	EC	06762
3 Corral	Quadrangle	125 g/l	EC	06647
4 Fusilade 250 EW	Zeneca	250 g/l	EW	06531
5 Standon Fluazifop-P	Standon	125 g/l	EC	06060
6 Stefes Slayer	Stefes	125 g/l	EC	06277

Uses

Annual grasses, barren brome, perennial grasses, volunteer cereals in SETASIDE *(off-label)* [4]. Annual grasses, barren brome, volunteer cereals in FIELD MARGINS *(off-label)* [4]. Annual grasses, blackgrass, volunteer cereals, wild oats in FIELD BEANS, HOPS [1, 2, 4]. Annual grasses, perennial grasses, volunteer cereals, wild oats in FODDER BEET, SUGAR BEET, WINTER OILSEED RAPE [1-6]. Annual grasses, perennial grasses, volunteer cereals, wild oats in BLACKCURRANTS, CARROTS, GOOSEBERRIES, KALE *(stockfeed only)*, ONIONS, PEAS, RASPBERRIES, STRAWBERRIES, SWEDES *(stockfeed only)* [1-4, 6]. Annual grasses, perennial grasses in FARM FORESTRY *(off-label)* [4].

Notes

Efficacy
- Best results achieved by application when weed growth active under warm conditions with adequate soil moisture. Agral wetter must always be added to spray
- Spray weeds from 2-expanded leaf stage to fully tillered, couch from 4 leaves when majority of shoots have emerged, with a second application if necessary
- Control of perennial grasses may be reduced under dry conditions. Do not cultivate for 2 wk after spraying couch
- Annual meadow grass is not controlled
- May also be used to remove grass cover crops

Crop Safety/Restrictions
- Maximum number of treatments 2 per crop for sugar and fodder beet; 2 per crop (1 per crop [3, 5]) for winter oilseed rape; 2 per yr for strawberries and farm forestry; 1 per crop or yr for other crops
- Apply to sugar and fodder beet from 1-true leaf to 50% ground cover
- Apply to winter oilseed rape from 1-true leaf to established plant stage
- Apply in fruit crops after harvest. See label for timing details on other crops
- Before using on onions or peas use crystal violet test to check that leaf wax is sufficient
- Off-label uses on legumes refer to seed crops only
- Do not sow cereals for at least 8 wk after application of high rate or 2 wk after low rate
- Do not apply through CDA sprayer, with hand-held equipment or from air

Special precautions/Environmental safety
- Irritating to skin [4] and eyes [1, 2, 5, 6]
- Flammable
- Harmful to fish or other aquatic life. Do not contaminate surface waters or ditches with chemical or used container

Protective clothing/Label precautions
- A, C, H
- 6a, 12c [1, 2, 5, 6]; 6b [1, 2, 4-6]; 18, 21, 29, 36, 37, 53, 63, 66, 70, 78 [1-6]

Latest application/Harvest Interval(HI)
* Before flowering for fruit crops and hops; before flower bud visible for peas, field beans, broad beans and beans harvested dry; before 31 Dec in yr of sowing for winter oilseed rape; before 50% ground cover for swedes.
* HI beet crops, carrots, kale, 8 wk; onions 4 wk

Approval
* Off-label Approval unlimited for use in farm forestry (OLA 1302/94)[4]; to Oct 1998 for use on non-cropped field margins (OLA 1301/94)[4]; to Oct 1998 for use on land temporarily removed from production (OLA 1301/94)[4]

---

280   fluazinam
A dinitroaniline fungicide for use in potatoes

Products					
	1 Salvo	Zeneca	500 g/l	SC	07092
	2 Shirlan	Zeneca	500 g/l	SC	07091

Uses   Blight in POTATOES.

Notes

Efficacy
* Commence treatment at the first blight warning and before blight enters the crop
* In the absence of a warning treatment should start before foliage of adjacent plants meets in the rows
* Spray at 7-14 d intervals depending on severity of risk
* Ensure complete coverage of the foliage and stems, increasing volume as haulm growth progresses, in dense crops and if blight risk increases

Crop Safety/Restrictions
* Maximum number of treatments 10 per crop

Special precautions/Environmental safety
* Irritant. May cause sensitization by skin contact
* Do not contaminate surface waters or ditches with chemical or used container
* Do not allow direct spray to fall within 6 m of surface waters or ditches. Direct spray away from water

Protective clothing/Label precautions
* A, C, H
* 10a, 14, 16, 18, 21, 29, 36, 37, 52, 54a, 63, 66, 78

Latest application/Harvest Interval(HI)
* HI 7 d before harvest

FOR FULL CONDITIONS OF USE ALWAYS READ THE PRODUCT LABEL

281   fludioxonil

A phenylpyrrole fungicide seed treatment for wheat and barley

Products	Beret Gold	Ciba Agric.	25 g/l	SC	07531

Uses   Bunt, fusarium foot rot and seedling blight, snow mould in WINTER WHEAT. Covered smut, foot rot, fusarium foot rot and seedling blight, snow mould in SPRING BARLEY, WINTER BARLEY. Leaf stripe in SPRING BARLEY *(partial control)*, WINTER BARLEY *(partial control)*.

Notes   **Efficacy**
- Apply direct to seed using conventional seed treatment equipment. Continuous flow treaters should be calibrated using product before use
- Effective against benzimidazole-resistant strains of *Fusarium nivale*
- Product may reduce flow rate of seed through drill. Recalibrate with treated seed before drilling

Crop Safety/Restrictions
- Maximum number of treatments 1 per seed batch
- Do not apply to cracked, split or sprouted seed
- Sow treated seed within 6 mth

Special precautions/Environmental safety
- Harmful to fish or other aquatic life. Do not contaminate surface waters or ditches with chemical or used container
- Treated seed not to be used for food or feed. Do not re-use sack for food or feed
- Treated seed must not be applied from the air

Protective clothing/Label precautions
- A, H
- 29, 53, 60, 67, 72, 73, 74, 75, 76, 77

Latest application/Harvest Interval(HI)
- Before drilling

282   fluoroglycofen-ethyl

A diphenyl ether herbicide available only in mixtures

283   fluoroglycofen-ethyl + mecoprop-P

A contact and translocated herbicide mixture for winter cereals

Products	Estrad	BASF	20% w/w:500 g/l	KK	07196

Uses   Annual dicotyledons, cleavers, speedwells in WINTER BARLEY, WINTER WHEAT.

Notes   **Efficacy**
- Best results achieved by treatment when weeds are actively growing

- Later applications may result in reduced weed control and failure to achieve optimum yield
- Increase water volume for cleavers larger than 4-6 whorls and for dense weed infestations or crop cover

Crop Safety/Restrictions
- Maximum number of treatments 1 per crop
- Do not use late applications in mixture with any product
- Do not apply if crops are wet or stressed from poor fertility or previous pesticide treatment
- Do not apply during cold weather, drought, or if rain or frost expected
- Do not open or touch the water soluble bag with wet gloves or hands
- Avoid damage by spray drift onto susceptible crops (e.g. beans, beet, brassicae, oilseed rape, fruit, glasshouse crops, hops, lettuce, ornamentals, peas, potatoes, tomatoes, vines)
- Cultivate to 10 cm before drilling following broad-leaved crops
- Wash equipment thoroughly after use to remove traces which may damage susceptible crops sprayed later

Special precautions/Environmental safety
- Harmful if swallowed
- Risk of serious damage to eyes
- Extremely dangerous to fish or other aquatic life. Do not contaminate surface waters or ditches with chemical or used container
- Do not allow direct spray from ground based/vehicle mounted sprayers to fall within 6 m, or from hand-held sprayers to within 2 m, of surface waters or ditches. Direct spray away from water

Protective clothing/Label precautions
- A, C, H
- 18, 21, 29, 36, 37, 43, 51, 54a, 63, 66 (container B), 67 (container A), 70, 78

Withholding period
- Keep livestock out of treated areas until foliage of poisonous weeds such as ragwort has died and become unpalatable

Latest application/Harvest Interval(HI)
- Before third node detectable stage (GS 33)

---

284   fluroxypyr

A post-emergence pyridyloxyacetic herbicide for use in cereals

See also *bromoxynil + clopyralid + fluroxypyr*
        bromoxynil + fluroxypyr
        bromoxynil + fluroxypyr + ioxynil
        clopyralid + fluroxypyr + ioxynil
        clopyralid + fluroxypyr + triclopyr

Products					
1 Barclay Hurler	Barclay	200 g/l	EC	06952
2 Standon Fluroxypyr	Standon	200 g/l	EC	07055
3 Starane 2	DowElanco	200 g/l	EC	05496
4 Stefes Fluroxpyr	Stefes	200 g/l	EC	05977

FOR FULL CONDITIONS OF USE ALWAYS READ THE PRODUCT LABEL

Uses   Annual dicotyledons, black bindweed, chickweed, cleavers, docks, forget-me-not, hemp-nettle, volunteer potatoes in MAIZE [3]. Annual dicotyledons, black bindweed, chickweed, cleavers, docks, forget-me-not, hemp-nettle, volunteer potatoes in BARLEY, DURUM WHEAT, ESTABLISHED GRASSLAND, OATS, RYE, SEEDLING LEYS, TRITICALE, WHEAT [1-4]. Annual dicotyledons, volunteer potatoes in BULB ONIONS *(off-label)* [3].

Notes   **Efficacy**
- Best results achieved under good growing conditions in a strongly competing crop
- A number of tank mixtures with HBN and other herbicides are recommended for use in autumn and spring to extend range of species controlled. See label for details
- Spray is rainfast in 1 h
- Do not spray if frost imminent

Crop Safety/Restrictions
- Maximum number of treatments 1 per crop or yr
- Apply to new leys from 3 expanded leaf stage
- Timing varies in tank mixtures. See label for details
- Do not use on crops undersown with clovers or other legumes
- Crops undersown with grass may be sprayed provided grasses are tillering
- Do not treat crops suffering stress caused by any factor
- Do not roll or harrow for 7 d before or after treatment

Special precautions/Environmental safety
- Flammable
- Harmful to fish or other aquatic life. Do not contaminate surface waters or ditches with chemical or used container

Protective clothing/Label precautions
- 12c, 21, 28, 29, 43, 53, 63, 66

Withholding period
- Keep livestock out of treated areas for at least 3 d and until foliage of poisonous weeds such as ragwort has died and become unpalatable

Latest application/Harvest Interval(HI)
- Before flag leaf sheath opening (GS 47) for winter wheat and barley; before flag leaf sheath extending (GS 41) for spring wheat and barley; before second node detectable (GS 32) for oats, rye, triticale and durum wheat; before buttress roots appear for maize

Approval
- Off-label approval to Oct 1998 for use in bulb onions (OLA 1339/93)[3]

285   fluroxypyr + mecoprop-P

A post-emergence herbicide for broadleaved weeds in amenity turf

Products   Bastion T   Rigby Taylor   72:300 g/l   EC   06011

Uses   Annual dicotyledons, perennial dicotyledons, slender speedwell in AMENITY TURF, SPORTS TURF.

Notes   **Efficacy**
- Best results achieved when soil moist and weeds in active growth, normally Apr-Sep

- Do not apply in drought period unless irrigation applied
- Avoid mowing for 3 d before or after spraying (5 d before for young turf)
- Do not spray if turf wet

Crop Safety/Restrictions
- Maximum number of treatments 2 per yr
- Young turf must only be treated in spring provided that at least 2 mth have elapsed between sowing and application
- After mowing young turf allow 5 d before treatment
- Do not treat turf under stress from any cause, if night temperatures are low, if ground frost imminent or during prolonged cold weather

Special precautions/Environmental safety
- Harmful in contact with skin or if swallowed, irritating to eyes
- Harmful to fish or other aquatic life. Do not contaminate surface waters or ditches with chemical or used container

Protective clothing/Label precautions
- A, C
- 5a, 5c, 6a, 18, 21, 28, 29, 36, 37, 53, 63, 66, 70, 78

286   fluroxypyr + triclopyr
A foliar acting herbicide for docks in grassland

Products	Doxstar	DowElanco	100:100 g/l	EC	06050

Uses   Docks in ESTABLISHED GRASSLAND.

Notes   **Efficacy**
- Apply in spring or autumn or, at lower dose, in spring and autumn
- Treat docks in rosette stage, up to 200 mm high or across
- Grass and weeds must be growing actively to ensure best results
- Do not spray in drought, very hot or very cold weather
- Allow 2-3 wk growth after cutting or grazing to allow sufficient regrowth to occur before spraying
- Control may be reduced if rain falls within 2 h of application
- Do not cut grass for 28 d after spraying

Crop Safety/Restrictions
- Maximum number of treatments 1 per yr (2 per yr at lower dose)
- Do not spray grass less than 1 yr old
- Clover will be killed or severely checked by treatment
- Do not roll or harrow for 10 d before or 7 d after spraying
- Do not sow kale, turnips, swedes or grass mixtures containing clover by direct drilling or minimum cultivation techniques within 6 wk of application

---

FOR FULL CONDITIONS OF USE ALWAYS READ THE PRODUCT LABEL

Special precautions/Environmental safety
* Irritating to eyes
* Dangerous to fish or other aquatic life. Do not contaminate surface waters or ditches with chemical or used container
* Do not allow drift to come into contact with crops, amenity plantings, gardens, ponds, lakes or watercourses

Protective clothing/Label precautions
* A, C
* 6a, 14, 18, 21, 28, 29, 34, 36, 37, 43, 52, 57, 63, 66, 70

Withholding period
* Keep livestock out of treated areas for at least 7 d and until foliage of any poisonous weeds such as ragwort has died and become unpalatable

Latest application/Harvest Interval(HI)
* 7 d before grazing or harvest of grass

---

287 flusilazole

A systemic, protective and curative conazole fungicide for cereals

See also carbendazim + flusilazole
fenpropimorph + flusilazole
fenpropimorph + flusilazole + tridemorph

Products					
1	DUK 747	DuPont	400 g/l	EC	05775
2	Genie	DuPont	400 g/l	EC	05272
3	Lyric	DuPont	250 g/l	EC	06543
4	Sanction	DuPont	400 g/l	EC	04773

Uses

Brown rust, eyespot, net blotch, powdery mildew, rhynchosporium, septoria, yellow rust in SPRING BARLEY, WINTER BARLEY [1-4]. Brown rust, eyespot, powdery mildew, septoria, yellow rust in WINTER WHEAT [1-4]. Light leaf spot in OILSEED RAPE [4].

Notes

Efficacy
* Use as a routine preventative spray or when disease first develops
* Best control of eyespot achieved by spraying between leaf-sheath erect and second node detectable stages (GS 30-32)
* Product active against both MBC-sensitive and MBC-resistant strains of eyespot
* Treat oilseed rape in autumn and/or spring at stem extension stage [4]
* Rain occurring within 2 h after application may reduce effectiveness
* See label for recommended tank-mixes

Crop Safety/Restrictions
* Maximum number of treatments (including other products containing flusilazole) 3 per crop for winter wheat; 2 per crop for barley. Do not use high rate more than once in any crop [1-3]
* Maximum number of treatments (including other products containing flusilazole) 1 per crop at full dose and/or 2 per crop at reduced dose on winter wheat and barley; 1 at full dose or 2 at half dose on oilseed rape [4]
* Lower rate may be applied at any time up to and including grain watery ripe (GS 71). Higher rate may be applied up to and including second node detectable (GS 32)
* Do not apply to crops under stress or during frosty weather

Special precautions/Environmental safety
* Harmful if swallowed
* Flammable
* Dangerous to fish or other aquatic life. Do not contaminate surface waters or ditches with chemical or used container

Protective clothing/Label precautions
* A, C
* 5c, 12c, 18, 24, 28, 29, 36, 37, 52, 63, 66, 70, 78

Latest application/Harvest Interval(HI)
* High dose before 3rd node detectable (GS 33) plus reduced dose before early milk stage (GS 72) on winter wheat and barley [1-4]; before first flowers open (GS 4,0) on oilseed rape [4]

288 flusilazole + tridemorph
A systemic protective and curative fungicide for cereals

Products	Fusion	DuPont	160:350 g/l	EC	04908

Uses Brown rust, net blotch, powdery mildew, rhynchosporium, yellow rust in SPRING BARLEY, WINTER BARLEY. Brown rust, powdery mildew, septoria diseases, yellow rust in WINTER WHEAT.

Notes

Efficacy
* Apply at start of disease attack. Treatment most effective when disease levels low at application
* Rain within 2 h of application may reduce effectiveness

Crop Safety/Restrictions
* Maximum number of treatments (including other products containing flusilazole) 3 per crop for winter wheat, 2 per crop for winter or spring barley
* Some crop scorch may occur if treatment applied during frosty weather, high temperatures or drought

Special precautions/Environmental safety
* Irritating to eyes
* Flammable
* Dangerous to fish or other aquatic life. Do not contaminate surface waters or ditches with chemical or used container

Protective clothing/Label precautions
* A, C
* 6a, 12c, 18, 24, 28, 29, 36, 37, 43, 52, 63, 66

Withholding period
* Keep livestock out of treated areas for at least 14 d

FOR FULL CONDITIONS OF USE ALWAYS READ THE PRODUCT LABEL

Latest application/Harvest Interval(HI)
* Before early milk stage of grain (GS 73)

289   flutriafol

A conazole fungicide for broad spectrum disease control in wheat and barley

See also carbendazim + flutriafol
                 chlorothalonil + flutriafol
                 ethirimol + flutriafol + thiabendazole

Products					
1	Landgold Flutriafol	Landgold	125 g/l	SC	06244
2	Pointer	Zeneca	125 g/l	SC	06695

Uses

Brown rust, leaf blotch, net blotch, powdery mildew, yellow rust in SPRING BARLEY. Brown rust, powdery mildew, septoria diseases, yellow rust in SPRING WHEAT, WINTER WHEAT. Leaf blotch, net blotch, powdery mildew in WINTER BARLEY.

Notes

Efficacy
* Best results obtained from application as protectant or at early stage of disease development
* See label for details of recommended timing, need for repeat treatments and tank mixes

Crop Safety/Restrictions
* Maximum number of treatments 2 per crop (including any other products containing flutriafol)
* Under conditions of stress some wheat varieties can exhibit flag leaf tip scorch which may be increased by fungicide treatment

Special precautions/Environmental safety
* Irritating to eyes and skin
* Harmful to fish or other aquatic life. Do not contaminate surface waters or ditches with chemical or used container

Protective clothing/Label precautions
* A, C
* 6a, 6b, 14, 18, 22, 28, 29, 36, 37, 53, 63, 66, 70, 78

Latest application/Harvest Interval(HI)
* Up to grain watery ripe (GS 71)

290   fomesafen

A diphenyl ether herbicide only available in mixtures

291   fomesafen + terbutryn

A residual herbicide for use in leguminous crops

Products					
Reflex T		Zeneca	80:400 g/l	SC	07423

Uses	Annual dicotyledons in PEAS *(spring sown)*, SPRING BROAD BEANS, SPRING FIELD BEANS.

Notes	**Efficacy**

* Apply after planting, but before the crop emerges
* Best results achieved when applied to moist, fine, firm tilth
* Rain soon after application is essential for optimum weed control
* Results may be unsatisfactory in dry, or excessively wet, soil conditions
* Weed control may be reduced on soils with more than 10% organic matter

Crop Safety/Restrictions
* Maximum number of treatments 1 per crop
* Use only once every 5 yr
* All varieties of spring sown peas may be treated but forage varieties may be damaged from which recovery may not be complete. All varieties of spring sown field and broad beans may be treated
* Emerged crop leaves may be severely scorched or killed
* Do not use on Sands
* Plough or cultivate to at least 150 mm before drilling or planting another crop. Only cereals should be planted in the calendar year of use with a minimum interval of 4 mth after treatment

Special precautions/Environmental safety
* Irritant. May cause sensitisation by skin contact
* Harmful to fish or other aquatic life. Do not contaminate surface waters or ditches with chemical or used container

Protective clothing/Label precautions
* A, C, H
* 6, 10a, 16, 18, 21, 28, 30, 36, 37, 53, 63, 65, 70

Latest application/Harvest Interval(HI)
* Pre-emergence of crop

292   fonofos

An organophosphorus insecticide for soil and seed treatment

Products	1 Cudgel	Zeneca	433 g/l	CS	06648
	2 Fonofos Seed Treatment	Zeneca	433 g/l	FS	06664

Uses	Cabbage root fly in LEAF BRASSICAS, ROOT BRASSICAS [1]. Frit fly, wheat bulb fly, wireworms in SPRING WHEAT, WINTER BARLEY, WINTER WHEAT [2]. Sciarid flies, vine weevil in HARDY ORNAMENTALS [1].

Notes	**Efficacy**

* May be applied either by incorporation into compost, as a compost or soil drench, or as a hygiene spray treatment under glass against adults and larvae of sciarid flies [1]

FOR FULL CONDITIONS OF USE ALWAYS READ THE PRODUCT LABEL

- Should be incorporated into compost before striking cuttings, pricking out or potting on and followed with a drench treatment 6 wk later [1]
- When drenching use sufficient water to ensure thorough wetting of the growing medium but without excessive run-through [1]
- When used for cabbage root fly control in modular raised brassicas, slowly maturing crops should be re-treated in the field with a suitable granular insecticide to prevent re-infestation [1]
- Apply seed treatment with suitable liquid seed treatment machinery [2]
- For best results drill into a well-prepared seed-bed no more than 25mm deep. Do not broadcast treated seed [2]
- Recommended for all crops sown from early Sep until mid-Mar [2]
- Check drill calibration after each seed lot [2]
- Where wheat bulb fly egg counts are high, insecticide sprays at egg-hatch or dead-heart may be necessary [2]

Crop Safety/Restrictions
- Maximum number of treatments 1 per batch of seed [2]; 3 per season for container nursery stock [1]
- Do not drench under hot sunny conditions and rinse off plant foliage immediately with plain water [1]
- Check susceptibility of any one variety by treating a small number of plants first [1]
- Do not treat seed with more than 16% moisture [2]
- Treated seed may be stored for up to 3 mth in cool, dry, well ventilated place [2]
- All varieties of winter wheat, spring wheat and winter barley seed may be treated [2]

Special precautions/Environmental safety
- Fonofos is subject to the Poisons Rules 1982 and the Poisons Act 1972. See notes in Section 1
- Contains an anticholinesterase organophosphorus compound. Do not use if under medical advice not to work with such compounds
- Harmful in contact with skin [1]
- Toxic in contact with skin; may cause sensitisation by skin contact [2]
- Harmful if swallowed [2]
- Dangerous to game, wild birds and animals [1]
- Treated seed is dangerous to game and wildlife. Bury spillages [2]
- Dangerous to livestock [2]
- Extremely dangerous to fish or other aquatic life. Do not contaminate surface waters or ditches with chemical or used container
- Treated foliage must not be handled until the spray deposit has dried [1]
- Do not harvest crops for human or animal consumption for at least 6 wk after sowing treated seed [2]

Protective clothing/Label precautions
- A, C, D, H, K, M [2]; B, C, H, J, K, L, M [1]
- 1, 14, 16, 18, 22, 25, 28, 29, 36, 37, 45, 51, 64, 65, 70, 78 [1, 2]; 4a, 10a, 35, 40, 5c, 74, 75, 77 [2]; 5a, 26, 27 [1]

Withholding period
- Keep all livestock out of treated areas for at least 6 wk. Bury or remove spillages [2]

Latest application/Harvest Interval(HI)
- HI 6 wk

293 formaldehyde (commodity substance)

An agricultural/horticultural and animal husbandry fungicide

Products	1 formaldehyde	various	38-40%	SL
	2 paraformaldehyde	various		

Uses Fungus diseases in FLOWER BULBS *(dip)*, GLASSHOUSES, MUSHROOM HOUSES [1]. Fungus diseases in LIVESTOCK HOUSES [2]. Soil-borne diseases in SOIL AND COMPOST [1].

Notes **Efficacy**
* Use as a dip to sterilize flower bulbs [1]
* Use as drench to sterilize soil and compost, indoors and outdoors [1]
* Use as spray or fumigant in mushroom houses [1]
* Use as spray, dip or fumigant for glasshouse hygiene [1]
* Use as fumigant to sterilize animal houses [2]

Crop Safety/Restrictions
* Observe maximum permitted concentrations. See PR 12, 1990, pp 9-10

Special precautions/Environmental safety
* Formaldehyde is subject to the Poisons Rules 1982 and the Poisons Act 1972. See notes in Section 1
* Operators must observe Occupational Exposure Standard as set out in HSE Guidance Note EH40/90 and ACOP 30. Control of Substances Hazardous to Health in Fumigation
* Operators must be supplied with a Section 6 (HSW) Safety Data Sheet before commencing work

Protective clothing/Label precautions
* A, D, P

294 fosamine-ammonium

A contact phosphonic acid herbicide for woody weed control

Products	Krenite	DuPont	480 g/l	SL	01165

Uses Woody weeds in FORESTRY, NON-CROP AREAS, WATERSIDE AREAS.

Notes **Efficacy**
* Apply as overall spray to foliage during Aug-Oct when growth ceased but before leaves have started to change colour. Effects develop in following spring
* Thorough coverage of leaves and stems needed for effective control
* Addition of non-ionic wetter or Actipron recommended. Rain within 24 h may reduce effectiveness
* Most deciduous species controlled or suppressed but evergreens resistant

---

FOR FULL CONDITIONS OF USE ALWAYS READ THE PRODUCT LABEL

• Little effect on underlying herbaceous vegetation

Crop Safety/Restrictions
• Do not use in conifer plantations

Protective clothing/Label precautions
• 21, 29, 54, 60, 63, 65

Approval
• May be used alongside river, canal and reservoir banks but should not be sprayed into water. See notes in Section 1 on use of herbicides in or near water

---

295   fosetyl-aluminium

A systemic phosphonic acid fungicide for various horticultural crops

Products	Aliette	RP Agric.	80% w/w	WP	05648

Uses   Ascochyta, downy mildew in COMBINING PEAS *(off-label)*, VINING PEAS *(off-label)*. Collar rot, crown rot in APPLES. Damping off, downy mildew in BRASSICAS *(off-label)*. Downy mildew in BROAD BEANS, HOPS, LETTUCE *(off-label)*, PROTECTED LETTUCE. Phytophthora in CAPILLARY BENCHES, HARDY ORNAMENTAL NURSERY STOCK, PROTECTED POT PLANTS. Red core in STRAWBERRIES, STRAWBERRIES *(spring treatment, off-label)*.

Notes   **Efficacy**
• Spray young orchards for crown rot protection after blossom and repeat after 3-6 wk. Apply as paste to bark of apples to control collar rot
• Apply to broad beans at early flowering and 14 d later
• Spray autumn-planted strawberries 2-3 wk after planting or use dip treatment at planting. Spray established crops in late summer/early autumn after picking
• Apply to hops as early season band spray and as foliar spray every 10-14 d
• May be used incorporated in blocking compost for protected lettuce
• Apply as drench to rooted cuttings of hardy nursery stock after first potting and repeat mthly. Up to 6 applications may be needed
• See label for details of application to capillary benches and list of tolerant plants

Crop Safety/Restrictions
• Maximum number of treatments 1 per batch of compost for lettuce; 1 per yr for strawberries (root dip or foliar spray); 1 per yr for apples (bark paste), 6 per yr for apples (foliar spray); 2 per crop for broad beans, peas; 2 per yr for hops (basal spray); 6 per yr for hops (foliar spray); 1 per crop for brassicas; 1 per crop during propagation, 2 per crop after planting out for lettuce
• When using on strawberries do not mix with any other product
• Use in protected lettuce only from Sep to Apr
• Do not apply to ornamentals in mixture with nutrient solutions
• Apply test treatment on lettuce cultivars or ornamentals of unknown tolerance before committing whole batches

Protective clothing/Label precautions
• A, C, H
• 30, 35, 36, 37, 54, 63, 67

Latest application/Harvest Interval(HI)
* Up to 31 Dec for autumn treatment on strawberries; pre-emergence for brassicas.
* HI hops 4 d; lettuce 14 d; broad beans, peas 17 d; apples 14 d; strawberries 3 mth; applied as bark paste 5 mth

Approval
* Off-label approval to Sep 1997 for use on leaf brassicas, peas (OLA 0221/95); to Sep 1997 for use on strawberries (OLA 0222/95); to Sep 1997 for use on outdoor and protected lettuce (OLA 0223/95)

296 fuberidazole

A benzimidazole (MBC) fungicide available only in mixtures

See also bitertanol + fuberidazole

297 fuberidazole + triadimenol

A broad spectrum systemic fungicide seed treatment for cereals

Products					
1 Baytan Flowable	Bayer	22.5:187.5 g/l	FS	02593
2 Baytan Flowable	Zeneca	22.5:187.5 g/l	FS	06635

Uses
Blue mould, brown foot rot and ear blight, brown rust, bunt, loose smut, mildew, septoria, yellow rust in WHEAT. Blue mould, brown foot rot and ear blight in TRITICALE. Brown foot rot and ear blight, brown rust, covered smut, loose smut, mildew, rhynchosporium, yellow rust in BARLEY. Brown foot rot and ear blight, crown rust, loose smut, pyrenophora leaf spot in OATS. Brown foot rot and ear blight, stripe smut in RYE. Leaf stripe in SPRING BARLEY.

Notes   **Efficacy**
* Apply through suitable seed treatment machinery. See label for details
* Calibrate drill for treated seed and drill at 2.5-4 cm into firm, well prepared seedbed
* Ferrax recommended as preferable seed treatment for barley
* In addition to seed-borne diseases early attacks of various foliar, air-borne diseases are controlled or suppressed. See label for details
* Product may be used in conjunction with Gammasan 30. No other combinations recommended [1]

Crop Safety/Restrictions
* Maximum number of treatments 1 per batch of seed
* Do not use on naked oats
* Do not drill treated winter wheat or rye seed after end of Nov. Seed rate should be increased as drilling season progresses
* Treated spring wheat may be drilled in autumn up to end of Nov or from Feb onward
* Germination tests should be done on all batches of seed to be treated to ensure seed viability and suitability for treatment
* Do not use on seed with moisture content above 16%, on sprouted, cracked or skinned seed or on seed already treated with another seed treatment

FOR FULL CONDITIONS OF USE ALWAYS READ THE PRODUCT LABEL

- Store treated seed in cool, dry, well-ventilated store and drill as soon as possible, preferably in season of purchase
- Treatment may accentuate effects of adverse seedbed conditions on crop emergence

Special precautions/Environmental safety
- Treated seed not to be used as food or feed. Do not re-use sack
- Dangerous to fish or other aquatic life. Do not contaminate surface waters or ditches with chemical or used container

Protective clothing/Label precautions
- A
- 14, 29, 52, 60, 63, 70, 72, 73, 74, 76, 77

Latest application/Harvest Interval(HI)
- Before drilling

298   furalaxyl

A protective and curative phenylamide (acylalanine) fungicide for ornamentals

Products	Fongarid 25 WP	Ciba Agric.	25% w/w	WP	03595

Uses   Damping off in BEDDING PLANTS. Phytophthora, pythium in HARDY ORNAMENTAL NURSERY STOCK, POT PLANTS.

Notes   **Efficacy**
- Apply by incorporation into compost or as a drench to obtain at least 12 wk protection
- Apply drench within 3 d of seeding or planting out as a protective treatment or as soon as first signs of root disease appear as curative treatment
- Do not apply to field or border soil

Crop Safety/Restrictions
- Manufacturer's Guide lists 86 genera which have been treated successfully. With all subjects of unknown susceptibility treat a few plants before committing whole batches

Special precautions/Environmental safety
- Irritating to eyes and skin
- Harmful to fish or other aquatic life. Do not contaminate surface waters or ditches with chemical or used container

Protective clothing/Label precautions
- A, C
- 6a, 6b, 14, 18, 22, 28, 29, 36, 37, 53, 63, 66, 70

299 gamma-HCH
A contact, ingested and fumigant organochlorine insecticide

See also carboxin + gamma-HCH + thiram
fenpropimorph + gamma-HCH + thiram

Products					
1 Atlas Steward	Atlas	560 g/l	SC	03062
2 Fumite Lindane 10	Hortichem	21.1 g a.i.	FU	00933
3 Fumite Lindane 40	Hortichem	84.4 g a.i.	FU	00934
4 Fumite Lindane Pellets	Hortichem	2.7 g or 7.6 g a.i.	FU	00937
5 Gamma-Col	Zeneca	800 g/l	SC	06670
6 Gamma-Col Turf	ICI Professional	800 g/l	SC	05450
7 Gamma-Col Turf	Zeneca Prof.	800 g/l	SC	06826
8 Gammasan 30	Zeneca	30% w/w	FS	06671
9 Kotol FS	RP Agric.	125 g/l	FS	06968
10 Lindane Flowable	PBI	800 g/l	SC	02610
11 Unicrop Leatherjacket Pellets	Unicrop	1.8% w/w	GB	02272

Uses

Adelgids, aphids, capsids, chrysanthemum midge, leaf miners, leafhoppers, rhododendron bug, springtails, symphylids in ORNAMENTALS [1, 5]. Adelgids, aphids, leafhoppers, rhododendron bug in WOODY ORNAMENTALS [1]. Adelgids, large pine weevil in FORESTRY PLANTATIONS [10]. Ambrosia beetle, elm bark beetle, great spruce bark beetle in CUT TIMBER [10]. Ants, aphids, capsids, earwigs, leaf miners, mushroom flies, thrips, whitefly, woodlice in PROTECTED CROPS [4]. Aphids, capsids, leaf miners, leafhoppers, springtails, symphylids, thrips in TOMATOES [1, 5]. Aphids, capsids, leaf miners, leafhoppers in VEGETABLES [1]. Aphids in LETTUCE [1]. Apple blossom weevil, sawflies, suckers in APPLES, PEARS [1, 5]. Cabbage stem flea beetle, cabbage stem weevil, cutworms, flea beetles, leatherjackets, pollen beetles, weevils, wireworms in BRASSICAS [1, 5, 10]. Cabbage stem flea beetle, cabbage stem weevil, pollen beetles, seed weevil, weevils in OILSEED RAPE [1, 5, 10]. Cabbage stem flea beetle, cabbage stem weevil, pollen beetles, seed weevil in BRASSICA SEED CROPS [10]. Chafer grubs, leatherjackets, millipedes, wireworms in TURF [6, 7]. Chafer grubs, leatherjackets, wireworms in STRAWBERRIES [1, 5]. Clay-coloured weevil, cutworms, poplar leaf beetles, strawberry root weevil in FOREST NURSERY BEDS [10]. Cutworms, flea beetles, leatherjackets, millipedes, pygmy mangold beetle, springtails, symphylids, wireworms in SUGAR BEET [1, 5, 10, 11]. Cutworms, leatherjackets, wireworms in GRASSLAND [1, 5, 10, 11]. Cutworms, leatherjackets, wireworms in CEREALS [1, 5, 8-10]. Cutworms, leatherjackets, wireworms in MAIZE [1, 5]. Grain beetles, grain storage pests, grain weevils in GRAIN STORES *(empty)* [2, 3]. Leatherjackets in OATS, RYE, SPRING BARLEY, SPRING WHEAT, TRITICALE, WINTER BARLEY, WINTER WHEAT [11]. Raspberry cane midge in CANE FRUIT [1, 5]. Springtails, symphylids, thrips in CUCUMBERS [5].

Notes

Efficacy
- Apply as foliar spray, soil spray with incorporation, soil drench, dip, bran-based bait, fumigant or seed treatment as appropriate
- Method of application dose, timing and number of applications vary with formulation, crop and pest. See label for details

FOR FULL CONDITIONS OF USE ALWAYS READ THE PRODUCT LABEL

Crop Safety/Restrictions
* Maximum number of treatments 1 per crop for cereals, maize, sugar beet, strawberries, tomatoes, cucurbits; 1 per yr for grassland; 2 per yr for brassicas, apples, pears, cane fruit; 4 per crop for oilseed rape; 1 per batch of seed [9]
* Do not use if potatoes or carrots are to be planted within 18 mth [1, 10]
* Do not apply to potatoes, carrots or cucurbits or to blackcurrants or soft fruit after flowering
* In spring and summer apply in early morning or late evening to reduce hazard to bees
* Do not apply as soil/compost drench when roots are dry
* Do not fumigate newly pricked-out seedlings until root action has started again. Advisable to cut blooms before fumigating. Do not fumigate rare or unusual plants without first testing on small scale

Special precautions/Environmental safety
* Toxic if swallowed [6]; harmful if swallowed [7-9, 11]; harmful in contact with skin and if swallowed [1]; irritating to eyes and respiratory system [2, 3]
* Flammable [11]
* Dangerous to bees. Do not apply to crops in flower or to those in which bees are actively foraging. Do not apply when flowering weeds are present [1, 5, 6, 8, 9]
* Dangerous to fish [5, 6], harmful to fish [1-4, 7-11] or aquatic life. Do not contaminate surface waters or ditches with chemical or used container
* Treated seed not to be used as food or feed. Do not re-use sack [8, 9]
* Do not handle treated seed without protective gloves [8, 9]

Protective clothing/Label precautions
* A, C [2-11]; A [1]
* 4c [5, 6]; 5a, 30, 79 [1, 9]; 5b, 72, 77 [9]; 5c [1, 8-10]; 6a, 6c [2-4]; 14, 22, 28, 29, 35, 41, 52, 60, 66, 67, 78 [2-11]; 16, 74, 75 [8, 9]; 18, 36, 37, 48, 53, 63, 70 [1-11]; 21 [1, 8, 9]; 35 (14 d), 43 (14 d), 65 [1]; 73, 76 [8]

Withholding period
* Keep all livestock out of treated areas for at least 14 d [1, 5, 10, 11]

Latest application/Harvest Interval(HI)
* HI 2 d [4]; 14 d [1, 8, 10, 11]

Maximum Residue Levels (mg residue/kg food)
* strawberries, blackberries, loganberries, raspberries, bilberries, cranberries, currants, gooseberries 3; tomatoes, peppers, aubergines, cauliflowers, Brussels sprouts, head cabbages, lettuce, meat (sheep meat) 2; citrus fruits, pome fruits, apricots, peaches, nectarines, plums, bananas, swedes, turnips, garlic, onions, shallots, cucumbers, gherkins, courgettes, beans (with pods), celery, leeks, rhubarb, cultivated mushrooms, meat (except sheep meat) 1; grapes 0.5; carrots, horseradish, parsnips, parsley root, salsify, tea 0.2; peas (with pods), cereals, eggs 0.1; potatoes 0.05

300 gamma-HCH + thiabendazole + thiram
An insecticide and fungicide seed dressing for brassica crops

Products   Hysede FL   Agrichem   400:120:140 g/l   FS   02863

Uses   Canker, damping off, flea beetles, millipedes in BROCCOLI, BRUSSELS SPROUTS, CAULIFLOWERS, OILSEED RAPE, SWEDES, TURNIPS. Flea beetles, millipedes in CABBAGES.

Notes

Efficacy
- Apply with automated seed dresser if treating large quantities of seed. For smaller quantities use batch-type equipment or other forms of closed container and ensure even coverage by not treating batches of more than 25 kg
- May be applied diluted with water, in which case treated seed may require drying

Crop Safety/Restrictions
- Maximum number of treatments 1 per batch

Special precautions/Environmental safety
- Harmful in contact with skin or if swallowed
- Irritating to eyes, skin and respiratory system
- May cause sensitisation by skin contact
- Treated seed not to be used as food or feed. Do not re-use sack. Bury spillages
- Dangerous to fish or other aquatic life. Do not contaminate surface waters or ditches with chemical or used container

Protective clothing/Label precautions
- A, C, D, H
- 5a, 5c, 6a, 6b, 6c, 10b, 18, 21, 28, 29, 36, 37, 52, 63, 65, 70, 74, 75, 76, 77, 78

Latest application/Harvest Interval(HI)
- Pre-drilling

Maximum Residue Levels (mg residue/kg food)
- see gamma-HCH and thiabendazole entries

---

301   gamma-HCH + thiophanate-methyl
An insecticide and lumbricide for use on turf

Products   Castaway Plus   RP Environ.   60:500 g/l   SC   05327

Uses   Earthworms, leatherjackets in TURF.

Notes

Efficacy
- Apply by spraying in spring or autumn. Drenching is not required
- Do not mow for 2 d after spraying. If mown beforehand leave clippings
- Do not collect first clippings after treatment
- Do not spray during drought or if ground frozen
- Effectiveness is not impaired by rain or irrigation immediately after application
- Do not mix with any other product

Special precautions/Environmental safety
- Irritating to eyes and skin
- Harmful to bees. Do not apply to crops in flower or to those in which bees are actively foraging. Do not apply when flowering weeds are present
- Harmful to fish or other aquatic life. Do not contaminate surface waters or ditches with chemical or used container

---

FOR FULL CONDITIONS OF USE ALWAYS READ THE PRODUCT LABEL

Protective clothing/Label precautions
• A, C, H, M
• 6a, 6b, 14, 18, 21, 28, 30, 36, 37, 53, 63, 70, 78

Withholding period
• Keep livestock out of treated areas for at least 14 d

Maximum Residue Levels (mg residue/kg food)
• see gamma-HCH entry

---

302   gamma-HCH + thiram
An insecticide and fungicide seed dressing for brassica crops

Products	Hydraguard	Agrichem	533:200 g/l	LS	03278

Uses   Damping off, flea beetles in BRUSSELS SPROUTS, CABBAGES, KALE, OILSEED RAPE, SWEDES, TURNIPS.

Notes   **Efficacy**
• Apply in batch or continuous flow type seed treatment machinery
• Normally used undiluted but may be diluted with up to an equal volume of water in which case seed may require drying before bagging
• Seed should be evenly treated for best pest and disease control

Crop Safety/Restrictions
• Maximum number of treatments 1 per batch

Special precautions/Environmental safety
• Harmful in contact with skin and if swallowed
• Irritating to skin, eyes and respiratory system
• May cause sensitisation by skin contact
• Treated seed not to be used as food or feed. Do not re-use sacks
• Harmful to fish or other aquatic life. Do not contaminate surface waters or ditches with chemical or used container
• For use only by agricultural contractors
• Must be dispensed from 20 l container with a suitable probe and pump mechanism

Protective clothing/Label precautions
• A, C
• 5a, 5c, 6a, 6b, 6c, 10a, 18, 21, 28, 29, 36, 37, 53, 63, 65, 70, 73, 74, 75, 77, 78

Latest application/Harvest Interval(HI)
• Pre-drilling

Maximum Residue Levels (mg residue/kg food)
• See gamma-HCH entry

303   gibberellins
A plant growth regulator for use in apples, pears etc

Products					
1	Berelex	Whyte	1 g/tablet	TB	00231
2	Berelex	Whyte	1 g/tablet	TB	06637
3	Regulex	Zeneca	10 g/l	SL	06704

Uses

Growth regulation in WATERCRESS *(off-label)* [1]. Increasing fruit set in PEARS [1, 2]. Increasing fruit set in APPLES *(off-label)* [1]. Increasing germination, increasing yield in CELERY [1-3]. Increasing germination in NOTHOFAGUS [3]. Increasing quality in HOPS *(off-label)* [1]. Increasing yield in RHUBARB [1, 2]. Reducing fruit russeting in APPLES [3].

Notes

Efficacy
* Apply to apples at completion of petal fall and repeat 3 times at 10 d intervals. Good coverage of fruitlets is essential for successful results. Best results achieved by spraying under humid, slow drying conditions [3]
* Useful in pears when blossom sparse, setting conditions poor or where frost has killed many flowers. Apply as single or split application. See label for details. Resulting fruit may be seedless [1, 2]
* May be used on pears if 80% of flowers frosted but not effective on very severely frosted blossom. Conference pear responds well, Beurre Hardy only in some seasons. Young trees generally less responsive [1, 2]
* Apply to seed as a soak treatment, 24 h for nothofagus, 48 h at 5°C for celery, and sow immediately
* Apply to celery 3 wk before harvest to increase head size, to rhubarb crowns on transfer to forcing shed or as drench at first signs of growth in field [1, 2]

Crop Safety/Restrictions
* Maximum number of treatments 1 per crop or yr for watercress, hops; 4 per yr for apples
* Good results achieved on Cox's Orange Pippin, Discovery, Golden Delicious and Karmijn apples. For other cultivars test on a small number of trees [3]
* Do not apply to pears after petal fall [1, 2]

Protective clothing/Label precautions
* 29, 54, 60, 61, 63, 65, 66, 70

Latest application/Harvest Interval(HI)
* HI watercress 5 d

Approval
* Off-label Approval to Jan 1997 for use on watercress, hops (OLA 0022/92) [1]; to Jan 1996 for use on apples (OLA 0023/92) [1]

FOR FULL CONDITIONS OF USE ALWAYS READ THE PRODUCT LABEL

304   glufosinate-ammonium

A non-selective, non-residual phosphinic acid contact herbicide

Products					
	1 Challenge	AgrEvo	150 g/l	SL	07306
	2 Challenge	Hoechst	150 g/l	SL	05176
	3 Dash	Nomix-Chipman	120 g/l	SL	05177
	4 Harvest	AgrEvo	150 g/l	SL	07321
	5 Harvest	Hoechst	150 g/l	SL	06337

Uses   Annual dicotyledons, annual grasses, perennial dicotyledons, perennial grasses in NURSERY STOCK, WOODY ORNAMENTALS [3]. Annual dicotyledons, annual grasses, perennial dicotyledons, perennial grasses in FIELD CROPS *(pre-cropping situations)*, FIELD CROPS *(sward destruction)*, SETASIDE [1, 2, 4, 5]. Annual dicotyledons, annual grasses in BARLEY *(pre-harvest)*, COMBINING PEAS *(pre-harvest)*, FIELD BEANS *(pre-harvest)*, LINSEED *(pre-harvest)*, OILSEED RAPE *(pre-harvest)*, POTATOES *(pre-emergence)*, POTATOES *(pre-harvest)*, SUGAR BEET *(pre-emergence)*, VEGETABLES *(pre-emergence)*, WHEAT *(pre-harvest)* [1, 2, 4, 5]. Annual weeds, perennial weeds in APPLES, CANE FRUIT, CHERRIES, CURRANTS, DAMSONS, FORESTRY, GRAPEVINES, HEADLANDS, NON-CROP AREAS, PEARS, PLUMS, STRAWBERRIES, TREE NUTS [1, 2, 4, 5]. Pre-harvest desiccation in BARLEY, COMBINING PEAS, FIELD BEANS, LINSEED, OILSEED RAPE, POTATOES, WHEAT [1, 2, 4, 5].

Notes   **Efficacy**
- Activity quickest under warm, moist conditions. Light rainfall 3-4 h after application will not affect activity. Do not spray wet foliage or if rain imminent
- For weed control uses apply between 1 Mar and 30 Sep when weeds growing actively
- Ploughing or other cultivations can follow 4 h after spraying
- On uncropped headland apply in May/Jun to prevent weeds invading field
- Crops can normally be sown/planted immediately after spraying or sprayed post-drilling. On sand, very light or immature peat soils allow at least 3 d before sowing/planting or expected emergence
- For weed control in potatoes apply pre-emergence or up to 10% emergence on earlies and seed crops, up to 40% on maincrop, on plants up to 15 cm high
- In sugar beet and vegetables apply just before crop emergence, using stale seedbed technique
- In top and soft fruit, grapevines, woody ornamentals, nursery stock and forestry apply up to 3 treatments between 1 Mar and 30 Sep as directed sprays
- For grass destruction apply before winter dormancy occurs. Heavily grazed fields should show active regrowth. Plough from the day after spraying
- Apply pre-harvest desiccation treatments 10-21 d before harvest (14-21 d for oilseed rape). See label for timing details on individual crops
- For potato haulm desiccation apply to listed varieties (not seed crops) at onset of senescence, 14-21 d before harvest

Crop Safety/Restrictions
- Maximum number of treatments 4 per crop for potatoes (including 2 desiccant uses); 2 per crop (including 1 desiccant use) for oilseed rape, dried peas, field beans, linseed, wheat and barley; 3 per yr for fruit and forestry; 2 per yr for strawberries, non-crop land and setaside; 1 per crop for sugar beet, vegetables and other crops; 1 per yr for grassland destruction
- Pre-harvest desiccation sprays should not be used on seed crops of wheat, barley, peas or potatoes but may be used on seed crops of oilseed rape, field beans and linseed

- Do not desiccate potatoes in exceptionally wet weather or in saturated soil. See label for details
- Do not spray potatoes after emergence if grown from small or diseased seed or under very dry conditions
- Do not spray hedge bottoms
- Treated pea haulm may be fed to livestock from 7 d after spraying, treated grain from 14 d

Special precautions/Environmental safety
- Harmful if swallowed and in contact with skin. Irritating to eyes
- Do not use straw from treated crops as animal feed
- Harmful to fish or other aquatic life. Do not contaminate surface waters or ditches with chemical or used container

Protective clothing/Label precautions
- A, C, H, M
- 5a, 5c, 6a, 16, 18, 21, 25, 26, 27, 28, 29, 36, 37, 43, 44a, 53, 63, 70, 78

Withholding period
- Keep livestock out of treated areas until foliage of any poisonous weeds such as ragwort has died and become unpalatable

Latest application/Harvest Interval(HI)
- 30 Sep for use on non-crop land, top fruit, forestry, pre-drilling, pre-planting, pre-emergence in sugar beet, vegetables and other crops; before winter dormancy for grassland destruction.
- HI potatoes, oilseed rape, drying peas, field beans, linseed 7 d; wheat, barley 14 d

---

FOR FULL CONDITIONS OF USE ALWAYS READ THE PRODUCT LABEL

305 glyphosate (agricultural)

A translocated non-residual phosphonic acid herbicide

See also diuron + glyphosate

Products

1 Apache	Zeneca	320 g/l	SL	06748
2 Barbarian ✓	Barclay	360 g/l ✓	SL	07625
3 Barclay Dart	Barclay	180 g/l ✓	SL	05129
4 Barclay Gallup ✓	Barclay	360 g/l ✓	SL	05161
5 Clarion	DuPont	320 g/l	SL	06749
6 Glyphogan ✓	PBI	360 g/l ✓	SL	05784
7 Helosate	Helm	360 g/l	SL	06499
8 I T Glyphosate	I T Agro	360 g/l	SL	07212
9 Landgold Glyphosate 360	Landgold	360 g/l	SL	05929
10 Roundup ✓	Monsanto	360 g/l ✓	SL	01828
11 Roundup Biactive	Monsanto	360 g/l ✓	SL	06941
12 Roundup Biactive Dry	Monsanto	42% w/w ✓	WG	06942
13 Roundup Four 80	Monsanto	480 g/l ✓	SL	03176
14 Stacato	Sipcam	360 g/l ✓	SL	05892
15 Stampede	Zeneca	480 g/l ✓	SL	06327
16 Standon Glyphosate 360	Standon	360 g/l	SL	05582
17 Stefes Complete	Stefes	120 g/l ✓	SL	06084
18 Stefes Glyphosate ✓	Stefes	360 g/l ✓	SL	05819
19 Stefes Kickdown ✓	Stefes	360 g/l ✓	SL	06329
20 Stefes Kickdown 2	Stefes	360 g/l	SL	06548
21 Sting CT	Monsanto	120 g/l ✓	SL	04754
22 Sulfosate	Zeneca	640 g/l	SL	06750
23 Touchdown	Zeneca	480 g/l ✓	SL	06326

Uses

Annual and perennial weeds, annual dicotyledons, annual grasses, bracken, perennial dicotyledons, perennial grasses, rushes in GRASSLAND *(sward destruction)* [1, 2, 5, 15, 22, 23]. Annual and perennial weeds, annual grasses, black bent, couch, creeping bent, perennial dicotyledons, volunteer cereals, volunteer potatoes in FIELD CROPS *(stubble treatment)* [1-23]. Annual and perennial weeds, annual weeds, perennial weeds in NON-CROP FARM AREAS [2, 4, 7, 10-12, 14, 18-21]. Annual and perennial weeds, black bent, couch, creeping bent, perennial dicotyledons in FIELD BEANS *(pre-harvest)* [2, 4, 6, 7, 9-12, 16, 18-20]. Annual and perennial weeds, black bent, couch, creeping bent, perennial dicotyledons in OILSEED RAPE *(pre-harvest)* [2, 4, 6, 7, 9-12, 14, 16, 18-20]. Annual and perennial weeds, black bent, couch, creeping bent, perennial dicotyledons in BARLEY *(pre-harvest)*, WHEAT *(pre-harvest)* [1, 2, 4-16, 18-20, 22, 23]. Annual and perennial weeds, bracken, couch, rushes, woody weeds in FORESTRY *(directed spray)* [4, 6, 7, 10-12, 18-20]. Annual and perennial weeds, couch, volunteer cereals in FIELD CROPS *(sward destruction/direct drilling)* [3, 4, 6-14, 16-21]. Annual and perennial weeds, couch in LINSEED *(pre-harvest)* [10-12]. Annual and perennial weeds, couch in COMBINING PEAS *(pre-harvest)* [4, 6, 7, 9-12, 16, 18-20]. Annual and perennial weeds, couch in OATS *(pre-harvest)* [4, 6-14, 16, 18-20]. Annual and perennial weeds, couch in MUSTARD *(pre-harvest)* [6, 7, 10-12]. Annual and perennial weeds, green cover in NON-CROP AREAS [1, 5, 15, 22, 23]. Annual and perennial weeds, green cover in SETASIDE [5, 10-12]. Annual and perennial weeds, sucker control in APPLES, CHERRIES, DAMSONS, PEARS, PLUMS [4, 6, 7, 10-12, 18-20]. Annual and perennial weeds in BROADLEAVED TREES *(pre-planting)*, CONIFERS *(pre-planting)* [10-12]. Annual and perennial weeds in ASPARAGUS *(off-label)*, GRAPEVINES *(off-label)*, TREE NUTS *(off-label)* [10]. Annual and perennial weeds in GRASSLAND *(pre-cut/graze)* [4, 6, 10, 13, 18]. Annual and perennial weeds in FALLOWS, HEADLANDS [4, 7, 10-12, 14, 18-21]. Annual and perennial weeds in DURUM WHEAT *(pre-harvest)* [4]. Annual weeds, chemical thinning, perennial weeds, woody weeds in FORESTRY [2, 6, 7, 10-12]. Annual weeds, destruction of short term leys in GRASSLAND [3, 4, 6, 7, 9-14, 17-

21]. Annual weeds, perennial weeds in ORCHARDS [2]. Annual weeds, volunteer cereals in FIELD CROPS *(pre-drilling/pre-emergence)*, LEEKS *(pre-drilling/pre-emergence)*, ONIONS *(pre-drilling/pre-emergence)*, PEAS *(pre-drilling/pre-emergence)*, SPRING CEREALS *(pre-drilling/pre-emergence)*, SUGAR BEET *(pre-drilling/pre-emergence)*, SWEDES *(pre-drilling/pre-emergence)*, TURNIPS *(pre-drilling/pre-emergence)* [3, 11, 12, 17, 21]. Black bent, couch, creeping bent, perennial dicotyledons in PEAS *(pre-harvest)*[2]. Bracken, brambles, perennial grasses, woody weeds in TOLERANT CONIFERS *(overall dormant spray)* [10-12]. Perennial dicotyledons in FORESTRY *(wiper application)*, GRASSLAND *(wiper application)*, ORCHARDS *(wiper application)* [4, 10-12, 18]. Perennial grasses, reeds, rushes, sedges, waterlilies in AQUATIC SITUATIONS [6, 7, 10-12]. Pre-harvest desiccation in BARLEY, OATS, WHEAT [4, 6, 7, 9, 10, 14, 18-20]. Pre-harvest desiccation in OILSEED RAPE [4, 6-12, 14, 16, 18-20]. Pre-harvest desiccation in MUSTARD [6-8, 10-12]. Total vegetation control in NON-CROP AREAS *(amenity situations)* [7, 18]. Weed beet in SUGAR BEET *(wiper application)*[4, 10-12, 18-20]. Wild oats in CEREALS *(wiper glove)* [4, 10].

Notes

Efficacy
* For best results apply to actively growing weeds with enough leaf to absorb chemical
* Annual weed grasses should have at least 5 cm of leaf and annual broad-leaved weeds at least 2 expanded true leaves
* Perennial grass weeds should have 4-5 new leaves and be at least 10 cm long when treated. Perennial broad-leaved weeds should be treated at or near flowering but before onset of senescence
* Volunteer potatoes and polygonums are not controlled by harvest-aid rates
* In order to allow translocation, do not cultivate before treating perennials and do not apply other pesticides, lime, fertiliser or farmyard manure within 5 d of treatment
* Recommended intervals after treatment and before cultivation vary - see label
* A rainfree period of at least 6 h (preferably 24 h) should follow spraying
* Do not tank-mix with other pesticides or fertilisers as such mixtures may lead to reduced control. Adjuvants are obligatory for some products recommended for some uses with others - see label
* Do not spray weeds affected by drought, waterlogging, frost or high temperatures

Crop Safety/Restrictions
* Maximum number of applications 1 per crop (pre-harvest), 1 per yr (grassland destruction, non-crop land, green cover treatment), 1 in other situations
* Do not treat cereals grown for seed or undersown crops
* Disperse decaying matter by thorough cultivation before sowing or planting a following crop
* Consult grain merchant before treating crops grown on contract or intended for malting
* With wiper application weeds should be at least 10 cm taller than crop

Special precautions/Environmental safety
* Irritating to eyes [22] and skin [2, 3, 6-10, 13, 14, 16, 18-20]
* Harmful if swallowed [1, 3, 5, 15, 19, 22, 23]
* Dangerous [3, 7], harmful [1, 2, 4-10, 13-23] to fish or other aquatic life. Do not contaminate surface waters or ditches with chemical or used container
* Do not mix, store or apply in galvanized or unlined mild steel containers or spray tanks
* Do not leave spray in spray tanks for long period and make sure tanks are well vented
* Take extreme care to avoid drift
* Treated poisonous plants must be removed before grazing or conserving [4, 6, 10-13, 18]

FOR FULL CONDITIONS OF USE ALWAYS READ THE PRODUCT LABEL

- Do not use in covered areas such as greenhouses or under polythene
- For field edge treatment direct spray away from hedge bottoms

Protective clothing/Label precautions
- A, C (+ H, M when using ULVA equipment) [1]; A, C, F, H, M, N [7, 10]; A, C, H (+M for ULVA equipment) [22]; A, C, H, M, N [20]; A, C, H, M [5, 11, 12, 15, 23]; A, C [2-4, 6, 8, 9, 13, 14, 16, 18, 19]
- 5c [1, 3, 5, 15, 19, 22, 23]; 6a [2-4, 6-10, 13, 14, 16, 18-20, 22]; 6b, 21, 28 [2-4, 6-10, 13, 14, 16, 18-20]; 9 [3, 19]; 14 [3, 4, 6-10, 13, 14, 16, 18-20]; 18 [1-10, 13-16, 18-20, 22, 23]; 29, 36, 37 [1-16, 18-20, 22, 23]; 30 [21]; 43 [1, 2, 5, 15, 22, 23]; 44a [5, 22]; 52, [3]; 53 [1, 2, 4-10, 13-23]; 54 [11, 12]; 55 [7, 14]; 60 [3, 4, 6, 7, 9, 14, 16, 18-21]; 63 [1-16, 18-23]; 66 [2, 7, 8, 10-13, 21]; 67 [1, 5, 15, 22, 23]; 70 [1, 3-7, 9, 14-16, 18-23]; 78 [1, 3, 5, 15, 19, 21-23]

Latest application/Harvest Interval(HI)
- 5 d before harvest, grazing or drilling for grassland; stubbles 24 h-5 d depending on product and dose - see label for details; pre-emergence for autumn crops and spring cereals; 72 h post-drilling for sugar beet, peas, swedes, turnips, onions and leeks; post-leaf fall but before white bud or green cluster for top fruit.
- HI wheat, barley, oats, field beans, combining peas, linseed 7 d; mustard 8 d; oilseed rape 14 d

Approval
- May be applied through CDA equipment. See label for details [4, 6, 7, 10-13, 17, 21]. See notes in Section 1 on ULV application

Maximum Residue Levels (mg residue/kg food)
- wild mushrooms 50; barley, oats 20; linseed, rape seed 10; wheat, rye, triticale 5; meat (cattle, goat, sheep kidney) 2; meat (pig kidney) 0.5; all other products (except beans, peas) 0.1

---

306   glyphosate (horticulture, forestry, amenity etc.)
A translocated non-residual phosphonic acid herbicide

See also diuron + glyphosate

Products					
1	Barclay Gallup	Barclay	360 g/l	SL	05161
2	Barclay Gallup Amenity	Barclay	360 g/l	SL	06753
3	CDA Spasor	RP Environ.	174 g/l	EW	06414
4	Glyphogan	PBI	360 g/l	SL	05784
5	Helosate	Helm	360 g/l	SL	06499
6	Hilite	Nomix-Chipman	144 g/l	UL	06261
7	Roundup	Monsanto	360 g/l	SL	01828
8	Roundup Biactive	Monsanto	360 g/l	SL	06941
9	Roundup Biactive Dry	Monsanto	42% w/w	WG	06942
10	Roundup Pro	Monsanto	360 g/l	SL	04146
11	Roundup Pro Biactive	Monsanto	360 g/l	SL	06954
12	Spasor	RP Environ.	360 g/l	SL	07211
13	Stefes Glyphosate	Stefes	360 g/l	SL	05819
14	Stefes Kickdown	Stefes	360 g/l	SL	06329
15	Stefes Kickdown 2	Stefes	360 g/l	SL	06548
16	Stirrup	Nomix-Chipman	144 g/l	UL	06132
17	Touchdown LA	Zeneca Prof.	480 g/l	SL	06444

Uses	Annual and perennial weeds, bracken, couch, rushes, woody weeds in FORESTRY *(directed spray)* [1, 2, 4-11, 13-16]. Annual and perennial weeds, bracken in AMENITY TREES AND SHRUBS, FENCELINES, NON-CROP AREAS, PATHS AND DRIVES [17]. Annual and perennial weeds, couch, volunteer cereals, volunteer potatoes in FIELD CROPS *(stubble treatment)* [1, 4, 5, 7-9, 13-15]. Annual and perennial weeds, couch, volunteer cereals in FIELD CROPS *(sward destruction/direct drilling)* [1, 4, 5, 7-9, 13-15]. Annual and perennial weeds, couch in LINSEED *(pre-harvest)* [7-9]. Annual and perennial weeds, couch in BARLEY *(pre-harvest)*, COMBINING PEAS *(pre-harvest)*, FIELD BEANS *(pre-harvest)*, OATS *(pre-harvest)*, OILSEED RAPE *(pre-harvest)*, WHEAT *(pre-harvest)* [1, 4, 5, 7-9, 13-15]. Annual and perennial weeds, couch in MUSTARD *(pre-harvest)* [4, 5, 7-9]. Annual and perennial weeds, sucker control in PLUMS [1, 4, 5, 7-9, 13-16]. Annual and perennial weeds, sucker control in APPLES, CHERRIES, DAMSONS, PEARS [1, 4-9, 13-16]. Annual and perennial weeds in BROADLEAVED TREES *(pre-planting)* [7-9, 16]. Annual and perennial weeds in SETASIDE [7-9]. Annual and perennial weeds in ASPARAGUS *(off-label)*, GRAPEVINES *(off-label)*, TREE NUTS *(off-label)* [7]. Annual and perennial weeds in GRASSLAND *(pre-cut/graze)* [1, 4, 7, 13]. Annual and perennial weeds in FALLOWS, HEADLANDS, NON-CROP FARM AREAS [1, 5, 7-9, 13-15]. Annual and perennial weeds in DURUM WHEAT *(pre-harvest)* [1]. Annual and perennial weeds in AMENITY TREES AND SHRUBS *(directed spray)*, HEDGES *(directed spray)* [6, 16]. Annual and perennial weeds in AMENITY GRASS *(directed spray)* [6]. Annual and perennial weeds in AMENITY GRASS *(pre-planting/sowing)*, AMENITY TREES AND SHRUBS *(pre-planting/sowing)*, WOODY ORNAMENTALS *(directed spray)* [3, 6, 10-12, 16]. Annual and perennial weeds in AMENITY GRASS *(wiper application)*, AMENITY TREES AND SHRUBS *(wiper application)* [2, 10-12]. Annual and perennial weeds in CONIFERS *(pre-planting)* [2, 6-11, 16]. Annual and perennial weeds in BROAD LEAVED TREES *(pre-planting)* [2, 6, 10, 11]. Annual weeds, destruction of short term leys in GRASSLAND [1, 4, 5, 7-9, 13-15]. Annual weeds, volunteer cereals in FIELD CROPS *(pre-drilling/pre-emergence)*, LEEKS *(pre-drilling/pre-emergence)*, ONIONS *(pre-drilling/pre-emergence)*, PEAS *(pre-drilling/pre-emergence)*, SPRING CEREALS *(pre-drilling/pre-emergence)*, SUGAR BEET *(pre-drilling/pre-emergence)*, SWEDES *(pre-drilling/pre-emergence)*, TURNIPS *(pre-drilling/pre-emergence)* [8, 9]. Bracken, brambles, perennial grasses, woody weeds in TOLERANT CONIFERS *(overall dormant spray)* [7-11]. Chemical thinning in FORESTRY [2, 4, 5, 7-11]. Perennial dicotyledons in FORESTRY *(wiper application)*, GRASSLAND *(wiper application)*, ORCHARDS *(wiper application)* [1, 7-11, 13]. Perennial grasses, reeds, rushes, sedges, waterlilies in AQUATIC SITUATIONS [2, 4, 5, 7-12]. Pre-harvest desiccation in BARLEY, OATS, WHEAT [1, 4, 5, 7, 13-15]. Pre-harvest desiccation in OILSEED RAPE [1, 4, 5, 7-9, 13-15]. Pre-harvest desiccation in MUSTARD [4, 5, 7-9]. Total vegetation control in NON-CROP AREAS *(amenity situations)* [2, 3, 5, 6, 10-13, 16]. Weed beet in SUGAR BEET *(wiper application)* [1, 7-9, 13-15]. Wild oats in CEREALS *(wiper glove)* [1, 7].
Notes	**Efficacy** • For best results apply to actively growing weeds with enough leaf to absorb chemical • Annual weed grasses should have at least 5 cm of leaf and annual broad-leaved weeds at least 2 expanded true leaves • Perennial grass weeds should have 4-5 new leaves and be at least 10 cm long when treated. Perennial broad-leaved weeds should be treated at or near flowering but before onset of senescence • Bracken must be treated at full fron expansion • In order to allow translocation, do not cultivate before treating perennials and do not apply other pesticides, lime, fertiliser or farmyard manure within 5 d of treatment • Recommended intervals after treatment and before cultivation vary - see label • A rainfree period of at least 6 h (preferably 24 h) should follow spraying

FOR FULL CONDITIONS OF USE ALWAYS READ THE PRODUCT LABEL

- Do not tank-mix with other pesticides or fertilisers as such mixtures may lead to reduced control. Adjuvants are obligatory for some products recommended for some uses with others - see label
- Do not spray weeds affected by drought, waterlogging, frost or high temperatures
- Fruit tree suckers best treated in late spring
- Chemical thinning treatment can be applied as stump spray or stem injection
- For rhododendron control apply to stumps or regrowth. Addition of adjuvant Mixture B recommended for application to foliage by knapsack sprayer [7, 10]
- Product formulated ready for use without dilution. Apply only through specified applicator (see label for details) [3, 6, 16]

Crop Safety/Restrictions
- Maximum number of treatments unrestricted in forestry and aquatic situations [7, 10] or for non-crop use [10, 17]
- Decaying remains of plants killed by spraying must be dispersed before direct drilling
- Do not use treated straw as a mulch or growing medium for horticultural crops
- For use in orchards, grapevines and tree nuts care must be taken to avoid contact with the trees. Do not use in orchards established less than 2 yr and keep off low-lying branches
- Do not spray root suckers in orchards in late summer or autumn
- Do not use under glass or polythene as damage to crops may result
- With wiper application weed should be at least 10 cm taller than crop
- Do not use wiper techniques in soft fruit crops
- Certain conifers may be sprayed overall in dormant season. See label for details [7-11]
- Use a tree guard when spraying in established forestry plantations

Special precautions/Environmental safety
- Irritating to eyes and skin [2-5, 7, 10, 12-16]. Harmful if swallowed [14, 17]
- Dangerous [5], harmful [2-5, 7, 10, 12-17] to fish or other aquatic life. Do not contaminate surface waters or ditches with chemical or used container
- Maximum permitted concentration in treated water 0.2 ppm [2, 4, 5, 7, 10]
- The National Rivers Authority or Local River Purification Authority must be consulted before use in or near water
- Do not mix, store or apply in galvanised or unlined mild steel containers or spray tanks
- Do not leave spray in spray tanks for long period and make sure that tanks are well vented
- Take extreme care to avoid drift onto other crops
- For field edge treatment direct spray away from hedge bottoms

Protective clothing/Label precautions
- A, C, F, H, M, N [5, 7, 10]; A, C, H, M, N [15]; A, C, H, M [8, 9, 11, 17]; A, C, M, N [6, 12, 16]; A, C [1, 2, 4, 13, 14]; A, H, M [3]
- 5c, 78 [14, 17]; 6a [1-5, 7, 10, 12, 14]; 6b [1-5, 7, 10, 12]; 14, 21, 28 [1-7, 10, 12-16]; 18 [1-7, 10, 12-17]; 29, 36, 37, 63 [1-17]; 53 [1-5, 7, 10, 12-15, 17]; 54 [8, 9, 11]; 55 [5]; 60 [1-6, 12-16]; 66 [5, 7-11]; 67 [17]; 70 [1-6, 12-17]

Latest application/Harvest Interval(HI)
- 5 d before drilling for grassland; stubbles 24 h-5 d depending on product and dose - see label for details; post-leaf fall but before white bud or green cluster for top fruit

Approval
- Approved for aquatic weed control. See notes in Section 1 on use of herbicides in or near water [2, 4, 5, 7-12]
- May be applied through CDA equipment. See label for details [2-12, 16, 17]. See notes in Section 1 on ULV application
- Off-label approval unlimited for use on grapevines and tree nuts (OLA 0337/92)[7]; unlimited for use on outdoor asparagus (OLA 0584/94)[7]

Maximum Residue Levels (mg residue/kg food)
• see glyphosate (agriculture) entry

307 guazatine

A guanidine fungicide seed dressing for cereals

Products					
	Panoctine	RP Agric.	300 g/l	LS	06207

Uses Bunt, septoria seedling blight in WHEAT. Fusarium foot rot and seedling blight in BARLEY *(reduction)*, OATS *(reduction)*, WHEAT *(reduction)*.

Notes **Efficacy**
• Apply with conventional seed treatment machinery
• After treating, bag seed immediately and keep in dry, draught-free store

Crop Safety/Restrictions
• Maximum number of treatments 1 per batch
• Do not treat grain with moisture content above 16% and do not allow moisture content of treated seed to exceed 16%
• Do not apply to cracked, split or sprouted seed
• Treatment may lower germination capacity, particularly if seed grown, harvested or stored under adverse conditions

Special precautions/Environmental safety
• Harmful if swallowed and in contact with skin. Risk of serious damage to eyes
• Dangerous to fish or other aquatic life. Do not contaminate surface waters or ditches with chemical or used container

Protective clothing/Label precautions
• A, C, H
• 5a, 5c, 9, 18, 21, 29, 36, 37, 52, 63, 65, 70, 72, 73, 76, 78

Latest application/Harvest Interval(HI)
• Pre-drilling

308 guazatine + imazalil

A fungicide seed treatment for barley and oats

Products					
	Panoctine Plus	RP Agric.	300:25 g/l	LS	06208

Uses Brown foot rot, fusarium foot rot and seedling blight in BARLEY *(reduction)*, OATS *(reduction)*. Foot rot, leaf stripe, net blotch in BARLEY. Pyrenophora leaf spot in OATS.

FOR FULL CONDITIONS OF USE ALWAYS READ THE PRODUCT LABEL

Notes

Efficacy
• Apply with conventional seed treatment machinery
• After treating, bag seed immediately and keep in dry, draught-free store

Crop Safety/Restrictions
• Maximum number of treatments 1 per batch
• Do not treat grain with moisture content above 16% and do not allow moisture content of treated seed to exceed 16%
• Do not apply to cracked, split or sprouted seed
• Do not store treated seed for more than 6 mth
• Treatment may lower germination capacity, particularly if seed grown, harvested or stored under adverse conditions

Special precautions/Environmental safety
• Harmful if swallowed and in contact with skin. Risk of serious damage to eyes
• Dangerous to fish or other aquatic life. Do not contaminate surface waters or ditches with chemical or used container

Protective clothing/Label precautions
• A, C, H
• 5a, 5c, 9, 18, 21, 29, 36, 37, 52, 63, 65, 70, 72, 73, 76, 78

Latest application/Harvest Interval(HI)
• Pre-drilling

Maximum Residue Levels (mg residue/kg food)
• see imazalil entry

309   heptenophos

A contact, systemic and fumigant anti-cholinesterase organophosphorus insecticide

See also deltamethrin + heptenophos

Products

1 Hostaquick	AgrEvo	550 g/l	EC	07326
2 Hostaquick	Hoechst	550 g/l	EC	01079

Uses

American serpentine leaf miner in NON-EDIBLE ORNAMENTALS *(off-label)*, PROTECTED CUCURBITS *(off-label)*, PROTECTED VEGETABLES *(off-label)*. Aphids, leafhoppers in TOMATOES. Aphids, thrips in PROTECTED POT PLANTS. Aphids, woolly aphid in APPLES. Aphids in BRASSICAS, BROAD BEANS, CALABRESE *(off-label)*, CELERY, CEREALS, COURGETTES, FENNEL *(off-label)*, FIELD BEANS, FRENCH BEANS, LETTUCE, MARROWS, PEARS, PEAS, PROTECTED VEGETABLE SEEDLINGS *(off-label)*, RADICCHIO *(off-label)*, RUNNER BEANS, STRAWBERRIES. Leafhoppers, thrips in CUCUMBERS.

Notes

Efficacy
• Spray when pests first seen and repeat as necessary. Knock-down effect is rapid
• Use sufficient water for good spray coverage. On glasshouse crops apply at high volume
• May be used in integrated control programmes with Phytoseiulus or Encarsia in glasshouses. Allow at least 4 d after treatment before introducing any new beneficial insects

Crop Safety/Restrictions
* In glasshouses do not apply in early morning following a low night temperature nor in excessively high daytime temperatures. Do not spray crops at temperatures above 30°C
* Do not apply through a fogging or ULV machine in glasshouse crops

Special precautions/Environmental safety
* This product contains an anti-cholinesterase organophosphorus compound. Do not use if under medical advice not to work with such compounds
* Toxic if swallowed, harmful in contact with skin
* Allow 24 h before handling treated crops
* Flammable
* Dangerous to bees. Do not apply to crops in flower or to those in which bees are actively foraging except as directed on peas. Do not apply when flowering weeds are present
* Harmful to fish or other aquatic life. Do not contaminate surface waters or ditches with chemical or used container

Protective clothing/Label precautions
* A, C, M
* 1, 4c, 5a, 12c, 14, 16, 18, 21, 25, 28, 29, 35, 36, 37, 48, 53, 60, 63, 66, 70, 79

Latest application/Harvest Interval(HI)
* HI 24 h

Approval
* Approved for aerial application on cereals, peas, brassicas, beans (expires 31 Jul 97). See notes in Section 1. Consult firm for details
* Off-label Approval unlimited for use on a wide range of protected vegetables and non-edible ornamentals for American serpentine leaf miner control (OLA 0220/95)[1], (1083/92)[2]; for aphid control on outdoor calabrese, fennel, radicchio and protected vegetable seedlings (OLA 1169/92)[2]

---

310   hymexazol
A systemic isoxazole fungicide for pelleting sugar beet seed

Products					
Tachigaren 70 WP	Sumitomo	70% w/w	WP	02649

Uses   Aphanomyces cochlioides in BEETROOT *(off-label)*. Black leg, damping off in SUGAR BEET.

Notes
Efficacy
* Incorporate into pelleted seed using suitable seed pelleting machinery

Crop Safety/Restrictions
* Maximum number of treatments 1 per batch of seed

Special precautions/Environmental safety
* Irritating to eyes, skin and respiratory system
* Wear protective gloves when handling treated seed
* Treated seed not to be used as food or feed

FOR FULL CONDITIONS OF USE ALWAYS READ THE PRODUCT LABEL

• Harmful to fish of aquatic life. Do not contaminate surface waters or ditches with chemical or used container

Protective clothing/Label precautions
• A, C, F
• 6a, 6b, 6c, 18, 29, 36, 37, 53, 63, 67, 72, 73, 74, 76

Approval
• Off-label Approval unlimited for use on pelleted beetroot seed to be sown outdoors (OLA 0356/92)

---

311   imazalil

A systemic and protectant conazole fungicide

See also *carboxin + imazalil + thiabendazole*
       guazatine + imazalil

Products					
1	Fungaflor	Hortichem	200 g/l	EC	05967
2	Fungaflor Smoke	Hortichem	15% w/w	FU	05968
3	Fungazil 100 SL	RP Agric.	100 g/l	SL	06202

Uses

Dry rot, gangrene, silver scurf, skin spot in SEED POTATOES [3]. Powdery mildew in CUCUMBERS, PROTECTED ORNAMENTALS, PROTECTED ROSES [1, 2]. Powdery mildew in COURGETTES *(off-label)*, GHERKINS *(off-label)*, ORNAMENTALS, ROSES [1].

Notes

Efficacy
• Treat cucurbits before or as soon as disease appears and repeat every 10-14 d or every 7 d if infection pressure great or with susceptible cultivars [1, 2]
• Apply to potatoes post-harvest before putting into store or at first grading. A further treatment may be applied in early spring before planting

Crop Safety/Restrictions
• Maximum number of treatments 2 per batch of seed potatoes [3]
• Do not spray cucurbits or ornamentals in full, bright sunshine. When spraying in the evening the spray should dry before nightfall. May cause damage if open flowers are sprayed. Do not use on rose cultivar Dr A.J. Verhage [1]
• With ornamentals of unknown tolerance test on a few plants in first instance [1]

Special precautions/Environmental safety
• Harmful if swallowed [1, 3]
• Irritating to eyes and skin [1, 3]; irritating to eyes [2]
• Flammable [1], highly flammable [3]
• Keep unprotected persons out of treated areas for at least 2 h [2]
• Harmful to bees. Do not apply to crops in flower or to those in which bees are actively foraging. Do not apply when flowering weeds are present [1]
• Harmful to fish or other aquatic life. Do not contaminate surface waters or ditches with chemical or used container [1, 3]

Protective clothing/Label precautions
• A, C, D [3]; A, C, H [1]
• 5c, 6b, 28, 29, 53, 70 [1, 3]; 6a, 63 [1-3]; 12b, 38 [2]; 12c, 21, 50 [1]; 16, 65, 72, 73, 74, 78 [3]; 18, 36, 37 [2, 3]; 35, 67 [1, 2]

Latest application/Harvest Interval(HI)
• During storage and/or prior to chitting for potatoes [3].
• HI cucumbers 1 d [1, 2]

Approval
• Off-label approval unlimited for use on courgettes and gherkins (OLA 0939/95)[1]

Maximum Residue Levels (mg residue/kg food)
• citrus fruits; pome fruits, ware potatoes 5; bananas 2; cucumbers, gherkins, courgettes 0.2; tea, hops 0.1; all other products 0.02

---

312   imazalil + pencycuron
A fungicide mixture for treatment of seed potatoes

Products					
1 Monceren IM	Bayer	0.6:12.5% w/w	DP	06259
2 Monceren IM Flowable	Bayer	20:250 g/l	SC	06731

Uses   Black scurf in SEED POTATOES. Silver scurf, stem canker in SEED POTATOES *(reduction)*.

Notes   **Efficacy**
• Apply to seed tubers during planting (see label for suitable method) or sprinkle over tubers in chitting trays before loading into planter. It is essential to obtain an even distribution over tubers [1]
• If seed tubers become damp from light rain distribution of product should not be affected. Tubers in the hopper should be covered if a shower interrupts planting
• Apply to clean seed tubers at any time, into or out of store [2]
• Use suitable misting equipment. See label for details [2]
• Treatment usually most conveniently carried out over roller table at the end of grading out [2]

Crop Safety/Restrictions
• Maximum number of treatments 1 per batch
• Do not use on tubers which have previously been treated with other imazalil fungicides or where other dry powder seed treatments or hot water treatment are used
• Do not tank mix with other potato fungicides or storage products [2]
• Treated tubers may only be used as seed. They must not be used for human or animal consumption [2]

Special precautions/Environmental safety
• Operators must wear a dust mask when handling product and when riding on planter [2]
• Irritating to eyes [2]
• Harmful to fish or other aquatic life. Do not contaminate surface waters or ditches with chemical or used container

Protective clothing/Label precautions
• A, C [2]; A, F [1]
• 6a [2]; 29, 53, 63, 67 [1, 2]

FOR FULL CONDITIONS OF USE ALWAYS READ THE PRODUCT LABEL

Latest application/Harvest Interval(HI)
* Immediately before planting

Maximum Residue Levels (mg residue/kg food)
* see imazalil entry

---

313   imazalil + thiabendazole
A fungicide mixture for treatment of seed potatoes

Products   Extratect Flowable   MSD AGVET   100:300 g/l   FS   05507

Uses   Dry rot, gangrene, silver scurf, skin spot in SEED POTATOES *(reduction)*. Silver scurf, skin spot, stem canker in POTATOES *(reduction)*.

Notes   **Efficacy**
* Apply immediately after lifting with suitable spinning disc or low volume hydraulic applicator for disease reduction during storage
* Apply within 2 mth of planting for maximum disease reduction in the progeny crop

Crop Safety/Restrictions
* Maximum number of treatments 1 per batch
* Do not mix with any other product
* Delayed emergence noted under certain conditions but with no final effect on yield

Special precautions/Environmental safety
* Irritating to respiratory system. Risk of serious damage to eyes
* Harmful to fish or other aquatic life. Do not contaminate surface waters or ditches with chemical or used container

Protective clothing/Label precautions
* A, C, D, P
* 6c, 9, 18, 22, 24, 28, 29, 36, 37, 53, 72, 73, 74, 76

Latest application/Harvest Interval(HI)
* Pre-planting

Maximum Residue Levels (mg residue/kg food)
* see imazalil and thiabendazole entries

---

314   imazamethabenz-methyl
A post-emergence grass weed herbicide for use in winter cereals

Products   Dagger   Cyanamid   300 g/l   SC   03737

Uses   Blackgrass, charlock, loose silky bent, onion couch, volunteer oilseed rape, wild oats in WINTER BARLEY, WINTER WHEAT.

Notes

Efficacy
- Has contact and residual activity. Wild oats controlled from pre-emergence to 4 leaves + 3 tillers stage, charlock and volunteer oilseed rape to 10 cm. Good activity on onion couch up to 7.5 cm and useful suppression of cleavers
- When used alone add Agral for improved contact activity
- Best results achieved when applied to fine, firm, clod-free seedbed when soil moist
- Do not use on soils with more than 10% organic matter
- Effects of autumn/winter treatment normally persist to control spring flushes of weeds
- Tank mixture with isoproturon recommended for control of tillered blackgrass, with pendimethalin for general grass and dicotyledon control

Crop Safety/Restrictions
- Maximum number of treatments 2 per crop (as split dose treatment)
- Apply from 2-fully expanded leaf stage of crop to before second node detectable (GS 12-32)
- Do not use on durum wheat
- Do not use on soils where surface water is likely to accumulate
- See label for restrictions on following crops. In case of crop failure land must be ploughed to 15 cm and re-drilled in spring. Sugar beet, oilseed rape or other brassicas should not be sown in these circumstances

Special precautions/Environmental safety
- Irritating to eyes

Protective clothing/Label precautions
- A, C
- 6a, 18, 21, 29, 35, 37, 54, 60, 63, 66, 70, 78

Latest application/Harvest Interval(HI)
- Before 2nd node detectable (GS 32)

315   imazapyr

A non-selective translocated and residual imidazolinone herbicide

Products

1 Arsenal	Nomix-Chipman	250 g/l	SL	05537
2 Arsenal	Cyanamid	250 g/l	SL	04064
3 Arsenal 50	Nomix-Chipman	50 g/l	SL	05567
4 Arsenal 50	Cyanamid	50 g/l	SL	04070
5 Arsenal 50 F	Cyanamid	50 g/l	SL	05856

Uses

Annual weeds, perennial dicotyledons, perennial grasses in CONIFER PLANTATIONS [5]. Bracken, total vegetation control in NON-CROP AREAS [1-4].

Notes

Efficacy
- Chemical is absorbed through roots and foliage, kills underground storage organs and gives long term residual control. Complete kill may take several weeks
- May be applied before weed emergence but gives best results from application at any time of year when weeds are growing actively

FOR FULL CONDITIONS OF USE ALWAYS READ THE PRODUCT LABEL

• As a conifer site preparation treatment apply from beginning Jul to end Oct

Crop Safety/Restrictions
• Maximum number of treatments 1 per yr [1-4]; 1 per planting [5]
• Avoid drift onto desirable plants
• Do not apply to soil which may later be used to grow desirable plants
• Do not apply on or near desirable trees or plants or on areas where their roots may extend or in locations where the chemical may be washed or move into contact with their roots
• At least 5 mth must elapse between application and planting of named conifers only (Sitka spruce, Lodgepole pine, Corsican pine)
• Conifer plants must be at least 2 yr old [5]
• Do not use on conifer nursery bed

Special precautions/Environmental safety
• Irritating to eyes
• Not to be used on food crops

Protective clothing/Label precautions
• A, C, M [1-4]; A, C [5]
• 6a, 14, 18, 21, 28, 36, 37, 54, 63, 66 [1-5]

Approval
• May be applied through CDA equipment. See label for details. See notes in Section 1 [1-4]

316   imazaquin

An imidazolinone herbicide and plant growth regulator available only in mixtures

See also chlormequat + choline chloride + imazaquin

317   imidacloprid

A nitroimidazolidinylideneamine insecticide for seed treatment of sugar beet and spray treatment of hops

Products	1 Admire	Bayer	70% w/w	WG	07481
	2 Gaucho	Bayer	70% w/w	WS	06590

Uses Beet virus yellows vectors, flea beetles, mangold fly, millipedes, pygmy beetle, springtails, symphylids in SUGAR BEET [2]. Damson-hop aphid in HOPS [1].

Notes **Efficacy**
• Apply to sugar beet seed as part of the normal commercial pelleting process using special treatment machinery [2]
• Treated seed should be drilled within the season of purchase [2]
• Apply as a directed stem base spray before most bines reach a height of 2 m. If necessary treat both sides of the crown at half the normal concentration [1]
• Base of hop plants should be free of weeds and debris at application [1]
• Bines emerging away from the main stock or adjacent to poles may require a separate application [1]
• Uptake and movement within hops requires soil moisture and good growing conditions [1]

- Control may be impaired in plantations greater than 3640 plants/ha [1]

Crop Safety/Restrictions
- Maximum number of treatments 1 per batch of seed [2], 1 per yr [1]
- To minimise likelihood of resistance do not treat all the crop in any one year with this product [1]

Special precautions/Environmental safety
- Irritant. May cause sensitization by skin contact
- Extremely dangerous to bees. Do not apply to crops in flower or those in which bees are actively foraging. Do not apply when flowering weeds are present [1]

Protective clothing/Label precautions
- A, H [2]; A [1]
- 6, 18, 36, 37, 48a, 70 [1]; 10a, 29, 54, 63, 67 [1, 2]; 6, 73, 74, 75, 76, 77 [2]

Latest application/Harvest Interval(HI)
- Before bines reach 2 m in length or before end of first week in Jun whichever is sooner [1]

318   indol-3-ylacetic acid
A plant growth regulator for promoting rooting of cuttings

Products					
1 Rhizopon A Powder	Fargro	1% w/w	DP	07131
2 Rhizopon A Tablets	Fargro	50 mg ai	TB	07132

Uses   Rooting of cuttings in ORNAMENTALS.

Notes   **Efficacy**
- Apply by dipping end of prepared cuttings into powder or dissolved tablet solution
- Shake off excess powder and make planting holes to prevent powder stripping off [1]
- Consult manufacturer for details of application by spray or total immersion

Crop Safety/Restrictions
- Maximum number of treatments 1 per cutting
- Store product in a cool, dark and dry place
- Use solutions once only. Discard after use [2]
- Use plastic, not metallic, container for solutions [2]

Protective clothing/Label precautions
- 28 [1]; 29, 54, 63, 67 [1, 2]

Latest application/Harvest Interval(HI)
- Before cutting insertion

FOR FULL CONDITIONS OF USE ALWAYS READ THE PRODUCT LABEL

319   4-indol-3-yl-butyric acid

A plant growth regulator promoting the rooting of cuttings

See also 4-indol-3-yl-butyric acid + 2-(1-naphthyl)acetic acid with dichlorophen

Products					
1	Chryzotek Beige	Fargro	0.4% w/w	DP	07125
2	Chryzotop Green	Fargro	0.25% w/w	DP	07129
3	Rhizopon AA Powder (0.5%)	Fargro	0.5% w/w	DP	07126
4	Rhizopon AA Powder (1%)	Fargro	1% w/w	DP	07127
5	Rhizopon AA Powder (2%)	Fargro	2% w/w	DP	07128
6	Rhizopon AA Tablets	Fargro	50 mg ai	TB	07130

Uses   Rooting of cuttings in ORNAMENTALS.

Notes   **Efficacy**
* Dip base of cuttings into powder immediately before planting
* Powders or solutions of different concentration are required for different types of cutting. Lowest concentration for softwood, intermediate for semi-ripe, highest for hardwood
* See label for details of concentration and timing recommended for different species

Crop Safety/Restrictions
* Use of too strong a powder or solution may cause injury to cuttings

Protective clothing/Label precautions
* 28, 29, 54, 63, 67

320   4-indol-3-yl-butyric acid + 2-(1-naphthyl)acetic acid with dichlorophen

A plant growth regulator for promoting rooting of cuttings

Products				
Synergol	Hortichem	0.025% w/w:5:5 g/l SL		04594

Uses   Rooting of cuttings in ORNAMENTALS.

Notes   **Efficacy**
* Dip base of cuttings into diluted concentrate immediately before planting
* Suitable for hardwood and softwood cuttings
* See label for details of concentration and timing for different species

Special precautions/Environmental safety
* Harmful if swallowed

Protective clothing/Label precautions
* 5c

321 iodofenphos

A contact and ingested organophosphorus insecticide

Products	Nuvanol N 500 SC	Ciba Agric.	39.1% w/w	SC	H4951

Uses Crickets, flies in REFUSE TIPS. Ectoparasites, hide beetle, lesser mealworm in POULTRY HOUSES. Flies in AGRICULTURAL PREMISES. Grain storage pests in GRAIN STORES.

Notes

Efficacy
* Used as indoor spray treatment pests controlled for 2-3 mth
* May be mixed with approved disinfectant for terminal disinfection of poultry houses

Special precautions/Environmental safety
* This product contains an anticholinesterase organophosphorus compound. Do not use if under medical advice not to work with such compounds
* Irritating to eyes and skin
* Do not apply to surfaces on which food or feed is stored, prepared or eaten
* Remove or cover all foodstuffs before application. Protect food preparing equipment and eating utensils from contamination
* Remove exposed milk and collect eggs before application. Protect milk machinery and containers from contamination
* Do not apply directly to livestock
* Dangerous to bees. Do not apply to crops in flower or to those in which bees are actively foraging. Do not apply when flowering weeds are present
* Dangerous to fish or other aquatic life. Do not contaminate surface waters or ditches with chemical or used container

Protective clothing/Label precautions
* A, C
* 1, 6a, 6b, 14, 18, 23, 28, 29, 36, 37, 52, 63, 66, 70

322 ioxynil

A contact acting HBN herbicide for use in turf and onions

See also benazolin + bromoxynil + ioxynil
bromoxynil + ethofumesate + ioxynil
bromoxynil + fluroxypyr + ioxynil
bromoxynil + ioxynil
bromoxynil + ioxynil + mecoprop-P
bromoxynil + ioxynil + triasulfuron
clopyralid + fluroxypyr + ioxynil

Products	1 Actrilawn 10	RP Environ.	100 g/l	SL	05247
	2 Totril	RP Agric.	225 g/l	EC	06116

FOR FULL CONDITIONS OF USE ALWAYS READ THE PRODUCT LABEL

Uses   Annual dicotyledons in NEWLY SOWN TURF, TURF [1]. Annual dicotyledons in CHIVES *(off-label)*, GARLIC, LEEKS, ONIONS, SHALLOTS [2].

Notes   ### Efficacy
* Best results on seedling to 4-leaf stage weeds in active growth during mild weather
* In newly sown turf apply after first flush of weed seedlings, in spring normally 4 wk after sowing
* May be used on established turf from May to Sep under suitable conditions. Do not mow within 7 d of treatment

Crop Safety/Restrictions
* Maximum number of treatments 1 per yr for turf; 1 per crop (more at split doses) for spring or autumn sown bulb onions, spring sown leeks and autumn sown salad onions; 1 per crop (more at split doses) for spring sown pickling and salad onions, transplanted onions, leeks, garlic and shallots. See label for details of split applications
* Apply to sown onion crops as soon as possible after plants have 3 true leaves or to transplanted crops when established
* Apply to newly sown turf after the 2-leaf stage of grasses
* Do not use on crested dogstail

Special precautions/Environmental safety
* Harmful if swallowed [1], if swallowed or in contact with skin [2]
* Irritating to eyes and skin [2]
* Do not apply by knapsack sprayer or at concentrations higher than those recommended
* Dangerous to fish or other aquatic life. Do not contaminate surface waters or ditches with chemical or used container
* Harmful to bees. Do not apply to crops in flower or to those in which bees are actively foraging. Do not apply when flowering weeds are present

Protective clothing/Label precautions
* A, C [1, 2]
* 5a, 6a, 6b [2]; 5c, 18, 21, 25, 28, 29, 36, 37, 43, 50, 52, 60, 63, 66, 70, 78 [1, 2]

Withholding period
* Keep livestock out of treated areas for at least 6 wk after treatment and until foliage of any poisonous weeds such as ragwort has died and become unpalatable

Latest application/Harvest Interval(HI)
* HI onions, shallots, garlic, leeks 14 d [2]

Maximum Residue Levels (mg residue/kg food)
* garlic, onions, shallots 0.1

---

323   iprodione

A protectant dicarboximide fungicide with some eradicant activity

See also carbendazim + iprodione

Products					
1	Rovral Flo	RP Agric.	255 g/l	SC	06328
2	Rovral Green	RP Environ.	250 g/l	SC	05702
3	Rovral Liquid FS	RP Agric.	500 g/l	FS	06366
4	Rovral WP	RP Agric.	50% w/w	WP	06091
5	Turbair Rovral	Fargro	5 g/l	UL	02248

Uses

Alternaria, botrytis, glume blotch in WINTER WHEAT [1]. Alternaria, botrytis, sclerotinia stem rot in OILSEED RAPE [1]. Alternaria, botrytis in STORED CABBAGES [4]. Alternaria, botrytis in BRASSICA SEED CROPS [1, 4]. Alternaria, botrytis in KALE SEED CROPS, MUSTARD [1]. Alternaria in LEAF BRASSICAS, STUBBLE TURNIPS [1]. Alternaria in LINSEED *(seed treatment)*, OILSEED RAPE *(seed treatment)* [3]. Black scurf in SEED POTATOES *(seed treatment)* [3]. Botrytis, sclerotinia in OUTDOOR LETTUCE *(off-label)*, RADICCHIO *(off-label)* [1]. Botrytis in AUBERGINES *(off-label)*, CHRYSANTHEMUMS, LETTUCE, PEPPERS *(off-label)*, PROTECTED ORNAMENTALS *(off-label)*, TOMATOES [5]. Botrytis in CUCUMBERS, OUTDOOR LETTUCE, OUTDOOR TOMATOES, POT PLANTS, PROTECTED LETTUCE, PROTECTED TOMATOES, RASPBERRIES [4]. Botrytis in PEARS *(off-label)*, STRAWBERRIES [1, 4]. Botrytis in GRAPEVINES *(off-label)* [1]. Brown patch, dollar spot, fusarium patch, grey snow mould, melting out, red thread in AMENITY GRASS, TURF [2]. Chocolate spot in FIELD BEANS [1]. Collar rot, leaf rot in ONIONS, SALAD ONIONS [1]. Fungus diseases in BORAGE *(off-label)* [4]. Grey mould, powdery mildew in CUCURBITS *(off-label)* [5]. Net blotch in BARLEY [1]. Root rot in CHICORY *(off-label)* [4].

Notes

Efficacy

* Many diseases require a programme of 2 or more sprays at intervals of 2-4 wk
* Recommended spray schedules vary with disease and crop. See label for details
* Use a drench to control Rhizoctonia on bedding plants and cabbage storage diseases [4]
* Some flower seed may be soaked in a solution of product and sown or packed after drying. See label for details [4]
* Apply turf treatments after mowing and do not mow again for 24 h [1]
* Treatment harmless to Phytoseiulus and Encarsia being used for integrated pest control
* Apply to dry rapeseed and linseed prior to sowing [3]
* Best results on seed potatoes achieved using hydraulic sprayer with solid or hollow cone jets with air or liquid pressure atomisation [3]
* Apply to seed potatoes after harvest or after grading out and before traying out for chitting [3]

Crop Safety/Restrictions

* Maximum number of treatments 1 per batch for seed treatments [3]; 1 per crop on cereals, vining peas, cabbage (as drench), chicory, borage; 2 per crop on combining peas, field beans and stubble turnips; 3 per crop on brassicas (including seed crops), oilseed rape, protected winter lettuce (Oct-Feb); 4 per crop on strawberries, grapevines, salad onions, cucumbers; 5 per crop on raspberries; 6 per crop on bulb onions, tomatoes, curcurbits, peppers, aubergines and turf; 7 per crop on lettuce (Mar-Sep)
* Not to be used on protected lettuce or radicchio [1]
* A minimum of 3 wk must elapse between treatments on leaf brassicas [1]
* See label for pot plants showing good tolerance. Check other species before applying on a large scale [4]. Check safety on new chrysanthemum cultivars [5]
* Do not treat oilseed rape seed that is cracked or broken or of low viability [3]
* Do not excessively wet seed potato skins [3]
* Do not treat oats [1]

Special precautions/Environmental safety

* Irritating to eyes and skin [2]
* Harmful to fish or other aquatic life. Do not contaminate surface waters or ditches with chemical or used container

FOR FULL CONDITIONS OF USE ALWAYS READ THE PRODUCT LABEL

Protective clothing/Label precautions
- A [1, 2]
- 6a, 6b, 16, 29 [2]; 18, 21, 36, 37, 66 [1, 2]; 28 [4, 5]; 30 [1, 3, 4]; 35, 65 [5]; 53, 63 [1-5]; 67 [3, 4]; 70 [2-4]; 72, 73, 74, 76, 77 [3]

Latest application/Harvest Interval(HI)
- Before grain watery ripe (GS 69) for wheat and barley; pre-planting for seed treatments.
- HI strawberries, protected tomatoes 1 d; outdoor tomatoes, cucurbits, peppers, aubergines 2 d; lettuce (Mar-Sep), onions, raspberries 7 d; grapevines 2 wk; brassicas, brassica seed crops, oilseed rape, beans, stubble turnips, borage 21 d; protected winter lettuce (Oct-Feb) 4 wk; cabbage (drench treatment) 2 mth

Approval
- Approved for aerial application on oilseed rape, peas [1]. See notes in Section 1
- Approved for ULV application [5]. See label for details. See notes in Section 1
- Off-label Approval to Feb 1998 for use on borage (0169/93)[4]; unlimited for use on protected peppers, aubergines and ornamentals (OLA 0089/93)[5]; unlimited for use on outdoor crops of lettuce, cress, lamb's lettuce, scarole, radicchio (OLA 0715/95)[1]; to Jan 1997 for use on protected cucurbits (OLA 0090/93)[5]; unlimited for use on grapevines (OLA 0478/93)[1]; to Sep 1998 for use on pears (OLA 1059/93)[1]; to Sep 1998 for use on pears as a post-harvest dip (OLA 1442/94)[4]; to Oct 1998 for use on chicory roots (conveyor belt treatment) (OLA 1306/93)[4]

Maximum Residue Levels (mg residue/kg food)
- pome fruits, bilberries, currants, gooseberries, lettuces, herbs 10; stone fruits, cane fruits, kiwi fruit, garlic, onions, shallots, tomatoes, peppers, aubergines, head cabbages 5; cucumbers, gherkins, courgettes 2; witloof 1; rape seed, wheat 0.5; pulses 0.2; kohlrabi, tea, hops 0.1; animal products 0.05; citrus fruits, tree nuts, cranberries, wild berries, miscellaneous fruit (except bananas and kiwi fruit), root and tuber vegetables (except carrots, celeriac, radishes), squashes, water melons, sweetcorn, spinach, beet leaves, water cress, stem vegetables (except celery, fennel), mushrooms, oilseeds (except linseed, rape seed, mustard seed), rye, oats, triticale, maize 0.02

324   iprodione + thiophanate-methyl
A protectant and systemic fungicide for oilseed rape, field beans and peas

Products	Compass	RP Agric.	167:167 g/l	SC	06190

Uses   Alternaria, crown rot in CARROTS *(off-label)*. Alternaria, grey mould, light leaf spot, sclerotinia stem rot, stem canker in OILSEED RAPE. Chocolate spot in FIELD BEANS. Grey mould, leaf and pod spot, stem rot in COMBINING PEAS, VINING PEAS.

Notes   **Efficacy**
- Timing of sprays on oilseed rape varies with disease, see label for details
- Apply to field beans at early flowering stage before disease enters aggressive stage and repeat after 3 wk if necessary
- Apply to peas at mid flowering stage. For combining peas repeat treatment in 2-4 wk

Crop Safety/Restrictions
- Maximum number of treatments (refers to total sprays containing benomyl, carbendazim and thiophanate methyl) 2 per crop for winter oilseed rape, field beans and combining peas; 1 per crop for vining peas

• Under conditions of crop stress, e.g. after prolonged dry conditions on light soils do not spray winter oilseed rape from aircraft in less than 100 l/ha in strong direct sunlight. Under these conditions aerial application should be made in early morning or evening

Special precautions/Environmental safety
• Treated haulm must not be fed to livestock
• Harmful to fish or other aquatic life. Do not contaminate surface waters or ditches with chemical or used container

Protective clothing/Label precautions
• A, C, H, M
• 30, 53, 63, 66

Latest application/Harvest Interval(HI)
• Before end of flowering for winter oilseed rape and field beans.
• HI 3 wk

Approval
• Provisional Approval for aerial application to a limited area of winter oilseed rape and field beans. Contact manufacturer for allocation
• Off-label Approval to Aug 1999 for use on carrots (OLA 1264/94)

Maximum Residue Levels (mg residue/kg food)
• see iprodione entry

---

325   isoproturon

A residual urea herbicide for use in cereals

See also diflufenican + isoproturon

Products					
1 Alpha Isoproturon 650	Makhteshim	650 g/l	SC	07034
2 Arelon	AgrEvo	538 g/l	SC	06716
3 Arelon	Hoechst	538 g/l	SC	04544
4 Auger	RP Agric.	500 g/l	SC	06581
5 Barclay Guideline	Barclay	500 g/l	SC	06743
6 Isoproturon 500	AgrEvo	500 g/l	SC	06718
7 Luxan Isoproturon 500 Flowable	Luxan	500 g/l	SC	07437
8 MSS Iprofile	Mirfield	500 g/l	SC	06341
9 Sabre	AgrEvo	538 g/l	SC	06717
10 Sabre	Schering	538 g/l	SC	04148
11 Stefes IPU	Stefes	500 g/l	SC	05776
12 Tripart Pugil	Tripart	500 g/l	SC	06153

Uses   Annual dicotyledons, annual grasses, blackgrass, rough meadow grass, wild oats in AUTUMN SOWN SPRING WHEAT [4]. Annual dicotyledons, annual grasses, blackgrass, rough meadow grass, wild oats in WINTER RYE [2-5, 9-12]. Annual dicotyledons, annual grasses, blackgrass, rough meadow grass, wild oats in TRITICALE [2-6, 9-12]. Annual dicotyledons, annual grasses, blackgrass, rough meadow grass, wild oats in SPRING WHEAT [2, 3, 5, 6, 9-12]. Annual dicotyledons, annual grasses, blackgrass, rough meadow grass, wild oats in RYE [6]. Annual

---

FOR FULL CONDITIONS OF USE ALWAYS READ THE PRODUCT LABEL

dicotyledons, annual grasses, blackgrass, rough meadow grass, wild oats in WINTER BARLEY, WINTER WHEAT [1-12].

Notes

Efficacy
* May be applied in autumn or spring (but only post-emergence on wheat or barley). Best control normally achieved by early post-emergence treatment
* See label for details of rates, timings and tank mixes for different weed problems
* Apply to moist soils. Effectiveness may be reduced in seasons of above average rainfall or by prolonged dry weather
* Residual activity reduced on soils with more than 10% organic matter. Only use on such soils in spring

Crop Safety/Restrictions
* Maximum total dose 2.5 kg a.i./ha for any crop. Where possible reduce to 1.5 kg a.i./ha using mixtures or sequences with other herbicides
* Recommended timing of treatment varies depending on crop to be treated, method of sowing, season of application, weeds to be controlled and product being used. See label for details
* Do not apply pre-emergence to wheat or barley. On triticale and winter rye use pre-emergence only and on named varieties. Do not use on durum wheats, undersown cereals or crops to be undersown
* Do not apply to very wet or waterlogged soils, or when heavy rainfall is forecast
* Do not use on very cloddy soils or if frost is imminent or after onset of frosty weather
* Crop damage may occur on free draining, stony or gravelly soils if heavy rain falls soon after spraying. Early sown crops may be damaged if spraying precedes or coincides with a period of rapid growth
* Do not roll for 1 wk before or after treatment or harrow for 1 wk before or any time after treatment

Special precautions/Environmental safety
* Harmful to fish or other aquatic life. Do not contaminate surface waters or ditches with chemical or used container [1, 5, 8, 11, 12]
* Do not spray where soils are cracked, to avoid run-off through drains
* Avoid direct or indirect contamination of water by use of buffer strips

Protective clothing/Label precautions
* A [5, 11]
* 16 [11, 12]; 18, 36 [1, 8-10]; 21 [1-4, 6, 8-11]; 28 [1, 4, 8, 11]; 29, 63, 66 [1-12]; 37 [1, 8]; 53 [1, 5, 7, 8, 11, 12]; 54 [2-4, 6, 10]; 60 [1, 11]; 70 [9-12]

Latest application/Harvest Interval(HI)
* Not later than second node detectable stage (GS 32)

Approval
* May be applied through CDA equipment. See label for details. See notes in Section 1 [4, 11]
* All approvals for aerial use of isoproturon were revoked in 1995

326 isoproturon + isoxaben
A residual herbicide mixture for cereals

Products Ipso DowElanco 450:19 g/l SC 05736

Uses Annual dicotyledons, annual grasses, blackgrass in AUTUMN SOWN SPRING WHEAT, TRITICALE, WINTER BARLEY, WINTER RYE, WINTER WHEAT.

Notes
Efficacy
- May be applied in autumn or spring (but only post-emergence on wheat or barley). Best control normally achieved by early post-emergence treatment
- See label for details of rates and timings for different weed problems
- Apply to moist soils. Prolonged dry weather after application reduces effectiveness
- Residual activity reduced on soils with more than 10% organic matter

Crop Safety/Restrictions
- Maximum number of treatments 1 per crop. Maximum total dose of isoproturon 2.5 kg a.i./ha per crop
- Do not apply pre-emergence to wheat or barley. On triticale and winter rye use pre-emergence only and on named varieties. Do not use on durum wheats, undersown cereals or crops to be undersown
- Treat winter wheat or barley up to 31 Mar or 2nd node detectable (GS 32), whichever comes first
- Treat autumn sown spring wheat in spring
- Do not use on waterlogged or very cloddy soils or if heavy rain is forecast. Crop damage may occur on free draining, stony or gravelly soils if heavy rain falls soon after spraying
- Broadcast crops should be treated post-emergence
- Do not treat if frost imminent or after onset of frosty weather
- Early sown crops may be damaged if spraying precedes or coincides with a period of rapid growth
- Do not roll for 1 wk before or after treatment or harrow for 1 wk before or at any time after treatment
- Following crops of beet, oilseed rape and other brassicas are sensitive to isoxaben if direct drilled or minimally cultivated. Land must be ploughed to at least 20 cm before drilling subsequent crops

Special precautions/Environmental safety
- Harmful to fish or other aquatic life. Do not contaminate surface waters or ditches with chemical or used container
- Do not spray where soils are cracked, to avoid run-off through drains
- Avoid direct or indirect contamination of water by use of buffer strips

Protective clothing/Label precautions
- A
- 21, 28, 29, 53, 60, 63, 65

Latest application/Harvest Interval(HI)
- 31 Dec in yr of sowing, 31 Mar in yr of harvest or before 3rd node detectable (GS 33), whichever is sooner, for winter wheat and barley; before 31 Mar in yr of harvest or before 3rd node detectable (GS 33), whichever is sooner, for autumn sown spring wheat; pre-emergence of crop for winter rye and triticale

FOR FULL CONDITIONS OF USE ALWAYS READ THE PRODUCT LABEL

327   isoproturon + pendimethalin
A contact and residual herbicide for use in winter cereals

Products

1 Encore	Cyanamid	125:250 g/l	SC	04737
2 Jolt	Cyanamid	125:250 g/l	SC	05488
3 Trump	Cyanamid	236:236 g/l	SC	03687

Uses

Annual dicotyledons, annual grasses, blackgrass, wild oats in TRITICALE, WINTER BARLEY, WINTER RYE, WINTER WHEAT.

Notes

Efficacy
- May be applied in autumn or spring (but only post-emergence on wheat or barley)
- Annual grasses controlled from pre-emergence to 3-leaf stage, extended to 3-tiller stage with blackgrass by tank mixing with additional isoproturon. Best results on wild oats post-emergence in autumn
- Annual dicotyledons controlled pre-emergence and up to 4-leaf stage [1, 2], up to 8-leaf stage [3]. See label for details
- Apply to moist soils. Best results achieved by application to fine, firm seedbed. Trash, ash or straw should have been incorporated evenly
- Contact activity is reduced by rain within 6 h of application
- Activity may be reduced on soil with more than 6% organic matter or ash. Do not use on soils with more than 10% organic matter

Crop Safety/Restrictions
- Maximum number of treatments 1 per crop. Maximum total dose of isoproturon 2.5 kg a.i. /ha per crop
- Apply before first node detectable (GS 31) [1, 2], to before leaf sheath erect (GS 30) [3]
- Do not apply pre-emergence to wheat or barley. On winter rye and triticale use pre-emergence only on named cultivars
- Do not use on durum wheat, spring cereals or spring cultivars drilled in autumn
- Do not use pre-emergence on winter rye or triticale drilled after 30 Nov or unless crop seed covered with at least 32 mm settled soil. May be applied to shallowly drilled crops after crop has emerged
- Do not use on crops suffering stress from disease, drought, waterlogging, poor seedbed conditions or other causes or apply post-emergence when frost imminent
- Do not apply to very wet or waterlogged soils, or when heavy rain is forecast
- Do not undersow treated crops
- Do not roll or harrow after application
- In the event of autumn crop failure spring wheat, spring barley, maize, potatoes, beans or peas may be grown following ploughing to at least 150 mm

Special precautions/Environmental safety
- Dangerous to fish or other aquatic life. Do not contaminate surface waters or ditches with chemical or used container
- Do not spray where soils are cracked, to avoid run-off through drains
- Avoid direct or indirect contamination of water by use of buffer strips

Protective clothing/Label precautions
- A, C [1, 2]; A [3]
- 21, 25, 28, 29, 52, 63, 66, 70 [1-3]

Latest application/Harvest Interval(HI)
- Pre-emergence for winter rye and triticale; before leaf sheath erect (GS 30)[3] or before 1st node detectable (GS 31)[1, 2] for winter wheat and barley

328   isoproturon + simazine
A contact and residual herbicide for winter wheat and barley

Products	Harlequin 500 SC	Ciba Agric.	450:50 g/l	SC	05847

Uses   Annual dicotyledons, annual grasses, blackgrass in WINTER BARLEY, WINTER WHEAT.

Notes   **Efficacy**
- May be applied in autumn or spring (but only post-emergence of the crop
- Annual grasses controlled up to early tillering, annual dicotyledons to 50 mm
- Apply to fine, firm, even seedbed. Any trash or burnt straw should be dispersed in preparing seedbed
- Apply to moist soils. Weed control may be reduced in excessively wet autumns or if prolonged dry weather follows application to dry soil

Crop Safety/Restrictions
- Maximum number of treatments 1 per crop. Maximum total dose of isoproturon 2.5 hg a.i./ha per crop
- Apply from 2-leaf stage of crop to before leaf sheath erect (GS 12-30)
- Do not use on durum wheat, undersown crops or those due to be undersown
- With direct drilled crops soil surface should be broken by surface cultivation and seed covered by 12-25 mm soil
- Do not use on sand or on soils with more than 10% organic matter
- On stony or gravelly soils there is risk of crop damage, especially with heavy rain soon after application
- Do not apply when frost or heavy rain is forecast or to crops severely checked by frost, waterlogging, pest or disease attack.
- Early sown crops may be prone to damage if spraying precedes or coincides with period of rapid growth in autumn
- Do not harrow for 7 d before treatment or roll for 7 d before or after treatment in spring
- In the event of crop failure land should be inverted by mouldboard ploughing to at least 150 mm and harrowed before drilling/planting another crop
- Do not harrow after application

Special precautions/Environmental safety
- Irritating to skin
- Do not spray where soils are cracked, to avoid run-off through drains
- Avoid direct or indirect contamination of water by use of buffer strips

Protective clothing/Label precautions
- 6b, 18, 21, 28, 29, 36, 37, 54, 63, 66, 70

FOR FULL CONDITIONS OF USE ALWAYS READ THE PRODUCT LABEL

Latest application/Harvest Interval(HI)
• Leaf sheath erect (GS 30)

329   isoproturon + trifluralin

A residual early post-emergence herbicide for cereals

Products	1 Autumn Kite	AgrEvo	300:200 g/l	EC	07119
	2 Autumn Kite	Schering	300:200 g/l	EC	03830

Uses   Annual dicotyledons, annual grasses, blackgrass in WINTER BARLEY, WINTER WHEAT.

Notes   **Efficacy**
• Provides contact and residual control. Best results achieved by application to dicotyledons pre-emergence or up to 2-leaf stage, to blackgrass up to 2-3 tillers
• Apply post-emergence when leaves dry, weeds are actively growing and rain not expected for 2 h.
• Effectiveness may be reduced by prolonged dry or sunny weather after application
• Do not use on soils with more than 10% organic matter
• If used in conjunction with minimum cultivation ensure that all trash and burnt straw is removed, buried or dispersed before spraying

Crop Safety/Restrictions
• Maximum number of treatments 1 per crop. Maximum total dose of isoproturon 2.5 kg a.i./ha per crop
• Apply post-emergence to 4-leaf and 2 tillers stage and to broadcast crops after 3-leaf stage (GS 13)
• Do not use on durum wheat or crops to be undersown
• Do not use on Sands and do not incorporate in soil. On very stony, gravelly or other free draining soils crops may be damaged if heavy rain falls soon after treatment
• Do not spray when heavy rain is forecast or to crops stressed by frost, waterlogging, deficiency or pest attack
• Do not roll after treatment until following spring
• In case of crop failure only sow carrots, peas or sunflowers within 5 mth and plough to at least 15 cm. Do not sow sugar beet in spring following treatment
• Do not harrow treated crops

Special precautions/Environmental safety
• Irritating to skin and eyes. Flammable
• Keep in original container, tightly closed, in a safe place, under lock and key
• Harmful to fish or other aquatic life. Do not contaminate surface waters or ditches with chemical or used container
• Do not spray where soils are cracked, to avoid run-off through drains
• Avoid direct or indirect contamination of water by use of buffer strips

Protective clothing/Label precautions
• A, C
• 6a, 6b, 12c, 18, 21, 25, 28, 29, 36, 37, 53, 60, 63, 66

Latest application/Harvest Interval(HI)
• Before 5 leaf stage (GS 15)

330   isoxaben

A soil-acting amide herbicide for use in grass and fruit

See also isoproturon + isoxaben

Products

1 Flexidor 125	DowElanco	125 g/l	SC	05104
2 Gallery 125	Rigby Taylor	125 g/l	SC	06889
3 Knot Out	Vitax	125 g/l	SC	05163

Uses

Annual dicotyledons in FORESTRY TRANSPLANTS [1, 2]. Annual dicotyledons in APPLES, BLACKBERRIES, BLACKCURRANTS, CHERRIES, CONTAINER-GROWN WOODY ORNAMENTALS, GOOSEBERRIES, GRAPEVINES, HARDY ORNAMENTAL NURSERY STOCK, HOPS, PEARS, PLUMS, PROTECTED ORNAMENTALS *(off-label)*, RASPBERRIES, STRAWBERRIES [1]. Annual dicotyledons in AMENITY GRASS [2, 3]. Annual dicotyledons in AMENITY TREES AND SHRUBS [2].

Notes

Efficacy
* When used alone apply pre-weed emergence
* Effectiveness is reduced in dry conditions. Weed seeds germinating at depth are not controlled
* Activity reduced on soils with more than 10% organic matter. Do not use on peaty soils
* Various tank mixtures are recommended for early post-weed emergence treatment (especially for grass weeds). See label for details
* Best results on turf achieved by applying to firm, moist seedbed within 2 d of sowing. Avoid disturbing soil surface after application [3]

Crop Safety/Restrictions
* Maximum number of treatments 1 per crop, sowing or yr for amenity grass, ornamentals, strawberries, 2 per yr for hardy ornamental nursery stock and forestry transplants
* See label for details of crops which may be sown in the event of failure of a treated crop

Special precautions/Environmental safety
* Do not contaminate surface waters or ditches with chemical or used container
* Dangerous to livestock

Protective clothing/Label precautions
* A, C
* 54, 60, 63, 66, 70

Withholding period
* Keep all livestock out of treated areas for at least 50 d

Latest application/Harvest Interval(HI)
* Before 1 Feb for strawberries; before 1 Apr for bush, cane, top fruit and conifers and in farm forestry; before 1 May for hardy ornamental nursery stock

Approval
* Off-label approval unlimited for use in protected ornamentals (OLA 0605/94)[1]

FOR FULL CONDITIONS OF USE ALWAYS READ THE PRODUCT LABEL

331 isoxaben + terbuthylazine
A contact and residual herbicide for use in peas

Products   Skirmish 495 SC   Ciba Agric.   75:420 g/l   SC   05692

Uses   Annual dicotyledons in COMBINING PEAS, VINING PEAS.

Notes   **Efficacy**
* May be applied pre- or post-emergence of crop but before second node stage (GS 102)
* Product will give residual control of germinating weeds on mineral soils for up to 8 wk
* Product is slow acting and control may not be evident for 7-10 d or more after spraying
* Best results achieved when soil surface damp and with a fine, firm tilth. Do not use on very cloddy or stony soil
* Rain after spraying will normally improve weed control but excessive rainfall, very dry conditions or unusually low soil temperatures may lead to unsatisfactory control

Crop Safety/Restrictions
* Maximum number of treatments 1 per crop
* Product may be used on all varieties of spring sown vining and combining peas but Vedette and Printana may be damaged
* Do not use on forage peas
* Do not use on soils lighter than coarse sandy loam, on very stony soils or soils with more than 10% organic matter
* Pea seed should be covered by at least 25 mm of soil
* Heavy rain after application may cause some crop damage, especially on light soils. Do not use on soils where surface water is likely to accumulate
* For post-emergence treatment a crystal violet test for cuticle wax is advised

Special precautions/Environmental safety
* Irritant. May cause sensitization by skin contact
* Harmful to fish or other aquatic life. Do not contaminate surface waters or ditches with chemical or used container

Protective clothing/Label precautions
* A
* 10a, 18, 29, 36, 37, 54, 63, 66

Latest application/Harvest Interval(HI)
* Before second node stage (GS 102)

332 lambda-cyhalothrin
A quick-acting contact and ingested pyrethroid insecticide

Products
1 Hallmark	Zeneca	50 g/l	EC	06434
2 Landgold Lambda-C	Landgold	50 g/l	EC	06097
3 Standon Lambda-C	Standon	50 g/l	EC	05672

Uses   Aphids, apple leaf midge, apple sucker in APPLES [2, 3]. Aphids, cabbage stem flea beetle, virus vectors in WINTER OILSEED RAPE [2, 3]. Aphids, cabbage stem flea beetle in OILSEED RAPE [1]. Aphids in BARLEY, DURUM WHEAT, WHEAT [1]. Barley yellow dwarf virus vectors in

WINTER BARLEY, WINTER WHEAT [1-3]. Carrot fly, insect pests in CARROTS *(off-label)*, PARSNIPS *(off-label)* [1]. Caterpillars in BROCCOLI, BRUSSELS SPROUTS, CABBAGES, CALABRESE, CAULIFLOWERS [1-3]. Damson-hop aphid, red spider mites in HOPS [1-3]. Pea and bean weevils, pea moth in PEAS [1-3]. Pea and bean weevils in FIELD BEANS [1-3]. Pear sucker in PEARS [1-3]. Silver y moth in DWARF BEANS *(off-label)*, NAVY BEANS *(off-label)*, RUNNER BEANS *(off-label)* [1].

Notes

Efficacy
* Apply at first signs of pest attack or as otherwise recommended and repeat after 3-4 wk (10-14 d for pea moth) if necessary
* Add Agral wetter to spray for use on brassicas

Crop Safety/Restrictions
* Maximum number of applications 1 per crop for winter wheat and barley; 2 per crop for winter oilseed rape, peas and field beans; 3 per crop for apples and pears; 4 per crop for brassicas; 6 per crop for hops

Special precautions/Environmental safety
* Harmful in contact with skin or if swallowed. Irritating to eyes and skin
* Flammable
* Keep in original container, tightly closed, in a safe place, under lock and key
* Extremely dangerous to bees. Do not apply to crops in flower or to those in which bees are actively foraging, except as directed in peas and winter oilseed rape. Do not apply when flowering weeds are present.
* Do not apply to a cereal crop if any product containing a pyrethroid insecticide or dimethoate has been applied to the crop after the start of ear emergence (GS 51)
* Extremely dangerous to fish or other aquatic life. Do not contaminate surface waters or ditches with chemical or used container
* Do not allow direct spray from vehicle mounted/drawn hydraulic sprayers to fall within 6 m, or from hand-held sprayers to within 2 m, of surface waters or ditches. Direct spray away from water
* Do not spray cereals in spring/summer within 6 m of the edge of the crop

Protective clothing/Label precautions
* A, B, C, H, J, K, L, M [3]; A, C, H, J, K, L, M [2]; A, C, H, J, K, M [1]
* 5a, 5c, 6a, 6b, 12c, 14, 18, 21, 29, 36, 37, 48a, 51, 66, 70, 78 [1-3]; 54a, 63 [2]; 64 [1, 3]

Latest application/Harvest Interval(HI)
* Before 31 Mar for oilseed rape [1, 3] and for winter wheat, barley [1]

Approval
* Off-label approval unlimited (OLA 1192/93) or to Sep 1997 (OLA 1194/93) for use on potatoes for Colorado beetle control as required by Statutory Notice [1]; to Apr 2000 for use on carrots and parsnips (OLA 0575/95)[1]; to Sep 2000 for use on dwarf beans, navy beans and runner beans (OLA 1247/94)[1]

FOR FULL CONDITIONS OF USE ALWAYS READ THE PRODUCT LABEL

Maximum Residue Levels (mg residue/kg food)

* hops 10; cress, lamb's lettuce, lettuce, scarole, chervil, chives, parsley, celery leaves, tea 1; meat (except poultry) 0.5; apricots, peaches, grapes, head cabbages, beans (with pods), peas (with pods) 0.2; pome fruits, cherries, plums, currants, gooseberries,cucumbers, gherkins, courgettes 0.1; tree nuts, Brussels sprouts, barley, milk and dairy produce 0.05; blackberries, dewberries, loganberries, raspberries, bilberries, cranberries, wild berries, miscellaneous fruits, root and tuber vegetables, garlic, onions, shallots, sweet corn, watercress, beans (without pods), peas (without pods), asparagus, wild mushrooms, pulses, oilseed, potatoes, wheat, rye, oats, triticale, maize, rice, poultry meat, eggs 0.02

333   lenacil

A residual, soil-acting uracil herbicide for horticultural crops

See also chloridazon + lenacil

Products

1 Stefes Lenacil	Stefes	80% w/w	WP	07103
2 Venzar	DuPont	80% w/w	WP	02293
3 Venzar Flowable	DuPont	440 g/l	SC	06907
4 Vizor	DuPont	440 g/l	SC	06572

Uses

Annual dicotyledons, annual grasses in SPINACH BEET *(off-label)*, SPINACH *(off-label)* [2, 4]. Annual dicotyledons, annual grasses in OUTDOOR LEAF HERBS *(off-label)* [2]. Annual dicotyledons, annual meadow grass in FODDER BEET, MANGELS, RED BEET, SUGAR BEET [1-4]. Annual dicotyledons, annual meadow grass in BLACKCURRANTS, FLOWER BULBS, GOOSEBERRIES, HERBACEOUS PERENNIALS, MINT, NURSERY STOCK, RASPBERRIES, STRAWBERRIES [1].

Notes

Efficacy

* Weeds controlled as they germinate, not after emergence. Rain or irrigation necessary after spraying to activate chemical. Effectiveness may be reduced by dry conditions
* May be used in strawberries on soils with less than 10% organic matter, in beet crops on more highly organic soils with incorporation [1]
* Treat bush and cane fruit after planting, after each day's work if planting takes place over a period of time [1]
* Treat flower bulbs before or shortly after crop emergence [1]

Crop Safety/Restrictions

* Maximum number of treatments 1 per crop for red and fodder beet, mangels, spinach, spinach beet and strawberries; 1 per crop pre- and 3 per crop post-emergence for sugar beet; 1 per yr for strawberries, blackcurrants, gooseberries, raspberries [1]; 2 per yr for mint
* Apply to beet crops pre-drilling incorporated [1], pre- or post-emergence [1-4]
* Recommended in tank mixture with other beet herbicides. See label for details
* See label for soil type restrictions
* Heavy rain after application may cause damage to beet and maiden strawberries
* Strawberries may be treated immediately after planting (including cold-stored runners), as maidens, established crops or runner bed. Rainfall and soil may influence safety of treatment
* Do not spray strawberries when flowering or fruiting [1]
* Do not apply other residual herbicides within 3 mth before or after spraying
* Do not treat crops under stress from drought, low temperatures, nutrient deficiency, pest or disease attack, or waterlogging

- Blackcurrants, gooseberries and raspberries must only be treated between the end of harvest and the start of flowering [1]
- Succeeding crops should not be planted or sown for at least 4 mth after treatment following ploughing to at least 150 mm

Special precautions/Environmental safety
- Irritating to eyes, skin [1-4] and respiratory system [1, 2]
- Harmful to fish or other aquatic life. Do not contaminate surface waters or ditches with chemical or used container

Protective clothing/Label precautions
- A, C
- 6a, 6b, 6c, 18, 21, 28, 29, 36, 37, 53, 63, 65, 67

Latest application/Harvest Interval(HI)
- Pre-emergence for red beet, fodder beet, spinach, spinach beet, mangels; before leaves meet over rows for sugar beet; before start of flowering for strawberries, blackcurrants, gooseberries, raspberries

Approval
- Off-label approval to March 1996 for use on spinach and spinach beet (OLA 1327/93)[2], (OLA 1330/93)[3], (OLA 1329/93)[4]; unlimited for use on outdoor and protected herbs (see OLA notice for list of species)(OLA 0704/93)[2]
- Approval expiry: 31 Dec 96 [02293]

---

334   lenacil + phenmedipham
A contact and residual herbicide for use in sugar beet

Products	DUK-880	DuPont	440:114 g/l	KL	04121

Uses   Annual dicotyledons in SUGAR BEET.

Notes   **Efficacy**
- Apply at any stage of crop when weeds at cotyledon stage
- Germinating weeds controlled by root uptake for several weeks
- Best results achieved under warm, moist conditions on a fine, firm seedbed
- Treatment may be repeated up to a maximum of 3 applications on later weed flushes
- Residual activity may be reduced on highly organic soils or under very dry conditions
- Rain falling 1 h after application does not reduce activity

Crop Safety/Restrictions
- Maximum number of treatments 3 per crop
- Do not apply when temperature above or likely to exceed 21°C on day of spraying or under conditions of high light intensity
- Do not spray any crop under stress from drought, waterlogging, cold, wind damage or any other cause
- Heavy rain after application may reduce stand of crop particularly in very hot weather. In severe cases yield may be reduced

---

FOR FULL CONDITIONS OF USE ALWAYS READ THE PRODUCT LABEL

- Do not sow or plant any crop within 4 mth of treatment. In case of crop failure only sow or plant beet crops, strawberries or other tolerant horticultural crop within 4 mth

Special precautions/Environmental safety
- Irritating to eyes, skin and respiratory system
- Harmful to fish or other aquatic life. Do not contaminate surface waters or ditches with chemical or used container

Protective clothing/Label precautions
- A, C
- 6a, 6b, 6c, 18, 21, 28, 29, 36, 37, 53, 60, 63, 66

---

335 lindane
See gamma-HCH

---

336 linuron

A contact and residual urea herbicide for various field crops
See also 2,4-DB + linuron + MCPA
chlorpropham + linuron

Products

1 Afalon	AgrEvo	450 g/l	SC	07299
2 Afalon	Hoechst	450 g/l	SC	04665
3 Alpha Linuron 50 SC	Makhteshim	500 g/l	SC	06967
4 Alpha Linuron 50 WP	Makhteshim	50% w/w	WP	04870
5 Ashlade Linuron FL	Ashlade	480 g/l	SC	06221
6 Atlas Linuron	Atlas	370 g/l	SC	03054
7 Campbell's Linuron 45% Flowable	MTM Agrochem.	450 g/l	SC	00408
8 Linuron Flowable	PBI	450 g/l	SC	02965
9 MSS Linuron 50	Mirfield	500 g/l	SC	07862
10 UPL Linuron 45% Flowable	United Phosphorus	450 g/l	SC	07435

Uses

Annual dicotyledons, annual meadow grass, black bindweed, chickweed, corn marigold, fat-hen, redshank in SPRING CEREALS [3-7, 10]. Annual dicotyledons, annual meadow grass, black bindweed, chickweed, corn marigold, fat-hen, redshank in CELERY [1, 2, 6-8]. Annual dicotyledons, annual meadow grass, black bindweed, chickweed, corn marigold, fat-hen, redshank in PARSNIPS [1-4, 6-10]. Annual dicotyledons, annual meadow grass, black bindweed, chickweed, corn marigold, fat-hen, redshank in CARROTS, POTATOES [1-10]. Annual dicotyledons, annual meadow grass, black bindweed, chickweed, corn marigold, fat-hen, redshank in PARSLEY [1-7, 9, 10]. Annual dicotyledons, annual meadow grass, black bindweed, chickweed, corn marigold, fat-hen, redshank in BARLEY, OATS, SPRING WHEAT [9]. Annual dicotyledons, annual meadow grass, black bindweed, chickweed, redshank in WINTER BARLEY, WINTER WHEAT [6]. Annual weeds in CELERIAC *(off-label)*, GARLIC *(off-label)*, HERBS *(off-label)*, LEEKS *(off-label)*, ONIONS *(off-label)* [8].

Notes

Efficacy
- Many weeds controlled pre-emergence or post-emergence to 2-3 leaf stage, some (annual meadow grass, mayweed) only susceptible pre-emergence. See label for details
- Best results achieved by application to firm, moist soil of fine tilth
- Little residual effect on soil with more than 10% organic matter

Crop Safety/Restrictions
* Maximum number of treatments 1 per crop [3, 9, 10]; 1 per crop for cereals [4-7], potatoes [1, 2, 4-8], celery [1, 2, 6-8], onions, leeks, garlic [9]; 2 per crop for carrots [1, 2, 4-8], parsnips [1, 2, 4, 6-8], parsley [1, 2, 4, 6, 7]; 3 per crop for celeriac, herbs [8]
* Apply only pre-crop emergence [10]
* Drill spring cereals at least 3 cm deep and apply pre-emergence of crop or weed
* Do not use on undersown cereals or crops grown on Sands or Very Light soils or soils heavier than Sandy Clay Loam or with more than 10% organic matter
* Apply to potatoes well earthed up to a rounded ridge pre-crop emergence and do not cultivate after spraying. May be used in tank mix with glufosinate-ammonium [1, 2]
* Apply within 4 d of drilling carrots or post-emergence after first rough leaf stage.
* Do not apply to emerged crops of carrots, parsnips or parsley under stress
* Recommendations for parsnips, parsley and celery vary. See label for details
* Potatoes, carrots and parsnips may be planted at any time after application. Lettuce should not be grown within 12 mth of treatment. Transplanted brassicas may be grown from 3 mth after treatment

Special precautions/Environmental safety
* Irritating to eyes, skin and respiratory system [4-7]. Irritating to eyes and skin [10]
* Dangerous to fish or other aquatic life. Do not contaminate surface waters or ditches with chemical or used container
* Do not allow direct spray from vehicle mounted/drawn hydraulic sprayers to fall within 6 m of surface waters or ditches. Direct spray away from water
* Do not apply by hand-held sprayers

Protective clothing/Label precautions
* A, C, H [1-5, 7-10]; A, C [6]
* 6a, 6b [4-7, 10]; 6c [4-7]; 18, 36, 37 [3-8, 10]; 21, 29, 63 [1-10]; 22, 60, 70, 78 [10]; 28 [3-10]; 43 (5 mth) [5]; 52 [1-5, 7, 8, 10]; 53 [6, 9]; 54a [1-5, 7-10]; 65 or 66 [7, 8]; 66 [1, 2, 5, 6; 10]; 67 [3, 4, 9]

Latest application/Harvest Interval(HI)
* Pre-emergence for cereals and linseed; 2 rough leaf stage for carrots, parsnips, parsley and celery; pre- or up to 40% emergence for potatoes (products differ, see label for details).
* HI onions, garlic, herbs 7 d (pre-emergence [8]), leeks 21 d

Approval
* Off-label approval to Mar 1998 for use on celeriac, onions, leeks, garlic (OLA 0208/93)[8]; unlimited for use on herbs (OLA 0209/93)[8]
* Approval expiry: 31 Dec 96 [00408]

337   linuron + trifluralin
A residual pre-emergence herbicide for use in winter cereals

Products					
	1 Campbell's Trifluron	MTM Agrochem.	250:480 g/l	KL	02682
	2 Linnet	PBI	106:192 g/l	EC	01555

FOR FULL CONDITIONS OF USE ALWAYS READ THE PRODUCT LABEL

Uses   Annual dicotyledons, annual grasses, annual meadow grass, perennial ryegrass, rough meadow grass in WINTER BARLEY, WINTER WHEAT.

Notes   **Efficacy**
* Effective against weeds germinating near soil surface. Best results achieved by application to fine, firm, moist seedbed free of clods, crop residues or established weeds
* Effectiveness reduced by long dry periods after application or on waterlogged soil
* Do not use on peaty soils or where organic matter exceeds 10%
* With autumn application residual effects normally last until spring but further herbicide treatment may be needed on thin or backward crops
* With loose seedbed on lighter soils results improved by rolling after drilling

Crop Safety/Restrictions
* Maximum number of treatments 1 per crop
* Apply without incorporation as soon as possible after drilling and before crop emergence, within 3 d on early drilled crops
* Do not treat durum wheat or undersown crops
* Crop seed must be well covered, minimum depth specified varies from 12 to 30 mm
* Do not use on soils classed as Sands. Do not harrow after treatment
* Only spring barley or spring wheat may be sown within 6 mth of application

Special precautions/Environmental safety
* Irritating to eyes, skin and respiratory system
* Dangerous to fish or other aquatic life. Do not contaminate surface waters or ditches with chemical or used container
* Do not allow direct spray from vehicle mounted/drawn hydraulic sprayers to fall within 6 m of surface waters or ditches. Direct spray away from water
* Do not apply by hand-held sprayers

Protective clothing/Label precautions
* A, C, H
* 6a, 6b, 6c, 14, 18, 21, 25, 28, 29, 36, 37, 52, 54a, 60, 61, 63, 66, 70

Latest application/Harvest Interval(HI)
* Pre-emergence of crop

338   malathion
A broad-spectrum contact organophosphorus insecticide and acaricide

Products   MTM Malathion 60   MTM Agrochem.   600 g/l   EC   05714

Uses   Aphids, bryobia mites, codling moth, leafhoppers, red spider mites, suckers, woolly aphid in APPLES. Aphids, bryobia mites, leafhoppers, red spider mites, suckers, woolly aphid in PEARS. Aphids, celery fly in CARROTS, CELERY, PARSNIPS. Aphids, leafhoppers, mealy bugs, scale insects, thrips, whitefly in PROTECTED FLOWERS, PROTECTED LETTUCE, PROTECTED ROSES, TOMATOES. Aphids, red spider mites, sawflies in GOOSEBERRIES. Aphids, red spider mites in CHERRIES, DAMSONS, PLUMS. Aphids, thrips in FLOWERS. Aphids in APRICOTS, BROAD BEANS, CANE FRUIT, CURRANTS, FIELD BEANS, FRENCH BEANS, KALE, LETTUCE, NECTARINES, PEACHES, PEAS, POTATOES, RED BEET, STRAWBERRIES, SWEDES, TURNIPS. Cabbage seed weevil, cabbage stem weevil, pollen beetles in MUSTARD, OILSEED RAPE. Flea beetles, midges, mustard beetle in WATERCRESS *(off-label)*. Leafhoppers, raspberry beetle in RASPBERRIES. Scale insects in ROSES. Sciarid flies in MUSHROOMS. Thrips in LEEKS, ONIONS.

Notes

Efficacy
* Spray when pest first seen and repeat as necessary, usually at 7-14 d intervals
* Number and timing of sprays vary with crop and pest. See label for details
* Repeat spray routinely for scale insect and whitefly control in glasshouses
* Apply as drench to control sciarid larvae in mushroom casing
* Strains of aphids and red spider mite resistant to organophosphorus compounds have developed in some areas
* Addition of wetting agent recommended against certain pests eg woolly aphids and leaf-curling aphids

Crop Safety/Restrictions
* On carrots the maximum total dose applied per crop must not exceed the equivalent of 3 (on mineral soils) or 4 (on organic soils) full dose applications
* Do not use on antirrhinums, crassula, ferns, fuchsias, gerberas, petunias, pileas, sweet peas or zinnias
* Pick mushrooms hard before treatment and do not harvest for 4 d afterwards

Special precautions/Environmental safety
* This product contains an anticholinesterase organophosphorus compound. Do not use if under medical advice not to work with such compounds
* Dangerous to bees. Do not apply to crops in flower or to those in which bees are actively foraging. Do not apply when flowering weeds are present
* Extremely dangerous to fish or other aquatic life. Do not contaminate surface waters or ditches with chemical or used container

Protective clothing/Label precautions
* A, H, J, M
* 1, 2, 21, 28, 29, 50, 53, 63, 65, 70

Latest application/Harvest Interval(HI)
* HI 4 d; crops for processing 7 d

Approval
* Off-label Approval to Dec 1998 (see approval notice) for use on watercress beds (OLA 0172/95)[1]

Maximum Residue Levels (mg residue/kg food)
* cereals 8; garlic, onions, shallots, tomatoes, peppers, aubergines, cucumbers, gherkins, courgettes, cauliflowers, Brussels sprouts, head cabbages, lettuce, beans (with pods), peas (with pods), celery, leeks, rhubarb, cultivated mushrooms 3; citrus fruits 2; pome fruits, apricots, peaches, nectarines, plums, grapes, cane fruits, bilberries, cranberries, currants, gooseberries, bananas, carrots, horseradish, parsnips, parsley root, salsify, swedes, turnips, potatoes 0.5

FOR FULL CONDITIONS OF USE ALWAYS READ THE PRODUCT LABEL

339   maleic hydrazide

A pyridazinone plant growth regulator suppressing sprout and bud growth
See also dicamba + maleic hydrazide + MCPA

Products

1 Burtolin	RP Environ.	185 g/l	SL	05399
2 Fazor	DowElanco	80% w/w	SG	05558
3 Mazide 25	Vitax	250 g/l	SL	02067
4 MSS MH 18	Mirfield	180 g/l	SL	03065
5 Regulox K	RP Environ.	250 g/l	SL	05405
6 Royal MH 180	Uniroyal	180 g/l	SC	07043
7 Source	Chiltern	80% w/w	SG	07427

Uses

Growth retardation in HEDGES [3]. Growth suppression in AMENITY GRASS [3-6]. Growth suppression in GRASS NEAR WATER [5, 6]. Growth suppression in HARD SURFACES, INDUSTRIAL SITES [5]. Sprout suppression, volunteer suppression in POTATOES [2, 7]. Sprout suppression in ONIONS [2-4, 6, 7]. Sucker inhibition in AMENITY TREES AND SHRUBS [1, 3, 5].

Notes

Efficacy
* Apply to grass at any time of yr when growth active, best when growth starting in Apr-May and repeated when growth recommences
* Uniform coverage and 8 h dry weather necessary for effective results
* Mow 2-3 d before and 5-10 d after spraying for best results. Need for mowing reduced for up to 6 wk
* Apply to onions at 10% necking and not later than 50% necking stage when the tops are still green. Rain or irrigation within 24 h may reduce effectiveness
* Apply to second early or maincrop potatoes at least 3 wk before haulm destruction. Accurate timing essential. See label for details [2]
* To control suckers wet trunks thoroughly, especially pruned and basal bud areas [1, 3]
* Spray hawthorn hedges in full leaf, privet 7 d after cutting, in Apr-May [3]

Crop Safety/Restrictions
* Maximum number of treatments 2 per yr on amenity grass, land not intended for cropping and land adjacent to aquatic areas; 1 per crop on onions, potatoes and on or around tree trunks
* Do not apply in drought or when crops are suffering from pest, disease or herbicide damage. Do not treat fine turf or grass seeded less than 8 mth previously
* May be applied to grass along water courses but not to water surface [5, 6]
* Do not treat onions more than 2 wk before maturing. Treated onions may be stored until Mar but must then be removed to avoid browning
* Only use on potatoes of good keeping quality, not on seed, first earlies or crops for seed
* Do not treat potatoes within 3 wk of applying a haulm desiccant
* Avoid drift onto nearby vegetables, flowers or other garden plants

Special precautions/Environmental safety
* Irritant. Risk of serious damage to eyes [7]
* Only apply to grass not to be used for grazing
* Do not use treated water for irrigation purposes within 3 wk of treatment or until concentration in water falls below 0.02 ppm [5, 6]
* Maximum permitted concentration in water 2 ppm [5, 6]
* Do not contaminate surface waters or ditches with chemical or used container
* Do not apply by knapsack sprayer [2]

Protective clothing/Label precautions
- A, C [7]; A, H [2]
- 6, 9, 14, 18, 25. 26, 27, 28, 32, 36, 37 [7]; 21 [1-5]; 29 [1, 3-5, 7]; 30 [2, 6]; 54, 63 [1-7]; 55, 56 [5, 6]; 60, 66, 70 [1, 3-5]; 65 [1, 3-6]; 67 [2]

Latest application/Harvest Interval(HI)
- 3 wk before haulm destruction for potatoes; before 50% necking for onions [7]
- HI onions 4 d; potatoes 3 wk

Approval
- Approved for use on grass near water. See notes in Section 1 on use of herbicides in or near water [5, 6]

Maximum Residue Levels (mg residue/kg food)
- ware potatoes 50; garlic, onions, shallots 10; all other products 1

---

340   mancozeb

A protective dithiocarbamate fungicide for potatoes and other crops

See also benalaxyl + mancozeb
carbendazim + mancozeb
cymoxanil + mancozeb
cymoxanil + mancozeb + oxadixyl
dimethomorph + mancozeb

Products					
1 Agrichem Mancozeb 80	Agrichem	80% w/w	WP	06354
2 Ashlade Mancozeb FL	Ashlade	412 g/l	SC	06226
3 Barclay Manzeb 80	Barclay	80% w/w	WP	05944
4 Dithane 945	PBI	80% w/w	WP	04017
5 Dithane Dry Flowable	PBI	75% w/w	WG	04255
6 Headland Zebra Flo	Headland	455 g/l	SC	07442
7 Headland Zebra WP	Headland	80% w/w	WP	07441
8 Karamate Dry Flo	Rohm & Haas	75% w/w	WG	04252
9 Luxan Mancozeb Flowable	Luxan	455 g/l	SC	06812
10 Manzate 200	DuPont	80% w/w	WP	01281
11 Unicrop Flowable Mancozeb	Unicrop	412 g/l	SC	04700
12 Unicrop Mancozeb	Unicrop	80% w/w	WP	05467

Uses   Black spot, rust in ROSES [8, 12]. Blight in POTATOES [1-3, 6, 7, 9-12]. Brown rust, net blotch, rhynchosporium, septoria diseases, sooty moulds, yellow rust in BARLEY [1, 12]. Brown rust, net blotch in SPRING BARLEY *(moderate control)*, WINTER BARLEY *(moderate control)* [3]. Brown rust, septoria diseases, sooty moulds, yellow rust in WHEAT [1, 4, 5, 9, 12]. Currant leaf spot in BLACKCURRANTS, GOOSEBERRIES [8]. Downy mildew in LETTUCE, PROTECTED LETTUCE [8, 12]. Downy mildew in WINTER OILSEED RAPE [1, 12]. Fire in TULIPS [8, 12]. Net blotch, rhynchosporium, sooty moulds in WINTER BARLEY [4-7, 9]. Net blotch in WINTER BARLEY *(reduction)* [6, 7, 9]. Ray blight in CHRYSANTHEMUMS [8, 12]. Rust in PELARGONIUMS, PROTECTED CARNATIONS [8, 12]. Scab in APPLES, PEARS [8]. Septoria, septoria diseases, septoria leaf spot, sooty moulds, yellow rust in WINTER WHEAT [3-7, 9]. Septoria, septoria leaf spot, sooty moulds in SPRING WHEAT [3, 6, 7, 9]. Yellow rust in SPRING WHEAT *(reduction)* [9]. Yellow rust in WINTER WHEAT *(reduction)* [6, 7, 9].

FOR FULL CONDITIONS OF USE ALWAYS READ THE PRODUCT LABEL

Notes

Efficacy
- In cereals may be used in tank mixture with carbendazim to improve disease control. This mixture obligatory for some products - see label. Do not mix with other fungicides
- Apply to cereals before any disease is well established. See label for recommended timing
- May be recommended for suppression or control of mildew in cereals depending on product and tank mix. See label for details
- Persistence on cereals may be slightly reduced in rapid growing conditions
- Apply to potatoes before haulm meets across rows (usually mid-Jun) or at earlier blight warning, and repeat every 10-14 d depending on conditions
- May be used on potatoes up to desiccation of haulm
- On oilseed rape apply as soon as disease develops between cotyledon and 5-leaf stage (GS 1,0-1,5)

Crop Safety/Restrictions
- Maximum number of treatments 1 per crop for oilseed rape [1], 2 per crop for cereals; 8 per crop (10 per crop [10]) for potatoes
- A minimum of 10 d must elapse between applications to potatoes
- Apply to cereals from 4-leaf stage to before early milk stage-GS 71. Recommendations vary, see labels for details
- Use alone on cereals not permitted; must be mixed with carbendazim only [9]
- Treat winter oilseed rape before 6 true leaf stage (GS 1,6) and before 31 Dec
- On protected lettuce only 2 post-planting applications of mancozeb or of any combination of products containing EBDC fungicide (mancozeb, maneb, thiram, zineb) either as a spray or a dust are permitted within 2 wk of planting out and none thereafter.
- Avoid treating wet cereal crops or those suffering from drought or other stress

Special precautions/Environmental safety
- Irritating to respiratory system [1, 3-5, 7, 8, 10-12]
- May cause sensitization by skin contact
- Harmful to fish or other aquatic life. Do not contaminate surface waters or ditches with chemical or used container

Protective clothing/Label precautions
- A, C, D [3]; A, C [10]; A [1, 2, 4-9, 11, 12]
- 6a, 6b [2-5, 8, 10-12]; 6c [1-5, 7, 8, 10-12]; 10a [1, 6, 7, 9]; 14 [3]; 18, 21, 29, 36, 37, 63, 67 [1-11]; 28 [2-6, 8-11]; 53 [1-12]; 60, 70 [2, 4-6, 8-11]

Latest application/Harvest Interval(HI)
- Before early milk stage (GS 73) [1, 3, 6, 7, 9] for cereals; before 6 true leaf stage and before 31 Dec for winter oilseed rape.
- HI potatoes 7 d; outdoor lettuce 14 d; protected lettuce 21 d; apples, pears, blackcurrants, gooseberries 4 wk

Approval
- Approved for aerial application on potatoes [1-5, 9-12]; on cereals [2]. See notes in Section 1

Maximum Residue Levels (mg residue/kg food)
- hops 25; lettuces, herbs 5; oranges, apricots, peaches, nectarines, grapes, strawberries, peppers, aubergines 2; cherries, plums 1; garlic, onions, shallots, cucumbers 0.5; celeriac, witloof 0.2; tree nuts, oilseeds (except rape seed), tea 0.1; bilberries, cranberries, wild berries, miscellaneous fruit, horseradish, Jerusalem artichokes, parsley root, sweet potatoes, swedes, turnips, yams, spinach, beet leaves, asparagus, cardoons, fennel, rhubarb, mushrooms, potatoes, maize, rice, animal products 0.05

341 mancozeb + metalaxyl

A systemic and protectant fungicide for potatoes and some other crops

Products

1 Fubol 58 WP	Ciba Agric.	48:10% w/w	WP	00927
2 Fubol 75 WP	Ciba Agric.	67.5:7.5% w/w	WP	03462
3 Osprey 58 WP	Ciba Agric.	48:10% w/w	WP	05717

Uses

Blight in POTATOES [2, 3]. Cavity spot in CARROTS [1]. Damping off in WATERCRESS *(off-label)* [1]. Downy mildew in ONIONS *(off-label)* [1, 2]. Fungus diseases in FLOWERHEAD BRASSICAS *(off-label)*, KOHLRABI *(off-label)*, LEAF BRASSICAS *(off-label)*, SWEDES *(off-label)*, TURNIPS *(off-label)* [1]. Phytophthora fruit rot in APPLES *(off-label)* [1, 2]. Root rot in RASPBERRIES *(off-label)* [1]. White blister in BROCCOLI *(off-label)*, BRUSSELS SPROUTS *(off-label)*, CABBAGES *(off-label)*, CALABRESE *(off-label)*, CAULIFLOWERS *(off-label)* [1, 2]. White blister in RADISHES *(off-label)* [2]. White tip in LEEKS *(off-label)* [1, 2].

Notes

Efficacy

* Apply as protectant spray on potatoes immediately risk of blight in district or as crops begin to meet in (not across) rows and repeat every 10-21 d according to blight risk (every 10-14 d on irrigated crops) [2, 3]
* Use for first part of spray programme to mid-Aug. Later use a protectant fungicide (preferably tin-based) up to complete haulm destruction or harvest [2, 3]
* Do not treat potato crops showing active blight infection
* Recommended for potato blight control in Northern Ireland but specific Northern Ireland label must be consulted [3]
* Best results on carrots and parsnips achieved by application to damp soil. Do not use where these crops have been grown on same site during either of 2 previous years [1]
* Do not apply on potatoes for 2-3 h before rainfall or when raining

Crop Safety/Restrictions

* Maximum number of treatments 5 per crop for potatoes [2]; 5 per crop for maincrop, 3 per crop for early potatoes [3]; 3 per crop on Brussels sprouts, cabbages, cauliflowers, broccoli, calabrese, onions, leeks; 2 per crop on oilseed rape, field and broad beans, apple orchards; 1 per crop on carrots
* Do not use other acylalanine products on potatoes after final spray [2, 3]

Special precautions/Environmental safety

* Irritating to skin, eyes and respiratory system
* Harmful to fish or other aquatic life. Do not contaminate surface waters or ditches with chemical or used container

Protective clothing/Label precautions

* A
* 6a, 6b, 6c, 18, 21, 29, 35, 36, 37, 53, 63, 67

Latest application/Harvest Interval(HI)

* Within 6 wk of drilling for carrots and parsnips; pre-planting for seedling brassicas.
* HI early and maincrop potatoes 1 d [2]; early potatoes 7 d, maincrop potatoes 14 d [3]; Brussels sprouts, onions 14 d; field and broad beans, apples 4 wk; carrots 8 wk; parsnips 16 wk; raspberries 3 mth

FOR FULL CONDITIONS OF USE ALWAYS READ THE PRODUCT LABEL

Approval
- Approved for aerial application on potatoes. See notes in Section 1[2, 3]
- Off-label Approval to Feb 1997 for use on apples (orchard floor), cabbages, cauliflowers, broccoli, calabrese, Brussels sprouts, leeks and onions (OLA 0201/92)[1], (OLA 0202/92)[2]; to Apr 1998 for use on outdoor and protected seedlings of leaf and flowerhead brassicas, swedes, turnips and kohlrabi (OLA 0618/93)[1]; to Dec 1998 for use on protected and outdoor watercress (OLA 1210/93)[1]; to Aug 1997 for use as drench on raspberries (OLA 0886/93)[1]; to Jul 2000 for use on outdoor radishes (OLA 0997/95)[2]; unlimited for use on protected radishes (OLA 0996/95)[2]

Maximum Residue Levels (mg residue/kg food)
- see mancozeb and metalaxyl entries

342   mancozeb + ofurace
A systemic and protectant fungicide mixture for potatoes

Products	Patafol	Mirfield	67:5.8% w/w	WP	07397

Uses   Blight in POTATOES.

Notes   **Efficacy**
- Apply to potatoes at first blight warning or just before foliage meets in the row and repeat at 10-14 d intervals
- Do not apply after the end of Aug
- Complete blight programme up to haulm destruction with organotin products
- Always apply before the crop is infected
- Do not apply if rain likely within 2-3 h of treatment
- After rain or irrigation wait until foliage is dry before spraying

Crop Safety/Restrictions
- Maximum number of treatments 5 per crop

Special precautions/Environmental safety
- Irritating to eyes, skin and respiratory system
- Harmful to fish or other aquatic life. Do not contaminate surface waters or ditches with chemical or used container

Protective clothing/Label precautions
- A, C
- 6a, 6b, 6c, 18, 21, 28, 35 (7 d), 36, 37, 53, 63, 67

Latest application/Harvest Interval(HI)
- III 7 d

Maximum Residue Levels (mg residue/kg food)
- see mancozeb entry

343 mancozeb + oxadixyl

A systemic and contact protectant fungicide for potato blight control

Products	Recoil	Sandoz	56:10% w/w	WP	04039

Uses   Blight in POTATOES. Root rot in BLACKBERRIES *(off-label)*, CANE FRUIT *(off-label)*, DEWBERRIES *(off-label)*, LOGANBERRIES *(off-label)*, RASPBERRIES *(off-label)*.

Notes

Efficacy
* Commence treatment for potato blight early in season as soon as there is risk of infection
* In absence of a blight warning treatment should start before potatoes meet within row
* Repeat treatments every 10-14 d depending on degree of blight infection risk
* Use as a protectant treatment. Potatoes showing blight should not be treated
* Do not spray within 2-3 h of rainfall and only apply to dry foliage
* To complete blight spray programme after end of Aug make subsequent treatments up to haulm destruction with another fungicide, preferably fentin based

Crop Safety/Restrictions
* Maximum number of treatments 5 per crop for potatoes; 2 per yr for raspberries
* Do not treat raspberries on light soils with less than 2% organic matter

Special precautions/Environmental safety
* Irritating to skin, eyes and respiratory system. May cause sensitization by skin contact
* Do not spray within 2 m of surface water courses (raspberries, blackberries, dewberries, loganberries and cane fruit only)

Protective clothing/Label precautions
* A, H
* 6a, 6b, 6c, 14, 18, 24, 24a, 28, 29, 36, 37, 54, 63, 67

Latest application/Harvest Interval(HI)
* Before end of Aug.
* HI zero

Approval
* Approved for aerial application on potatoes. See notes in Section 1. Consult firm for details
* Off-label approval to Dec 1996 for use as a soil drench treatment for raspberries (OLA 0623/94)[1]; unlimited for use as a soil drench on raspberries, blackberries, dewberries, loganberries, cane fruit (other than wild) (OLA 0750/95)[1]

Maximum Residue Levels (mg residue/kg food)
* see mancozeb entry

FOR FULL CONDITIONS OF USE ALWAYS READ THE PRODUCT LABEL

344   mancozeb + propamocarb hydrochloride
A systemic and contact protectant fungicide for potato blight control

Products	1 Tattoo	AgrEvo	301.6:248 g/l	SC	07293
	2 Tattoo	Schering	301.6:248 g/l	SC	06745

Uses

Blight in POTATOES.

Notes

Efficacy
- Commence treatment as soon as there is risk of blight infection
- In the absence of a warning treatment should start before the crop meets within the row
- Repeat treatments every 10-14 d depending on degree of infection risk
- Irrigated crops should be regarded as at high risk and treated every 10 d
- Do not use once blight has become readily visible
- Do not spray when rainfall is imminent and apply only to dry foliage
- To complete the spray programme after the end of Aug, make subsequent treatments up to haulm destruction with a protectant fungicide, preferably fentin based.

Crop Safety/Restrictions
- Maximum number of treatments 5 per crop

Special precautions/Environmental safety
- Irritant. May cause sensitization by skin contact
- Harmful to fish or other aquatic life. Do not contaminate surface waters or ditches with chemical or used container

Protective clothing/Label precautions
- A, H, M
- 10a, 18, 21, 26, 28, 29, 36, 37, 53, 63, 66

Latest application/Harvest Interval(HI)
- Before end of Aug.
- HI 14 d

Maximum Residue Levels (mg residue/kg food)
- see mancozeb entry

345   maneb
A protectant dithiocarbamate fungicide

See also *carbendazim + chlorothalonil + maneb*
      carbendazim + maneb
      carbendazim + maneb + sulfur
      carbendazim + maneb + tridemorph
      copper oxychloride + maneb + sulfur
      fentin acetate + maneb
      ferbam + maneb + zineb

Products

1	Agrichem Maneb 80	Agrichem	80% w/w	WP	05474
2	Amos Maneb 80	Luxan	80% w/w	WP	06560
3	Ashlade Maneb Flowable	Ashlade	480 g/l	SC	06477
4	Campbell's X-Spor SC	MTM Agrochem.	480 g/l	SC	03252
5	Headland Spirit	Headland	480 g/l	SC	04548
6	Luxan Maneb 80	Luxan	80% w/w	WP	06570
7	Luxan Maneb 800 WP	Luxan	80% w/w	WP	06561
8	Maneb 80	Rohm & Haas	80% w/w	WP	01276
9	Stefes Maneb DF	Stefes	75% w/w	WP	06418
10	Unicrop Flowable Maneb	Unicrop	480 g/l	SC	05546
11	Unicrop Maneb 80	Unicrop	80% w/w	WP	06926
12	X-Spor SC	United Phosphorus	480 g/l	SC	06310

Uses

Blight in POTATOES [1-12]. Brown rust, eyespot, late ear diseases, rhynchosporium, yellow rust in WINTER BARLEY [1, 12]. Brown rust, late ear diseases, septoria, septoria diseases, sooty moulds, yellow rust in WINTER WHEAT [1, 2, 4-9, 11, 12]. Brown rust, net blotch, rhynchosporium, yellow rust in BARLEY [2, 4-9, 11]. Didymella in TOMATOES *(off-label)* [11]. Eyespot in WINTER WHEAT *(late infections)* [12]. Late ear diseases in SPRING BARLEY [1]. Powdery mildew in WINTER WHEAT *(partial suppression)* [12].

Notes

Efficacy
* Apply to potatoes before blight infection occurs, at blight warning or before haulms meet in row and repeat every 10-14 d
* Most effective on cereals as a preventative treatment before disease established, an early application at about first node stage (GS 31), a late application after flag leaf emergence (GS 37) and during ear emergence before watery ripe stage (GS 71) [5]
* In cereals may be used in tank mixture with carbendazim to improve disease control. This mixture obligatory for some products - see label. Do not mix with other fungicides
* Apply to tomatoes at first sign of disease and repeat every 7-14 d

Crop Safety/Restrictions
* Maximum number of treatments 2 per crop in barley and winter wheat [4, 5, 7, 8, 11, 12]; 2 per crop in winter wheat [2, 6]; 5 per crop in all crops [1]
* Do not apply if frost or rain expected, if crop wet or suffering from drought or physical or chemical stress
* A minimum of 10 d must elapse between applications to potatoes

---

FOR FULL CONDITIONS OF USE ALWAYS READ THE PRODUCT LABEL

Special precautions/Environmental safety
* Harmful if swallowed [4, 5, 12]. Irritating to skin, eyes and respiratory system [2-6, 10, 12], to respiratory system [1, 7-9, 11]
* May cause sensitization by skin contact [1, 2, 6-9, 11]
* Dangerous to fish or other aquatic life. Do not contaminate surface waters or ditches with chemical or used container [1, 2, 6-9, 11]

Protective clothing/Label precautions
* A, C [4, 5, 12]; A, D [1, 9]; A [2, 3, 6, 8, 10, 11]; C, D [7]
* 5c, 16, 35 [2, 4-8, 10-12]; 6a, 6b, 54, 70 [2-8, 10-12]; 6c, 18, 21, 29, 36, 37, 63 [1-12]; 10a, 52 [1, 9]; 28 [9]; 66 [3, 9]; 67 [1, 2, 4-8, 10-12]

Latest application/Harvest Interval(HI)
* Before grain milky-ripe (GS 71) for winter wheat [4, 5, 12]; before flag leaf sheath opening (GS 45) for barley [4, 5, 12]; full ear emergence (GS 59) for cereals [7, 11]; before early milk stage (GS 71) for cereals [1].
* HI protected lettuce 3 wk; outdoor celery 14 d; potatoes, outdoor crops 7 d; protected crops 2 d

Approval
* Approved for aerial application on potatoes [1, 3-5, 8, 9, 11]; winter cereals [5]. See notes in Section 1. Consult firm for details
* Approval expiry: 30 Sep 1995 [03252]; replaced by MAFF 06310

Maximum Residue Levels (mg residue/kg food)
* hops 25; lettuces, herbs 5; oranges, apricots, peaches, nectarines, grapes, strawberries, peppers, aubergines 2; cherries, plums 1; garlic, onions, shallots, cucumbers 0.5; celeriac, witloof 0.2; tree nuts, oilseeds (except rape seed), tea 0.1; bilberries, cranberries, wild berries, miscellaneous fruit, horseradish, Jerusalem artichokes, parsley root, sweet potatoes, swedes, turnips, yams, spinach, beet leaves, asparagus, cardoons, fennel, rhubarb, mushrooms, potatoes, maize, rice, animal products 0.05

346   maneb + zinc

A protectant fungicide for use in potatoes and other crops

Products					
1 Barclay Manzeb Flow	Barclay	37:1.25% w/w	SC	05872
2 Manex	Chiltern	480:16 g/l	SC	05731

Uses

Alternaria, downy mildew in BRASSICAS, OILSEED RAPE [1, 2]. Blight, leaf mould in TOMATOES [1, 2]. Blight in POTATOES [1, 2]. Brown foot rot and ear blight, glume blotch, powdery mildew, rust, septoria, sooty moulds in SPRING BARLEY, SPRING WHEAT, WINTER BARLEY, WINTER WHEAT [1, 2]. Rhynchosporium, rust in BARLEY [1, 2]. Scab in PEARS [2]. Scab in APPLES [1, 2].

Notes

Efficacy
* Apply spray before or at first appearance of disease and repeat at 10-14 d intervals (7-10 d on potatoes under severe blight conditions)
* Number and timing of sprays vary with crop and disease. See label for details
* Optimum time for treating cereals is boot stage (GS 45), repeat spray if necessary

Crop Safety/Restrictions
* Maximum number of treatments 1 per crop for oilseed rape

- Spray cereals from last leaf just visible to ear three-quarters emerged (GS 20-57)

Special precautions/Environmental safety
- Irritating to eyes, skin and respiratory system
- Do not feed treated sugar beet tops to livestock
- Do not graze livestock in treated orchards

Protective clothing/Label precautions
- A, C
- 6a, 6b, 6c, 18, 21, 29, 35, 36, 37, 54, 63, 66, 70

Latest application/Harvest Interval(HI)
- Before emergence of inflorescence completed (GS 59) for cereals.
- HI protected tomatoes 2 d; field crops 7 d

Maximum Residue Levels (mg residue/kg food)
- see maneb entry

---

347   maneb + zinc oxide
A protectant fungicide for use on potatoes

Products	Mazin	Unicrop	75:2.5% w/w	WP	06061

Uses   Blight in POTATOES. Powdery scab in SEED POTATOES.

Notes   **Efficacy**
- Apply as spray for blight control before disease appears or at blight warning and repeat at 10-14 d intervals. 3 or more sprays may be needed through season
- May also be used as dust application for reduction in tuber-borne powdery scab (not in situations of high disease pressure)

Special precautions/Environmental safety
- Irritating to eyes, skin and respiratory system
- Keep away from fire and sparks
- Use contents of opened container as soon as possible and do not store opened containers until following season. If product becomes wet, effectiveness may be reduced and inflammable vapour produced

Protective clothing/Label precautions
- A, C, F
- 6a, 6b, 6c, 18, 22, 28, 29, 35, 36, 37, 54, 67

Latest application/Harvest Interval(HI)
- At planting for dust treatment.
- HI 7 d for foliar spray

Approval
- Approved for aerial application on potatoes. See notes in Section 1

FOR FULL CONDITIONS OF USE ALWAYS READ THE PRODUCT LABEL

Maximum Residue Levels (mg residue/kg food)
• see maneb entry

348   MCPA

A translocated phenoxyacetic herbicide for cereals and grassland

See also 2,4-D + dichlorprop + MCPA + mecoprop
2,4-DB + linuron + MCPA
2,4-DB + MCPA
benazolin + 2,4-DB + MCPA
bentazone + MCPA + MCPB
clopyralid + diflufenican + MCPA
dicamba + dichlorprop + MCPA
dicamba + maleic hydrazide + MCPA
dicamba + MCPA + mecoprop-P
dichlorprop + ferrous sulfate + MCPA

Products

1	Agrichem MCPA-25	Agrichem	250 g/l	SL	00050
2	Agricorn 500	FCC	476 g/l	SL	00055
3	Atlas MCPA	Atlas	480 g/l	SL	03055
4	Barclay Meadowman	Barclay	500 g/l	SL	07639
5	BASF MCPA Amine 50	BASF	500 g/l	SL	00209
6	BH MCPA 75	RP Environ.	750 g/l	SL	05395
7	Campbell's MCPA 50	MTM Agrochem.	500 g/l	SL	00381
8	Headland Spear	Headland	500 g/l	SL	07115
9	HY-MCPA	Agrichem	500 g/l	SL	06293
10	Luxan MCPA 500	Luxan	500 g/l	SL	07470
11	MSS MCPA 50	Mirfield	500 g/l	SL	01412
12	Quad MCPA 50%	Quadrangle	500 g/l	SL	01669
13	Stefes Phenoxylene 50	Stefes	500 g/l	SL	07612

Uses

Annual dicotyledons, buttercups, charlock, dandelions, docks, fat-hen, hemp-nettle, perennial dicotyledons, thistles, wild radish in ESTABLISHED GRASSLAND [1-5, 7-12]. Annual dicotyledons, buttercups, dandelions, docks, perennial dicotyledons, thistles in PERMANENT PASTURE [13]. Annual dicotyledons, charlock, fat-hen, hemp-nettle, perennial dicotyledons, wild radish in WINTER RYE [1, 2, 5, 7, 8, 10-12]. Annual dicotyledons, charlock, fat-hen, hemp-nettle, perennial dicotyledons, wild radish in BARLEY, OATS, WHEAT [1-5, 7, 8, 10-13]. Annual dicotyledons, charlock, fat-hen, hemp-nettle, perennial dicotyledons, wild radish in YOUNG LEYS [3, 13]. Annual dicotyledons, charlock, fat-hen, hemp-nettle, perennial dicotyledons, wild radish in SPRING BARLEY, SPRING OATS, SPRING WHEAT, WINTER BARLEY, WINTER WHEAT [9]. Annual dicotyledons, charlock, fat-hen, hemp-nettle, perennial dicotyledons, wild radish in GRASS SEED CROPS [10, 12, 13]. Annual dicotyledons, charlock, fat-hen, hemp-nettle, perennial dicotyledons, wild radish in NEWLY SOWN GRASS [10]. Annual dicotyledons, charlock, fat-hen, hemp-nettle, perennial dicotyledons, wild radish in RYE [4, 13]. Annual dicotyledons, creeping thistle, daisies, docks, perennial dicotyledons in INDUSTRIAL SITES, ROAD VERGES, TURF [6, 11]. Annual dicotyledons, creeping thistle, daisies, docks, perennial dicotyledons in AMENITY GRASS [6]. Annual dicotyledons, perennial dicotyledons in ASPARAGUS [13].

Notes

Efficacy
• Best results achieved by application to weeds in seedling to young plant stage under good growing conditions when crop growing actively

- Spray perennial weeds in grassland before flowering. Most susceptible growth stage varies between species. See label for details
- Do not spray during cold weather, drought, if rain or frost expected or if crop wet

Crop Safety/Restrictions
- Maximum number of treatments 1 per crop or yr except grass (2 per yr) for some products. See label
- Apply to winter cereals in spring from fully tillered, leaf sheath erect stage to before first node detectable (GS 30-31)
- Apply to spring barley and wheat from 5-leaves unfolded (GS 15), to oats from 1-leaf unfolded (GS 11) to before first node detectable (GS 31)
- Apply to cereals undersown with grass after grass has 2-3 leaves unfolded
- Do not sow any crop within 1 mth of application [4]
- Do not use on cereals before undersowing
- Do not roll, harrow or graze for a few days before or after spraying; see label
- Recommendations for crops undersown with legumes vary. Red clover may withstand low doses after 2-trifoliate leaf stage, especially if shielded by taller weeds but white clover is more sensitive. See label for details
- Apply to linseed as soon as possible after 5 cm high and before 25 cm
- Do not use on grassland where clovers are an important part of the sward
- Apply to grass seed crops from 2-3 leaf stage to 5 wk before head emergence
- Do not use on any crop suffering from stress or herbicide damage
- Do not direct drill brassicas or legumes within 6 wk of spraying grassland
- Avoid spray drift onto nearby susceptible crops

Special precautions/Environmental safety
- Harmful if swallowed [1-9, 11-13]. Harmful in contact with skin [6, 9, 13]
- Irritating to skin. Risk of serious damage to eyes [5]
- Harmful to fish or other aquatic life. Do not contaminate surface waters or ditches with chemical or used container [2, 5, 9]

Protective clothing/Label precautions
- A, C [1-3, 5-13]
- 5a [6, 9-11, 13]; 5c, 18, 21, 29, 36, 37, 78 [1-13]; 6b, 9 [5, 10]; 14 [10]; 43 (2 wk) [4]; 43 [1-3, 5-13]; 53 [2, 5, 9]; 54 [1, 3, 4, 6-8, 10-13]; 60, 65 [1, 2, 5-9, 12]; 63 [1-10, 12, 13]; 66 [3, 4, 10, 11, 13]; 70 [1-12]

Withholding period
- Keep livestock out of treated areas for at least 2 wk and until foliage of poisonous weeds such as ragwort has died and become unpalatable

Latest application/Harvest Interval(HI)
- Before 1st node detectable (GS 31) for cereals; 6 wk before heading for grass

FOR FULL CONDITIONS OF USE ALWAYS READ THE PRODUCT LABEL

349 MCPA + MCPB
A translocated herbicide for undersown cereals and grassland

Products

1 Bellmac Plus	United Phosphorus	38:262 g/l	SL	07521
2 Campbell's Bellmac Plus	MTM Agrochem.	38:262 g/l	SL	00385
3 MSS MCPB + MCPA	Mirfield	25:275 g/l	SL	01413
4 Trifolex-Tra	Cyanamid	34:216 g/l	SL	07147

Uses

Annual dicotyledons, perennial dicotyledons in ESTABLISHED GRASSLAND [3, 4]. Annual dicotyledons, perennial dicotyledons in CEREALS [3]. Annual dicotyledons, perennial dicotyledons in PEAS [4]. Annual dicotyledons, perennial dicotyledons in LEYS, UNDERSOWN CEREALS [1-4]. Annual dicotyledons, perennial dicotyledons in SAINFOIN [1-3].

Notes

Efficacy
- Best results achieved by application to weeds in seedling to young plant stage under good growing conditions when crop growing actively. Spray perennials when adequate leaf surface before flowering. Retreatment often needed in following year
- Spray leys and sainfoin before crop provides cover for weeds
- Rain, cold or drought may reduce effectiveness

Crop Safety/Restrictions
- Maximum number of treatments 1 per crop or yr
- Apply to cereals from 2-expanded leaf stage to before jointing (GS 12-30) and, where undersown, after 1-trifoliate leaf stage of clover
- Apply to direct-sown seedling clover after 1-trifoliate leaf stage
- Apply to mature white clover for fodder at any stage. Do not spray red clover after flower stalk has begun to form
- Do not spray clovers for seed [3, 4]
- Apply to sainfoin after first compound leaf stage [1, 2], second compound leaf [3]
- Do not spray peas [1-3]. Only spray peas in tank mix with Fortrol from 4-leaf stage to before bud visible (GS 104-201) [4]
- Do not roll or harrow for a few days before or after spraying [1-4]
- Avoid spray drift onto nearby sensitive crops

Special precautions/Environmental safety
- Harmful to fish or other aquatic life. Do not contaminate surface waters or ditches with chemical or used container

Protective clothing/Label precautions
- 18, 21, 29, 36, 37, 43, 53, 60, 63, 66

Withholding period
- Keep livestock out of treated areas until foliage of any poisonous weeds such as ragwort has died and become unpalatable

Latest application/Harvest Interval(III)
- Before first node detectable (GS 31) for cereals

Approval
- Approval expiry: 28 Feb 96 [00385]; replaced by MAFF 06794

350 MCPA + mecoprop

A translocated selective herbicide for amenity grass

Products					
1 Fisons Greenmaster Extra	Levington	0.49:0.57% w/w	GR	03130
2 Greenmaster Extra	Levington	0.49:0.57% w/w	GR	07491

Uses   Annual dicotyledons, perennial dicotyledons in AMENITY GRASS.

Notes   **Efficacy**
* Apply from Apr to Sep, when weeds growing actively and have large leaf area for absorption of chemical
* Granules contain NPK fertilizer to encourage grass growth
* Do not apply when heavy rain expected or during prolonged drought. Irrigate after 1-2 d unless rain has fallen
* Do not mow for 2-3 d before or after treatment

Crop Safety/Restrictions
* Maximum number of treatments 2 per yr
* Avoid contact with cultivated plants
* Compost for not less than 6 mth before using first 4 mowings as compost or mulch
* Do not apply to newly sown or turfed areas for at least 6 mth or reseed bare patches for 8 wk

Special precautions/Environmental safety
* Do not contaminate surface waters or ditches with chemical or used container

Protective clothing/Label precautions
* 29, 43, 54, 63, 67

Withholding period
* Keep livestock out of treated areas until foliage of any poisonous weeds such as ragwort has died and become unpalatable

351 MCPA + mecoprop-P

A translocated selective herbicide for amenity grass

Products					
Greenmaster Extra	Levington	0.49:0.29 % w/w	GR	07594

Uses   Annual dicotyledons, perennial dicotyledons in AMENITY GRASS.

Notes   **Efficacy**
* Apply from April to September, when weeds growing actively and have large leaf area available for chemical absorption
* Granules contain NPK fertiliser to encourage grass growth

FOR FULL CONDITIONS OF USE ALWAYS READ THE PRODUCT LABEL

- Do not apply when heavy rain expected or during prolonged drought. Irrigate after 1-2 d unless rain has fallen
- Do not mow within 2-3 d of treatment

Crop Safety/Restrictions
- Maximum number of treatments 2 per yr
- Avoid contact with cultivated plants
- Do not use first 4 mowings as compost or mulch within 6 mth
- Do not treat newly sown or turfed areas for at least 6 mth
- Do not reseed bare patches for 8 wk after treatment

Special precautions/Environmental safety
- Do not contaminate surface waters or ditches with chemical or used container

Protective clothing/Label precautions
- 29, 36, 37, 43, 54, 63, 67

Withholding period
- Keep livestock out of treated areas until foliage of any poisonous weeds such as ragwort has died and become unpalatable

---

352   MCPB

A translocated phenoxybutyric herbicide for use in peas etc.

See also bentazone + MCPA + MCPB
bentazone + MCPB
MCPA + MCPB

Products					
1 Bellmac Straight	United Phosphorus	400 g/l	SL	07522
2 Campbell's Bellmac Straight	MTM Agrochem.	400 g/l	SL	00386

Uses   Annual dicotyledons, perennial dicotyledons in EDIBLE PODDED PEAS *(off-label)*, LEYS [2]. Annual dicotyledons, perennial dicotyledons in BLACKCURRANTS, PEAS, UNDERSOWN CEREALS, WHITE CLOVER SEED CROPS [1, 2].

Notes   **Efficacy**
- Best results achieved by spraying young seedling weeds in good growing conditions
- Best results on perennials by spraying before flowering
- Effectiveness may be reduced by rain within 12 h, by very cold or dry conditions

Crop Safety/Restrictions
- Maximum number of treatments 1 per crop
- Apply to undersown cereals from 2-leaves unfolded to first node detectable (GS 12-31), and after first trifoliate leaf stage of clover
- Red clover seedlings may be temporarily damaged but later growth is normal
- Apply to white clover seed crops in Mar to early Apr, not after mid-May, and allow 3 wk before cutting and closing up for seed
- Apply to peas from 3-6 leaf stage but before flower bud detectable (GS 103-201). See label for details of cultivars which can be treated and requirements for leaf-wax testing
- Do not use on beans, vetches or established red clover
- Apply to blackcurrants in Aug after harvest and after shoot growth ceased but while bindweed still growing actively; direct spray onto weed as far as possible

459

- Do not roll or harrow for 7 d before or after treatment
- Avoid drift onto nearby sensitive crops

Special precautions/Environmental safety
- Harmful in contact with skin and if swallowed
- Harmful to fish or other aquatic life. Do not contaminate surface waters or ditches with chemical or used container

Protective clothing/Label precautions
- A, C
- 5a, 5c, 18, 21, 29, 36, 37, 43, 53, 60, 63, 66, 70, 78

Withholding period
- Keep livestock out of treated areas until foliage of any poisonous weeds such as ragwort has died and become unpalatable

Latest application/Harvest Interval(HI)
- First node detectable stage (GS 31) for cereals; before flower buds appear in terminal leaf (GS 201) for peas

Approval
- Off-label Approval to Apr 1997 for use on edible podsded peas (OLA 0357/92)[2]
- Approval expiry: 31 Dec 95 [00386]; replaced by MAFF 07522

353   mecoprop

A translocted phenoxypropionic herbicide for cereals and grassland

See also 2,4-D + dicamba + mecoprop
2,4-D + dichlorprop + MCPA + mecoprop
2,4-D + mecoprop
bifenox + isoproturon + mecoprop
dicamba + MCPA + mecoprop
dicamba + mecoprop
dicamba + mecoprop + triclopyr
diclorprop + mecoprop
MCPA + mecoprop

Products					
1	Amos CMPP	Luxan	560 g/l	SL	06556
2	Atlas CMPP	Atlas	570 g/l	SL	03050
3	Barclay Mecrop	Barclay	570 g/l	SL	07578
4	Campbell's CMPP	MTM Agrochem.	570 g/l	SL	02918
5	Clenecorn	FCC	570 g/l	SL	00542
6	Clovotox	RP Environ.	300 g/l	SL	05354
7	Headland Charge	Headland	570 g/l	SL	04495
8	Hymec	Agrichem	600 g/l	SL	01091
9	Luxan CMPP	Luxan	560 g/l	SL	06037
10	Mascot Cloverkiller	Rigby Taylor	285 g/l	SL	02438
11	MSS CMPP	Mirfield	600 g/l	SL	01405
12	MSS CMPP Amine 60	Mirfield	600 g/l	SL	02446

FOR FULL CONDITIONS OF USE ALWAYS READ THE PRODUCT LABEL

Uses   Annual dicotyledons, chickweed, cleavers, perennial dicotyledons in YOUNG LEYS [4, 5, 7]. Annual dicotyledons, chickweed, cleavers, perennial dicotyledons in ESTABLISHED GRASSLAND [1, 2, 9, 12]. Annual dicotyledons, chickweed, cleavers, perennial dicotyledons in BARLEY, OATS, WHEAT [1-5, 7-9, 11, 12]. Annual dicotyledons, chickweed, cleavers, perennial dicotyledons in NEWLY SOWN GRASS [3]. Annual dicotyledons, chickweed, clovers, perennial dicotyledons in TURF [6, 10].

Notes   **Efficacy**
- Best results achieved by application to weeds in seedling to young plant stage growing actively in a strongly competing crop
- Do not spray when rain imminent or when growth hard from cold or drought. Rainfall within 8-12 h may reduce effectiveness
- For dock control in grassland cut in mid-summer and spray regrowth in Aug-Sep
- Spray turf weeds in spring or early summer when in active growth. More resistant perennials may need repeat treatment after 4-6 wk

Crop Safety/Restrictions
- Maximum number of treatments 1 per crop for cereals (2 per crop for winter cereals not undersown [3]); 2 per yr for grassland and turf; 1 per yr for newly sown grass [3]
- Range of timings recommended for cereals varies with product. See label for details
- Do not spray rye, crops undersown with legumes or crops about to be undersown
- If hard frosts occur within 3-4 wk of application to barley of low vigour on light soils crop may be scorched or stunted
- Do not roll or harrow within a few days before or after treatment
- Apply to grasses after 3-fully expanded leaves and 1 tiller stage
- Clovers present in sprayed grass will be severely damaged
- Do not direct-drill brassicas or legumes within 6 wk after application
- Do not use first 4 mowings for mulching. Compost for at least 6 mth before use
- Avoid drift onto nearby fruit trees or other susceptible crops

Special precautions/Environmental safety
- Harmful in contact with skin and if swallowed [1-3, 5-12]
- Irritating to eyes and skin [1, 3, 5-12]
- Irritating to eyes [2]

Protective clothing/Label precautions
- A, C, H, M [10]; A, C [1-9, 11, 12]
- 5a, 5c [1, 2, 4-9, 11, 12]; 6a [1, 2, 4, 6-12]; 6b [1, 4, 6-12]; 18, 29, 36, 37, 43, 63 [1-12]; 21 [2, 3, 10]; 22 [1, 4-9, 11, 12]; 25, 53, 67 [10]; 54, 66, 78 [1-9, 11, 12]; 60 [2]; 70 [2, 3, 5]; 70, [1, 4, 6-9, 11, 12]

Withholding period
- Keep livestock out of treated areas until foliage of any poisonous weeds such as ragwort has died and become unpalatable

Latest application/Harvest Interval(HI)
- Varies between before first node detectable (GS 31) and before third node detectable (GS 33) for cereals. See product label

354 mecoprop-P

A translocated phenoxypropionic herbicide for cereals and grassland

See also bromoxynil + ioxynil + mecoprop-P
    dicamba + MCPA + mecoprop-P
    dicamba + mecoprop-P
    fluoroglycofen-ethyl + mecoprop-P
    fluroxypyr + mecoprop-P
    MCPA + mecoprop-P

Products					
1 Duplosan	BASF	600 g/l	SL	05889
2 MSS Optica	Mirfield	600 g/l	SL	04973
3 Standon Mecoprop-P	Standon	600 g/l	SL	05651
4 Stefes Mecoprop-P	Stefes	600 g/l	SL	05780
5 Stefes Mecoprop-P2	Stefes	600 g/l	SL	06239

Uses
Annual dicotyledons, chickweed, cleavers, perennial dicotyledons in ESTABLISHED GRASSLAND [3]. Annual dicotyledons, chickweed, cleavers, perennial dicotyledons in BARLEY, OATS, WHEAT [1-3, 5]. Annual dicotyledons, chickweed, cleavers, perennial dicotyledons in GRASSLAND [1, 2, 4, 5]. Annual dicotyledons, chickweed, cleavers, perennial dicotyledons in GRASS SEED CROPS [1, 2, 4]. Annual dicotyledons, chickweed, cleavers, perennial dicotyledons in YOUNG LEYS [1, 2, 5].

Notes   **Efficacy**
* Best results achieved by application to seedling weeds which have not been frost hardened, when soil warm and moist and expected to remain so for several days
* Do not spray during cold weather, periods of drought, if rain or frost expected or if crop wet

Crop Safety/Restrictions
* Maximum number of treatments 1 per crop for cereals [2, 5]; 1 per crop for spring cereals, 2 per crop for winter cereals [1, 3, 4]; 1 per yr for newly sown grass leys, 2 per yr for established grass [1, 3, 4]; 2 per yr for grassland [2, 5]; 1 per yr for grass seed crops
* Spray winter cereals from 1 leaf stage in autumn up to and including first node detectable in spring (GS 10-31) or up to second node detectable (GS 32) if necessary. Apply to spring cereals from first fully expanded leaf stage (GS 11) but before first node detectable (GS 31)
* Use maximum rate of 2.3 l/ha only after crop has reached GS 30
* Spray cereals undersown with grass after grass starts to tiller
* Do not spray cereals undersown with clovers or legumes or to be undersown with legumes or grasses
* Spray newly sown grass leys when grasses have at least 3 fully expanded leaves and have begun to tiller. Any clovers will be damaged
* Do not spray grass seed crops within 5 wk of seed head emergence
* Do not spray crops suffering from herbicide damage or physical stress
* Do not roll or harrow for 7 d before or after treatment
* Avoid drift onto beans, beet, brassicas, fruit and other sensitive crops

FOR FULL CONDITIONS OF USE ALWAYS READ THE PRODUCT LABEL

Special precautions/Environmental safety
* Harmful if swallowed and in contact with skin, irritating to eyes [5]. Harmful if swallowed, risk of serious damage to eyes [1-4]
* Harmful to fish or other aquatic life. Do not contaminate surface waters or ditches with chemical or used container

Protective clothing/Label precautions
* A, C
* 5a, 6a [5]; 5c, 18, 21, 29, 36, 37, 43, 53, 60, 63, 66, 70, 78 [1-5]; 9 [1-4]

Withholding period
* Keep livestock out of treated areas until foliage of any poisonous weed, such as ragwort, has died and become unpalatable

Latest application/Harvest Interval(HI)
* Before 1st node detectable (GS 31) for spring cereals; before 3rd node detectable (GS 33) for winter cereals [1, 3, 4]; before second node detectable (GS 32) for cereals [2, 5]; 5 wk before emergence of seed head for grass seed crops [1, 2, 4]

355   mefluidide

A sulfonanilide plant growth regulator for suppression of grass

Products	Embark	Intracrop	240 g/l	SL	04810

Uses   Growth suppression in AMENITY GRASS, ROADSIDE GRASS.

Notes

Efficacy
* Apply when grass growth active in Apr/May and repeat when growth re-commences
* Grass growth suppressed for up to 8 wk
* Mow (not below 2 cm) 2-3 d before or 5-10 d after spraying
* Only apply to dry foliage but do not use in drought
* A rainfree period of 8 h is needed after spraying
* May be tank-mixed with suitable 2,4-D or mecoprop product for control of dicotyledons
* Delay application for 3-4 wk after application of high-nitrogen fertilizer

Crop Safety/Restrictions
* Maximum number of treatments 2 per yr
* Do not use on fine turf areas or on turf established for less than 4 mth
* Do not use on grass suffering from herbicide damage
* Treated turf may appear less dense and be temporarily discoloured but normal colour returns in 3-4 wk
* Do not reseed within 2 wk after application

Special precautions/Environmental safety
* Do not allow animals to graze treated areas

Protective clothing/Label precautions
* 29, 54, 63, 66, 70

356   mephosfolan

A systemic organophosphorus insecticide for use on hops

Products	Cytro-Lane		Cyanamid	250 g/l	EC	00626

Uses   Damson-hop aphid in HOPS.

Notes   **Efficacy**
* Apply as accurately measured drench to lower part of bines and centre of hop plant at ground level in mid- to late May after most bines have reached 1.5-2 m. A second application may be made before mid-Jul. See label for details of suitable applicators
* Base of hop plants should be free of weeds and dead leaves
* Bines emerging away from main stock or near poles may require separate treatment
* Systemic uptake may be delayed by very dry conditions
* Product may be used in integrated control programmes. Predator insects not harmed

Crop Safety/Restrictions
* Do not apply as a spray
* Total dose per season should not exceed 8 l/1000 plants

Special precautions/Environmental safety
* Mephospholan is subject to the Poisons Rules 1982 and the Poisons Act 1972. See notes in Section 1
* This product contains an anticholinesterase organophosphorus compound. Do not use if under medical advice not to work with such compounds
* Very toxic in contact with skin, by inhalation and if swallowed. Irritating to skin
* Keep in original container, tightly closed, in a safe place, under lock and key
* Keep unprotected persons out of treated areas for at least 1 d
* Harmful to livestock
* Dangerous to game, wild birds and animals
* Harmful to bees. Do not apply to crops in flower or to those in which bees are actively foraging. Do not apply when flowering weeds are present
* Dangerous to fish or other aquatic life. Do not contaminate surface waters or ditches with chemical or used container

Protective clothing/Label precautions
* A, C, H + K or L, M
* 1, 3a, 3b, 3c, 6b, 14, 16, 18, 21, 25, 28, 29, 35, 36, 37, 38, 41, 45, 50, 52, 64, 65, 70, 79

Withholding period
* Keep all livestock out of treated areas for at least 10 d

Latest application/Harvest Interval(HI)
* HI 4 wk

FOR FULL CONDITIONS OF USE ALWAYS READ THE PRODUCT LABEL

357 mepiquat chloride

A quaternary ammonium plant growth regulator available only in mixtures
See also 2-chloroethylphosphonic acid + mepiquat chloride

358 mercuric oxide

An inorganic mercury fungicide paint all approvals for which were revoked with effect from 31 March 1992

359 mercurous chloride

A soil applied inorganic mercury fungicide all approvals for which were revoked with effect from 31 March 1992

360 metalaxyl

A phenylamide (acylalanine) fungicide available only in mixtures
See also carbendazim + metalaxyl
 chlorothalonil + metalaxyl
 copper oxychloride + metalaxyl
 mancozeb + metalaxyl

361 metalaxyl + thiabendazole + thiram

A protective fungicide seed treatment for peas and beans
See also thiabendazole

Products	Apron Combi FS	Ciba Agric.	233:120:100 g/l	FS	07203

Uses Ascochyta, damping off, downy mildew in BEANS *(seed treatment)*, COMBINING PEAS *(seed treatment)*, DWARF BEANS *(seed treatment)*, MANGE-TOUT PEAS *(seed treatment)*, RUNNER BEANS *(seed treatment)*, VINING PEAS *(seed treatment)*.

Notes **Efficacy**
• Apply through continuous flow seed treaters which should be calibrated before use

Crop Safety/Restrictions
• Ensure moisture content of treated seed satisfactory and store in a dry place
• Check calibration of seed drill with treated seed before drilling

Special precautions/Environmental safety
• Harmful if swallowed. Irritating to eyes
• Treated seed not to be used as food or feed. Do not re-use sack
• Harmful to fish or other aquatic life. Do not contaminate surface waters or ditches with chemical or used container

Protective clothing/Label precautions
• A, C, D, H, M
• 5c, 6a, 18, 22, 25, 29, 36, 37, 53, 63, 65, 72, 73, 74, 75, 76, 78

Maximum Residue Levels (mg residue/kg food)
* see metalaxyl and thiabendazole entries

362   metalaxyl + thiram
A systemic and protectant fungicide for use in lettuce

Products   Favour 600 SC   Ciba Agric.   100:500 g/l   SC   05842

Uses   Crown rot in PARSLEY *(off-label)*. Downy mildew in LETTUCE, PROTECTED LETTUCE.

Notes   **Efficacy**
* On lettuce apply from 100% emergence and repeat at 14 d intervals
* Resistant strains of lettuce downy mildew may develop against which product may be ineffective. See label for details

Crop Safety/Restrictions
* Maximum number of treatments; 2 post-planting applications of thiram or any combination of products containing thiram or other EBDC compound (mancozeb, maneb, zineb) either as a spray or a dust are permitted on protected lettuce within 2 wk of planting out and none thereafter. On winter lettuce, if only thiram based products are being used, 3 applications may be made within 3 wk of planting out and none thereafter
* Maximum number of treatments 5 per crop for outdoor lettuce; 3 per yr on outdoor parsley
* Do not use as a soil block treatment

Special precautions/Environmental safety
* Harmful if swallowed. Irritating to eyes, skin and respiratory system
* Harmful to fish or other aquatic life. Do not contaminate surface waters or ditches with chemical or used container

Protective clothing/Label precautions
* A, C
* 5a, 6a, 6b, 6c, 18, 21, 29, 35, 36, 37, 53, 63, 66, 70, 78

Latest application/Harvest Interval(HI)
* HI protected lettuce 3 wk; outdoor lettuce 2 wk; outdoor parsley 4 wk

Approval
* Off-label Approval to Aug 1998 for use on outdoor parsley (OLA 1020/93)

Maximum Residue Levels (mg residue/kg food)
* see metalaxyl entry

FOR FULL CONDITIONS OF USE ALWAYS READ THE PRODUCT LABEL

363 metaldehyde
A molluscicide bait for controlling slugs and snails

Products

1	Devcol Morgan 6% Metaldehyde Slug Killer	Nehra	6% w/w	PT	07422
2	Doff Agricultural Slug Killer with Animal Repellent	Doff Portland	6% w/w	PT	04898
3	Doff Horticultural Slug Killer Blue Mini Pellets	Doff Portland	3% w/w	PT	05688
4	Doff Metaldehyde Slug Killer Mini Pellets	Doff Portland	6% w/w	PT	00741
5	Escar-Go 6	Chiltern	6% w/w	PT	06076
6	Fisons Helarion	Levington	5% w/w	PT	02520
7	Gastratox 6G Slug Pellets	Truchem	6% w/w	PT	04066
8	Hardy	Chiltern	6% w/w	PT	06948
9	Luxan 9363	Luxan	6% w/w	PT	07359
10	Luxan Metaldehyde	Luxan	6% w/w	PT	06564
11	Metarex	De Sangosse	6% w/w	PT	04910
12	Metarex RG	De Sangosse	6% w/w	RB	06754
13	Mifaslug	FCC	6% w/w	PT	01349
14	Optimol	Zeneca	4% w/w	PT	06688
15	PBI Slug Pellets	PBI	6% w/w	PT	01558
16	Quad Mini Slug Pellets	Quadrangle	6% w/w	PT	01670
17	Superflor 6% Metaldehyde Slug Killer Mini Pellets	CMI	6% w/w	PT	05453
18	Tripart Mini Slug Pellets	Tripart	6% w/w	PT	02207
19	Unicrop 6% Mini Slug Pellets	Unicrop	6% w/w	PT	02275

Uses

Slugs, snails in BRASSICAS *(seed admixture)*, CEREALS *(seed admixture)* [4]. Slugs, snails in BRASSICAS, CEREALS, GRASSLAND, OILSEED RAPE, POTATOES [10]. Slugs, snails in FIELD CROPS, FRUIT CROPS, ORNAMENTALS, PROTECTED CROPS, VEGETABLES [1, 3-9, 11-19].

Notes

Efficacy
* Apply pellets by hand, fiddle drill, fertilizer distributor, by air or in admixture with seed. See labels for rates and timing.
* Best results achieved by application during mild, damp weather when slugs and snails most active. May be applied in standing crops
* Do not apply when rain imminent or water glasshouse crops within 4 d of application
* To prevent slug build up apply at end of season to brassicas and other leafy crops
* To reduce tuber damage in potatoes apply twice in Jul and Aug

Special precautions/Environmental safety
* Dangerous to game, wild birds and animals
* Harmful to fish or other aquatic life. Do not contaminate surface waters or ditches with chemical or used container
* Product contains proprietary cat and dog deterrent [9]

Protective clothing/Label precautions
* 29, 66 [1-4, 7, 10-16, 18, 19]; 30 [5, 6, 8, 9, 17]; 37, 45, 53, 63, 67 [1-19]; 42 (7 d) [5, 8, 9, 17]; 42, 70 [1-4, 6, 7, 10-16, 18, 19]; 62 [5, 8]

Withholding period
* Keep poultry out of treated areas for at least 7 d

Approval
* Approved for aerial application on all crops [2, 4, 12-19]. See notes in Section 1

364 metam-sodium

A methyl isothiocyanate producing sterilant for glasshouse, nursery and outdoor soils

Products

1	Campbell's Metham Sodium	MTM Agrochem.	400 g/l	SL	00412
2	Sistan	Unicrop	380 g/l	SL	01957

Uses

Dutch elm disease in ELM TREES *(off-label)* [2]. Nematodes, soil insects, soil-borne diseases, weed seeds in CHRYSANTHEMUMS, ORNAMENTALS, OUTDOOR TOMATOES *(Jersey)*, PROTECTED CARNATIONS, PROTECTED CROPS, TOMATOES [1]. Nematodes, soil pests, soil-borne diseases, weed seeds in GLASSHOUSE SOILS, NURSERY SOILS, OUTDOOR SOILS, POTTING SOILS [2].

Notes

Efficacy
* Product acts by breaking down in contact with soil to release methyl isothiocyanate
* Apply to glasshouse soils as a drench, inject undiluted to 20 cm at 30 cm intervals and seal with water or apply to surface, rotavate and seal
* After treatment allow sufficient time (several weeks) for residues to dissipate and aerate soil by forking. Time varies with season. See label for details
* Apply to outdoor soils by drenching or sub-surface application
* Apply when soil temperatures exceed 7°C, preferably above 10°C, between 1 Apr and 31 Oct. Soil must be of fine tilth and adequate moisture content
* Product may also be used by mixing into potting soils
* Avoid using in equipment incorporating natural rubber parts
* Use in control programme against dutch elm disease to sever root grafts which can transmit disease from tree to tree. See OLA notice for details [2]

Crop Safety/Restrictions
* No plants must be present during treatment. Do not treat glasshouses within 2 m of growing crops. Do not plant until soil is entirely free of fumes
* Cress test advised to check for absence of residues. See label for details

Special precautions/Environmental safety
* Harmful in contact with skin and if swallowed [2]
* Irritating to eyes, skin and respiratory system [1, 2]

Protective clothing/Label precautions
* A, C, D, H, M [1, 2]
* 5a, 5c, 65, [2]; 6a, 6b, 6c, 14, 16, 18, 21, 28, 29, 36, 37, 54, 63, 78. See label for additional precautions. [1, 2]; 65 [1]

FOR FULL CONDITIONS OF USE ALWAYS READ THE PRODUCT LABEL

Latest application/Harvest Interval(HI)
• Pre-planting of crop

Approval
• Off-label approval unlimited for use around elm trees infected with Dutch elm disease
(OLA 0592/93)[2]

---

365   metamitron

A contact and residual triazinone herbicide for use in beet crops

Products					
	1 Goltix WG	Bayer	70% w/w	WG	02430
	2 Stefes 7G	Stefes	70% w/w	WG	06350
	3 Stefes Metamitron	Stefes	70% w/w	WG	05821
	4 Tripart Accendo	Tripart	70% w/w	WG	06110

Uses   Annual dicotyledons, annual grasses, annual meadow grass, fat-hen in FODDER BEET,
MANGELS, RED BEET, SUGAR BEET [1-4]. Annual dicotyledons in OUTDOOR LEAF
HERBS *(off-label)* [1].

Notes   **Efficacy**
• Low dose programme (LDP). Apply post-weed emergence as a fine spray, including
adjuvant oil, timing treatment according to weed emergence and size. See label for details
and for recommended tank mixes and sequential treatments. On mineral soils the LDP
should be preceeded by pre-drilling or pre-emergence treatment
• Use up to 5 LDP sprays on mineral soils (up to 3 where also used pre-emergence) and up
to 6 sprays on organic soils
• Traditional application. Apply either pre-drilling before final cultivation with incorporation
to 8-10 cm, or pre-crop emergence at or soon after drilling into firm, moist seedbed to
emerged weeds from cotyledon to first true leaf stage
• On emerged weeds at or beyond 2-leaf stage addition of adjuvant oil advised
• Up to 3 post-emergence sprays may be used on soils with over 10% organic matter
• For control of wild oats and certain other weeds, tank mixes with other herbicides or
sequential treatments are recommended. See label for details

Crop Safety/Restrictions
• Using traditional method post-crop emergence on mineral soils do not apply before first
true leaves have reached 1 cm long
• Crop tolerance may be reduced by stress caused by growing conditions, effects of pests,
disease or other pesticides, nutrient deficiency etc
• Only sugar beet, fodder beet or mangels may be drilled within 4 mth after treatment.
Winter cereals may be sown in same season after ploughing, provided 16 wk passed since
last treatment

Protective clothing/Label precautions
• 29, 54, 63, 70

Approval
• Off-label approval unlimited for use on outdoor leafy herbs (OLA 0802/95)[1]

366   metazachlor

A residual anilide herbicide for use in brassicas and nursery stock

Products

1 Barclay Metaza	Barclay	500 g/l	SC	07354
2 Butisan S	BASF	500 g/l	SC	00357
3 Butisan S	Clifton	500 g/l	SC	05687
4 Standon Metazachlor 50	Standon	500 g/l	SC	05581

Uses

Annual dicotyledons, annual grasses in FARM FORESTRY *(off-label)* [2]. Annual dicotyledons, annual meadow grass, blackgrass, triazine resistant groundsel in WINTER OILSEED RAPE [2-4]. Annual dicotyledons, annual meadow grass, blackgrass, triazine resistant groundsel in HARDY ORNAMENTAL NURSERY STOCK, NURSERY FRUIT TREES AND BUSHES, ORNAMENTALS, SPRING OILSEED RAPE [2, 3]. Annual dicotyledons, annual meadow grass, blackgrass, triazine resistant groundsel in BROCCOLI, BRUSSELS SPROUTS, CABBAGES, CALABRESE, CAULIFLOWERS, SWEDES, TURNIPS [1-3]. Annual dicotyledons, annual meadow grass, blackgrass in OILSEED RAPE [1]. Annual dicotyledons in HONESTY *(off-label)*[2].

Notes

Efficacy

* Activity is dependent on root uptake. For pre-emergence use apply to firm, moist, clod-free seedbed
* Some weeds (chickweed, mayweed, blackgrass etc) susceptible up to 2- or 4-leaf stage. Moderate control of cleavers achieved provided weeds not emerged and adequate soil moisture present
* Split pre- and post-emergence treatments recommended for certain weeds in winter oilseed rape on light and/or stony soils
* Effectiveness is reduced on soils with more than 10% organic matter
* Various tank-mixtures recommended. See label for details

Crop Safety/Restrictions

* Maximum number of treatments 1 per crop for spring oilseed rape, swedes, turnips and brassicas; 2 per crop for winter oilseed rape and honesty (split dose treatment); 3 per yr for ornamentals, nursery stock, nursery fruit trees and farm forestry
* On winter oilseed rape may be applied pre-emergence from drilling until seed chits, post-emergence after fully expanded cotyledon stage (GS 1,0) or by split dose technique depending on soil and weeds. See label for details
* On spring oilseed rape, swedes, turnips and transplanted brassicas recommended as a pre-emergence sequential treatment following trifluralin
* On spring oilseed rape may also be used pre-weed-emergence from cotyledon to 10-leaf stage of crop (GS 1,0-1,10)
* With pre-emergence treatment ensure seed covered by 15 mm of well consolidated soil. Harrow across slits of direct-drilled crops
* Ensure brassica transplants have roots well covered and are well established
* In ornamentals and hardy nursery stock apply after plants established and hardened off as a directed spray or, on some subjects, as an overall spray. See label for list of tolerant subjects. Do not treat plants in containers
* Do not use on sand, very light or poorly drained soils
* Do not treat protected crops or spray overall on ornamentals with soft foliage

FOR FULL CONDITIONS OF USE ALWAYS READ THE PRODUCT LABEL

• Do not spray crops suffering from wilting, pest or disease
• Do not spray broadcast crops or if a period of heavy rain forecast
• Any crop can follow normally harvested treated oilseed rape. See label for details of crops which may be planted in event of crop failure

Special precautions/Environmental safety
• Harmful if swallowed. Irritating to eyes and skin
• Harmful to fish or other aquatic life. Do not contaminate surface waters or ditches with chemical or used container

Protective clothing/Label precautions
• A, C, H, M [2-4]; A, C, H [1]
• 5c, 6b, 18, 21, 28, 29, 36, 37, 63, 66, 70, 78 [1-4]; 6a, 35, 43.53 [2-4]; 43, 53 [1]

Withholding period
• Keep livestock out of treated areas until foliage of any poisonous weeds such as ragwort has died and become unpalatable
• Keep livestock out of treated areas of swede and turnip for at least 5 wk following treatment [1]

Latest application/Harvest Interval(HI)
• Pre-emergence for swedes and turnips; before 10 leaf stage for spring oilseed rape; before end of Jan for winter oilseed rape; before 6 pairs of true leaves for honesty.
• HI brassicas 6 wk. Do not use on nursery fruit trees within 1 yr of fruit production

Approval
• Off-label Approval unlimited for use in farm forestry (OLA 0227/94)[2], unlimited for honesty grown outdoors (OLA 0551/93)[2]

367   methabenzthiazuron

A contact and residual urea herbicide for cereals and ryegrass

Products	Tribunil	Bayer	70% w/w	WP	02169

Uses   Annual dicotyledons, annual meadow grass, blackgrass, rough meadow grass in WINTER BARLEY, WINTER WHEAT. Annual dicotyledons, annual meadow grass, rough meadow grass in DURUM WHEAT, RYEGRASS LEYS, RYEGRASS SEED CROPS, SPRING BARLEY, TRITICALE, WINTER OATS, WINTER RYE.

Notes   **Efficacy**
• Apply pre-weed emergence on fine, firm seedbed or post-emergence to 1-2 true leaf stage of weeds, blackgrass (at higher rates) not beyond 2-leaf stage or after end Nov
• Do not use lower rates on soils with more than 10% organic matter
• Do not use higher rate for blackgrass control on soils with more than 4% organic matter or where seedbed preparation has not been preceded by ploughing

Crop Safety/Restrictions
• Maximum number of treatments 1 per crop
• Apply in winter wheat or autumn sown spring wheat pre- or post-emergence up to 6 wk after drilling, provided weeds in correct stage
• Apply pre-emergence in winter barley, but not in crops drilled after mid-Oct on sand in E. Anglia

- Only use pre-emergence in spring barley drilled to at least 25 mm and not on sand
- Apply pre-emergence in perennial ryegrass drilled before mid-Oct; not on sand
- Do not use on crops to be undersown with clover
- Use lower rate on direct drilled crops and roll before spraying where harrowed or disced after drilling

Special precautions/Environmental safety
- Harmful to fish or other aquatic life. Do not contaminate surface waters or ditches with chemical or used container

Protective clothing/Label precautions
- 21, 28, 29, 43, 53, 63, 67

Withholding period
- Keep livestock out of treated areas for at least 14 d after application

Latest application/Harvest Interval(HI)
- Pre-emergence for spring barley, winter oats, winter rye, triticale and autumn sown spring wheat; post-emergence before 30 Oct for winter barley and durum wheat (pre-emergence before 30 Nov at higher rate); post-emergence before 1st node detectable (GS 31) for winter wheat (pre-emergence up to 6 wk after drilling but before 30 Nov at higher rate); pre-emergence but before 15 Oct for perennial ryegrass (post-emergence but before 15 Sep at higher rate on undersown crops)

Approval
- Approved for aerial application on cereals, grass leys. See notes in Section 1

368   methiocarb
A stomach acting carbamate molluscicide and insecticide

Products					
1 Decoy	Bayer	2% w/w	PT	06535
2 Draza	Bayer	4% w/w	PT	00765
3 Draza ST	Bayer	4% w/w	PT	05315
4 Exit	Bayer	4% w/w	PT	07632

Uses   Leatherjackets in CEREALS *(reduction)* [1, 2, 4]. Slugs, snails in PROTECTED CROPS *(off-label)* [2]. Slugs, snails in BRASSICAS, CEREALS, FIELD CROPS, FRUIT CROPS, GRASSLAND, OILSEED RAPE, ORNAMENTALS, POTATOES [1, 2, 4]. Slugs in CEREALS *(seed admixture)* [3]. Strawberry seed beetle in STRAWBERRIES [1, 2, 4].

Notes   **Efficacy**
- Use as a surface, overall application when pests active (normally mild, damp weather), pre-drilling or post-emergence. May also be used on cereals or ryegrass as admixture with seed only at time of drilling [1, 2, 4]
- Also reduces populations of cutworms and millipedes [1, 2, 4]
- Best on potatoes in late Jul to Aug, on peas before start of pods formation [1, 2, 4]
- Apply to strawberries before strawing down, to blackcurrants at grape stage to prevent snails contaminating crop [1, 2, 4]
- See label for details of suitable application equipment

FOR FULL CONDITIONS OF USE ALWAYS READ THE PRODUCT LABEL

Crop Safety/Restrictions
* Do not allow pellets to lodge in crops as this may affect marketability [1, 2, 4]
* Product not suitable for broadcast surface application. For use only by seed merchants as a bait for admixture with cereal seed to be drilled [3]

Special precautions/Environmental safety
* Harmful if swallowed
* This product contains an anticholinesterase carbamate compound. Do not use if under medical advice not to work with such compounds
* Keep poultry out of treated areas for at least 7 d
* Harmful to game, wild birds and animals. Keep away from animals
* Harmful to fish or other aquatic life. Do not contaminate surface waters or ditches with chemical or used container

Protective clothing/Label precautions
* A
* 2, 5c, 18, 29, 36, 37, 42, 46, 53, 63, 67, 70 [1-4]; 72, 73, 74, 75, 76, 77 [3]; 78 [1, 4]; 78 [2, 3]

Latest application/Harvest Interval(HI)
* HI 7 d

Approval
* Approved for aerial application on all crops. See notes in Section 1 [1, 2, 4, 5135]
* Off-label Approval unlimited for use on all protected edible and non-edible crops (OLA 0186/92)[2]

---

369   methomyl
A carbamoyloxime insecticide available only in mixtures

---

370   methomyl + (Z)-9-tricosene
An insecticide bait for fly control in animal houses

Products	Golden Malrin Fly Bait	Sanofi	1:0.025% w/w	GR	05579

Uses   Flies in LIVESTOCK HOUSES.

Notes   **Efficacy**
* Product contains a pheromone attractant and can be applied as a scatter bait.
* Alternatively sprinkle granules on to polystyrene ceiling tiles, old egg trays or similar coated with adhesive and hang in infested areas out of reach of animals
* Re-apply every 4-5 d, or when granules disappear, until fly population has been reduced to acceptable levels. then apply every 7 d

Crop Safety/Restrictions
* Prevent access to bait by children and livestock
* Do not contaminate animal feed or use in areas where bait might come into contact with food

Special precautions/Environmental safety
- Product contains an anticholinesterase carbamate compound. Do not use if under medical advice not to work with such compounds
- Dangerous to fish or other aquatic life. Do not contaminate surface waters or ditches with chemical or used container

Protective clothing/Label precautions
- A, H
- 2, 25, 29, 36, 52, 63, 65, 81, 82

371   methoprene
An insect growth regulator

Products	Apex 5E	Sandoz Speciality	600 g/l	EC	05730

Uses   Sciarid flies in MUSHROOM HOUSES.

Notes   **Efficacy**
- Use as surface spray either on the compost or the casing
- Use pre-casing treatment only when early pest infestation occurs
- Product controls sciarids at pupal stage. Treatment will not control the generation of sciarids present at application

Crop Safety/Restrictions
- Maximum number of treatments 1 per spawning on compost or casing
- Product prevents development of adults from eggs laid up to 14 d before or after treatment. Alternative treatments may be necessary if adult numbers increase subsequently
- Pre-casing treatment may not prevent development from larvae able to move into untreated casing prior to pupation

Special precautions/Environmental safety
- Dangerous to fish or other aquatic life. Do not contaminate surface waters or ditches with chemical or used conatiner

Protective clothing/Label precautions
- A, C, H
- 18, 28, 30, 36, 37, 52, 63, 67

Latest application/Harvest Interval(HI)
- Before casing on the compost or immediately after casing

372   2-methoxyethyl mercury acetate
An organomercury fungicide seed treatment all approvals for which were revoked with effect from 31 March 1992

FOR FULL CONDITIONS OF USE ALWAYS READ THE PRODUCT LABEL

373   methyl bromide

A highly toxic alkyl halide fumigant for food and non-food commodities

Products	1 Methyl Bromide 100	Bromine & Chem.	99.7% w/w	GA	01336
	2 Sobrom BM 100	Brian Jones	100%	GA	04381

Uses

Food storage pests in FOOD STORAGE AREAS [1, 2]. Grain storage pests in GRAIN STORES [1]. Insect pests in STORED PLANT MATERIAL [1, 2].

Notes

Efficacy
* Chemical is released from containers as a highly toxic, colourless gas with only a slight odour
* Dose depends on temperature. At lower temperatures higher doses and longer exposures should be used
* See label for dosage and exposure required in different situations

Crop Safety/Restrictions
* Maximum number of treatments 1 per infestation [2]; 1 for flour, 2 for other food and non-food commodities [1]
* Air animal feed for at least 5 d following treatment
* Some types of produce, cut flowers and plants are susceptible to damage by methyl bromide fumigation. Check before treating
* Methyl bromide can cause taint to flours and foodstuffs containing reactive sulfur compounds. See label for guidance

Special precautions/Environmental safety
* Methyl bromide is subject to the Poisons Rules 1982 and the Poisons Act 1972. See notes in Section 1
* Not for retail sale
* Very toxic by inhalation. Causes burns. Fatal if swallowed.
* Wear suitable protective clothing (See HSE Guidance Note CS 12)
* Wear suitable respiratory equipment during fumigation and while removing sheets. Do *not* wear gloves or rubber boots
* Do not use application equipment incorporating natural rubber, PVC or aluminium, zinc, magnesium or their alloys
* Extinguish all naked flames, including pilot lights when applying fumigant
* Keep in original container, tightly closed in a safe place under lock and key. Do not re-use container and return to supplier when empty
* Keep unprotected persons out of treated area until it is shown by test to be clear of methyl bromide
* Harmful to fish or other aquatic life. Do not contaminate surface waters or ditches with chemical or used container [2]
* Special equipment is needed for application and the chemical may only be used by professional operators trained in its use and familiar with the precautionary measures to be observed
* All operations must be in accordance with the HSE guidance note CS 12. A minimum of 2 operators are required

Protective clothing/Label precautions
* See HSE Guidance Note CS 12 [1, 2]
* 3a, 8, 24, 24a, 43, 53, 79 [2]; 3b, 16, 18, 36, 37, 38, 64, 70 [1, 2]; 69, 79 [1]

Withholding period
* Keep animals and livestock out of treated area until it is shown by test to be clear of methyl bromide

Latest application/Harvest Interval(HI)
* 3 d before loading grains; 7 d before use of all other commodities and spaces [1]

Maximum Residue Levels (mg residue/kg food)
* oilseeds, cereals 0.1; all other products (except tree nuts, stone fruits, grapes, figs, pulses, animal products) 0.05

---

374   methyl bromide with amyl acetate
A highly toxic alkyl halide soil fumigant

Products	Fumyl-O-Gas	Brian Jones	99.7%	GA	04833

Uses Food storage pests in FOOD STORAGE. Nematodes, soil insects, soil-borne diseases, weed seeds in FIELD CROPS, PROTECTED CROPS.

Notes

Efficacy
* Chemical is released from containers as a highly volatile, highly toxic gas and must be released under gas-proof sheeting
* Amyl acetate is added as a warning odourant
* Soil must be clear of trash and cultivated to at least 40 cm before treatment
* Gas must be confined in soil for 48-96 h

Crop Safety/Restrictions
* Maximum number of treatments 1 per yr
* Crops may be planted 3-21 d after removal of gas-proof sheeting. For some crops a leaching irrigation is essential. See firms literature for details

Special precautions/Environmental safety
* Methyl bromide is subject to the Poisons Rules 1982 and the Poisons Act 1972. See notes in Section 1
* Product not for retail sale
* Special equipment is needed for application and the chemical may only be used by professional operators trained in its use and familiar with the precautionary measures to be observed. Refer to HSE Guidance Note CS 12 before applying
* Very toxic by inhalation. Causes burns. Risk of serious damage to health by prolonged exposure
* Keep in original container, tightly closed, in a safe place, under lock and key
* Wear suitable respiratory equipment during fumigation and while removing sheeting
* Do not wear gloves or rubber boots
* Ventilate treated areas thoroughly when gas has cleared
* Keep unprotected persons out of treated areas for at least 24 h
* Dangerous to game, wild birds, animals and bees
* Harmful to fish or other aquatic life. Do not contaminate surface waters or ditches with chemical or used container

---

FOR FULL CONDITIONS OF USE ALWAYS READ THE PRODUCT LABEL

Protective clothing/Label precautions
• See HSE Guidance Note CS 12
• 3b, 8, 9, 11, 16, 18, 23, 24, 25, 26, 27, 29, 36, 37, 38, 41, 45, 48, 53, 61, 64, 78

Withholding period
• Keep livestock out of treated areas for at least 24 h

Latest application/Harvest Interval(HI)
• Before sowing or planting

Maximum Residue Levels (mg residue/kg food)
• see methyl bromide entry

375   methyl bromide with chloropicrin
A highly toxic alkyl halide fumigant for treatment of soil and stored products

Products					
1 Methyl Bromide 98	Bromine & Chem.	980:20 g/l	GA	01335
2 Sobrom BM 98	Brian Jones	980:20 g/l	GA	04189

Uses

Grain storage pests in GRAIN STORES [1]. Nematodes, soil insects, soil-borne diseases, weed seeds in BULBS/CORMS, FLOWERS, NURSERY STOCK, STRAWBERRIES, VEGETABLES [1]. Nematodes, soil insects, soil-borne diseases, weed seeds in FIELD CROPS, ORCHARDS, PROTECTED CROPS [2]. Nematodes, soil insects, weed seeds in TURF [1].

Notes

Efficacy
• Chemical is released from containers as a highly volatile, highly toxic gas and for soil fumigation must be released under a plastic tarpaulin
• Chloropicrin is added as a warning odourant tear gas
• Soil must be clear of trash and cultivated to at least 40 cm before treatment
• Gas must be confined in soil for 48-96 h

Crop Safety/Restrictions
• Maximum number of treatments 1 per yr
• Crops may be planted 3-21 d after removal of tarpaulin. For some crops a leaching irrigation is essential. See firms literature for details

Special precautions/Environmental safety
• Methyl bromide and chloropicrin are subject to the Poisons Rules 1982 and the Poisons Act 1972. See notes in Section 1
• Product not for retail sale
• Special equipment is needed for application and the chemical may only be used by professional operators trained in its use and familiar with the precautionary measures to be observed. Refer to HSE Guidance Note CS 12 before applying
• Very toxic by inhalation, in contact with skin and if swallowed
• Irritating to eyes, respiratory system and skin
• Keep in original container, tightly closed, in a safe place, under lock and key
• Wear suitable respiratory equipment during fumigation and while removing sheeting
• Do *not* wear gloves or rubber boots
• Ventilate treated areas thoroughly when gas has cleared
• Keep unprotected persons out of treated areas for at least 24 h

• Harmful to fish or other aquatic life. Do not contaminate surface waters or ditches with chemical or used container [2]

Protective clothing/Label precautions
• See HSE Guidance Note CS 12 [1, 2]
• 3a, 3b, 3c, 6a, 6b, 6c, 16, 18, 36, 37, 38, 43, 64, 70 [1, 2]; 53, 79 [2]; 69, 79 [1]

Withholding period
• Keep livestock out of treated areas for at least 24 h

Latest application/Harvest Interval(HI)
• Before sowing or planting

Maximum Residue Levels (mg residue/kg food)
• see methyl bromide entry

---

376   metoxuron
A contact and residual urea herbicide for cereals and carrots

Products	Dosaflo	Sandoz	500 g/l	SC	00754

Uses
Annual dicotyledons, annual grasses, barren brome, blackgrass in DURUM WHEAT, TRITICALE, WINTER BARLEY, WINTER WHEAT. Annual dicotyledons, annual grasses, mayweeds in CARROTS.

Notes

Efficacy
• Best results achieved by application to weeds in seedling to young plant stage
• Best control of barren brome in autumn or winter from 1-3 leaf stage
• Effects on blackgrass reduced on soils of high organic matter or low pH

Crop Safety/Restrictions
• Maximum number of treatments 2 per crop for cereals with minimum of 6 wk between applications
• Apply to named winter cereal varieties from 2-leaf unfolded stage to before first node detectable (GS 12-31). See label for lists of resistant and susceptible varieties
• Apply to carrots after 2-true leaf stage. Do not spray when soil very dry or wet
• Do not spray cereals on sand or very light soils, or carrots on soils with more than 80% sand or less than 1% organic matter
• Do not apply during prolonged frosty weather, when frost or heavy rain imminent or ground waterlogged
• If treatment repeated in spring at least 6 wk must elapse after autumn/winter spray
• Do not roll for 7 d before or after spraying
• Do not cultivate after spraying or harrow or cultivate within 14 d beforehand
• Do not spray crops checked by pests, wind, frost or waterlogging until recovered
• No crop should be sown within 6 wk of treatment

Protective clothing/Label precautions
• 21, 28, 29, 54, 63, 66

FOR FULL CONDITIONS OF USE ALWAYS READ THE PRODUCT LABEL

Latest application/Harvest Interval(HI)
• Before 1st node detectable (GS 31) for cereals

377 metribuzin

A contact and residual triazinone herbicide for use in potatoes

Products					
1 Lexone 70DF	DuPont	70% w/w	WG	04991
2 Sencorex WG	Bayer	70% w/w	WG	03755

Uses Annual dicotyledons, annual grasses, volunteer oilseed rape in POTATOES.

Notes **Efficacy**
• May be applied pre- or post-emergence of crop. Best results achieved on weeds at cotyledon to 1-leaf stage
• Apply to moist soil with well-rounded ridges and few clods
• Activity reduced by dry conditions and on soils with high organic matter content
• Do not cultivate after treatment
• On fen and moss soils pre-planting incorporation to 10-15 cm gives increased activity
• With named maincrop and second early varieties on soils with more than 10% organic matter shallow pre- or post-planting incorporation may be used. See label for details
• Effective control using a programme of reduced doses is made possible by using a spray of smaller droplets, thus improving retention. See label for details

Crop Safety/Restrictions
• Maximum number of treatments 3 per crop subject to maximum permitted dose
• Apply pre-emergence only on named first earlies, pre- or post-emergence on named second earlies. On named maincrop varieties apply pre-emergence (except for certain varieties on sand or very light soils) or post-emergence before longest shoots reach 15 cm. See label for details
• On stony or gravelly soils there is risk of crop damage, especially if heavy rain falls soon after application
• When days are hot and sunny delay spraying until evening
• Some varieties may be sensitive to post-emergence treatment if crop under stress
• Ryegrass, cereals or winter beans may be sown in same season provided at least 16 wk elapsed after treatment and ground ploughed to 15 cm and thoroughly cultivated soon after harvest. Other crops may be sown normally in spring of next year
• Do not drill lettuce or radish in season following treatment

Protective clothing/Label precautions
• 21, 25, 28, 29, 54, 63, 67

Latest application/Harvest Interval(HI)
• Pre-emergence for earlies; before most advanced shoots reach 15 cm for maincrop

378 metsulfuron-methyl

A contact and residual sulfonylurea herbicide for use in cereals

Products

1 Ally	DuPont	20% w/w	WG	02977
2 Ally WSB	DuPont	20% w/w	WB	06588
3 Jubilee (WSB)	DuPont	20% w/w	WB	06082
4 Jubilee 20 DF	DuPont	20% w/w	WG	06136
5 Landgold Metsulfuron	Landgold	20% w/w	WG	06280
6 Lorate 20 DF	DuPont	20% w/w	WG	06135
7 Standon Metsulfuron	Standon	20% w/w	WG	05670

Uses   Annual dicotyledons, chickweed, mayweeds in LINSEED, TRITICALE [1-4, 6]. Annual dicotyledons, chickweed, mayweeds in BARLEY, OATS, WHEAT [1-7].

Notes   **Efficacy**
* Best results achieved on small, actively growing weeds up to 6-true leaf stage. Good spray cover is important
* Commonly used in tank-mixture on wheat and barley with other cereal herbicides to improve control of resistant dicotyledons (cleavers, fumitory, ivy-leaved speedwell), larger weeds and grasses. See label for recommended mixtures
* Always add metsulfuron-methyl product to tank first

Crop Safety/Restrictions
* Maximum number of treatments 1 per crop
* Product must be applied only after 1 Feb
* Apply to wheat and oats from 2-leaf (GS 12), to barley, and triticale from 3-leaf stage (GS 13) until flag-leaf fully emerged (GS 39). Do not spray Igri barley before leaf sheath erect stage (GS 30)
* Apply to linseed from first pair of true leaves unfolded up to 30 cm high or before flower bud visible, whichever is the sooner
* Recommendations for oats, triticale and linseed apply to product alone
* Do not use on cereal crops undersown with grass or legumes
* Do not use on any crop suffering stress from drought, waterlogging, frost, deficiency, pest or disease attack or apply within 7 d of rolling
* Do not spray in tank mixture, or in sequence, with a product containing any other sulfonylurea
* Take extreme care to avoid drift onto broad-leaved crops or contamination of land planted or to be planted with any crops other than cereals
* Use recommended procedure to clean out spraying equipment
* Only cereals, oilseed rape, field beans or grass may be sown in same calendar year after treating cereals with the product alone. Other restrictions apply to tank mixtures. See label for details. In the event of crop failure sow only wheat within 3 mth after treatment. Only cereals should be planted within 16 mth of applying to a linseed crop

Special precautions/Environmental safety
* Extremely dangerous to aquatic higher plants. Do not contaminate surface waters or ditches with chemical or used container

FOR FULL CONDITIONS OF USE ALWAYS READ THE PRODUCT LABEL

- Do not allow direct spray from field mounted/drawn hydraulic sprayers to fall within 6 m, or from hand-held sprayers to within 2 m, of surface waters or ditches. Direct spray away from water
- Take extreme care to avoid damage by drift onto broad-leaved plants outside the target area, onto surface waters or ditches or onto land intended for cropping
- Do not open the water soluble bag or touch it with wet hands or gloves [2, 3]

Protective clothing/Label precautions
- 18, 29 [1, 3-7]; 21, 28, 36, 37, 51a, 54a, 63, 67 [1-7]; 32 [2, 3]

Latest application/Harvest Interval(HI)
- Before flag leaf sheath extending stage for cereals (GS 41). Before flower bud visible or up to 30 cm tall, whichever is earlier, for linseed

---

379 metsulfuron-methyl + thifensulfuron-methyl

A contact residual and translocated sulfonylurea herbicide mixture for use in cereals

Products Harmony M DuPont 7:68% w/w WG 03990

Uses Annual dicotyledons, cleavers, field pansy, polygonums, speedwells in SPRING BARLEY, SPRING WHEAT, WINTER WHEAT.

Notes **Efficacy**
- Best results by application to small, actively growing weeds up to 6-true leaf stage
- Effectiveness may be reduced if heavy rain occurs within 4 h of application or if soil conditions very dry
- Tank-mixture with mecoprop, mecoprop-P or reduced rate of fluroxypyr improves control of cleavers and other problem weeds
- When used in tank-mixes Harmony M must be added to spray tank first and fully dispersed before adding other product

Crop Safety/Restrictions
- Maximum number of treatments 1 per crop
- Apply in spring to crops from 3-leaf stage to flag-leaf fully emerged (GS 13-39)
- Do not use on durum wheat or winter barley or on any crop suffering stress from drought, waterlogging, frost, deficiency, pest or disease attack
- Do not spray in tank mixture, or in sequence, with a product containing any other sulfonylurea herbicide
- Do not apply within 7 d of rolling
- Take extreme care to avoid drift onto broad-leaved crops or contamination of land planted or to be planted with any crops other than cereals
- Use recommended procedure to clean out spraying equipment
- Only cereals, oilseed rape, field beans or grass may be sown in same calendar year after treatment. In the event of crop failure sow only wheat within 3 mth after treatment

Special precautions/Environmental safety
- Extremely dangerous to aquatic higher plants. Do not contaminate surface waters or ditches with chemical or used container

- Do not allow direct spray from field mounted/drawn hydraulic sprayers to fall within 6 m, or from hand-held sprayers to within 2 m, of surface waters or ditches. Direct spray away from water

Protective clothing/Label precautions
- 21, 28, 29, 51a, 54a, 63, 67

Latest application/Harvest Interval(HI)
- Before flag leaf ligule/collar just visible (GS 39)

380   monolinuron

A residual urea herbicide for potatoes, leeks and French beans

Products					
1 Arresin	AgrEvo	200 g/l	EC	07303
2 Arresin	Hoechst	200 g/l	EC	00118

Uses

Annual dicotyledons, annual grasses, annual meadow grass, fat-hen, polygonums in FRENCH BEANS, LEEKS, POTATOES.

Notes

Efficacy
- Best results achieved by application on moist, firm, clod-free seedbed pre-weed emergence or on seedlings up to 2-3 leaf stage
- Light rain after application can improve control
- In potatoes or French beans on fen or peat soils apply post-weed emergence but before emergence of crop

Crop Safety/Restrictions
- Maximum number of treatments 1 per crop
- May be used in tank-mixture with Challenge on potatoes up to 10% emergence on earlies, 40% emergence on maincrop. Do not apply alone post-emergence
- On early sown French beans apply at least 5 d pre-emergence
- Some bean cultivars may be sensitive on light soils, with application close to emergence with heavy rain after treatment or in unevenly drilled crop. See label for list of cultivars
- Apply to transplanted or direct-sown leeks after established and 180 mm high
- Only apply pre-emergence to crops grown under polythene
- Do not use on very light soils
- Do not sow or plant other crops within 2-3 mth, depending on rate applied. Lettuce must not be sown in same season

Special precautions/Environmental safety
- Irritating to skin, eyes and respiratory system
- Flammable
- Extremely dangerous to fish or other aquatic life. Do not contaminate surface waters or ditches with chemical or used container

Protective clothing/Label precautions
- A, C, H, J, M
- 6a, 6b, 6c, 12c, 16, 18, 21, 28, 29, 37, 51, 63, 66, 70

FOR FULL CONDITIONS OF USE ALWAYS READ THE PRODUCT LABEL

Latest application/Harvest Interval(HI)
* Pre-emergence for French beans; 3 wk after transplanting for leeks; pre-emergence or up to 10% emergence in seed or early potatoes, pre-emergence or up to 40% emergence (only in mixture with Challenge) in maincrop potatoes

Approval
* Approved for aerial application on potatoes, French beans and leeks. See notes in Section 1

381   monolinuron + paraquat

A contact and residual herbicide for potatoes and French beans

Products					
	Gramonol Five	Zeneca	154:110 g/l	SC	06673

Uses
Annual dicotyledons, annual grasses, annual meadow grass, blackgrass, wild oats in FRENCH BEANS, POTATOES.

Notes

Efficacy
* Intended for use where some weeds have emerged and further germination expected
* Best results by application to fine, firm seedbed with weeds up to 2-3 leaf stage
* Residual activity assisted by moist soil conditions and light rain after application
* Residual action greatest on light soils, very little on highly organic soils

Crop Safety/Restrictions
* Maximum number of treatments 1 per crop
* Do not use on potatoes if any plants are 15 cm or more high
* Apply to certain cultivars of French bean after sowing but at least 3 d before crop emergence. See label for details of cultivars which may be treated
* Do not use on weak or unhealthy crops
* Crops may be sensitive on light, sandy soils, with application close to emergence, with heavy rain after treatment or in unevenly drilled crop

Special precautions/Environmental safety
* Paraquat is subject to the Poisons Rules 1982 and the Poisons Act 1972. See notes in Section 1
* Toxic if swallowed
* Harmful in contact with skin. Irritating to eyes, skin and respiratory system
* Keep in original container, tightly closed, in a safe place, under lock and key
* Harmful to animals
* Extremely dangerous to fish or other aquatic life. Do not contaminate surface waters or ditches with chemical or used container

Protective clothing/Label precautions
* A, C, H, J, M
* 4c, 5a, 6a, 6b, 6c, 17, 18, 21, 28, 29, 36, 37, 41, 51, 64, 65, 70, 79

Withholding period
* Keep all livestock out of treated areas for at least 24 h

Latest application/Harvest Interval(HI)
* 3 d pre-emergence of crop for French beans; before 10% emergence for early potatoes, 20% emergence for maincrop

Maximum Residue Levels (mg residue/kg food)
* see paraquat entry

382   myclobutanil
A systemic, protectant and curative conazole fungicide

Products

1 Systhane 6 Flo	Promark	60 g/l	EW	07334
2 Systhane 6 W	PBI	6% w/w	WP	04571

Uses

American gooseberry mildew in BLACKCURRANTS, GOOSEBERRIES [1]. Black spot, mildew, rust in ROSES [2]. Powdery mildew, scab in APPLES [1]. Powdery mildew in HOPS *(off-label)*, STRAWBERRIES [1]. Scab in PEARS [1].

Notes

Efficacy
* Best results achieved when used as part of routine preventive spray programme from bud burst to end of flowering in apples and pears and from just before the signs of mildew infection in blackcurrants and gooseberries [1]
* In strawberries commence spraying at, or just prior to, first flower [1]
* Spray at 7-14 d intervals depending on disease pressure and dose applied [1]
* For improved scab control in post-blossom period tank-mix with Karamate Dry Flo (mancozeb) or captan [1]
* Apply alone from mid-Jun for control of secondary mildew on apples and pears [1]
* On roses spray at first signs of disease and repeat every 2 wk. In high risk areas spray when leaves emerge in spring, repeat 1 wk later and then continue normal programme [2]

Crop Safety/Restrictions
* Total application for season in apples and pears should not exceed 15 l/ha [1] (ie 10 sprays at highest recommended rate) and 9 l/ha in soft fruit [1]

Special precautions/Environmental safety
* Harmful to fish or other aquatic life. Do not contaminate surface waters or ditches with chemical or used container [1]
* Do not harvest apples, pears, blackcurrants or gooseberries for at least 14 d, and strawberries for at least 3 d, after last application [1]

Protective clothing/Label precautions
* A, H, J [1]
* 21, 35(strawberries 3 d; remainder 14 d), 36, 37 [1]; 29, 53, 67 [1, 2]; 63 [2]

Latest application/Harvest Interval(HI)
* HI 14 d (apples, pears, blackcurrants, gooseberries); 3 d (strawberries)

Approval
* Off-label approval unlimited for use on hops (OLA 0849/95)[1]

FOR FULL CONDITIONS OF USE ALWAYS READ THE PRODUCT LABEL

383   2-(1-naphthyl)acetic acid

A plant growth regulator and herbicide for sucker control

See also 4-indol-3-yl-butyric acid + 2-(1-naphthyl)acetic acid with dichlorophen

Products					
1 Rhizopon B Powder (0.1%)	Fargro	0.1% w/w	SP	07133
2 Rhizopon B Powder (0.2%)	Fargro	0.2% w/w	SP	07134
3 Rhizopon B Tablets	Fargro	25 mg ai	TB	07135
4 Tipoff	Unicrop	57.8 g/l	EC	05878

Uses   Control of water shoots, sucker control in APPLES, PEARS [1-4]. Rooting of cuttings in ORNAMENTALS [1-3]. Sucker control in CHERRIES, PLUMS, RASPBERRIES [1-4].

Notes   **Efficacy**
- Dip moistened base of cuttings into powder immediately before planting [1, 2]
- Spray root suckers of top fruit trees when 10-15 cm high. Best results obtained on sucker regrowth cut back the previous winter. A second application can be made later in year (Jul) if there are any signs of regrowth
- Spray water shoots of apple and pear when about 10 cm long [4]
- Spray raspberry suckers in alleys when 10-20 cm high and repeat if necessary [4]
- It is important that foliage of suckers is thoroughly wetted [4]
- See label for details of concentrations recommended for promotion of rooting in cuttings of different species [3]
- Dip prepared cuttings in solution for 4-24 h depending on species [3]

Crop Safety/Restrictions
- Maximum number of treatments 2 per yr
- Avoid spray drift on to non-target areas of fruit trees
- In raspberries only spray suckers in alleys, not replacement or fruiting canes or suckers in the crop rows

Protective clothing/Label precautions
- 21, 28, 29, 70 [1-3]; 30 [4]; 54, 63, 67 [1-4]

384   (2-naphthyloxy)acetic acid

A plant growth regulator for setting tomato fruit

Products				
Betapal Concentrate	Vitax	16 g/l	SL	00234

Uses   Increasing fruit set in TOMATOES.

Notes   **Efficacy**
- Apply with any fine sprayer or syringe when first half dozen flowers are open
- Spray actual trusses in flower and repeat every 2 wk

Crop Safety/Restrictions
- Do not spray growing head of tomato plants

385 napropamide

A soil applied amide herbicide for oilseed rape, fruit and woody ornamentals

Products	Devrinol	RP Agric.	450 g/l	SC	06195

Uses

Annual dicotyledons, annual grasses, cleavers, groundsel in BLACKCURRANTS, CONTAINER-GROWN WOODY ORNAMENTALS, ESTABLISHED WOODY ORNAMENTALS, GOOSEBERRIES, NEWLY PLANTED WOODY ORNAMENTALS, RASPBERRIES, STRAWBERRIES, WINTER OILSEED RAPE.

Notes

Efficacy

* Apply to winter oilseed rape as pre-drilling incorporated treatment in tank-mixture with trifluralin. Addition of TCA recommended where volunteer cereals serious
* Incorporate to 25 mm within 30 min. See label for details of recommended procedure
* With minimum cultivation spray onto stubble (allow 24 h after burning) and incorporate as part of surface cultivation
* Apply around established fruit crops as pre-weed emergence treatment in Nov-Feb period, in woody ornamentals from Nov to Apr
* Do not use on soils with more than 10% organic matter
* Do not cultivate after treatment

Crop Safety/Restrictions

* Maximum number of treatments 1 per crop or yr
* Apply up to 14 d prior to drilling winter oilseed rape
* Do not use on sand. Do not use on strawberries on very light soil
* Bush and cane fruit must be established for at least 10 mth before spraying
* Apply to named strawberry cultivars established for at least one season or to maiden crops as long as planted carefully and no roots exposed
* Apply immediately after planting strawberries in Nov-Feb. Do not treat before Nov if planted in other mth. Do not use on protected crops
* Only oilseed rape, swedes, fodder turnips, brassicas or potatoes should be sown within 12 mth of application
* Soil should be mould-board ploughed to a depth of at least 200 mm before drilling or planting a following crop
* Many woody ornamentals have been treated successfully, but not yellow or golden ornamentals or container-grown alpines. If in doubt treat small area initially to test safety

Special precautions/Environmental safety

* Irritating to skin and eyes
* Harmful to fish or other aquatic life. Do not contaminate surface waters or ditches with chemical or used container

Protective clothing/Label precautions

* A, C
* 6a, 6b, 18, 21, 30, 36, 37, 53, 63, 65, 70

FOR FULL CONDITIONS OF USE ALWAYS READ THE PRODUCT LABEL

Latest application/Harvest Interval(HI)
- Pre-sowing for winter oilseed rape; before end Feb for strawberries, bush and cane fruit; before end of Apr for field and container grown ornamental trees and shrubs

386 nicotine

A general purpose, non-persistent, contact, alkaloid insecticide

Products					
1 Campbell's Nicosoap	MTM Agrochem.	75 g/l	LI	00416
2 Nico Soap	United Phosphorus	75 g/l	LI	07517
3 Nicotine 40% Shreds	DowElanco	40% w/w	FU	05725
4 XL-All Insecticide	Vitax	70 g/l	LI	02369
5 XL-All Nicotine 95%	Vitax	950 g/l	LI	07402

Uses Aphids, capsids, glasshouse whitefly, leaf miners, leafhoppers, sawflies, thrips, woolly aphid in PROTECTED CROPS [3-5]. Aphids, capsids, leaf miners, leafhoppers, sawflies, thrips, woolly aphid in CORMS, FLOWER BULBS, FLOWERS, FRUIT CROPS, VEGETABLES [4, 5]. Aphids, leaf miners in CHRYSANTHEMUMS [1, 2]. Aphids in APPLES, BROAD BEANS, DWARF BEANS, LEAF BRASSICAS, LETTUCE, ROSES, STRAWBERRIES, TOMATOES [1, 2]. General insect control in FRENCH BEANS *(off-label)* [4]. Insect pests in CHICORY *(off-label)*[3]. Leaf miners in CELERY [1, 2]. Leaf miners in ORNAMENTALS [1]. Mushroom flies in MUSHROOMS [3]. Potato virus vectors in CHITTING POTATOES [3]. Sawflies in GOOSEBERRIES [1, 2].

Notes **Efficacy**
- Apply as foliar spray, taking care to cover undersides of leaves and repeat as necessary or dip young plants, cuttings or strawberry runners before planting out
- Best results achieved by spraying at air temperatures above 16°C
- Fumigate glasshouse crops and mushrooms at temperatures of at least 16°C. See label for details of recommended fumigation procedure [3]
- Potatoes may be fumigated in chitting houses to control virus-spreading aphids [3]
- May be used in integrated control systems to give partial control of organophosphorus-, organochlorine- and pyrethroid-resistant whitefly

Crop Safety/Restrictions
- Maximum number of treatments 3 per crop for protected vegetables and hops [1, 2]
- On plants of unknown sensitivity test first on a small scale

Special precautions/Environmental safety
- Nicotine is subject to the Poisons Rules 1982 and the Poisons Act 1972. See notes in Section 1
- Harmful in contact with skin, by inhalation or if swallowed [1-4]. Toxic in contact with skin, by inhalation or if swallowed [2370]
- Keep unprotected persons out of treated glasshouses for at least 12 h
- Wear overall, hood, rubber gloves and respirator if entering glasshouse within 12 h of fumigating [3]
- Highly flammable [3]
- Dangerous to livestock.
- Dangerous to game, wild birds and animals
- Harmful to bees. Do not apply to crops in flower or to those in which bees are actively foraging. Do not apply when flowering weeds are present

- Dangerous to fish or other aquatic life. Do not contaminate surface waters or ditches with chemical or used container

Protective clothing/Label precautions
- A, C, H, K, M [5]; A, C [2]; A, D, H, J [3]; A [1, 4]
- 4a, 4b, 4c, 79 [5]; 5a, 5b, 5c, 14, 16, 18, 25, 28, 29, 36, 37, 38, 45, 50, 52, 63, 66 [1-5]; 21, 23, 27, 35, 41, 65, 67, 70 [1, 3-5]; 22, 40 [2]; 64 [3, 5]; 78 [1-4]

Withholding period
- Keep all livestock out of treated areas for at least 12 h

Latest application/Harvest Interval(HI)
- HI 2 d [1, 2, 4, 5]; 24 h (2 d for chicory) [3]

Approval
- Off-label Approval unlimited for use on French beans (OLA 0080/92)[4]; to Mar 1996 for use on chicory in forcing shed (OLA 1334/93)[3]

---

387   nuarimol

A systemic protectant and eradicant pyrimidine fungicide

See also captan + nuarimol

Products					
	Flarepath	Chemtech	90 g/l	EC	05502

Uses
Net blotch, powdery mildew, rhynchosporium in BARLEY. Powdery mildew, septoria in WHEAT. Powdery mildew in OATS.

Notes

Efficacy
- Apply prior to or at first signs of disease, usually from first node detectable stage (GS 31), with a second application if necessary
- In tank-mixes products complement action of other fungicides against leaf and ear disease complexes. See label for details
- Mildew control may be reduced if rain falls before product has dried on leaves

Crop Safety/Restrictions
- Maximum number of treatments 1 per crop for oats and wheat; 2 per crop for barley

Special precautions/Environmental safety
- Irritating to eyes and skin
- Flammable
- Harmful to fish or other aquatic life. Do not contaminate surface waters or ditches with chemical or used container

Protective clothing/Label precautions
- A, C
- 6a, 6b, 12c, 18, 21, 28, 29, 36, 37, 53, 60, 63, 66

FOR FULL CONDITIONS OF USE ALWAYS READ THE PRODUCT LABEL

Latest application/Harvest Interval(HI)
* Before flowering (GS 60) for oats and wheat; before boots swollen (GS 45) for barley

388 octhilinone

An isothiazolone fungicide for treating canker and tree wounds

Products Pancil T Rohm & Haas 1% w/w PA 01540

Uses Canker, pruning wounds, silver leaf in APPLES, PEARS, TREE FRUIT, WOODY ORNAMENTALS. Pruning wounds, silver leaf in CHERRIES, PLUMS.

Notes **Efficacy**
* Treat canker by removing infected shoots, cutting back to healthy tissue and painting evenly over wound
* Apply to pruning wounds immediately after cutting. Cover all cracks in bark and ensure that coating extend beyond edges of wound
* Best results achieved by application in dry conditions

Crop Safety/Restrictions
* Only apply after harvest and before bud burst
* Do not apply to grafting cuts or bud. Do not apply under freezing conditions

Special precautions/Environmental safety
* Irritating to skin, risk of serious damage to eyes
* May cause sensitization by skin contact
* Dangerous to fish or other aquatic life. Do not contaminate surface waters or ditches with chemical or used container

Protective clothing/Label precautions
* A, C
* 6b, 9, 10a, 18, 22, 25, 28, 29, 36, 37, 52, 60, 63, 65, 70, 78

389 ofurace

A phenylamide (acylalanine) fungicide available only in mixtures
See also mancozeb + ofurace

390 oxadiazon

A residual and contact oxadiazolone herbicide for fruit and ornamentals

Products
1 Ronstar 2G RP Agric. 2% w/w GR 06492
2 Ronstar Liquid RP Agric. 250 g/l EC 06493

Uses Annual dicotyledons, annual grasses, bindweeds, cleavers, knotgrass in APPLES, BLACKCURRANTS, GOOSEBERRIES, GRAPEVINES, HOPS, PEARS, RASPBERRIES, WOODY ORNAMENTALS [2]. Annual dicotyledons, annual grasses in CONTAINER-GROWN ORNAMENTALS [1].

Notes

Efficacy
* Rain or overhead watering is needed soon after application for effective results
* Pre-emergence activity reduced on soils with more than 10% organic matter and, in these conditions, post-emergence treatment is more effective [2]
* Best results on bindweed when first shoots are 10-15 cm long
* Do not cultivate after treatment [2]

Crop Safety/Restrictions
* See label for list of ornamental species which may be treated with granules. Treat small numbers of other species to check safety. Do not treat hydrangea or spiraea
* Do not treat container stock under glass or use on plants rooted in media with high sand or non-organic content. Do not apply to plants with wet foliage
* Apply spray to apples and pears from Jan to Jul, avoiding young growth
* Treat bush fruit from Jan to bud-break, avoiding bushes, grapevines in Feb/Mar before start of new growth or in Jun/Jul avoiding foliage
* Treat hops cropped for at least 2 yr in Feb or in Jun/Jul after deleafing
* Treat woody ornamentals from Jan to Jun, avoiding young growth. Do not spray container stock overall
* Do not use more than 8 l/ha in any 12 mth period [2]

Special precautions/Environmental safety
* Irritating to eyes
* Flammable [2]
* Dangerous to fish or other aquatic life. Do not contaminate surface waters or ditches with chemical or used container

Protective clothing/Label precautions
* A, C [1, 2]
* 6a, 18, 21, 26, 27, 30, 36, 37, 52, 63, 67, 70 [1, 2]; 12c [2]

391   oxadixyl

A phenylamide (acylalanine) fungicide available only in mixtures

See also cymoxanil + mancozeb + oxadixyl
        mancozeb + oxadixyl

392   oxamyl

A soil-applied, systemic carbamoyloxime nematicide and insecticide

Products	Vydate 10G	DuPont	10% w/w	GR	02322

Uses   American serpentine leaf miner, non-indigenous leaf miners, stem nematodes in GARLIC *(off-label)*. American serpentine leaf miner, non-indigenous leaf miners in AUBERGINES *(off-label)*, BEETROOT *(off-label)*, BROAD BEANS *(off-label)*, CUCURBITS *(off-label)*, ORNAMENTALS *(off-label)*, PEPPERS *(off-label)*, SHALLOTS *(off-label)*, SOYA BEAN *(off-label)*, TOMATOES *(off-label)*. Aphids, docking disorder vectors, mangold fly, millipedes, pygmy beetle in SUGAR BEET. Aphids, free-living nematodes, potato cyst nematode, spraing

FOR FULL CONDITIONS OF USE ALWAYS READ THE PRODUCT LABEL

vectors in POTATOES. Pea and bean weevils, pea cyst nematode in PEAS. Stem and bulb nematodes in ONIONS. Stem nematodes in PROTECTED GARLIC *(off-label)*, PROTECTED ONIONS *(off-label)*.

Notes

Efficacy
* Apply granules with suitable applicator before drilling or planting. See label for details of recommended machines
* In potatoes incorporate thoroughly to 10 cm and plant within 3-4 d
* In sugar beet apply in seed furrow at drilling
* In onions apply in seed furrow or use bow-wave technique with higher rates
* In peas apply during final seed-bed preparation and incorporate to 10-15 cm immediately after treatment. Drill within 3-4 d

Crop Safety/Restrictions
* Maximum number of treatments 1 per crop or yr

Special precautions/Environmental safety
* This product contains an anticholinesterase carbamate compound. Do not use if under medical advice not to work with such compounds
* Harmful by inhalation or if swallowed
* Keep in original container, tightly closed, in a safe place, under lock and key
* Dangerous to fish or other aquatic life. Do not contaminate surface waters or ditches with chemical or used container
* Dangerous to game and wildlife. Cover granules completely. Bury spillages
* Wear protective gloves if handling treated compost or soil within 2 wk after treatment
* Allow at least 12 h, followed by at least 1 h ventilation, before entry of unprotected persons into treated glasshouses

Protective clothing/Label precautions
* A, B, C or D, H, K, M
* 2, 5b, 5c, 14, 16, 18, 22, 25, 28, 29, 35, 36, 37, 45, 52, 64, 67, 70, 79

Latest application/Harvest Interval(HI)
* At drilling/planting for vegetables; before drilling/planting for potatoes and peas.
* HI tomatoes, ornamentals 2 wk

Approval
* Off-label Approval to Jan 1998 for use on protected garlic, shallots, beetroot and cucurbits to control American serpentine leaf miner and other non-indigenous leaf miners (OLA 0021/93); unlimited for use on outdoor and protected ornamentals, protected tomatoes, aubergines, peppers, soya beans and broad beans (OLA 0020/93); unlimited for use on protected onions and garlic (OLA 0925/94); unlimited for use on outdoor garlic for stem nematode control (OLA 0163/92)

393   oxycarboxin

A protectant, eradicant and systemic anilide fungicide

Products					
1 Plantvax 20	Uniroyal	200 g/l	EC	01600
2 Plantvax 75	Fargro	75% w/w	WP	01601
3 Ringmaster	RP Environ.	200 g/l	EC	05334

Uses Fairy rings in AMENITY GRASS, SPORTS TURF [3]. Rust in CHRYSANTHEMUMS, PELARGONIUMS, PROTECTED CARNATIONS, ROSES [2]. Yellow rust in WHEAT [1].

Notes **Efficacy**
- Apply in wheat at first signs of infection. Where heavy infections occur a second application may be made after a minimum interval of 21 d
- Add wetter for first spray on carnations, geraniums and roses if rust already established. May be applied with fogging machine mixed with suitable carrier [2]
- Can be used as a drench on carnations grown in peat bags [2]
- Apply to fairy rings after spiking. On dry soil water before spraying or add wetter. Best results when fungus in active growth [3]

Crop Safety/Restrictions
- On chrysanthemums apply alone, not within 2 d of other sprays. Do not apply spraying oil for 14 d before or after. Can scorch chrysanthemum leaves when taken up by roots. Avoid spraying when blooms open [2]

Special precautions/Environmental safety
- Harmful if swallowed. Irritating to eyes, skin and respiratory system [1, 3]
- Not to be used on food crops [3]
- Dangerous to fish or other aquatic life. Do not contaminate surface waters or ditches with chemical or used container

Protective clothing/Label precautions
- A, C, G, J, L, M [2]; A, C [1, 3]
- 5c, 6a, 6b, 6c, 18, 66, 78 [1, 3]; 14 [1, 2]; 16 [1]; 21, 28, 29, 36, 37, 52, 63, 70 [1-3]; 34 [3]; 67 [2]

Latest application/Harvest Interval(HI)
- HI 21 d [1]

Approval
- Approved for application with thermal fogging machine. See notes in Section 1 on ULV application [2]

---

394   paclobutrazol

A conazole plant growth regulator for ornamentals and fruit

See also dicamba + paclobutrazol

Products

1 Bonzi	Zeneca	4 g/l	SC	06640
2 Cultar	Zeneca	250 g/l	SC	06649

Uses Controlling vigour, increasing fruit set in APPLES, PEARS [2]. Improving colour in POINSETTIAS [1]. Increasing flowering, stem shortening in AZALEAS, BEDDING PLANTS, BEGONIAS, CHRYSANTHEMUMS, KALANCHOES, LILIES, ROSES, TULIPS [1].

Notes **Efficacy**
- Chemical is active via both foliage and root uptake

FOR FULL CONDITIONS OF USE ALWAYS READ THE PRODUCT LABEL

- Apply as spray to produce compact pot plants and to improve bract colour of poinsettias [1]
- Apply as drench to reduce flower stem length of potted tulips [1]
- Timing is critical and varies with species. See label for details [1]
- Apply to apple and pear trees under good growing conditions as pre-blossom spray (apples only) and post-blossom at 7-14 d intervals [2]
- Timing and dose of orchard treatment vary with species and cultivar. See label [2]

Crop Safety/Restrictions
- Maximum number of treatments 4 per crop for pot-grown ornamentals [1]; 7 per crop for apples; 6 per crop for pears [2]
- Do not use on border grown chrysanthemums [1]
- Do not use on trees of low vigour or under stress [2]
- Do not tank mix with or apply on same day as Thinsec [2]
- With 7 d spray programme on fruit trees use up to 6 sprays, with 10 d programme up to 4 sprays with 14 d programme up to 3 sprays [2]
- Do not use on trees from green cluster to 2 wk after full petal fall [2]
- Do not use in underplanted orchards, nor for 3 yr before grubbing [2]

Special precautions/Environmental safety
- Do not use on food crops [1]
- Harmful to fish or other aquatic life. Do not contaminate surface waters or ditches with chemical or used container

Protective clothing/Label precautions
- A, C [2]
- 21, 34 [1]; 29, 53, 60, 63, 66, 70 [1, 2]; 35, 43 [2]

Withholding period
- Keep livestock out of treated areas for at least 2 yr after treatment [2]

Latest application/Harvest Interval(HI)
- HI apples, pears 6 wk [2]

395   paraffin oil (commodity substance)
An agent for the control of birds by egg treatment

Products	paraffin oil	various	OL

Uses   Birds in MISCELLANEOUS SITUATIONS *(egg treatment)*.

Notes   **Efficacy**
- Egg treatment should be undertaken as soon as clutch is complete
- Eggs should be treated by complete immersion in liquid paraffin

Crop Safety/Restrictions
- Treat eggs once only
- Use to control eggs of birds covered by licences issued by the Agriculture and Environment Departments under Section 16(1) of the Wildlife and Countryside Act (1981)

Protective clothing/Label precautions
- A, C

Approval
• Only to be used where a licence has been approved in accordance with Section 16(1) of the Wildlife and Countryside Act (1981)

396   paraquat

A non-selective, non-residual contact bipyridilium herbicide

See also *diquat + paraquat*
          diuron + paraquat
          monolinuron + paraquat

Products

1 Barclay Total	Barclay	200 g/l		SL	05260
2 Dextrone X	Nomix-Chipman	200 g/l		SL	00687
3 Gramoxone 100	Zeneca	200 g/l		SL	06674
4 Scythe LC	Cyanamid	200 g/l		SC	05877
5 Speedway	ICI Professional	8% w/w		WG	02732
6 Speedway	Zeneca Prof.	8% w/w		WG	06861
7 Standon Paraquat	Standon	200 g/l		SL	05621
8 Stefes Paraquat	Stefes	200 g/l		SL	05134

Uses   Annual dicotyledons, annual grasses, barren brome, creeping bent, volunteer cereals, wild oats in FIELD CROPS *(stubble treatment)* [1, 3, 4, 7, 8]. Annual dicotyledons, annual grasses, chemical stripping, creeping bent, perennial ryegrass, rough meadow grass in HOPS [1, 3, 4]. Annual dicotyledons, annual grasses, creeping bent, firebreak desiccation in FORESTRY [1-3]. Annual dicotyledons, annual grasses, creeping bent, perennial ryegrass, rough meadow grass in NON-CROP AREAS [1-8]. Annual dicotyledons, annual grasses, creeping bent, perennial ryegrass, rough meadow grass in BLACKCURRANTS, GOOSEBERRIES, ORCHARDS, RASPBERRIES [1, 3, 4]. Annual dicotyledons, annual grasses, creeping bent, volunteer cereals in FIELD CROPS *(minimum cultivation)* [1, 3, 4, 7, 8]. Annual dicotyledons, annual grasses, creeping bent in WOODY ORNAMENTALS [2, 5, 6]. Annual dicotyledons, annual grasses, creeping bent in FLOWER BULBS [1-4]. Annual dicotyledons, annual grasses, creeping bent in FORESTRY TRANSPLANT LINES [1-3]. Annual dicotyledons, annual grasses, volunteer cereals in POTATOES, ROW CROPS, SUGAR BEET [1, 3, 4, 7, 8]. Annual dicotyledons, annual grasses in FOREST NURSERY BEDS *(stale seedbed)* [1-3]. Annual dicotyledons, annual grasses in GRAPEVINES [3, 4]. Annual dicotyledons, annual grasses in HERBS *(off-label)*, LUCERNE *(off-label)*, MINT *(off-label)* [3]. Annual grasses, creeping bent, perennial ryegrass, rough meadow grass in FIELD CROPS *(sward destruction/direct drilling)* [1, 3, 4, 7, 8]. Runner desiccation in STRAWBERRIES [1, 3, 4].

Notes   **Efficacy**
• Apply to green weeds preferably less than 15 cm high. Spray is rainfast after 10 min (15 min [7])
• Addition of wetter recommended with lower application rates [1, 3-8]
• Only use clean water for mixing up spray
• Allow at least 4 h before cultivating, leave overnight if possible
• Spray in autumn to suppress couch when shoots have 2 leaves, repeat as necessary and plough after last treatment

FOR FULL CONDITIONS OF USE ALWAYS READ THE PRODUCT LABEL

- For direct-drilling land should be free of perennial weeds and protection against slugs should be provided. Allow 7-10 d before drilling into sprayed grass
- Best time for use in top fruit is Nov-Apr
- For forestry fire-break use apply in Jul-Aug and fire 7-10 d later

Crop Safety/Restrictions
- Maximum number of treatments 1 per yr for lucerne, outdoor lettuce, protected vegetables and ornamentals; 2 per yr for mint and leafy herbs
- Chemical is inactivated on contact with soil. Crops can be sown or planted soon after spraying on most soils 24 h [2], 4 h [3, 7], after 3 d on sandy or immature peat soils
- Do not use on straw or other artificial growing media
- Apply to lucerne in late Feb/early Mar when crop dormant
- Do not use on potatoes under hot, dry conditions or on hops under drought conditions
- For interrow use in row crops apply with guarded, no-drift sprayer
- Use as a carefully directed spray in blackcurrants, gooseberries, grapevines and other fruit crops, in raspberries only when dormant
- For stawberry runner control use guarded sprayer, not when flowers or fruit present
- Apply to bulbs pre-emergence (at least 3 d pre-emergence on sandy soils) or at end of season, provided no attached foliage and bulbs well covered (not on sandy soils)
- If using around glasshouses ensure vents and doors closed
- In forestry seedbed apply up to 3 d before seedling emergence

Special precautions/Environmental safety
- Paraquat is subject to the Poisons Rules 1982 and the Poisons Act 1972. See notes in Section 1
- Toxic if swallowed. Paraquat can kill if swallowed
- Harmful in contact with skin. Irritating to eyes and skin
- Keep in original container, tightly closed, in a safe place, under lock and key
- Do not put in a drinks bottle
- Not for use by amateur gardeners
- Harmful to animals
- Paraquat may be harmful to hares. Where possible spray stubbles early in the d

Protective clothing/Label precautions
- A, C
- 4c, 5a, 6a, 6b, 17, 18, 21, 28, 29, 36, 37, 40, 47, 54, 60, 64, 65, 70, 79

Withholding period
- Keep all livestock out of treated areas for at least 24 h

Latest application/Harvest Interval(HI)
- Before shoots 15 cm high and before 10% of shoots emerged for early potatoes, before 40% of shoots emerged for maincrop potatoes [1, 3, 4, 8]; before 31 Mar for lucerne.
- HI mint, leafy herbs 8 wk [3]

Approval
- Off-label Approval to Aug 1998 for use on leafy herbs (OLA 1000/93)[3]; to Aug 1997 for use on dormant lucerne (OLA 1037/93)[3]

Maximum Residue Levels (mg residue/kg food)
- tea, hops 0.1; all other products (except cereals, animal products) 0.05

397 penconazole

A protectant conazole fungicide with antisporulant activity

See also captan + penconazole
dithianon + penconazole

Products				
Topas 100 EC	Ciba Agric.	100 g/l	EC	03231

Uses

American gooseberry mildew in BLACKCURRANTS *(restricted area)*. Powdery mildew, scab in ORNAMENTAL TREES. Powdery mildew in APPLES, HOPS. Rust in ROSES.

Notes

Efficacy
- Apply every 10-14 d (every 7-10 d in warm, humid weather) at first sign of infection or as a protective spray ensuring complete coverage. See label for details of timing
- Increase dose and volume with growth of hops but do not exceed 2000 l/ha
- Antisporulant activity reduces development of secondary mildew in apples

Crop Safety/Restrictions
- Maximum number of treatments 10 per yr for apples, 6 per yr for hops, 4 per yr for blackcurrants
- Check for varietal susceptibility in roses. Some defoliation may occur after repeat applications on Dearest

Special precautions/Environmental safety
- Irritating to eyes and skin
- Harmful to fish or other aquatic life. Do not contaminate surface waters or ditches with chemical or used container

Protective clothing/Label precautions
- A, C
- 6a, 6b, 14, 18, 21, 29, 35, 36, 37, 53, 63, 66, 70

Latest application/Harvest Interval(HI)
- HI apples, hops 14 d

Approval
- For use on blackcurrants apply to Ciba Agriculture for an area allocation

398 pencycuron

A non-systemic urea fungicide for use on seed potatoes

See also imazalil + pencycuron

Products				
1 Monceren DS	Bayer	12.5% w/w	DP	04160
2 Monceren Flowable	Bayer	250 g/l	FS	04907

Uses

Black scurf in SEED POTATOES.

FOR FULL CONDITIONS OF USE ALWAYS READ THE PRODUCT LABEL

Notes

Efficacy
* Provides control of tuber-borne disease and useful, though variable, control of soil-borne disease. Also gives some reduction of stem canker
* Apply to seed tubers in chitting trays, to bulk bins immediately before planting or in hopper at planting [1]
* If rain interrupts planting cover tubers in hopper
* Apply to clean seed tubers at any time, into or out of store [2]

Crop Safety/Restrictions
* Maximum number of treatments 1 per batch of seed potatoes

Special precautions/Environmental safety
* Use of suitable dust mask is mandatory when applying dust, filling the hopper or riding on planter

Protective clothing/Label precautions
* F
* 29, 54, 63, 67

Latest application/Harvest Interval(HI)
* At planting

Approval
* May be applied by misting equipment mounted over roller table. See label for details. See notes in Section 1 [2]

---

399   pendimethalin

A residual dinitroaniline herbicide for cereals and other crops

See also chlorotoluron + pendimethalin
  isoproturon + pendimethalin

Products

1 Sovereign 400SC	Ciba Agric.	400 g/l	SC	07181
2 Stomp 400 SC	Cyanamid	400 g/l	SC	04183

Uses

Annual dicotyledons, annual grasses, annual meadow grass, blackgrass, cleavers, speedwells, wild oats in COMBINING PEAS, DURUM WHEAT, POTATOES, SPRING BARLEY, TRITICALE, WINTER BARLEY, WINTER RYE, WINTER WHEAT [2]. Annual dicotyledons, annual grasses in EVENING PRIMROSE *(off-label)*, FARM FORESTRY *(off-label)*, PARSLEY *(off-label)*, SAGE *(off-label)* [2]. Annual dicotyledons, annual grasses in HERBS *(off-label)*, OUTDOOR LEAF HERBS *(off-label)* [1, 2]. Annual dicotyledons, annual grasses in APPLES, BLACKCURRANTS, CARROTS, CHERRIES, GOOSEBERRIES, HOPS, LEEKS, LEEKS *(off-label)*, ONIONS, ONIONS *(off-label)*, PARSLEY, PARSNIPS, PEARS, PLUMS, STRAWBERRIES, TRANSPLANTED BRASSICAS [1]. Annual dicotyledons, annual meadow grass, speedwells in FODDER MAIZE [2]. Annual grasses in FODDER MAIZE *(off-label)*[2].

Notes

Efficacy
* Apply as soon as possible after drilling. Weeds are controlled as they germinate
* For effective blackgrass control apply not more than 2 d after final cultivation and before weed seeds germinate
* Tank mixes with approved formulations of isoproturon or chlorotoluron recommended for improved pre- and post-emergence control of blackgrass, with isoxaben for additional broad-leaved weeds [2]

- Tank mixture with atrazine recommended for weed control in fodder maize
- Best results by application to fine firm, moist, clod-free seedbeds when rain follows treatment. Effectiveness reduced by prolonged dry weather after treatment
- Do not use on spring barley after end Mar (mid-Apr in Scotland) because dry conditions likely. Do not apply to dry seedbeds in spring unless rain imminent
- Effectiveness reduced on soils with more than 6% organic matter. Do not use where organic matter exceeds 10%
- Any trash, ash or straw should be incorporated evenly during seedbed preparation
- Do not disturb soil after treatment
- On peas drilled after end Mar (mid-Apr in Scotland) tank-mix with Fortrol [2]
- Apply to potatoes as soon as possible after planting and ridging in tank-mix with Fortrol or Sencorex [2]

Crop Safety/Restrictions
- Maximum number of treatments 1 per crop or yr
- May be applied pre-emergence of cereal crops sown before 30 Nov provided seed covered by at least 32 mm soil, or post-emergence to early tillering stage (GS 23) [2]
- Do not undersow treated crops
- Do not use on crops suffering stress due to disease, drought, waterlogging, poor seedbed conditions or chemical treatment or on soils where water may accumulate
- Apply to combining peas as soon as possible after sowing, not when plumule within 13 mm of surface [2]
- Apply to potatoes up to 7 d before first shoot emerges [2]
- In the event of crop failure specified crops may be sown after at least 2 mth following ploughing to 150 mm. See label for details
- After a dry season land must be ploughed to 150 mm before drilling ryegrass
- Apply in top fruit, bush fruit and hops from autumn to early spring when crop dormant [1]
- Apply in strawberries from autumn to early spring (not before Oct on newly planted bed). Do not apply pre-planting or during flower initiation period (post-harvest to mid-Sep) [1]
- Apply pre-emergence in drilled onions or leeks as tank-mixture with propachlor, not on sand, very light, organic or peaty soils or when heavy rain forecast [1]
- Apply to brassicas after final plant-bed cultivation but before transplanting. Avoid unnecessary soil disturbance after application and take care not to introduce treated soil into the root zone when transplanting [1]
- Do not use on protected crops or in greenhouses [1]

Special precautions/Environmental safety
- Dangerous to fish or other aquatic life. Do not contaminate surface waters or ditches with chemical or used container

Protective clothing/Label precautions
- 18, 21, 25, 28, 29, 36, 37, 52, 63, 66 [1, 2]; 70 [2]

Latest application/Harvest Interval(HI)
- Pre-emergence for spring barley, carrots, lettuce, fodder maize, parsnips, parsley, sage, peas, runner beans, potatoes, onions and leeks; before transplanting for brassicas; before main shoot and 3 tillers stage (GS 23) for winter cereals; before bud burst for blackcurrants, gooseberries; before flower trusses emerge for strawberries.
- HI evening primrose, outdoor parsley and sage 5 mth

FOR FULL CONDITIONS OF USE ALWAYS READ THE PRODUCT LABEL

Approval
* Off-label Approval unlimited for use on onions (OLA 0760/95)[1]; unlimited for use on outdoor parsley and sage (OLA 0175/93)[2]; unlimited for use in farm forestry (OLA 0226/94)[2]; unlimited for use on evening primrose (pre-emergence) (OLA 0319/92, 0660/92)[2]; unlimited for use on herbs and outdoor leaf herbs (OLA 1214/95)[1], (OLA 1215/95)[2]; unlimited for use on outdoor leaf herbs grown under covers (OLA 1225/95)[1]

400   pendimethalin + prometryn
A residual herbicide for peas, potatoes and field beans

Products | Monarch | Cyanamid | 264:170 g/l | SC | 05160

Uses

Annual dicotyledons, annual meadow grass in COMBINING PEAS, POTATOES, SPRING FIELD BEANS.

Notes

Efficacy
* Product acts through roots and shoots of germinating seedlings but also has contact action on cotyledon stage weeds
* Control improved by light rain after application but before weed emergence
* Do not disturb soil after application
* Deeper germinating weeds such as volunteer oilseed rape may not be completely controlled
* Weed control may be reduced after prolonged dry conditions

Crop Safety/Restrictions
* Maximum number of treatments 1 per crop
* Apply after drilling but not after plumule within 13 mm of soil surface (peas, beans) or within 7 d of emergence of first shoot (potatoes)
* All varieties of combining peas, spring field beans and early or maincrop potatoes (including crops grown for seed) may be treated
* Do not use on potatoes grown under polythene
* Do not use on peas or beans on Sands, very light, stony or gravelly soils
* In the event of crop failure or when following early potatoes, plough or cultivate to 15 cm before sowing or planting another crop. See label for list of crops which may be sown after various intervals

Special precautions/Environmental safety
* Irritating to eyes and skin
* Dangerous to fish or other aquatic life. Do not contaminate surface waters or ditches with chemical or used container

Protective clothing/Label precautions
* A, C
* 6a, 6b, 19, 21, 25, 28, 30, 36, 37, 52, 63, 66, 70

Latest application/Harvest Interval(HI)
* Pre-crop emergence

401   pendimethalin + simazine
A contact and residual herbicide for use in winter cereals

Products	Merit	Cyanamid	300:100 g/l	SC	04976

Uses   Annual dicotyledons, meadow grasses in WINTER BARLEY, WINTER WHEAT.

Notes

Efficacy
* Apply from pre-emergence to early tillering stage of crop
* Best results achieved on fine, firm, moist seedbed
* Do not use on soils with more than 10% organic matter
* Any straw residues should be incorporated before spraying

Crop Safety/Restrictions
* Maximum number of treatments 1 per crop
* Do not use on durum wheat
* Ensure crop seed is evenly covered with at least 32 mm of settled soil
* Do not use on Sands, very light, stony or gravelly soils
* Do not apply to crops under stress or on waterlogged soils
* Avoid overlapping spray swathes

Special precautions/Environmental safety
* Irritating to eyes and skin
* Dangerous to fish or other aquatic life. Do not contaminate surface waters or ditches with chemical or used container

Protective clothing/Label precautions
* A, C
* 6a, 6b, 18, 21, 25, 28, 29, 36, 37, 52, 60, 63, 66, 70

Latest application/Harvest Interval(HI)
* Before main shoot and 2 tillers (GS 21)

402   pentanochlor
A contact anilide herbicide for various horticultural crops
See also chlorpropham + pentanochlor

Products					
1 Atlas Solan 40	Atlas	400 g/l	EC	03834
2 Croptex Bronze	Hortichem	400 g/l	EC	04087

Uses   Annual dicotyledons, annual meadow grass in ANEMONES, ANNUAL FLOWERS, CARROTS, CHRYSANTHEMUMS, FREESIAS, NURSERY STOCK, PARSNIPS, SWEET PEAS [1, 2]. Annual dicotyledons, annual meadow grass in APPLES, CELERIAC, CELERY, CHERRIES, CONIFER SEEDLINGS, CURRANTS, FENNEL, FOXGLOVES, GOOSEBERRIES, LARKSPUR, PARSLEY, PEARS, PLUMS, PROTECTED CARNATIONS, ROSES, SWEET WILLIAMS, TOMATOES, UMBELLIFEROUS HERBS *(off-label)*, WALLFLOWERS [1].

FOR FULL CONDITIONS OF USE ALWAYS READ THE PRODUCT LABEL

Notes

Efficacy
- Best results achieved on young weed seedlings under warm, moist conditions
- Weeds most susceptible in cotyledon to 2-leaf stage, up to 3 cm high, redshank, fat-hen, fumitory and some others also controlled at later stages
- Some residual effect in early spring when adequate soil moisture and growing conditions good. Effectiveness reduced by very cold weather or drought

Crop Safety/Restrictions
- Maximum number of treatments 1 or 2 per crop depending on dose applied. See label for details
- Apply pre-emergence in anemones, freesias, foxgloves, larkspur
- Not recommended for flower crops on light sandy soils
- Apply pre-emergence or after fully expanded cotyledon stage of carrots and related crops
- Apply as directed spray in fruit crops, nursery stock, perennial flowers and tomatoes
- Any crop may be planted after 4 wk following ploughing and cultivation

Special precautions/Environmental safety
- Irritating to skin and eyes. Harmful if swallowed

Protective clothing/Label precautions
- A, C
- 6a, 6b, 18, 21, 28, 29, 36, 37, 54, 63, 66

Latest application/Harvest Interval(HI)
- HI 28 d for parsnips, carrots

Approval
- Off-label Approval unlimited for use on chervil, coriander, dill, bronze and green fennel, wild celery leaf, lovage, sweet cicely (OLA 0465/92)[1]

---

403   permethrin

A broad spectrum, contact and ingested pyrethroid insecticide

See also bioallethrin + permethrin
fenitrothion + permethrin + resmethrin

Products

1 Coopex Maxi Smoke Generators	Roussel Uclaf	13.5% w/w	FU	H5131
2 Coopex Mini Smoke Generators	Roussel Uclaf	13.5% w/w	FU	H5130
3 Coopex WP	AgrEvo Environ.	25% w/w	WP	H5096
4 Fumite Permethrin Smoke	Hortichem	2 and 3.5 g a.i.	FU	00940
5 Permasect 10 EC	Mitchell Cotts	90 g/l	EC	03920
6 Permasect 25 EC	Mitchell Cotts	230 g/l	EC	01576
7 Turbair Permethrin	Fargro	5 g/l	UL	02246

Uses

Aphids, caterpillars, cutworms, diamond-back moth, leaf miners, whitefly in BRASSICAS [5, 7]. Aphids, caterpillars, fruitlet mining tortrix, suckers in APPLES, PEARS [5]. Aphids, caterpillars, leaf miners, tomato fruitworm, whitefly in CUCUMBERS, TOMATOES [4-7]. Aphids, caterpillars, leaf miners, whitefly in ORNAMENTALS [4, 7]. Aphids, caterpillars, leaf miners, whitefly in CHRYSANTHEMUMS, FUCHSIAS, POT PLANTS [7]. Aphids, caterpillars, leaf miners, whitefly in PEPPERS [4, 5, 7]. Aphids, caterpillars, whitefly in PROTECTED ORNAMENTALS [5, 6]. Aphids, tomato fruitworm, whitefly in AUBERGINES [4-6]. Aphids, tomato fruitworm, whitefly in CHILLIES [5, 6]. Black pine beetle, pine weevil in CONIFER

SEEDLINGS *(off-label)*, FORESTRY TRANSPLANTS *(off-label)* [6]. Browntail moth in AMENITY TREES AND SHRUBS [7]. Colorado beetle in POTATOES *(off-label)* [6]. Flies in MANURE HEAPS [3]. Grain storage pests in GRAIN STORES [1-3]. Phorid flies, sciarid flies in MUSHROOMS [7]. Pine weevil in CONIFERS *(off-label)*[6]. Whitefly in CELERY, PROTECTED LETTUCE [4].

Notes

Efficacy
- Sprays give rapid knock-down effect with persistent protection on leaf surfaces
- Apply as soon as pest or damage appears, or as otherwise recommended and repeat as necessary usually at 14 d intervals
- Number and timing of sprays vary with crop and pest. See label for details
- Addition of non-ionic wetter recommended for use on brassicas
- Do not mix Turbair formulation with any other product
- For whitefly or mushroom fly control use at least 3 fumigations at 5-7 d intervals
- Apply in grain stores by fumigation or surface spraying. Foodstuffs should not come into contact with treated surfaces unless otherwise directed

Crop Safety/Restrictions
- Do not use Turbair treatment on Chinese cabbage, coleus, exacum, peperomia, red calceolaria or protected roses. Check safety on chrysanthemums, fuchsias, pot plants and ornamentals and any plants of unknown sensitivity before treating on a large scale
- Cut any open blooms before fumigating flower crops
- Do not fumigate young seedlings or plants being hardened off
- Allow 7 d after fumigating before introducing Encarsia or Phytoseiulus

Special precautions/Environmental safety
- Irritating to eyes, skin and respiratory system [7]; to eyes and respiratory system [4]; to eyes and skin [5, 6]
- Flammable [1-4, 7]
- Dangerous to bees. Do not apply to crops in flower or to those in which bees are actively foraging. Do not apply when flowering weeds are present
- Extremely dangerous to fish or other aquatic life. Do not contaminate surface waters or ditches with chemical or used container

Protective clothing/Label precautions
- A, C, D, H [3]; A, C [5, 6]; A, D, E, H [1, 2, 4]
- 6a [4-7]; 6b [5-7]; 6c [4, 7]; 12c, 66, 67, 70 [1, 2, 4, 7]; 16, 65 [5, 6]; 18, 28, 29, 36, 48, 51, 63 [1, 2, 4-7]; 21, 37 [1-7]; 32, 76 [3]; 57 [7]

Latest application/Harvest Interval(HI)
- HI zero

Approval
- Approved for ULV application. See label for details. See notes in Section 1 [7]
- Off-label Approval to May 1997 for ground or aerial application to potatoes as required by Statutory Notice to control Colorado beetle (OLA 0588)[6], (0589/92)[6]; to Jul 1995 for use as dip for forestry transplants (OLA 0815/92)[6] - see OLA notice for restrictions; unlimited for use on forestry trees to control pine weevil (OLA 0135/92)[6] - see OLA notice for restrictions; unlimited for use on containerised forestry seedlings (OLA 0451/94)[6]

FOR FULL CONDITIONS OF USE ALWAYS READ THE PRODUCT LABEL

Maximum Residue Levels (mg residue/kg food)
- lettuces, herbs, celery, rhubarb, cereals (except maize) 2; pome fruits, stone fruits, grapes, kiwi fruit, Chinese cabbage, kale, spinach, beet leaves 1; citrus fruits, tomatoes, peppers, aubergines, beans (with pods), leeks, meat 0.5; cotton seed, maize 0.2; almond, celeriac, radishes, cucumbers, gherkins, courgettes, melons, squashes, watermelons, sweetcorn, cauliflowers, peas (with pods), peanuts, rape seed, mustard seed 0.1; tree nuts (except almond), pome fruits, stone fruits, grapes, cane fruits, bilberries, cranberries, wild berries, miscellaneous fruit (except kiwi fruit, olives), root and tuber vegetables (except celeriac, radishes, turnips), garlic, kohlrabi, watercress, witloof, beans (without pods), peas (without pods), asparagus, cardoons, globe artichokes, mushrooms, pulses, linseed, poppy seed, sesame seed, soya bean, potatoes, milk, eggs 0.05

404   permethrin + thiram
An insecticide/fungicide mixture for chrysanthemums and pot plants

Products	Combinex	Fargro	4:25 g/l	EC	00562

Uses   Capsids, caterpillars, earwigs in BEGONIAS, CHRYSANTHEMUMS, CYCLAMENS, HYDRANGEAS, PRIMULAS. Leaf miners in BEGONIAS *(partial control)*, CHRYSANTHEMUMS *(partial control)*, CYCLAMENS *(partial control)*, HYDRANGEAS *(partial control)*, PRIMULAS *(partial control)*.

Notes

Efficacy
- Spray chrysanthemums as soon as plants established and repeat every 14 d, if powdery mildew infection particularly severe every 7 d
- Spray pot plants as soon as established in final pots and repeat as necessary

Crop Safety/Restrictions
- Spray up to stage immediately before new florets unfold. Later applications to unfolded petals may result in petal damage, particularly under adverse growing conditions
- Do not apply to carnations or vines
- Do not apply in bright sunshine or when plants are flagging
- Do not apply within 21 d before or after application of sulfur or dinocap
- Product not compatible with dinocap, magnesium sulfate, manganese sulfate or sulfur
- Do not mix with other chemicals when spraying at low volume

Special precautions/Environmental safety
- Irritating to skin, eyes and respiratory system
- Dangerous to bees. Do not apply to crops in flower or to those in which bees are actively foraging. Do not apply when flowering weeds are present
- Extremely dangerous to fish or other aquatic life. Do not contaminate surface waters or ditches with chemical or used container

Protective clothing/Label precautions
- A
- 6a, 6b, 6c, 14, 18, 21, 28, 29, 34, 36, 37, 48, 51, 63, 65

Latest application/Harvest Interval(HI)
- Immediately before the outer florets become unfolded for chrysanthemums

Maximum Residue Levels (mg residue/kg food)
- see permethrin entry

405   petroleum oil

An insecticidal and acaricidal hydrocarbon oil

Products	Hortichem Spraying Oil	Hortichem	710 g/l	EC	03816

Uses   Mealy bugs, red spider mites, scale insects in CUCUMBERS, GRAPEVINES, POT PLANTS, TOMATOES.

Notes   **Efficacy**
* Spray at 1% (0.5% on tender foliage) to wet plants thoroughly, particularly the underside of leaves, and repeat as necessary
* Apply under quick drying conditions but not in bright sun unless glass well shaded

Crop Safety/Restrictions
* Treat grapevines before flowering
* On plants of unknown sensitivity test first on a small scale
* Mixtures with certain pesticides may damage crop plants. If mixing, spray a few plants to test for tolerance before treating larger areas
* Do not mix with sulfur or use sulfur sprays within 28 d of treatment

Special precautions/Environmental safety
* Harmful if swallowed. Irritating to eyes, skin and respiratory system

Protective clothing/Label precautions
* A, C
* 5c, 6a, 6b, 6c, 18, 21, 28, 36, 37, 63, 65, 70, 78

Latest application/Harvest Interval(HI)
* HI zero

FOR FULL CONDITIONS OF USE ALWAYS READ THE PRODUCT LABEL

406 phenmedipham

A contact carbamate herbicide for beet crops and strawberries

See also *ethofumesate + phenmedipham*
lenacil + phenmedipham

Products					
1 Atlas Protrum K	Atlas	114 g/l	EC	03089
2 Beetup	MTM Agrochem.	114 g/l	EC	05299
3 Beetup	United Phosphorus	114 g/l	EC	07520
4 Betanal E	AgrEvo	114 g/l	EC	07248
5 Betanal E	Schering	114 g/l	EC	03862
6 Betosip	Sipcam	114 g/l	EC	06787
7 Cirrus	Quadrangle	114 g/l	EC	06367
8 Headland Dephend	Headland	114 g/l	EC	04925
9 Herbasan	Mirfield	160 g/l	SC	07161
10 Hickson Phenmedipham	Hickson & Welch	118 g/l	EC	02825
11 Kemifam E	PBI	114 g/l	EC	02935
12 Luxan Phenmedipham	Luxan	118 g/l	EC	06933
13 Stefes Forte	Stefes	114 g/l	EC	06427
14 Stefes Medipham	Stefes	114 g/l	EC	07152
15 Tripart Beta	Tripart	118 g/l	EC	03111
16 Tripart Beta 2	Tripart	114 g/l	EC	04831
17 Vangard	FCC	114 g/l	EC	02743

Uses
Annual dicotyledons in RED BEET [1-6, 8-11, 13-15, 17]. Annual dicotyledons in FODDER BEET, MANGELS, SUGAR BEET [1-17]. Annual dicotyledons in SPINACH (*off-label*) [5]. Annual dicotyledons in STRAWBERRIES [2-6, 8, 17].

Notes

Efficacy
- Best results achieved by application to young seedling weeds, preferably cotyledon stage, under good growing conditions when low doses are effective
- 2-3 repeat applications at 7-10 d intervals using a low dose are recommended on mineral soils, 3-5 applications may be needed on organic soils
- Do not spray wet foliage or if rain imminent
- Addition of adjuvant oil may improve effectiveness on some weeds
- Various tank-mixtures with other beet herbicides recommended. See label for details
- Use of certain pre-emergence herbicides is recommended in combination with post-emergence treatment. See label for details

Crop Safety/Restrictions
- Maximum number of treatments 2 per crop (higher dose), 5 per crop (lower dose) for beet crops; 2 per yr for strawberries; 1 per crop for spinach
- Apply to beet crops at any stage as low dose/low volume spray or from fully developed cotyledon stage with full rate. Apply to red beet after fully developed cotyledon stage
- At high temperatures (above 21°C) reduce rate and spray after 5 pm
- Do not apply immediately after frost or if frost expected
- Do not spray crops stressed by wind damage, nutrient deficiency, pest or disease attack etc. Do not roll or harrow for 7 d before or after treatment
- Apply to strawberries at any time when weeds in susceptible stage, except in period from start of flowering to picking
- Do not use on strawberries under cloches or polythene tunnels

Special precautions/Environmental safety
- Harmful if swallowed [1, 3, 13-17]; in contact with skin, by inhalation and if swallowed [2, 8, 12]
- Irritating to eyes, skin and respiratory system [1-8, 10-12, 15-17]
- Harmful (dangerous [9]) to fish or other aquatic life. Do not contaminate surface waters or ditches with chemical or used container [1-8, 10-17]

Protective clothing/Label precautions
- A, C, H [12]; A, C, P [8, 15]; A, C [1-7, 11, 13, 14, 17]
- 5a, 5b [8]; 5c [1-3, 8, 13-16]; 6a, 6b, 6c [1-8, 10, 11, 15-17]; 18, 21, 28, 29, 36, 37, 63, 66 [1-17]; 25 [2, 3]; 52 [9]; 53, 70, 78 [1-8, 10-17]; 60 [2-8, 10-17]

Latest application/Harvest Interval(HI)
- Before crop leaves meet between rows for beet crops; before flowering for strawberries; before 5 leaf stage for spinach

Approval
- Off-label Approval to Jul 1997 for use on outdoor spinach (OLA 0778/93) [5]
- Approval expiry: 31 Aug 95 [04831]; replaced by MAFF 06510

407   phenothrin
A non-systemic contact and ingested pyrethroid insecticide

Products					
	Sumithrin 10 Sec	Sumitomo	103 g/l	EC	H3762

Uses   Flies in LIVESTOCK HOUSES.

Notes

Efficacy
- Spray walls and other structural surfaces of milking parlours, dairies, cow byres and other animal housing

Crop Safety/Restrictions
- Do not apply directly to livestock
- Remove exposed milk and collect eggs before application. Protect milk machinery and containers from contamination

Special precautions/Environmental safety
- Irritating to eyes
- Wear approved respiratory equipment and eye protection (goggles) when applying with Microgen equipment
- Not to be sold or supplied to amateur users
- Dangerous to fish or other aquatic life. Do not contaminate surface waters or ditches with chemical or used container

Protective clothing/Label precautions
- A, C, D, E, F, H
- 6a, 18, 21, 29, 36, 37, 52, 63, 67

FOR FULL CONDITIONS OF USE ALWAYS READ THE PRODUCT LABEL

Approval
* May be applied with Microgen ULV equipment. See notes in Section 1

408 phenothrin + tetramethrin
A pyrethroid insecticide mixture for control of flying insects

Products

1 Deosan Fly Spray	Deosan	0.066:0.028% w/w	AE	H5246
2 Killgerm ULV 500	Killgerm	40:20 g/l	UL	H4647
3 Sorex Fly Spray RTU	Sorex	0.02:0.10% w/w	RH	H5718
4 Sorex Super Fly Spray	Sorex		AE	H4468

Uses Flies, mosquitoes, wasps in AGRICULTURAL PREMISES.

Notes **Efficacy**
* Close doors and windows and spray in all directions for 3-5 sec. Keep room closed for at least 10 min
* May be used in the presence of poultry and livestock

Crop Safety/Restrictions
* Do not use space sprays containing pyrethrins or pyrethroid more than once per week in intensive or controlled environment animal houses in order to avoid development of resistance. If necessary, use a different control method or product [2]

Special precautions/Environmental safety
* Do not spray directly on food, livestock or poultry
* Remove exposed milk and collect eggs before application. Protect milk machinery and containers from contamination
* Flammable [1, 2, 4]
* Dangerous (extremely dangerous [3]) to fish or other aquatic life. Do not contaminate surface waters or ditches with chemical or used container

Protective clothing/Label precautions
* A, H [3]
* 14, 22, 39, 51, 63 [3]; 28, 29 [1-4]; 36, 37, 52 [1, 2, 4]

409 phenylmercury acetate
An organomercury fungicide seed treatment approvals for which were revoked with effect from 31 March 1992

410 phorate
A systemic organophosphorus insecticide with vapour-phase activity

Products MTM Phorate MTM Agrochem. 10% w/w GR 06609

Uses Aphids, capsids, leafhoppers, wireworms in POTATOES. Aphids, capsids, pea and bean weevils in BROAD BEANS, FIELD BEANS, PEAS. Aphids, capsids in BEET CROPS. Aphids, carrot fly in CARROTS. Aphids, lettuce root aphid in LETTUCE. Aphids in BRASSICAS, PARSNIPS,

STRAWBERRIES. Cabbage root fly, cabbage stem flea beetle, rape winter stem weevil in OILSEED RAPE. Carrot fly in CELERY. Frit fly in MAIZE, SWEETCORN. Lettuce root aphid in ICEBERG LETTUCE *(higher rate, off-label)*.

Notes

Efficacy
- May be applied as a soil or foliar treatment. Application method, rate and timing vary with crop, pest and soil type. See label for details
- May be applied broadcast, with row crop applicators or in combine drills
- Effectiveness of soil application reduced by hot, dry weather, excessive rainfall or on soils with more than 10% organic matter content
- Foliar application recommended for aphid or capsid control in beet crops, peas, beans and strawberries
- Effectiveness of foliar application reduced on plants suffering from drought stress

Crop Safety/Restrictions
- Maximum number of treatments 2 per crop for beet crops (1 per crop for sugar beet seed crops); 1 per crop for other crops
- On carrots the maximum total dose applied per crop must not exceed the equivalent of 3 (on mineral soils) or 4 (on organic soils) full dose applications
- When applied at time of sowing granules must not be in direct contact with seed

Special precautions/Environmental safety
- This product contains an anticholinesterase organophosphorus compound. Do not use if under medical advice not to work with such compounds
- Toxic in contact with skin, by inhalation or if swallowed
- Keep in original container, tightly closed in a safe place, under lock and key
- Harmful to livestock
- Dangerous to game, wild birds and animals
- Dangerous to fish or other aquatic life. Do not contaminate surface waters or ditches with chemical or used container

Protective clothing/Label precautions
- A, B, C, D, E, H, J, K, M
- 1, 2, 4a, 4b, 4c, 14, 16, 18, 23, 24, 25, 28, 29, 35, 36, 37, 40, 45, 52, 64, 67, 70, 79

Withholding period
- Keep all livestock out of treated areas for at least 6 wk. Bury or remove spillages

Latest application/Harvest Interval(HI)
- Foliar application: before end of flowering for broad beans; before start of flowering for broad beans and strawberries. Soil treatment: at sowing or transplanting for spring field and broad beans, potatoes, maize, sweetcorn, celery, transplanted cabbage, cauliflower, broccoli, Brussels sprouts.
- HI beet crops, spring field and broad beans, peas (broadcast treatment), outdoor lettuce 6 wk; Iceberg lettuce (high rate) 10 wk; carrots, parsnips 21 wk

Approval
- Approved for aerial application on spring field beans and broad beans. See notes in Section 1
- Off-label Approval to June 1997 for root aphid control on Iceberg lettuce (OLA 0707/92)

FOR FULL CONDITIONS OF USE ALWAYS READ THE PRODUCT LABEL

411 picloram

A persistent, translocated picolinic herbicide for non-crop areas

See also 2,4-D + picloram
bromacil + picloram

Products	Tordon 22K	Nomix-Chipman	240 g/l	SL	05790

Uses

Annual dicotyledons, bracken, japanese knotweed, perennial dicotyledons, woody weeds in NON-CROP AREAS, NON-CROP GRASS.

Notes

Efficacy
* May be applied at any time of year. Best results achieved by application as foliage spray in late winter to early spring
* For bracken control apply 2-4 wk before frond emergence
* Clovers are highly sensitive and eliminated at very low doses
* Persists in soil for up to 2 yr

Crop Safety/Restrictions
* Maximum number of treatments 1 per yr
* Do not apply around desirable trees or shrubs where roots may absorb chemical
* Do not apply on slopes where chemical may be leached onto areas of desirable plants

Special precautions/Environmental safety
* Irritating to eyes
* Harmful to fish or other aquatic life. Do not contaminate surface waters or ditches with chemical or used container

Protective clothing/Label precautions
* 6a, 21, 29, 43, 53, 63, 66

Withholding period
* Keep livestock out of treated areas until foliage of any poisonous weeds such as ragwort has died and become unpalatable

412 pirimicarb

A carbamate insecticide for aphid control

Products					
1	Aphox	Zeneca	50% w/w	SG	06633
2	Barclay Pirimisect	Barclay	50% w/w	WG	06929
3	Phantom	Bayer	50% w/w	SG	04519
4	Pirimor	Zeneca	50% w/w	WP	06694
5	Standon Pirimicarb 50	Standon	50% w/w	WG	05622
6	Standon Pirimicarb H	Standon	50% w/w	WP	05669
7	Stefes Pirimicarb	Stefes	50% w/w	SG	05758

Uses

Aphids in SUGARBEET [5]. Aphids in PROTECTED CARNATIONS, PROTECTED CHRYSANTHEMUMS, PROTECTED LETTUCE [6]. Aphids in CHICORY *(off-label)*[1, 4]. Aphids in BROAD BEANS, BROCCOLI, BRUSSELS SPROUTS, CABBAGES, CALABRESE, CAULIFLOWERS, CEREALS, CHINESE CABBAGE, COLLARDS, FIELD BEANS, KALE, PEAS, POTATOES [1-3, 5, 7]. Aphids in APPLES, PEARS [1-3, 5]. Aphids in CARROTS,

CELERY, DWARF BEANS, GRASSLAND, MAIZE, OILSEED RAPE, PARSNIPS, RUNNER BEANS, SUGAR BEET, SWEDES, SWEETCORN, TURNIPS [1-3, 7]. Aphids in STRAWBERRIES [1-6]. Aphids in BLACKCURRANTS, CHERRIES, GOOSEBERRIES, RASPBERRIES, REDCURRANTS [1-3]. Aphids in HONESTY *(off-label)* [1]. Aphids in CUCUMBERS, LETTUCE, TOMATOES [4, 6]. Aphids in FOREST NURSERIES, ORNAMENTALS, PEPPERS [4].

Notes

Efficacy
* Chemical has contact, fumigant and translaminar activity
* Best results achieved under warm, calm conditions when plants not wilting and spray does not dry too rapidly. Little vapour activity at temperatures below 15°C
* Apply as soon as aphids seen or warning issued and repeat as necessary
* Addition of non-ionic wetter recommended for use on brassicas
* Chemical has little effect on bees, ladybirds and other insects and is suitable for use in integrated control programmes on apples and pears
* On cucumbers and tomatoes a root drench is preferable to spraying when using predators in an integrated control programme
* Aphids resistant to carbamate insecticides have developed in certain areas

Crop Safety/Restrictions
* Maximum number of treatments 2 per crop for honesty

Special precautions/Environmental safety
* This product contains an anticholinesterase carbamate compound. Do not use if under medical advice not to work with such compounds
* Harmful if swallowed
* Harmful to livestock
* Dangerous to fish or other aquatic life. Do not contaminate surface waters or ditches with chemical or used container

Protective clothing/Label precautions
* A, C, D, F, H, J, M [6]; A, C, D, H, J, M [1-5, 7]
* 2, 5c, 18, 21, 29, 36, 37, 54, 63, 67, 70, 78 [1-6]; 35 [1, 3, 4]; 41 (7 d) [1-5]; 41(7 d) [6]

Withholding period
* Keep all livestock out of treated areas for at least 7 d. Bury or remove spillages

Latest application/Harvest Interval(HI)
* HI oilseed rape, cereals, maize, sweetcorn, lettuce under glass 14 d; grassland 7 d; cucumbers, tomatoes and peppers under glass 2 d; protected courgettes and gherkins 24 h; other edible crops 3 d; flowers and ornamentals zero

Approval
* Approved for aerial application on cereals, maize, oilseed rape, swedes, turnips, brassicas, beans, sugar beet, carrots, peas, potatoes [1, 7]; cereals [2, 3]; sweetcorn [7]. See notes in Section 1
* Off-label approval unlimited for use on honesty (OLA 0584/93)[1]; to Mar 2000 for use on chicory (OLA 0500)[1], (0501/94)[4]

FOR FULL CONDITIONS OF USE ALWAYS READ THE PRODUCT LABEL

413   pirimiphos-methyl

A contact, fumigant and translaminar organophosphorus insecticide

Products					
1 Actellic D	Zeneca	250 g/l	EC	06930
2 Actellic Dust	Zeneca	2% w/w	DP	06931
3 Actellic Smoke Generator No 20	Zeneca	20 g a.i.	FU	06627
4 Actellifog	Hortichem	100 g/l	HN	06628
5 Blex	Zeneca	500 g/l	EC	06639
6 Fumite Pirimiphos Methyl Smoke	Hortichem	10 g a.i.	FU	00941

Uses

Aphids, apple sawfly, apple sucker, capsids, codling moth, rust mite, tortrix moths in APPLES [5]. Aphids, caterpillars, rust mite in PEARS [5]. Cabbage stem flea beetle in OILSEED RAPE [5]. Carrot fly in CARROTS, PARSLEY *(off-label)* [5]. Caterpillars, whitefly in FLOWERHEAD BRASSICAS, LEAF BRASSICAS [5]. Flour beetles, flour moths, grain beetles, grain storage mites, grain weevils, warehouse moth in STORED GRAIN [1-3]. French fly in CUCUMBERS [5]. Frit fly, wheat bulb fly in CEREALS [5]. Grain storage pests in GRAIN STORES [4]. Leaf miners in SUGAR BEET [5]. Storage pests in STORED LINSEED, STORED OILSEED RAPE [1, 2].

Notes

Efficacy

- Chemical acts rapidly and has short persistence in plants, though spray or dust persists for long period on inert surfaces
- Apply spray when pest first seen or at time of egg hatch and repeat as necessary
- For wheat bulb fly control apply on receipt of ADAS warning
- Rate, number and timing of sprays vary with crop and pest. See label for details
- For use on Brussels sprouts use of drop-legged sprayer is recommended
- For protection of stored grain disinfect empty stores by spraying surfaces and/or fumigation and treat grain by full or surface admixture or spray bagged grain. See label for details of treatment and suitable application machinery
- Best results obtained when grain stored at 15% moisture or less. Dry and cool moist grain coming into store before treatment
- For control of whitefly and other glasshouse pests apply as smoke or by thermal fogging. See label for details of techniques and suitable machines
- Actellifog may be used to clean up houses before introducing whitefly parasites but do not use when Encarsia present

Crop Safety/Restrictions

- On carrots the maximum total dose applied per crop must not exceed the equivalent of 3 (on mineral soils) or 4 (on organic soils) full dose applications
- Grain treated by admixture as specified may be consumed by humans and livestock
- Do not fumigate glasshouses in bright sunshine or when foliage is wet or roots dry
- Do not fumigate young seedlings or plants being hardened off
- Do not fog open flowers of ornamentals without first consulting firm
- Do not fog mushrooms when wet as slight spotting may occur
- Consult processors before using on crops for processing [5]

Special precautions/Environmental safety

- This product contains an organophosphorus, anticholinesterase compound. Do not use if under medical advice not to work with such compounds
- Irritating to eyes and skin

- Flammable [5]
- Ventilate fumigated or fogged spaces thoroughly before re-entry
- Dangerous to bees. Do not apply to crops in flower or to those in which bees are actively foraging. Do not apply when flowering weeds are present
- Harmful to fish or other aquatic life. Do not contaminate surface waters or ditches with chemical or used container

Protective clothing/Label precautions

- A, D, J, L, M [4]; A [1, 5]
- 1, 2, 18, 21, 28, 29, 35, 36, 37, 53, 63, 70 [1-6]; 6a, 6c, 12c, 49, 66, 67, 78 [4-6]; 16, 39, 65 [1-3]

Latest application/Harvest Interval(HI)

- HI zero [3, 4, 6];brassicas 3 d [5]; cereals, sugar beet, carrots, apples, pears 7 d [5]; cucumbers 3 wk [5]

Approval

- One product formulated for application by thermal fogging. See label for details. See introductory notes on ULV application [4]
- Off-label approval to Jan 1997 for use on outdoor parsley (OLA 0591/93)[5]

Maximum Residue Levels (mg residue/kg food)

- cereals (except rice) 5; citrus fruits 0.5

---

414   prochloraz

A broad-spectrum protectant and eradicant conazole fungicide

See also carbendazim + prochloraz
cyproconazole + prochloraz
fenpropidin + prochloraz
fenpropimorph + prochloraz

Products					
	1 Barclay Eyetak	Barclay	450 g/l	EC	06813
	2 Fisons Octave	Levington	46% w/w	WP	03416
	3 Levington Octave	Levington	46% w/w	WP	07505
	4 Mirage 40 EC	Makhteshim	400 g/l	EC	06770
	5 Sporgon 50WP	Darmycel	46% w/w	WP	03829
	6 Sportak 45	AgrEvo	450 g/l	EC	07286
	7 Sportak 45	Schering	450 g/l	EC	03815
	8 Stefes Poraz	Stefes	450 g/l	EC	07528

Uses
Alternaria, grey mould, light leaf spot, phoma, sclerotinia stem rot, white leaf spot in OILSEED RAPE [1, 6-8]. Botrytis in ORNAMENTALS [2, 3]. Cobweb, dry bubble, wet bubble in MUSHROOMS [5]. Eyespot, glume blotch, leaf spot, powdery mildew in WINTER WHEAT [1, 4, 6-8]. Eyespot, leaf spot, powdery mildew, rhynchosporium in WINTER RYE [1, 6-8]. Eyespot, net blotch, powdery mildew, rhynchosporium in WINTER BARLEY [1, 4, 6-8]. Eyespot, powdery mildew in SPRING WHEAT [4, 8]. Fungus diseases in CONTAINER-GROWN STOCK, HARDY ORNAMENTAL NURSERY STOCK [2, 3]. Fusarium bulb rot, penicillium rot in FLOWER BULBS *(off-label)* [6, 7]. Net blotch, powdery mildew, rhynchosporium in SPRING

FOR FULL CONDITIONS OF USE ALWAYS READ THE PRODUCT LABEL

BARLEY [1, 6-8]. Powdery mildew in SPRING WHEAT *(protection)*[1, 6, 7]. Ring spot in LETTUCE *(off-label)*, PROTECTED LETTUCE *(off-label)* [2].

Notes

Efficacy
* Spray cereals at first signs of disease. Protection of winter crops through season usually requires at least 2 treatments. See label for details of rates and timing. Treatment active against strains of eyespot resistant to benzimidazole fungicides
* Tank mixes with other fungicides recommended to improve control of rusts in wheat and barley. See label for details
* A period of at least 3 h without rain should follow spraying
* Apply as drench against soil diseases or as dip at propagation. Spray applications may be repeated at 7-14 d intervals. Under mist propagation use 7 d intervals [2]
* Apply to mushrooms as casing treatment or spray between flushes. Timing determined by anticipated disease occurrence [5]

Crop Safety/Restrictions
* Maximum number of treatments 2 at normal rate on cereals [6, 7] or 2 at split plus 1 at normal rate on oilseed rape [1, 6, 7]; 1 incorporated in casing plus 2 sprays or 3 spray treatments [5]
* See label for details of ornamental species which can be treated [2]

Special precautions/Environmental safety
* Harmful in contact with skin. Irritating to eyes and skin [1, 6-8]
* Irritating to skin and eyes [4]
* May cause sensitisation by skin contact [4]
* Flammable [1, 4, 6, 7]
* Dangerous to fish or other aquatic life. Do not contaminate surface waters or ditches with chemical or used container

Protective clothing/Label precautions
* A, C, H, J [5]; A, C [1, 4, 6-8]; A [2, 3]
* 5a, 70, 78 [1, 6-8]; 6a, 6b, 18, 36, 37 [1, 4, 6-8]; 10a [4]; 12c [1, 4, 6, 7]; 21, 29, 52, 63, 66 [1-8]; 60 [1, 6, 7]; 67 [2, 3, 5]

Latest application/Harvest Interval(HI)
* Milky ripe stage (GS 77) for cereals.
* HI 6 wk for oilseed rape and cereals; 21 d for lettuce [2]; 2 d for mushrooms [5]

Approval
* Off-label Approval unlimited for use on ornamental bulbs (OLA 1232/95)[7], (OLA 1231/95)[6]; to Dec 1998 for use on lettuce and protected lettuce (OLA 1420/94)[2]

415 prochloraz + tolclofos-methyl
A broad-spectrum fungicide mixture for use on seed potatoes

Products | Rizolex Nova | AgrEvo | 46% w/w:500 g/l | KK | 07434

Uses | Black scurf, stem canker in POTATOES. Gangrene, skin spot in SEED POTATOES *(reduction)*. Silver scurf, skin spot in POTATOES *(reduction)*.

Notes

Efficacy
- Treat seed potatoes as they are taken into, or out of, store. Tubers should be dry and free from soil deposits
- Apply pre-chitting or before the shoots protrude and become susceptible to damage by mechanical handling
- Apply using low volume hydraulic applicators, rotary or ultrasonic atomisers. Ensure even distribution of the spray over the whole tuber

Crop Safety/Restrictions
- Maximum number of treatments (including any other tolclofos-methyl product) 1 per seed lot
- Seed may be treated into store only if it is certain they will not be used for any other purpose than as seed
- Do not tank mix with any other potato fungicides or storage products
- Do not use on potatoes treated, or to be treated, with any sprout suppressant
- Consult processors before using on crops to be processed
- Not recommended on potatoes where hot water treatment has been or will be used

Special precautions/Environmental safety
- Product contains an anticholinesterase organophosphorus compound. Do not use if under medical advice not to work with such compounds
- Dangerous to fish or other aquatic life. Do not contaminate surface waters or ditches with chemical or used container
- Treated tubers to be used only as seed, not for human or animal consumption

Protective clothing/Label precautions
- A, C, D, E, H
- 2, 21, 28, 29, 52, 63, 65, 70,

Latest application/Harvest Interval(HI)
- At planting

416   prometryn

A contact and residual triazine herbicide for various field crops

See also pendimethalin + prometryn

Products

1 Alpha Prometryne 50WP	Makhteshim	50% w/w	WP	04871
2 Atlas Prometryne 50 WP	Atlas	50% w/w	WP	03502
3 Gesagard 50 WP	Ciba Agric.	50% w/w	WP	00981

Uses

Annual dicotyledons, annual grasses in DRILLED LEEKS *(off-label)*, GARLIC *(off-label)*, KOHLRABI *(off-label)*, ONIONS *(off-label)*, OUTDOOR HERBS *(off-label)*, PROTECTED HERBS *(off-label)*, SWEDES *(off-label)*, TURNIPS *(off-label)* [3]. Annual dicotyledons, annual grasses in COMBINING PEAS, VINING PEAS [2, 3]. Annual dicotyledons, annual grasses in CARROTS, CELERY, PARSLEY, POTATOES [1-3]. Annual dicotyledons, annual grasses in TRANSPLANTED LEEKS [1, 2]. Annual dicotyledons, annual grasses in PEAS, TRANSPLANTED CELERY [1].

FOR FULL CONDITIONS OF USE ALWAYS READ THE PRODUCT LABEL

Notes

Efficacy
- Best results achieved by application to young seedling weeds up to 5 cm high (cotyledon stage for knotgrass, mayweed and corn marigold) on fine, moist seedbed when rain falls afterwards. Do not use on very cloddy soils
- On organic soils only contact action effective and repeat application may be needed (certain crops only)

Crop Safety/Restrictions
- Maximum number of treatments 1 per crop for early potatoes, peas and transplanted leeks; 2 per crop for transplanted celery
- Apply to peas pre-emergence up to 3 d before crop expected to emerge
- Spring sown vining and drying peas may be treated. Damage may occur with Vedette or Printana, especially if emerging under adverse conditions. Do not treat forage peas
- Do not use on peas on Very Light soils, Sands, gravelly or stony soils
- Apply to early potatoes up to 10% emergence
- Apply to carrots, celery, parsley or coriander post-emergence after 2-rough leaf stage or after transplants established
- Apply to transplanted leeks or celery after transplants established. Do not use on drilled leeks
- Excessive rain after treatment may check crop
- In the event of crop failure only plant recommended crops within 8 wk

Special precautions/Environmental safety
- Harmful to fish or other aquatic life. Do not contaminate surface waters or ditches with chemical or used container [1, 2]

Protective clothing/Label precautions
- 29, 35 (6 wk), 53, 63, 67 [1, 2]; 54 [3]

Latest application/Harvest Interval(HI)
- Pre-emergence for peas; 6 wk before harvest and before 10% emergence for early potatoes; before 10% crop emergence for swedes and turnips; 3 wk after planting out for onions and garlic.
- HI 6 wk

Approval
- Off-label approval unlimited for use on drilled leeks (OLA 0054/92)[3], outdoor and protected herbs, swedes, turnips, kohlrabi, onions and garlic (OLA 0185/93)[3]

417   prometryn + terbutryn

A residual, pre-emergence herbicide for use in peas, beans and potatoes

Products					
Peaweed	PBI	152:304 g/l	SC	03248

Uses   Annual dicotyledons, annual grasses in COMBINING PEAS, POTATOES, SPRING BROAD BEANS, SPRING FIELD BEANS, VINING PEAS.

Notes

Efficacy
- Weeds are controlled before or shortly after emergence
- Best results on fine, firm, moist seedbed. Do not use on cloddy or very stony soils
- Residual effects may be reduced on soils with 5-10% organic matter. Do not use where organic matter exceed 10%

- Effectiveness may be reduced if heavy rain, below average soil temperatures or dry conditions follow application
- Do not cultivate or ridge up potatoes after treatment

Crop Safety/Restrictions
- Maximum number of treatments 1 per crop
- On peas and beans apply within 4 d of drilling
- Do not use on Vedette or forage peas, seed potatoes or any crop under plastic
- Crop seed should be covered by at least 25 mm of settled soil at time of treatment
- On potatoes apply within 4 d after final ridging. See label for tank mixtures
- Do not use on Sands, Loamy Sand or Loamy Fine Sand soils
- Heavy rain after application may cause crop damage on light soils
- Any crop may be drilled or planted after harvesting following cultivation to 15 cm

Special precautions/Environmental safety
- Harmful to fish or other aquatic life. Do not contaminate surface waters or ditches with chemical or used container

Protective clothing/Label precautions
- 28, 29, 36, 37, 53, 60, 63, 66, 67, 70, 78

Latest application/Harvest Interval(HI)
- 4 d after drilling for peas and beans; before 10% emergence for potatoes

418 propachlor

A pre-emergence amide herbicide for various horticultural crops

See also chloridazon + propachlor
    chlorthal-dimethyl + propachlor

Products

1 Alpha Propachlor 50SC	Makhteshim	50% w/w	SC	04873
2 Atlas Propachlor	Atlas	500 g/l	SC	06462
3 Ramrod Flowable	Monsanto	480 g/l	SC	01688
4 Ramrod Granular	Monsanto	20% w/w	GR	01687
5 Tripart Sentinel	Tripart	500 g/l	SC	03250
6 Tripart Sentinel 2	Tripart	480 g/l	SC	05140

Uses

Annual dicotyledons, annual grasses in ORNAMENTALS, PERENNIALS [4]. Annual dicotyledons, annual grasses in RAPE [3-6]. Annual dicotyledons, annual grasses in BROWN MUSTARD, LETTUCE *(off-label)*, OUTDOOR LEAF HERBS *(off-label)*, SAGE, WHITE MUSTARD [3]. Annual dicotyledons, annual grasses in BROCCOLI, BRUSSELS SPROUTS, CABBAGES, CAULIFLOWERS, KALE, LEEKS, ONIONS, SWEDES, TURNIPS [1-6]. Annual dicotyledons, annual grasses in OILSEED RAPE [1-3, 5, 6]. Annual dicotyledons, annual grasses in STRAWBERRIES [1, 2, 5]. Annual dicotyledons, annual grasses in CALABRESE [2-6].

Notes

Efficacy
- Controls germinating (not emerged) weeds for 6-8 wk

FOR FULL CONDITIONS OF USE ALWAYS READ THE PRODUCT LABEL

- Best results achieved by application to fine, firm, moist seedbed free of established weeds in spring, summer and early autumn
- Effective results with granules depend on rainfall soon after application [4]
- Use higher rate on soils with more than 10% organic matter
- Recommended as tank-mix with Sting CT or chlorthal-dimethyl on onions, leeks, swedes, turnips; with chlorpropham on leeks, onions [4]. See label for details

Crop Safety/Restrictions
- Maximum number of treatments 1 or 2 per crop, varies with product and crop - see label for details
- Apply in brassicas from drilling to time seed chits or after 3-4 true leaf stage but before weed emergence, in swedes and turnips pre-emergence only
- Apply in transplanted brassicas within 48 h of planting in warm weather. Plants must be hardened off and special care needed with block sown or modular propagated plants
- Apply in onions and leeks pre-emergence or from post-crook to young plant stage
- Apply to newly planted strawberries soon after transplanting, to weed-free soil in established crops in early spring before new weeds emerge
- Granules may be applied to onion, leek and brassica nurseries and to most flower crops after bedding out and hardening off
- Do not use pre-emergence of drilled wallflower seed
- Do not use on crops under glass or polythene
- Do not use under extremely wet, dry or other adverse growth conditions
- In the event of crop failure only replant recommended crops in treated soil

Special precautions/Environmental safety
- Irritating to eyes, skin and respiratory system [2], to eyes and skin [1, 3, 5, 6], to skin [4].
- May cause sensitization by skin contact [3-6]

Protective clothing/Label precautions
- A, C [1-3, 5, 6]; A [4]
- 6a, 14, 18, 36, 37, 63 [1-6]; 6b, 16, 21, 25, 28, 66 [1-3, 5, 6]; 6c [2, 3, 5, 6]; 10a [3, 5, 6]; 29 [1, 3-6]; 54, 70 [2-6]; 60 [1, 3, 5, 6]; 67 [4]; 76 [2]

Approval
- Off-label Approval to Sep 2000 for use on outdoor lettuce, lamb's lettuce, frise, radicchio, cress, scarole (all under covers) (OLA 1228/95)[3]; unlimited for use on outdoor leaf herbs (see OLA notice for details) (OLA 0652/95, 0957/95)[3]

419   propamocarb hydrochloride

A translocated protectant carbamate fungicide

See also chlorothalonil + propamocarb hydrochloride
mancozeb + propamocarb hydrochloride

Products					
1 Fisons Filex	Levington	722 g/l	SL	00869
2 Levington Filex	Levington	722 g/l	SL	07631

Uses   Botrytis, downy mildew in PROTECTED LETTUCE *(off-label)* [1]. Damping off, downy mildew in BRASSICAS, ONIONS [1, 2]. Damping off, phytophthora, pythium in POT PLANTS [1, 2]. Damping off, phytophthora, root rot in TOMATOES [1, 2]. Damping off, phytophthora in BEDDING PLANTS, FLOWERS [1, 2]. Damping off, root rot in AUBERGINES, CUCUMBERS, PEPPERS, ROCKWOOL AUBERGINES, ROCKWOOL CUCUMBERS, ROCKWOOL PEPPERS, ROCKWOOL TOMATOES [1, 2]. Damping off, root rot in INERT

SUBSTRATE CUCUMBERS *(off-label)*, INERT SUBSTRATE TOMATOES *(off-label)*, NFT TOMATOES *(off-label)*, ROCKWOOL CUCUMBERS *(off-label)*, ROCKWOOL TOMATOES *(off-label)* [1]. Damping off in CELERY, LEEKS, MARROWS, VEGETABLE SEEDLINGS [1, 2]. Downy mildew, phytophthora in ORNAMENTALS [1, 2]. Downy mildew in RADISHES *(off-label)* [1, 2]. Phytophthora, pythium in FLOWER BULBS, HARDY ORNAMENTAL NURSERY STOCK [1, 2]. Phytophthora in CONTAINER-GROWN STOCK, NURSERY STOCK, PROTECTED CROPS, STRAWBERRIES [1, 2].

| Notes | **Efficacy** |

Efficacy
* Chemical is absorbed through roots and translocated throughout plant
* Incorporate in compost before use or drench moist compost or soil before sowing, pricking out, striking cuttings or potting up. Use treated compost within 2 wk
* Drench treatment can be repeated at 3-6 wk intervals
* Concentrated solution is corrosive to all metals other than stainless steel
* May also be applied in trickle irrigation systems or feed solution
* To prevent root rot in bulbs apply as dip for 20 min or as a pre-planting drench

Crop Safety/Restrictions
* Maximum number of treatments 4 per crop for listed edible crops (except brassicas); 1 per crop for listed brassicas
* When applied over established seedlings rinse off foliage with water and do not apply under hot, dry conditions
* Do not treat young seedlings with overhead drench
* On plants of unknown tolerance test first on a small scale

Protective clothing/Label precautions
* A, B, C, H, K, M
* 21, 29, 35, 54, 60, 61, 63, 67

Latest application/Harvest Interval(HI)
* 6 wk after transplanting for Brussels sprouts.
* HI 14 d for listed edible crops except brassicas; 4 wk for calabrese, cauliflower, sprouting broccoli, Chinese cabbage; 10 wk for cabbage

Approval
* Off-label unlimited for use on outdoor and protected radishes (OLA 1210/95)[1], (OLA 1211/95)[2]; to Aug 1997 for use on protected lettuce (OLA 0939/92)[1]; to Dec 1997 for use on NFT and inert substrate tomatoes, inert substrate cucumbers (OLA 1315/93)[1]

420 propaquizafop
A phenoxy alkanoic acid foliar acting grass herbicide

| Products | Falcon | Cyanamid | 100 g/l | EC | 07025 |

| Uses | Annual grasses, perennial grasses in BULB ONIONS, CARROTS, COMBINING PEAS, FARM FORESTRY, FODDER BEET, LINSEED, MUSTARD, OILSEED RAPE, PARSNIPS, SPRING FIELD BEANS, SUGAR BEET. |

FOR FULL CONDITIONS OF USE ALWAYS READ THE PRODUCT LABEL

Notes

Efficacy
- Apply to emerged weeds when they are growing actively with adequate soil moisture
- Activity is slower under cool conditions
- Broad-leaved weeds and any weeds germinating after treatment are not controlled
- Annual meadow grass up to 3 leaves checked at low doses and severely checked at highest dose
- Spray barley cover crops when risk of wind blow has passed and before there is serious competition with the crop
- Various tank mixtures and sequences recommended for broader spectrum weed control in oilseed rape, peas and sugar beet. See label for details
- Severe couch infestations may require a second application at reduced dose when regrowth has 3-4 leaves unfolded

Crop Safety/Restrictions
- Maximum total dose 2 l/ha per crop or per yr for forestry
- See label for earliest crop growth stages for treatment
- See label for list of tolerant tree species
- Application in high temperatures and/or low soil moisture content may cause chlorotic spotting especially on combining peas and field beans
- Overlaps at the highest dose can cause damage from early applications to carrots and parsnips
- Product contains surfactants. Tank mixing with adjuvants not required or recommended
- An interval of 4 wk must elapse before redrilling a failed treated crop. Only broad-leaved crops may be redrilled

Special precautions/Environmental safety
- Flammable
- Harmful to fish or other aquatic life. Do not contaminate surface waters or ditches with chemical or used container
- Do not allow direct spray from vehicle mounted/drawn sprayers to fall within 6 m, or from hand-held sprayers within 2 m, of surface waters or ditches. Direct spray away from water

Protective clothing/Label precautions
- 12c, 14, 18, 21, 26, 27, 28, 29, 36, 37, 53, 54a, 63, 66

Latest application/Harvest Interval(HI)
- Before crop flower buds visible for winter oilseed rape, linseed, spring field beans; before 8 fully expanded leaf stage for spring oilseed rape, mustard; before weeds are covered by the crop for sugar beet, fodder beet; when pods first set for peas
- HI peas 2 wk; carrots, parsnips, bulb onions 4 wk; sugar beet, fodder beet 8 wk; spring field beans 14 wk

421   propham

A pre-sowing carbamate herbicide available only in mixtures

See also chloridazon + chlorpropham + fenuron + propham
chlorpropham + diuron + propham
chlorpropham + fenuron + propham
chlorpropham + propham

422 propiconazole

A systemic, curative and protectant conazole fungicide

See also carbendazim + propiconazole
chlorothalonil + propiconazole
fenpropidin + propiconazole
fenpropimorph + propiconazole

Products

1 Mantis 250 EC	Ciba Agric.	250 g/l	EC	06240
2 Radar	ICI Agrochem.	250 g/l	EC	01683
3 Radar	Zeneca	250 g/l	EC	06747
4 Standon Propiconazole	Standon	250 g/l	EC	07037
5 Stefes Restore	Stefes	250 g/l	EC	06267
6 Tilt 250 EC	Ciba Agric.	250 g/l	EC	02138

Uses

Alternaria, light leaf spot in OILSEED RAPE [1-6]. Brown rust, net blotch, powdery mildew, rhynchosporium, yellow rust in BARLEY [1-6]. Brown rust, powdery mildew, rhynchosporium, septoria in RYE [1-6]. Brown rust, powdery mildew, septoria, sooty moulds, yellow rust in WHEAT [1-6]. Brown rust, yellow rust in TRITICALE [2, 3, 6]. Cladosporium leaf blotch, rust in LEEKS [6]. Crown rust, drechslera leaf spot, mildew, rhynchosporium in GRASS FOR ENSILING, GRASS SEED CROPS [1-6]. Eyespot in WINTER WHEAT *(low levels only)* [1-4]. Eyespot in WINTER BARLEY *(low levels only)* [4]. Fungus diseases in HONESTY *(off-label)* [6]. Mildew, ramularia leaf spots, rust in SUGAR BEET [1-6]. Powdery mildew in OATS [1-6]. Snow rot in WINTER BARLEY [1-3]. White rust in CHRYSANTHEMUMS *(off-label)* [2, 6].

Notes

Efficacy
- Best results achieved by applying at early stage of disease. Recommended spray programmes vary with crop, disease, season, soil type and product. See label for details
- On leeks apply a programme of 3 treatments at 2-3 wk intervals [6]

Crop Safety/Restrictions
- Maximum number of treatments 4 per crop for wheat (up to 3 in yr of harvest); 3 per crop for barley, oats, rye, triticale (up to 2 in yr of harvest), leeks; 2 per crop or yr for oilseed rape, sugar beet, grass seed crops, honesty; 1 per yr on grass for ensiling
- May be applied up to and including grain watery ripe stage of cereals (GS 71)
- On oilseed rape do not apply during flowering
- A minimum interval of 14 d must elapse between treatments on leeks [6]
- Avoid spraying crops under stress, e.g. during cold weather or periods of frost

Special precautions/Environmental safety
- Irritating to eyes and skin
- Dangerous to fish or other aquatic life. Do not contaminate surface waters or ditches with chemical or used container

Protective clothing/Label precautions
- A, C
- 6a, 6b, 14, 18, 21, 28, 29, 36, 37, 52, 63, 66, 70 [1-6]; 23, 24, 35 [1-3, 5, 6]

Latest application/Harvest Interval(HI)
- Before 4 pairs of true leaves for honesty.

FOR FULL CONDITIONS OF USE ALWAYS READ THE PRODUCT LABEL

- HI cereals, grass and seed crops 5 wk; oilseed rape, sugar beet and grass for ensiling 4 wk; leeks 35 d [6]

Approval
- Approved for aerial application on wheat and barley [1-3, 5, 6], on wheat, barley, rye, oats [4]. See notes in Section 1
- Off-label approval unlimited for use only on *confirmed outbreaks* of white rust on chrysanthemums (OLA 0018/92)[2], (OLA 0019/92)[6]; unlimited for use on outdoor crops of honesty undersown in cereals (OLA 0481/93)[6]
- Approval expiry: 31 Oct 95 [01683]; replaced by MAFF 06747

Maximum Residue Levels (mg residue/kg food)
- grapes 0.5; apricots, peaches 0.2; bananas, tea, hops, ruminant liver 0.1; citrus fruits, tree nuts, pome fruits, strawberries, blackberries, dewberries, loganberries, raspberries, bilberries, cranberries, currants, gooseberries, wild berries, miscellaneous fruits (except bananas), root and tuber vegetables, bulb vegetables, tomatoes, aubergines, sweet corn, brassica vegetables, leaf vegetables and herbs, legume vegetables, asparagus, cardoons, fennel, leeks, rhubarb, fungi, pulses, peanuts, poppy seed, sesame seed, sunflower seed, soya beans, mustard seed, cotton seed, potatoes, cereals, meat (except ruminant liver) 0.05; milk and dairy produce 0.01

423 propiconazole + tebuconazole
A broad spectrum systemic fungicide mixture for cereals

Products					
1	Cogito	Ciba Agric.	250:250 g/l	EC	07384
2	Endeavour	Bayer	250:250 g/l	EC	07385

Uses

Brown rust, glume blotch, powdery mildew, septoria leaf spot, yellow rust in WHEAT. Brown rust, net blotch, powdery mildew, rhynchosporium, yellow rust in BARLEY.

Notes

Efficacy
- Best disease control and yield benefit obtained when applied at early stage of disease development before infection spread to new growth
- To protect the flag leaf and ear from Septoria diseases apply from flag leaf emergence to ear fully emerged (GS 37-59)
- Applications once foliar symptoms of *Septoria tritici* are already present on upper leaves will be less effective

Crop Safety/Restrictions
- Maximum total dose 1 l/ha/crop
- Occasional slight temporary leaf speckling may occur on wheat but this has not been shown to reduce yield response or disease control
- Do not treat durum wheat

Special precautions/Environmental safety
- Irritating to eyes and skin
- Dangerous to fish or other aquatic life. Do not contaminate surface waters or ditches with chemical or used container
- Do not allow direct spray from ground based/vehicle drawn sprayers to fall within 6 m, or from hand-held sprayers to within 2 m, of surface waters or ditches. Direct spray away from water

Protective clothing/Label precautions
* A, C, H
* 6a, 6b, 18, 22a, 29, 36, 37, 52, 54a, 63, 66

Latest application/Harvest Interval(HI)
* Before watery ripe stage (GS 71)

Maximum Residue Levels (mg residue/kg food)
* see propiconazole entry

424   propiconazole + tridemorph

A systemic, curative and protectant fungicide for cereals

Products	Tilt Turbo 475 EC	Ciba Agric.	125:350 g/l	EC	03476

Uses   Brown rust, leaf spot, powdery mildew, sooty moulds, yellow rust in WHEAT. Brown rust, yellow rust in WINTER BARLEY. Net blotch, powdery mildew, rhynchosporium in BARLEY.

Notes

Efficacy
* Apply at start of mildew development and repeat at or just before ear emergence. Treatment of particular benefit where established mildew present at time of spraying
* Sprays may be applied in autumn or spring. Optimum timing varies with crop and disease. See label for details
* Autumn treatment in barley suppresses Typhula snow rot and eyespot

Crop Safety/Restrictions
* Maximum number of treatments 3 per crop (1 autumn plus 2 spring/summer) for winter wheat and barley; 2 per crop for spring wheat and barley
* If used with other products containing propiconazole do not apply a total of more than 500 g propiconazole/ha in wheat or 375 g/ha in barley in any one season
* Only apply up to grain watery ripe stage (GS 71)

Special precautions/Environmental safety
* Irritating to eyes and skin
* Dangerous to fish or other aquatic life. Do not contaminate surface waters or ditches with chemical or used container

Protective clothing/Label precautions
* A, C
* 6a, 6b, 14, 16, 18, 21, 28, 29, 35, 36, 37, 52, 63, 66, 70

Latest application/Harvest Interval(HI)
* HI 5 wk

Approval
* Approved for aerial application on wheat, barley. See notes in Section 1

Maximum Residue Levels (mg residue/kg food)
* see propiconazole entry

FOR FULL CONDITIONS OF USE ALWAYS READ THE PRODUCT LABEL

425 propoxur

A carbamate insecticide for glasshouse and general use

Products   Fumite Propoxur Smoke   Hortichem   50% w/w   FU   00942

Uses   Aphids, whitefly in CHRYSANTHEMUMS, CUCUMBERS, PROTECTED CARNATIONS, TOMATOES.

Notes   **Efficacy**
* Fumigate glasshouse at first sign of whitefly infestation and, where serious, repeat after 7 d. Provides partial control of aphids
* Make house smoke tight, water plants several hours previously so that leaves are dry and damp down paths. Fumigate in late afternoon or evening in calm weather when temperature is 16°C or over. Do not fumigate in bright sunshine
* Keep house closed overnight and ventilate well next morning
* For aphid control in chrysanthemums treat when flower colour first appears in bud

Crop Safety/Restrictions
* Do not treat tomatoes until after third truss has set
* Allow 7 d after fumigation before introducing Encarsia or Phytoseiulus
* Open flowers of carnations and chrysanthemums should be cut before fumigating

Special precautions/Environmental safety
* This product contains an anticholinesterase carbamate compound. Do not use if under medical advice not to work with such compounds
* Irritating to eyes and respiratory system
* Harmful to bees. Do not apply to crops in flower or to those in which bees are actively foraging. Do not apply when flowering weeds are present
* Harmful to fish or other aquatic life. Do not contaminate surface waters or ditches with chemical or used container

Protective clothing/Label precautions
* D
* 18, 28, 29, 36, 37, 50, 53, 63, 70

Latest application/Harvest Interval(HI)
* HI 2 d

426 propyzamide

A residual amide herbicide for use in a wide range of crops
See also clopyralid + propyzamide

Products					
1 Barclay Piza 500	Barclay	50% w/w	WP	05283
2 Headland Judo	Headland	33.5% w/w	SC	07387
3 Kerb 50 W	PBI	50% w/w	WP	01133
4 Kerb Flo	PBI	400 g/l	SC	04521
5 Kerb Granules	PBI	4% w/w	GR	01135
6 Rapier	MTM Agrochem.	450 g/l	SC	05314
7 Standon Propyzamide 50	Standon	50% w/w	WP	05569
8 Stefes Pride	Stefes	50% w/w	WP	05616

Uses   Annual dicotyledons, annual grasses, horsetails, perennial grasses, sedges in FORESTRY [2-5]. Annual dicotyledons, annual grasses, perennial grasses, volunteer cereals, wild oats in WINTER OILSEED RAPE [1-4, 6-8]. Annual dicotyledons, annual grasses, perennial grasses, volunteer cereals, wild oats in WINTER FIELD BEANS [1-4, 7, 8]. Annual dicotyledons, annual grasses, perennial grasses in FARM WOODLAND [3-5]. Annual dicotyledons, annual grasses, perennial grasses in LETTUCE, RASPBERRIES, ROSES [3, 4, 7]. Annual dicotyledons, annual grasses, perennial grasses in RHUBARB, SEED BRASSICAS [3, 4]. Annual dicotyledons, annual grasses, perennial grasses in CHAMOMILE *(off-label)*, CHICORY *(off-label)*, EVENING PRIMROSE *(off-label)*, FENUGREEK *(off-label)*, HONESTY *(off-label)*, PROTECTED HERBS *(off-label)*, RADICCHIO *(off-label)*, SAGE *(off-label)*, TARRAGON *(off-label)*, VETCHES *(off-label)* [3]. Annual dicotyledons, annual grasses, perennial grasses in WOODY ORNAMENTALS [2-5]. Annual dicotyledons, annual grasses, perennial grasses in APPLES, BLACKBERRIES, BLACKCURRANTS, GOOSEBERRIES, LOGANBERRIES, PEARS, PLUMS, REDCURRANTS [2-4, 7]. Annual dicotyledons, annual grasses, perennial grasses in CLOVER SEED CROPS, LUCERNE, STRAWBERRIES, SUGAR BEET SEED CROPS [2-4]. Annual dicotyledons, annual grasses, perennial grasses in RASPBERRIES *(England only)*, RHUBARB *(outdoor)* [2]. Annual dicotyledons, annual grasses in BRASSICA SEED CROPS, OUTDOOR LETTUCE [2].

Notes   **Efficacy**
         • Active via root uptake. Weeds controlled from germination to young seedling stage, some species (including many grasses) also when established
         • Best results achieved by winter application to fine, firm, moist soil. Rain is required after application if soil dry
         • Do not use on soils with more than 10% organic matter except in forestry
         • Excessive organic debris or ploughed-up turf may reduce efficacy
         • For heavy couch infestations a repeat application may be needed in following winter
         • In lettuce lightly incorporate in top 25 mm pre-drilling or irrigate on dry soil

         Crop Safety/Restrictions
         • Maximum number of treatments 1 per crop or yr
         • Apply to most crops from 1 Oct to 31 Jan, to lettuce at any time
         • Apply as soon as possible after 3-true leaf stage of oilseed rape (GS 1,3) and seed brassicas, after 4-leaf stage of sugar beet for seed, within 7 d after sowing but before emergence for field beans, after perennial crops established for at least 1 season, strawberries after 1 yr
         • Only apply to strawberries on heavy soils, to field beans on medium and heavy soils, to established lucerne not less than 7 d after last cut
         • Do not treat protected crops or matted row strawberries
         • See label for lists of ornamental and forest species which may be treated
         • Lettuce may be sown or planted immediately after treatment, with other crops period varies from 5 to 40 wk. See label for details
         • Do not apply to the same land less than 9 mth after an earlier application

         Protective clothing/Label precautions
         • 29, 35, 37, 70 [1, 3-8]; 30 [2]; 54, 63, 67 [1-8]

FOR FULL CONDITIONS OF USE ALWAYS READ THE PRODUCT LABEL

Latest application/Harvest Interval(HI)
- Before 31 Dec for strawberries and winter field beans; before 31 Jan for oilseed rape,sugar beet seed crops, lucerne, seed crops of clover, fodder rape, kale and turnips, top, bush and cane fruit, rhubarb, woody ornamentals, hops, vines, evening primrose, tarragon, chamomile, strawberries, lettuce, radicchio, fenugreek; pre-emergence for chicory.
- HI strawberries, sage 6 wk; protected herbs 3 mth; winter vetches 4 mth

Approval
- May be applied through CDA equipment [4], through CDA equipment in forestry [3]. See label for details. See notes in Section 1
- Off-label approval unlimited for use on evening primrose (OLA 0489/92)[3]; tarragon, chamomile, radicchio, sage, fenugreek, protected herbs (OLA 0624/92)[3]; to June 1997 for use on winter vetches (OLA 0625/92)[3]; to May 1997 for use on chicory grown outdoors as a root crop (OLA 0429/93)[3]; unlimited for use on honesty (OLA 1364/93)[3]

427   pyrazophos
A systemic organophosphorus fungicide with insecticidal activity

Products					
1 Afugan	Promark	300 g/l	EC	00037
2 Afugan	Promark	300 g/l	EC	07301

Uses

Insect pests in BEETROOT *(off-label)*, CARROTS *(off-label)*, CELERIAC *(off-label)*, PARSNIPS *(off-label)*, SALSIFY *(off-label)*. Powdery mildew in APPLES, HOPS, POT PLANTS, ROSES.

Notes

Efficacy
- Spray apples at pink bud and repeat every 10-14 d until extension growth ceased
- Spray hops at first sign of disease or when shoots 10-12 cm long and repeat every 10-14 d. Treatment suppresses damson-hop aphid
- Spray pot plants and roses at first sign of infection and repeat every 10-14 d
- Spray Brussels sprouts from young plant stage to 2 wk before harvest. Use of pendant lances recommended to give good cover
- Temperatures above 30°C may reduce efficacy

Crop Safety/Restrictions
- Maximum number of treatments 4 per crop for beetroot, carrots, celeriac, parsnips, salsify; 7 per crop for protected marrows and melons; 10 per crop for cucumbers, other vegetable crops and ornamentals
- Do not tank-mix with sulfur, dinocap or other organophosphorus compounds
- Treatment may cause some yellowing of hop foliage but effect normally outgrown
- Do not spray roses under glass, aquilegia or scorzonera. Some outdoor roses may be slightly sensitive

Special precautions/Environmental safety
- Harmful if swallowed. Irritating to eyes and skin
- Flammable
- This product contains an anticholinesterase organophosphorous compound. Do not use if under medical advice not to work with such compounds
- Harmful to livestock
- Harmful to game, wild birds and animals

- Dangerous to bees. Do not apply to crops in flower or to those in which bees are actively foraging. Do not apply when flowering weeds are present
- Dangerous to fish or other aquatic life. Do not contaminate surface waters or ditches with chemical or used container

Protective clothing/Label precautions
- A, C
- 1, 5c, 6a, 6b, 12c, 14, 16, 18, 21, 24, 25, 28, 29, 35, 36, 37, 41, 46, 48, 52, 63, 66, 70, 78

Withholding period
- Keep all livestock out of treated areas for at least 2 wk

Latest application/Harvest Interval(HI)
- HI cucumbers 3 d; other vegetable crops 14 d

Approval
- Off-label approval unlimited for use on beetroot, carrots, salsify, celeriac, parsnips (OLA 1402/94)[1], (OLA1403/94)[2]

428   pyrethrins

A non-persistent, contact acting insecticide extracted from Pyrethrum

Products					
	1 Alfadex	Ciba Agric.	0.75 g/l	AL	00074
	2 Dairy Fly Spray	B H & B	0.75 g/l	AL	H5579
	3 Killgerm ULV 400	Killgerm	30 g/l	UL	H4838
	4 Multispray	AgrEvo Environ.	0.065% w/w	AL	H5165
	5 Pybuthrin 33	AgrEvo Environ.	3 g/l	AL	H5106

Uses
Flies in DAIRIES, FARM BUILDINGS, LIVESTOCK HOUSES, POULTRY HOUSES [1-3]. Flies in REFUSE TIPS [4, 5].

Notes

Efficacy
- For fly control close doors and windows and spray or apply fog as appropriate
- For best results outdoors spray during early morning or late afternoon and evening when conditions are still

Crop Safety/Restrictions
- Do not allow spray to contact open food products or food preparing equipment or utensils
- Remove exposed milk and collect eggs before application
- Do not treat plants
- Do not use space sprays containing pyrethrins or pyrethroid more than once per week in intensive or controlled environment animal houses in order to avoid development of resistance. If necessary, use a different control method or product [3]

Special precautions/Environmental safety
- Irritating to eyes, skin [4, 5] and respiratory system [1-3]
- Do not apply directly to livestock
- Harmful to fish or other aquatic life. Do not contaminate surface waters or ditches with chemical or used container [1-3, 5]

FOR FULL CONDITIONS OF USE ALWAYS READ THE PRODUCT LABEL

• Extremely dangerous to fish or other aquatic life and reptiles. Do not contaminate water courses or ground.
• Exclude all persons and animals during treatment [2]

Protective clothing/Label precautions
• A, B, E [2]; A, C, D, H [1, 3]; B, D, E, H [4, 5]
• 1, 23, 33, 51 [4]; 6a, 6b, 53 [1-3, 5]; 6c, 65 [1, 3]; 14, 39 [2, 4, 5]; 18, 28, 36, 37, 63 [1-5]; 21, 67 [2, 5]; 29 [1, 3-5]; 38 [5]

Approval
• Product formulated for ULV application [2]. May be applied through fogging machine or sprayer [1, 2, 4, 5]. See label for details. See notes in Section 1

---

429   pyrethrins + resmethrin
A contact insecticide for many glasshouse and horticultural crops

Products					
1 Pynosect 30 Fogging Solution	Mitchell Cotts	1.6:8.8 g/l	HN	01650
2 Pynosect 30 Water Miscible	Mitchell Cotts	1.4:9.1% w/w	EC	01653

Uses

American serpentine leaf miner in PROTECTED VEGETABLES *(off-label)*[1]. Aphids, thrips, whitefly in ORNAMENTALS [1, 2]. Aphids, whitefly in CUCUMBERS, TOMATOES [1, 2]. Phorid flies, sciarid flies in MUSHROOMS *(Off-label)* [2].

Notes

Efficacy
• Apply when pest first seen and repeat as necessary, for whitefly every 3-4 d
• Apply as high volume spray [2] or with fogging machine [1]. See label for details of suitable equipment
• Fog in late afternoon or early evening, not below 15°C or in bright sunshine

Crop Safety/Restrictions
• Do not spray when temperature above 24°C [2] or fog above 32°C [1]
• Care should be taken in spraying 'soft' cucumber plants grown during winter [2]
• Do not treat crops suffering from drought or stress
• On ornamentals a small test spraying is recommended before treating whole crop

Special precautions/Environmental safety
• Harmful by inhalation or if swallowed. Irritating to eyes, skin and respiratory system [1]
• Extremely dangerous to bees. Do not apply to crops in flower or to those in which bees are actively foraging. Do not apply when flowering weeds are present
• Dangerous [1], extremely dangerous [2], to fish or other aquatic life. Do not contaminate surface waters or ditches with chemical or used container

Protective clothing/Label precautions
• A, C, D, H, J, M [1]
• 5b, 6a, 6b, 14, 18, 22, 29, 36, 37, 52, 65, 78 [1]; 21, 28, 30, 51, 66 [2]; 48a, 63 [1, 2]

Latest application/Harvest Interval(HI)
• HI zero

Approval
• Product formulated for application by thermal fogging machines [1]. See label for details. See notes in Section 1

• Off-label approval unlimited for use on protected vegetables for control of American serpentine leaf miner (OLA 0019/93) [1]; unlimited for use on mushrooms for control of sciarid and phorid flies by cold fogging (OLA 0504/95) or by high volume spray (OLA 0505/95)[2]

430 pyridate
A contact pyridazine herbicide for cereals, oilseed rape and maize

Products

1 Barclay Pirate	Barclay	45% w/w	WP	07104
2 Lentagran WP	Sandoz	45% w/w	WB	07556

Uses

Annual dicotyledons, cleavers, dead nettle, speedwells in MAIZE, SWEETCORN [1, 2]. Annual dicotyledons, cleavers, dead nettle, speedwells in BARLEY, DURUM WHEAT, OATS, OILSEED RAPE, RYE, TRITICALE, WHEAT [1]. Annual dicotyledons, cleavers, dead nettle, speedwells in BRUSSELS SPROUTS, CABBAGES [2].

Notes

Efficacy
• Best results achieved by application to actively growing weeds at 6-8 leaf stage when temperatures are above 8°C before crop foliage forms canopy

Crop Safety/Restrictions
• Maximum number of treatments 1 per crop
• Apply to oilseed rape in winter after 6-true leaf stage (GS 1,6) but before mid-Dec, in spring after crop growth started but before 15 cm of stem extension
• Do not apply in mixture with or within 14 d of any other product which may result in dewaxing of crop foliage
• Apply to winter or spring cereals from first-tiller stage (GS 21)
• Apply to maize and sweetcorn after first-leaf stage. Do not use on cv. Meritos, Sunrise or Tainon 236
• Apply to cabbages and Brussels sprouts after 4 fully expanded leaf stage. Allow 2 wk after transplanting before treating [2]
• Do not use on crops suffering stress from frost, drought, disease or pest attack
• Do not apply to oilseed rape or cereals when night temperature consistently below 2°C or to oilseed rape when daytime temperature exceeds 16°C
• Do not apply to spring sown oilseed rape [2]

Special precautions/Environmental safety
• Irritating to eyes and skin
• May cause sensitisation by skin contact [2]
• Harmful to fish or other aquatic life. Do not contaminate surface waters or ditches with chemical or used container

Protective clothing/Label precautions
• A, C [1, 2]
• 6a, 6b, 18, 21, 29, 36, 37, 53, 63 [1, 2]; 10a, 32 [2]; 67 [1]

FOR FULL CONDITIONS OF USE ALWAYS READ THE PRODUCT LABEL

Latest application/Harvest Interval(HI)
- Before flag leaf ligule just visible (GS 39) for cereals; before flower bud visible (GS 3,1) for oilseed rape; before 7 leaf stage for maize, sweetcorn
- HI maize, sweetcorn 2 mth [1, 2]; cabbage, Brussels sprouts 1 mth [2]

431   pyrifenox
A systemic oxime fungicide for use on apples

Products   Dorado   Zeneca   200 g/l   EC   06657

Uses   Powdery mildew, scab in APPLES.

Notes   **Efficacy**
- Apply from bud-burst at 7-14 d intervals, depending on dose applied, until danger of scab infection ceases or extension growth completed
- Spray programmes which incorporate other suitable fungicides (such as Captan or Nimrod) will help to avoid development of tolerant strains

Crop Safety/Restrictions
- Maximum number of treatments 8 at high, 10 at intermediate and 16 at low dose
- Do not leave spray liquid in sprayer for long period

Special precautions/Environmental safety
- Harmful if swallowed. Irritating to eyes and skin
- Flammable
- Harmful to fish or other aquatic life. Do not contaminate surface waters or ditches with chemical or used container

Protective clothing/Label precautions
- A, C
- 5c, 6a, 6b, 17, 18, 21, 26, 27, 28, 29, 35, 36, 37, 53, 63, 70, 78

Latest application/Harvest Interval(HI)
- HI 14 d

432   quinalbarbitone-sodium
A stupefying agent for control of pest bird species

Products   Killgerm Seconal   Killgerm   100% w/w   SP   04715

Uses   Birds in FIELD CROPS.

Notes   **Efficacy**
- To be used under licence for making up baits

Special precautions/Environmental safety
- Under the Misuse of Drugs Act 1971 and Misuse of Drugs Regulations 1985 a licence is required to produce, possess or supply this product. Licences are issued by the Home Office, Drugs Branch

* Only to be used by operators trained in the use of stupefying baits and familiar with precautions to be observed
* Toxic by inhalation and if swallowed
* Keep in original container, tightly closed, in a safe place, under lock and key
* Any non-target species affected must be allowed to recover and then released
* Affected pest birds must be searched for and humanely killed. The bodies must be disposed of by burning or burial
* All remains of bait must be removed after treatment and burned or buried

Protective clothing/Label precautions
* A, D
* 4b, 4c, 14, 16, 18, 25, 29, 36, 37, 64, 67, 70, 79, 81, 82, 83

Approval
* Only to be used where a licence has been approved in accordance with section 16 (1) of the Wildlife and Countryside Act 1981

---

433   quinalphos

A broad-spectrum contact organophosphorus insecticide

See also disulfoton + quinalphos

Products	Savall	Sandoz	245 g/l	EC	01864

Uses   Carrot fly, cutworms in CARROTS, PARSNIPS. Celery fly in CELERY. Leatherjackets in CEREALS.

Notes

Efficacy
* Apply as soon as damage seen or after Ministry spray warning
* For late generation carrot fly apply in early Aug and early Sep, with further spray in early Oct if necessary

Crop Safety/Restrictions
* Maximum number of treatments 1 per crop for cereals, celery; 3 per crop for carrots, parsnips
* On carrots the maximum total dose applied per crop must not exceed the equivalent of 3 (on mineral soils) or 4 (on organic soils) full dose applications

Special precautions/Environmental safety
* Quinalphos is subject to the Poisons Rules 1982 and the Poisons Act 1972. See notes in Section 1
* This product contains an anticholinesterase organophosphorus compound. Do not use if under medical advice not to work with such compounds
* Harmful by inhalation and if swallowed. Irritating to eyes
* Keep in original container, tightly closed, in a safe place, under lock and key
* Harmful to livestock
* Dangerous to game, wild birds and animals

FOR FULL CONDITIONS OF USE ALWAYS READ THE PRODUCT LABEL

- Dangerous to bees. Do not apply to crops in flower or to those in which bees are actively foraging. Do not apply when flowering weeds are present
- Extremely dangerous to fish or other aquatic life. Do not contaminate surface waters or ditches with chemical or used container
- Do not allow spray from ground vehicle sprayers to fall within 6 m of surface waters or ditches; do not allow spray from hand-held sprayers to fall within 2 m of surface waters or ditches; direct spray away from water

Protective clothing/Label precautions
- A, C, H
- 1, 2, 5b, 5c, 6a, 12c, 14, 16, 18, 22, 25, 28, 29, 35, 36, 37, 41, 45, 48, 51, 64, 66, 70, 78

Withholding period
- Keep all livestock out of treated areas for at least 7 d

Latest application/Harvest Interval(HI)
- HI cereals, carrots, parsnips 3 wk; celery 7 d

434   quinomethionate

A non-systemic quinoxaline fungicide and acaricide for horticultural crops

Products	Morestan	Hortichem	25% w/w	WP	01376

Uses   Currant leaf spot, powdery mildew, red spider mites in GOOSEBERRIES. Currant leaf spot, powdery mildew in BLACKCURRANTS. Powdery mildew, red spider mites in MARROWS. Red spider mites in STRAWBERRIES.

Notes   **Efficacy**
- On recommended varieties of strawberry apply when mites appear in spring/early summer and repeat after harvest if necessary
- For control of mildew and leaf spot apply as soon as disease appears and repeat every 10-14 d if necessary (2 sprays at 2 wk intervals on marrows). Gives partial control of leaf spot on gooseberries

Crop Safety/Restrictions
- Do not apply near hop gardens, pears, alder windbreaks or Templar strawberry as drift may cause damage
- Establish crop tolerance on a few bushes before spraying large areas of blackcurrants of unknown tolerance

Protective clothing/Label precautions
- 21, 29, 54, 63, 67

Latest application/Harvest Interval(HI)
- HI blackcurrants, gooseberries, strawberries 2 wk; marrows (outdoor) 3 d

435 quintozene

A protectant soil applied chlorophenyl fungicide for horticultural crops

Products	1 Quintozene WP	RP Environ.	50% w/w	WP	05404
	2 Terraclor 20D	Hortichem	20% w/w	DP	06578

Uses   Botrytis, rhizoctonia, sclerotinia in BEDDING PLANTS, CHRYSANTHEMUMS, DAHLIAS, FUCHSIAS, LETTUCE, PELARGONIUMS, POT PLANTS, PROTECTED CARNATIONS, TOMATOES [2]. Dollar spot, fusarium patch, red thread in TURF [1]. Rhizoctonia, sclerotinia in ANEMONES, FLOWER BULBS, HYACINTHS, IRISES, NARCISSI, TULIPS [2]. Rhizoctonia in CUCUMBERS [2]. Sclerotinia in CHICORY *(forcing)* [2]. Wirestem in BRASSICAS [2].

Notes   **Efficacy**
* Apply to soil or compost and incorporate before planting or sowing [2]
* For use in turf apply drenching spray when fungal growth active [1]

Crop Safety/Restrictions
* Maximum number of treatments 1 per crop for ornamental bulbs, brassicas, chicory, cucumber, lettuce, tomato, bedding and pot plants [2]; 6 per yr for turf [1]
* In general leave treated soil or compost for 2-3 wk before planting unrooted cuttings, 4 wk before sowing tomatoes or cucumbers and 2-3 d before planting other crops

Special precautions/Environmental safety
* Irritating to skin, eyes and respiratory system

Protective clothing/Label precautions
* A [2]
* 6a, 6b, 6c, 18, 21, 28, 29, 36, 37, 54, 63, 67, 70 [1, 2]

Latest application/Harvest Interval(HI)
* Before sowing or planting for listed crops other than turf

Maximum Residue Levels (mg residue/kg food)
* lettuce 3; bananas 1; potatoes 0.2; tomatoes, peppers, aubergines 0.1; cauliflowers, head cabbages 0.02; beans (with pods) 0.01

436 quizalofop-ethyl

A phenoxypropionic post-emergence grass herbicide

Products	1 Pilot	AgrEvo	500 g/l	SC	07268
	2 Pilot	Schering	500 g/l	SC	03837
	3 Stefes Biggles	Stefes	500 g/l	SC	07083

Uses   Annual grasses, couch, perennial grasses, volunteer cereals in FODDER BEET, MANGELS, MUSTARD, RED BEET, SUGAR BEET, WINTER OILSEED RAPE.

FOR FULL CONDITIONS OF USE ALWAYS READ THE PRODUCT LABEL

Notes

Efficacy
- Apply to emerged weeds in mixture with suitable adjuvant oil, see label for details. Weeds emerging after treatment are not controlled. Effective on annual grasses from 2-leaf stage to fully tillered, on perennials from 4-6 leaf stage to before jointing
- Best results achieved by application to weeds growing actively in warm conditions with adequate soil moisture. Use split treatment to extend period of control
- Annual meadow-grass is not controlled
- Various spray programmes and tank-mixtures recommended to control mixed dicotyledon/grass weed populations. See label for details
- Leave at least 3 d before applying another herbicide
- For effective couch control do not hoe within 21 d after spraying
- At least 4 h rain free period required for effective results

Crop Safety/Restrictions
- Maximum number of treatments 2 per crop
- Apply to beet crops when weeds at appropriate stage and growing actively and not later than 31 Jul. Seed crops may be treated in autumn but consult seed house before use
- Apply to oilseed rape and mustard from expanded cotyledon stage (GS 1,0), before crop covers larger weeds and not later than 31 Jan
- Do not spray crops under stress from any cause or in frosty weather
- In the event of crop failure recommended crops may be resown at any time, broadleaved crops may be sown after 4 wk, cereals after 2-6 wk depending on dose

Special precautions/Environmental safety
- Harmful to fish or other aquatic life. Do not contaminate surface waters or ditches with chemical or used container

Protective clothing/Label precautions
- A
- 22, 29, 35, 53, 60, 63, 66, 70

Latest application/Harvest Interval(HI)
- 4 mth before harvest or 31 Jul, whichever earlier, for sugar beet, fodder beet, red beet and mangels; 6 mth before harvest or 31 Jan, whichever earlier, for oilseed rape and mustard

437   resmethrin

A contact acting pyrethroid insecticide

See also fenitrothion + permethrin + resmethrin
pyrethrins + resmethrin

Products   Turbair Resmethrin Extra   Fargro   6 g/l   UL   02247

Uses   American serpentine leaf miner in PROTECTED VEGETABLES *(off-label)*. Aphids, caterpillars in BRASSICAS. Aphids, sciarid flies, whitefly in AUBERGINES, CHRYSANTHEMUMS, CUCUMBERS, LETTUCE, PEPPERS, POT PLANTS, PROTECTED CARNATIONS, TOMATOES. Aphids in BEANS, SOFT FRUIT. Pine looper in PINE TREES. Sciarid flies in MUSHROOMS.

Notes

Efficacy
- Chemical has rapid contact effect and controls both adult and scale stages of whitefly
- Apply as ULV spray with suitable Turbair sprayer

- Spray when pest appears and repeat as needed, with whitefly every 2-3 d
- Spray when light intensity low. Chemical loses activity when exposed to light
- Do not mix with other products

Crop Safety/Restrictions
- If biological control of whitefly being practised only apply to top third of plants to control newly emerged adult whitefly
- Do not treat Chinese cabbage or red calceolaria

Special precautions/Environmental safety
- Irritating to eyes, skin and respiratory system
- Flammable
- Harmful to bees. Do not apply to crops in flower or to those in which bees are actively foraging. Do not apply when flowering weeds are present
- Harmful to fish or other aquatic life. Do not contaminate surface waters or ditches with chemical or used container

Protective clothing/Label precautions
- 6a, 6b, 6c, 12c, 18, 28, 29, 35, 36, 37, 50, 53, 63, 65, 70

Latest application/Harvest Interval(HI)
- HI zero

Approval
- Product formulated for application through ULV spraying equipment. See label for details. See notes in Section 1
- Off-label approval to Feb 1996 for use in protected fruit and vegetables (OLA 0183/92)[1], (OLA 0422/92)[1]

---

438   resmethrin + tetramethrin
A contact acting pyrethroid insecticide mixture

Products	Sorex Wasp Nest Destroyer	Sorex	0.10:0.05% w/w	AE	H4410

Uses   Wasps in MISCELLANEOUS SITUATIONS.

Notes   **Efficacy**
- Product acts by lowering temperature in and around nest and producing a stupefying vapour in addition to its contact insecticidal effect. Spray only on surfaces
- Apply by directing jet at nest from up to 3 m away and approach nest gradually until jet can be directed into entrance hole. Continue spraying until nest saturated

Special precautions/Environmental safety
- For use only by trained operators and owners of commercial and agricultural premises
- Harmful by inhalation. Contains perchloroethylene
- Keep off skin. Ensure adequate ventilation
- Harmful to caged birds and pets

FOR FULL CONDITIONS OF USE ALWAYS READ THE PRODUCT LABEL

- Dangerous to bees. Do not apply to crops in flower or to those in which bees are actively foraging. Do not apply when flowering weeds are present
- Extremely dangerous to fish or other aquatic life. Remove fish tanks before spraying

Protective clothing/Label precautions
- A, C, H
- 5b, 26, 28, 29, 36, 37, 46, 48, 51, 54, 63

439 rotenone
A natural, contact insecticide of low persistence

Products					
1 Devcol Liquid Derris	Nehra	50 g/l	EC	06063
2 FS Liquid Derris	Ford Smith	50 g/l	EC	01213

Uses   Aphids in FLOWERS, ORNAMENTALS, PROTECTED CROPS, SOFT FRUIT, TOP FRUIT, VEGETABLES. Raspberry beetle in CANE FRUIT. Sawflies in GOOSEBERRIES. Slug sawflies in PEARS, ROSES.

Notes   **Efficacy**
- Apply as high volume spray when pest first seen and repeat as necessary
- Spray raspberries at first pink fruit, loganberries when most of blossom over, blackberries as first blossoms open. Repeat 2 wk later if necessary
- Spray to obtain thorough coverage, especially on undersurfaces of leaves

Special precautions/Environmental safety
- Dangerous to fish or other aquatic life. Do not contaminate surface waters or ditches with chemical or used container
- Flammable

Protective clothing/Label precautions
- 12c, 29, 35, 52, 63, 65, 70

Latest application/Harvest Interval(HI)
- HI 1 d

440 sethoxydim
A post-emergence oxime grass weed herbicide

Products					
Checkmate	RP Agric.	193 g/l	EC	06129

Uses   Annual grasses, perennial grasses in BEDDING PLANTS *(off-label)*, FLOWER BULBS *(off-label)*, HERBACEOUS PERENNIALS *(off-label)*, TREES AND SHRUBS *(off-label)*, WOODY ORNAMENTALS *(off-label)*. Annual grasses, volunteer cereals in BRUSSELS SPROUTS, CABBAGES, CALABRESE, CAULIFLOWERS, COMBINING PEAS, FLAX, LINSEED, LUPINS, MUSTARD, PARSNIPS *(off-label)*, SPRING OILSEED RAPE, VINING PEAS, WINTER OILSEED RAPE. Black bent, couch, creeping bent, volunteer cereals, wild oats in BULB ONIONS, FODDER BEET, MANGELS, POTATOES, RASPBERRIES, STRAWBERRIES, SUGAR BEET. Blackgrass, volunteer cereals, wild oats in CLOVERS *(off-*

label), LUCERNE *(off-label)*, RED FESCUE SEED CROPS *(off-label)*, SAINFOIN *(off-label)*, TREFOIL *(off-label)*.

Notes

Efficacy
* Apply to emerged weeds in combination with Adder or Actipron adjuvant oil (except on peas). Weeds emerging after treatment are not controlled
* Best results when weeds growing actively. Action is faster under warm conditions
* Apply to annual grasses from 2-leaf stage to end of tillering, to couch when largest shoots at least 30 cm long. For effective couch control do not cultivate for 2 wk
* Annual meadow-grass is not controlled
* Apply when rain not expected for at least 2 h
* Various tank mixes recommended for grass/dicotyledon weed populations. See label for details of mixtures and sequential treatments
* Leave at least 7 d before applying another herbicide

Crop Safety/Restrictions
* Maximum number of treatments 1 per crop
* See label for details of crop growth stage and timing
* Do not mix with adjuvant oil on peas. Check pea leaf-wax by Crystal Violet test and do not spray where wax insufficient or damaged
* Do not spray crops suffering damage from other herbicides
* In raspberries apply to inter-row or to base of canes only
* Brassicas, field beans and onions may be sown 3 d after treatment, grass and cereal crops after 4 d

Special precautions/Environmental safety
* Irritating to eyes and skin
* Harmful to fish or other aquatic life. Do not contaminate surface waters or ditches with chemical or used container

Protective clothing/Label precautions
* A, C
* 6a, 6b, 14, 18, 21, 29, 35, 36, 37, 53, 63, 66, 70

Latest application/Harvest Interval(HI)
* HI vining peas, early potatoes, cabbages, Brussels sprouts, calabrese, cauliflowers, bulb onions, raspberries, strawberries 4 wk; seed potatoes 7 wk; maincrop potatoes, sugar beet 8 wk; fodder beet, mangels, parsnips 9 wk; combining peas 11 wk; winter oilseed rape, linseed, mustard, lupins 12 wk

Approval
* Off-label approval to Sep 1996 for use on parsnips, red fescue seed crops, clovers, lucerne, sainfoin, trefoil (OLA 0619/91); unlimited for use on bedding plants, bulbs, herbaceous perennials, ornamental trees and shrubs (OLA 1187/94)

FOR FULL CONDITIONS OF USE ALWAYS READ THE PRODUCT LABEL

441   simazine

A soil-acting triazine herbicide with selective and non-selective uses

See also isoproturon + simazine
         pendimethalin + simazine

Products					
1 Ashlade Simazine 50 FL	Ashlade	500 g/l	SC	06482
2 Atlas Simazine	Atlas	500 g/l	SC	05610
3 Gesatop 50 WP	Ciba Agric.	50% w/w	WP	00983
4 Gesatop 500 SC	Ciba Agric.	500 g/l	SC	05846
5 MSS Simazine 50 FL	Mirfield	500 g/l	SC	01418
6 Unicrop Flowable Simazine	Unicrop	500 g/l	SC	05447

Uses   Annual dicotyledons, annual grasses in MANGE-TOUT PEAS *(off-label)*, RUNNER BEANS *(off-label)* [3, 4]. Annual dicotyledons, annual grasses in MINT *(off-label)* [3]. Annual dicotyledons, annual grasses in BLACKCURRANTS, REDCURRANTS [6]. Annual dicotyledons, annual grasses in GRAPEVINES *(off-label)*, SAGE *(off-label)* [4]. Annual dicotyledons, annual grasses in FOREST NURSERY BEDS, FORESTRY TRANSPLANT LINES, SWEETCORN [1, 3-6]. Annual dicotyledons, annual grasses in APPLES, BROAD BEANS, CANE FRUIT, FIELD BEANS, GOOSEBERRIES, HOPS, PEARS, STRAWBERRIES [1-6]. Annual dicotyledons, annual grasses in CURRANTS [1-5]. Annual dicotyledons, annual grasses in ASPARAGUS, MAIZE, RHUBARB, ROSES, WOODY ORNAMENTALS [1, 2, 5, 6].

Notes   **Efficacy**
- Active via root uptake. Best results achieved by application to fine, firm, moist soil, free of established weeds, when rain falls after treatment
- Do not use on highly organic soils (soils with more than 10% organic matter [6])
- Following repeated use of simazine or other triazine herbicides resistant strains of groundsel and some other annual weeds may develop

Crop Safety/Restrictions
- Maximum number of treatments 1 per yr for maize, sweetcorn, broad and field beans, asparagus, rhubarb, sage, strawberries, hops, bush and cane fruit, apples (lower rate), forest transplant lines; 1 every 2 yr for apples (higher rate); 2 per yr for mint
- May be applied to spring beans up to 2-true leaf stage but control may be poor if weeds already germinated [3, 4]
- Do not spray beans on sandy or gravelly soils or cultivate after treatment. Do not treat varieties Beryl, Feligreen or Rowena
- Apply in top, bush and cane fruit and woody ornamentals established at least 12 mth in Feb-Mar. Roses may be sprayed immediately after planting. See label for lists of resistant and susceptible species
- Apply to strawberries in Jul-Dec, not in spring. Do not treat spring-planted crops established less than 6 mth, or winter-planted crops less than 9 mth. Do not spray varieties Huxley Giant, Madame Moutot or Regina
- Apply to hops overall in Feb-Apr before weeds emerge
- Apply to forest nursery seedbed in second yr or to transplant lines after plants 5 cm tall. Do not treat Norway spruce (Christmas trees)
- To reduce soil run-off on gradients especially in orchard and forest plantations, users are advised to plant grass strips or leave 6 m wide strips between treated areas and surface waters
- On sand, stony or gravelly soils there is risk of crop damage, especially with heavy rain
- Allow at least 7 mth before drilling or planting other crops, longer if weather dry
- Do not sow oats in autumn following spring application in maize

Special precautions/Environmental safety
* Dangerous to fish or other aquatic life and aquatic higher plants. Do not contaminate surface waters or ditches with chemical or used container
* Do not allow direct spray from vehicle-mounted/drawn hydraulic sprayers to fall within 6 m, or from hand-held sprayers to within 2 m, of surface waters or ditches. Direct spray away from water
* Use must be restricted to one product containing atrazine or simazine, and either to a single application at the maximum approved rate or (subject to any existing maximum permitted number of treatments) to several applications at lower doses up to the maximum approved rate for a single application

Protective clothing/Label precautions
* A, C, H [1, 2, 4-6]; B, C, D, H [3]
* 14, 21, 28 [4]; 29 [4, 5]; 30 [6]; 52, 54a, 63 [1-6]; 60 [1-4]; 65 [1-3, 6]; 66 [1-5]; 67 [1-3, 5]; 70 [1-3]

Latest application/Harvest Interval(HI)
* Before emergence of spears for asparagus; 7 d after drilling for maize and sweetcorn; before end of Feb (autumn planted), 10 d after drilling for spring planted field and broad beans; before end of Mar for top, bush and cane fruit; after harvest, before end of Nov for strawberries; after harvest, before 1 May for hops.
* HI mint and sage 4 mth

Approval
* Approvals for sale, supply and use of simazine on non-crop land by the approval holder or his agents were revoked with effect from 1 Sep 1992 and by persons other than the approval holder from 1 Sep 1993
* Off-label Approval to Jan 1997 for use on grapevines (OLA 0299/94)[4]; to May 1998 for use on sage (OLA 0293/94)[4] and mint (OLA 0300/94)[3]; unlimited for use on edible-podsded peas, runner beans (OLA 0296/94)[3], (OLA 0297/94)[4]

442 simazine + trietazine

A pre-emergence residual herbicide for peas, field and broad beans

Products					
1 Remtal SC	AgrEvo	57.5:402.5 g/l	SC	07270
2 Remtal SC	Schering	57.5:402.5 g/l	SC	03827

Uses

Annual dicotyledons in BROAD BEANS, COMBINING PEAS, EDIBLE PODDED PEAS *(off-label)*, FIELD BEANS, VINING PEAS.

Notes

Efficacy
* Chemical acts mainly via roots but some contact effect on cotyledon stage weeds. Gives residual control of germinating weeds for season
* Effectiveness increased by rain after spraying
* Weeds of intermediate susceptibility, including annual meadow grass and blackgrass, may be controlled if adequate rainfall occurs soon after spraying
* Control of deep germinating weeds on heavier soils may be incomplete
* Do not use on soils with more than 10% organic matter

FOR FULL CONDITIONS OF USE ALWAYS READ THE PRODUCT LABEL

- With repeated use of simazine or other triazine herbicides resistant strains of groundsel and some other annual weeds may develop

Crop Safety/Restrictions
- Maximum number of treatments 1 per crop
- Apply to spring peas, winter or spring field and broad beans between drilling and 5% emergence
- On early-drilled peas apply when seed chitting, on later crops as soon as possible after drilling. Do not spray winter sown peas. Vedette peas may be checked by treatment
- Do not spray any forage pea varieties
- Crop seed should be covered by at least 25 mm of settled soil
- Do not use on sand or on gravelly or cloddy soils
- Other crops may be sown or planted 10 wk or more after spraying. With drilled brassicas allow a minimum of 14 wk
- In the event of crop failure only drill peas, field or broad beans without ploughing but do not respray
- To reduce soil run-off users are advised to plant grass strips or leave strips 6 m wide between treated areas and surface waters

Special precautions/Environmental safety
- Harmful if swallowed
- Dangerous to fish or other aquatic life and aquatic higher plants. Do not contaminate surface waters or ditches with chemical or used container
- Do not allow direct spray from vehicle-mounted/drawn hydraulic sprayers to fall within 6 m, or from hand-held sprayers to within 2 m, of surface waters or ditches. Direct spray away from water
- Use must be restricted to one product containing atrazine or simazine either to a single application at maximum approved rate or (subject to any existing maximum permitted number of treatments) to several applications at lower doses up to the maximum approved rate for single application

Protective clothing/Label precautions
- A, B, C, D, H, M
- 5c, 18, 29, 36, 37, 52, 54a, 63, 66, 70, 78

Latest application/Harvest Interval(HI)
- Before 5% total crop emergence

Approval
- May be applied through CDA equipment. See label for details. See notes in Section 1 on ULV application
- Off-label approval to Apr 1997 for use on edible podded peas (OLA 0295/94)[2]

443 sodium chlorate

A non-selective inorganic herbicide for total vegetation control

Products					
1 Atlacide Soluble Powder	Nomix-Chipman	58.2% w/w	SP	00125
2 Centex	Chemsearch	6.4% w/w	SL	00456
3 Cooke's Weedclear	Nehra	50% w/w	SL	06512
4 Deosan Chlorate Weedkiller (Fire Suppressed)	Deosan	55% w/w	SP	00666
5 Doff Sodium Chlorate Weedkiller	Doff Portland	53% w/w	SP	06049
6 TWK Total Weedkiller	Yule Catto	89 g/l	SL	06393

Uses Total vegetation control in NON-CROP AREAS, PATHS AND DRIVES.

Notes **Efficacy**
* Active through foliar and root uptake
* Apply as overall spray at any time during growing season. Best results obtained from application in spring or early summer
* Do not apply before heavy rain

Crop Safety/Restrictions
* Clothing, paper, plant debris etc become highly inflammable when dry if contaminated with sodium chlorate
* Fire risk has been reduced by inclusion of fire depressant but product should not be used in areas of exceptionally high fire risk e.g. oil installations, timber yards

Special precautions/Environmental safety
* Harmful if swallowed
* Oxidizing - contact with combustible material may cause fire
* Wash clothing thoroughly after use
* If clothes become contaminated do not stand near an open fire

Protective clothing/Label precautions
* A, H, M
* 5c, 14, 18, 29, 36, 37, 43, 54, 63, 67, 70, 78

Withholding period
* Keep livestock out of treated areas until foliage of any poisonous weeds such as ragwort has died and become unpalatable

Approval
* Approval expiry: 28 Feb 95 [00666]; replaced by MAFF 05636

444 sodium cyanide

A poisonous gassing compound for control of rabbits and rats

Products				
Cymag	Zeneca	40% w/w	GE	06651

FOR FULL CONDITIONS OF USE ALWAYS READ THE PRODUCT LABEL

Uses   Rabbits, rats in OUTDOOR SITUATIONS.

Notes   **Efficacy**
* Hydrogen cyanide gas is produced when chemical is placed on moist earth. Use only in rabbit and rat holes out of doors and well away from farm or domestic buildings
* Place powder in burrows with a spoon or blow it in with a pump and seal openings with sod of grass. See label for details
* Do not use in wet or windy weather

Special precautions/Environmental safety
* Sodium cyanide is subject to the Poisons Rules 1982 and the Poisons Act 1972. See notes in Section 1
* Very toxic in contact with skin, by inhalation or if swallowed
* Cymag and the gas it evolves are deadly poisons. Use only in the presence of another person aware of the symptoms and first aid treatment for hydrogen cyanide poisoning and provided with amyl nitrite for use in an emergency
* Maximum exposure limits apply to this chemical. See HSE Approved Code of Practice for the Control of Substances Hazardous to Health
* Keep in original container, tightly closed, in a safe place, under lock and key
* Dangerous to people and livestock. Keep them out of treated areas during gassing operations
* Dangerous to fish or other aquatic life. Do not contaminate surface waters or ditches with chemical or used container

Protective clothing/Label precautions
* 3a, 3b, 3c, 18, 25, 26, 27, 29, 40, 52, 64, 68, 70, 79; because of its highly toxic nature a special set of precautions applies to this product and must be followed carefully. See label for details.

445   sodium monochloroacetate
A contact herbicide for various horticultural crops

Products
1 Atlas Somon	Atlas	96% w/w	SP	03045
2 Croptex Steel	Hortichem	95% w/w	SP	02418

Uses   Annual dicotyledons in BROCCOLI *(off-label)*, CALABRESE *(off-label)*, CAULIFLOWERS *(off-label)* [2]. Annual dicotyledons in APPLES, BLACKCURRANTS, BRUSSELS SPROUTS, CABBAGES, GOOSEBERRIES, KALE, LEEKS, ONIONS, PEARS, PLUMS, REDCURRANTS [1, 2]. Basal defoliation in HOPS *(off-label)* [2]. Basal defoliation in HOPS [1]. Sucker control in BLACKBERRIES *(off-label)*, HYBRID BERRIES *(off-label)*, LOGANBERRIES *(off-label)*, RASPBERRIES *(off-label)* [2].

Notes   **Efficacy**
* Best results achieved by application to emerged weed seedlings up to young plant stage under good growing conditions
* Effectiveness reduced by rain within 12 h
* May be used for hop defoliation in tank-mixture with tar-oil [1], with Wayfarer adjuvant [2]. Apply as a directed spray to the base of the plant when they are up to 20 cm high, after training when main shoots are at least 100 cm high

Crop Safety/Restrictions
- Maximum number of treatments 1 per crop or yr; 2 per yr for sucker control in cane fruit and hop defoliation
- Apply to brassicas from 2-4 leaf stage or after recovery from transplanting
- Do not spray cabbage that has begun to heart
- Apply to onions and leeks after crook stage but before 4-leaf stage
- Safety on brassicas, onions and leeks depends on presence of adequate leaf wax, check by crystal violet wax test. Do not add wetters, pesticides or nutrients to spray
- Apply in fruit crops established for at least 1 yr as a directed spray
- Do not spray if frost likely or if temperature likely to exceed 27°C
- Apply for sucker control in raspberries when canes 10-20 cm high with addition of Wayfarer adjuvant [2]

Special precautions/Environmental safety
- Harmful if swallowed, in contact with skin and by inhalation
- Irritating to respiratory system. Risk of serious damage to eyes
- Corrosive [1], causes burns [1, 2]
- Harmful to bees. Do not apply to crops in flower or to those in which bees are actively foraging. Do not apply when flowering weeds are present

Protective clothing/Label precautions
- A, C, D, E [1, 2]
- 5a, 5b, 5c, 7, 8, 9, 23, 24, 43, 70 [1]; 6a, 6b, 54 [2]; 6c, 18, 21, 28, 29, 36, 37, 42, 50, 63, 65 [1, 2]

Withholding period
- Keep livestock, especially poultry, out of treated areas for at least 2 wk

Latest application/Harvest Interval(HI)
- Before 4 true leaf stage for onions and leeks.
- HI 21 d for cabbage, kale, Brussels sprouts; 28 d for cane fruit; 14 wk for hops

Approval
- Off-label approval unlimited for use on cane fruit (OLA 0115/93)[2], hops (OLA 0357/93)[2]; to Apr 1997 for use on hops to within 6 wk of harvest (OLA 0358/93)[2]; unlimited for use on cauliflowers, broccoli, calabrese (OLA 0961/95)[2]

446   sodium silver thiosulfate
A plant-growth regulator used to extend life of flowers

Products	Argylene	Fargro	8% w/w	SP	03386

Uses   Prolonging flower life in GLASSHOUSE CUT FLOWERS, POT PLANTS.

Notes   **Efficacy**
- Acts by inhibiting production of ethylene
- Spray flowering pot plants to run off 8-14 d before shipment from glasshouse or at 3 wk intervals. Best time for spraying is late afternoon

FOR FULL CONDITIONS OF USE ALWAYS READ THE PRODUCT LABEL

• Vase life of cut flowers extended by dip treatment immediately after cutting

Crop Safety/Restrictions
• Spraying pot plants just before shipping may cause damage

Special precautions/Environmental safety
• Not to be used on food crops

Protective clothing/Label precautions
• 21, 29, 34, 54, 63, 67

---

447   strychnine hydrochloride (commodity substance)
A vertebrate control agent for destruction of moles underground

Products   strychnine   various

Uses   Moles in GRASSLAND *(areas of restricted public access)*, MISCELLANEOUS SITUATIONS *(areas of restricted public access)*.

Notes   **Efficacy**
• For use as poison bait against moles on commercial agricultural/horticultural land where public access restricted, on grassland associated with aircraft landing strips, horse paddocks, race and golf courses and other areas specifically approved by Agriculture Departments

Special precautions/Environmental safety
• Strychnine is subject to the Poisons Rules 1982 and the Poisons Act 1972. See notes in Section 1
• Must only be supplied to holders of an Authority to Purchase issued by appropriate Agricultural Department. Authorities to Purchase may only be issued to persons who satisfy the appropriate authority that they are trained and competent in its use
• Only to be supplied in original sealed pack in units up to 2 g
• Store in original container under lock and key and only on premises under control of holder of Authority to Purchase or a named individual
• A written COSHH assessment must be made before using
• Must be prepared for application with great care so that there is no contamination of the ground surface
• Any prepared bait remaining at the end of the day must be buried
• Operators must be supplied with a Section 6 (HSW) Safety Data Sheet before commencing work
• Other restrictions apply, see PR 1990, Issue 12

Protective clothing/Label precautions
• A
• 3

---

448   sulfonated cod liver oil
An animal repellent

Products	Scuttle	Fine	800 g/l	EC	06232

Uses	Deer, rabbits in BARLEY, CABBAGES, CAULIFLOWERS, FORESTRY, LETTUCE, OATS, OILSEED RAPE, WHEAT.

Notes

Efficacy
* Apply diluted as a spray before crop is being grazed or attacked or undiluted as a dip for forestry seedlings
* A rain-free period of about 5 h is required after application

Crop Safety/Restrictions
* Before using on vegetables users should consider risk of taint and consult processor before using on crops for processing

Special precautions/Environmental safety
* Do not apply directly to animals
* Dangerous to bees. Do not apply to crops in flower or to those in which bees are actively foraging. Do not apply when flowering weeds are present.
* Dangerous to fish or other aquatic life. Do not contaminate surface waters or ditches with chemical or used container

Protective clothing/Label precautions
* A, C
* 18, 29, 36, 37, 48, 52, 63, 66

---

449   sulfur

A broad-spectrum inorganic protectant fungicide and foliar feed
See also carbendazim + maneb + sulfur
       copper oxychloride + maneb + sulfur
       copper sulfate + sulfur

Products

1	Ashlade Sulphur FL	Ashlade	720 g/l	SC	06478
2	Atlas Sulphur 80 FL	Atlas	800 g/l	SC	03802
3	Headland Sulphur	Headland	800 g/l	SC	03714
4	Kumulus DF	BASF	80% w/w	SG	04707
5	Luxan Micro-Sulphur	Luxan	80% w/w	SG	06565
6	Microsul Flowable Sulphur	Stoller	960 g/l	SC	03907
7	Microthiol Special	PBI	80% w/w	MG	06268
8	MSS Sulphur 80	Mirfield	80% w/w	WG	05752
9	MTM Sulphur Flowable	MTM Agrochem.	800 g/l	SC	05312
10	Solfa	Atlas	80% w/w	MG	03529
11	Stoller Flowable Sulphur	Stoller	720 g/l	SC	03760
12	Sulphur Flowable	United Phosphorus	800 g/l	SC	07526
13	Thiovit	Sandoz	80% w/w	MG	02125
14	Thiovit	Sandoz	80% w/w	WG	05572
15	Tripart Imber	Tripart	720 g/l	SC	04050

Uses	Foliar feed, powdery mildew in WHEAT [3-7, 9-13, 15]. Foliar feed, powdery mildew in BARLEY [3-13, 15]. Foliar feed, powdery mildew in OATS [3, 6-8, 10, 11, 15]. Foliar feed, powdery mildew in TURNIPS [7, 10, 15]. Foliar feed, powdery mildew in HOPS [1-4, 6-15].

FOR FULL CONDITIONS OF USE ALWAYS READ THE PRODUCT LABEL

Foliar feed, powdery mildew in GRAPEVINES [1, 2, 4, 6-8, 10, 11, 15]. Foliar feed, powdery mildew in SWEDES [3, 5, 7, 9-15]. Foliar feed in OILSEED RAPE [3-5, 7, 10, 13, 15]. Foliar feed in GRASSLAND [2, 5, 7, 8, 10, 13]. Foliar feed in WINTER OILSEED RAPE, WINTER WHEAT [8]. Foliar feed in BRASSICAS, POTATOES, SOFT FRUIT, TOP FRUIT, TURF, VEGETABLES [7]. Gall mite in BLACKCURRANTS [1, 4, 8, 10, 15]. Powdery mildew, scab in APPLES [1-4, 6, 8-12, 15]. Powdery mildew, scab in PEARS [3, 6, 9, 11, 12]. Powdery mildew in CUCUMBERS *(off-label)*, PARSNIPS *(off-label)*, PROTECTED HERBS *(off-label)*, PROTECTED TOMATOES *(off-label)* [13]. Powdery mildew in ORNAMENTALS [10]. Powdery mildew in RYE [3]. Powdery mildew in STRAWBERRIES [1-4, 6, 8-12, 15]. Powdery mildew in SUGAR BEET [1-15]. Powdery mildew in GOOSEBERRIES [1, 2, 4, 8, 10, 15].

Notes

Efficacy
- Apply when disease first appears and repeat 2-3 wk later. Details of application rates and timing vary with crop, disease and product. See label for information
- Sulfur acts as foliar feed as well as fungicide and with some crops product labels vary in whether treatment recommended for disease control or growth promotion
- In grassland best at least 2 wk before cutting for hay or silage, 3 wk before grazing
- Treatment unlikely to be effective if disease already established in crop

Crop Safety/Restrictions
- Maximum number of treatments 2 per yr for sugar beet (3 per yr [1, 15]), cereals, parsnips, swedes, hops; 3 per yr for blackcurrants, gooseberries [3, 4, 6748]
- May be applied to cereals up to grain watery-ripe stage (GS 71) [1-4, 6, 8-11, 15], up to milky-ripe stage (GS 75) [13], to first node detectable (GS 31) [5]
- Do not use on sulfur-shy apples (Beauty of Bath, Belle de Boskoop, Cox's Orange Pippin, Lanes Prince Albert, Lord Derby, Newton Wonder, Rival, Stirling Castle) or pears (Doyenne du Comice)
- Do not use on gooseberry cultivars Careless, Early Sulphur, Golden Drop, Leveller, Lord Derby, Roaring Lion, or Yellow Rough
- Do not use on apples or gooseberries when young, under stress or if frost imminent
- Do not use on fruit for processing, on grapevines during flowering or near harvest on grapes for wine-making
- Do not use on hops at or after burr stage
- Do not spray top or soft fruit with oil or within 30 d of an oil-containing spray

Special precautions/Environmental safety
- Highly flammable [14]
- Harmful to fish or other aquatic life. Do not contaminate surface waters or ditches with chemical or used container [2]

Protective clothing/Label precautions
- 12b, 18 [14]; 29 [1, 12]; 30 [3-11, 13, 15]; 37 [2, 7]; 53, [3802]29 [2]; 54 [1, 3-15]; 60 [12]; 63 [1-15]; 65 [1]; 66 [2, 12]; 67 [3-11, 13-15]

Latest application/Harvest Interval(HI)
- Before 30 Sep in yr of harvest for sugar beet, swedes, parsnips; up to and including fruit swell for gooseberries; up to burr stage for hops.
- HI protected tomatoes 2 d [13]; grass for hay or silage 2 wk [7]

Approval
- Approved for aerial application on wheat, barley, winter oilseed rape [3, 4]; cereals, oilseed rape, sugar beet, swedes [13]. See notes in Section 1

• Off-label Approval unlimited for use on protected tomatoes (OLA 0720/91)[13], protected herbs (OLA 0491/92)[13], cucumbers (OLA 0928, 0929/92)[13], parsnips (OLA 0984/93)[13]

450   sulfuric acid (commodity substance)
A strong acid used as an agricultural desiccant

Products   sulfuric acid   various   77% w/w   SL

Uses   Haulm destruction in POTATOES. Pre-harvest desiccation in LINSEED, NARCISSI, ONIONS.

Notes   **Efficacy**
• Apply with suitable equipment between 1 Mar and 15 Nov for potatoes or narcissus bulbs, between 1 Aug and 15 Oct for linseed, between 1 Apr and 30 Aug for onions

Crop Safety/Restrictions
• Maximum number of treatments 3 per crop for potatoes; 2 per crop for linseed and onions, 1 per season for narcissus bulbs

Special precautions/Environmental safety
• Sulfuric acid is subject to the Poisons Rules 1982 and the Poisons Act 1972. See notes in Section 1
• Must only be used by suitably trained operators competent in use of equipment for applying sulfuric acid
• Not to be applied using hand-held or pedestrian controlled applicators
• A written COSHH assessment must be made before use. Operators should observe OES set out in HSE guidance note EH40/90 or subsequent issues
• Operators must have liquid suitable for eye irrigation immediately available at all times throughout spraying operation
• Spray must not be deposited within 1 m of public footpaths
• Written notice of any intended spraying must be given to owners of neighbouring land and warning notices posted. See PR 1990, Issue 12 for details
• Unprotected persons must be kept out of treated areas for at least 96 h after treatment

Protective clothing/Label precautions
• A, C, H, K, M
• 7, 8, 9

FOR FULL CONDITIONS OF USE ALWAYS READ THE PRODUCT LABEL

451   tar oils

Hydrocarbon and phenolic oils used as insecticidal and fungicidal winter washes

See also anthracene oil

Products

1 Sterilite Tar Oil Winter Wash 60% Stock Emulsion	Coventry Chemicals	636 g/l	EC	05061
2 Sterilite Tar Oil Winter Wash 80% Stock Emulsion	Coventry Chemicals	800 g/l	EC	05062

Uses

Aphids, scale insects, suckers, winter moth in APPLES. Aphids, scale insects, winter moth in CANE FRUIT, CHERRIES, CURRANTS, GOOSEBERRIES, NECTARINES, PEACHES, PEARS, PLUMS. Scale insects in GRAPEVINES.

Notes

Efficacy
* Spray kills hibernating insects and eggs as well as moss and lichens on trunk
* Spray winter wash over dormant branches and twigs. See labels for detailed timings
* Ensure that all parts of trees or bushes are completely wetted, especially cracks and crevices. Use higher concentration for cleaning up neglected orchard
* Treatment also partially controls winter and raspberry moths and reduces Botrytis on currants and powdery mildew on currants and gooseberries
* Combine with spring insecticide treatment for control of caterpillars
* When used in cold weather always add product to water and stir thoroughly before use
* Apply hop defoliant by spraying when bines 1.5-3 m high and direct spray downward at 45°

Crop Safety/Restrictions
* Maximum number of treatments 2 per yr for hops
* Only spray fruit trees or bushes when fully dormant
* Do not spray hops if temperature is above 21°C or after cones have formed. Do not drench rootstocks
* Do not spray on windy, wet or frosty days

Special precautions/Environmental safety
* Harmful if swallowed. Irritating to eyes, skin and respiratory system
* Dangerous to fish or other aquatic life. Do not contaminate surface waters or ditches with chemical or used container

Protective clothing/Label precautions
* A, C, H
* 5c, 6a, 6b, 6c, 18, 21, 28, 29, 36, 37, 52, 60, 61, 63, 65, 70, 78

Latest application/Harvest Interval(HI)
* Before any signs of bud swelling for fruit crops

452 tebuconazole

A systemic conazole fungicide for cereals and oilseed rape
See also fenpropidin + tebuconazole
propiconazole + tebuconazole

Products

	1 Folicur	Bayer	250 g/l	EW	06386
	2 Standon Tebuconazole	Standon	250 g/l	EC	07408

Uses

Alternaria, light leaf spot, powdery mildew, ring spot in BRUSSELS SPROUTS, CABBAGES [1]. Alternaria, light leaf spot, sclerotinia stem rot in OILSEED RAPE [1]. Alternaria in SPRING OILSEED RAPE, WINTER OILSEED RAPE [2]. Brown rust, fusarium ear blight, glume blotch, powdery mildew, septoria leaf spot, sooty moulds, yellow rust in WHEAT [1]. Brown rust, net blotch, powdery mildew, rhynchosporium, yellow rust in BARLEY [1]. Brown rust, net blotch, powdery mildew, rhynchosporium, yellow rust in SPRING BARLEY, WINTER BARLEY [2]. Brown rust, powdery mildew, rhynchosporium, yellow rust in RYE [1, 2]. Brown rust, powdery mildew, septoria diseases, yellow rust in SPRING WHEAT, WINTER WHEAT [2]. Chocolate spot, rust in FIELD BEANS [1]. Light leaf spot, sclerotinia in SPRING OILSEED RAPE *(reduction)*, WINTER OILSEED RAPE *(reduction)* [2]. Powdery mildew in SWEDES, TURNIPS [1]. Ring spot in OILSEED RAPE *(reduction)* [1].

Notes

Efficacy
- For best results apply at an early stage of disease development before infection spreads to new crop growth
- To protect flag leaf and ear from Septoria diseases apply from flag leaf emergence to ear fully emerged (GS 37-59). Earlier application may be necessary where there is a high risk of infection
- For light leaf spot control in oilseed rape apply in autumn/winter with a follow-up spray in spring/summer if required
- For control of most other diseases spray at first signs of infection with a follow-up spray 2-4 wk later if necessary. See label for details
- For disease control in Brussels sprouts and cabbages a 3-spray programme at 21-28 d intervals will give good control

Crop Safety/Restrictions
- Maximum total dose 2 l product/ha per crop for wheat, barley, rye, field beans, swedes, turnips; 2.25 l/ha per crop for Brussels sprouts, cabbages; 2.5 l/ha for oilseed rape
- Some transient leaf speckling may occur after application to wheat but this has not been shown to reduce yield response to disease control
- Do not treat durum wheat
- Do not apply before swedes and turnips have a root diameter of 2.5 cm, or before start of button formation in Brussels sprouts or heart formation in cabbages

Special precautions/Environmental safety
- Harmful if swallowed, risk of serious damage to eyes
- Harmful to fish or other aquatic life and aquatic plants. Do not contaminate surface waters or ditches with chemical or used container

Protective clothing/Label precautions
- A, C, H

FOR FULL CONDITIONS OF USE ALWAYS READ THE PRODUCT LABEL

• 5c, 9, 18, 24, 29, 36, 37, 53, 63, 66, 70, 78

Latest application/Harvest Interval(HI)
• Before grain milky-ripe for cereals (GS 71); when most seed green for oilseed rape (GS 6, 3)
• HI field beans, swedes, turnips 35 d; Brussels sprouts, cabbages 21 d

---

453   tebuconazole + triadimenol
A broad spectrum systemic fungicide for cereals

Products	Silvacur	Bayer	250:125 g/l	EC	06387

Uses   Brown rust, fusarium ear blight, glume blotch, leaf spot, powdery mildew, sooty moulds, yellow rust in WHEAT. Brown rust, net blotch, powdery mildew, rhynchosporium, yellow rust in BARLEY. Brown rust, powdery mildew, yellow rust in RYE.

Notes   **Efficacy**
• For best results apply at an early stage of disease development before infection spreads to new crop growth
• To protect flag leaf and ear from Septoria diseases apply from flag leaf emergence to ear fully emerged (GS 37-59). Earlier application may be necessary where there is a high risk of infection
• For control of rust, powdery mildew, leaf and net blotch apply at first signs of disease with a second application 2-3 wk later if necessary

Crop Safety/Restrictions
• A maximum dose of 2 l product/ha per crop must not be exceeded
• Do not use on durum wheat
• Some transient leaf speckling may occur after application to wheat but this has not been shown to reduce yield response or disease control

Special precautions/Environmental safety
• Irritating to eyes
• Harmful to fish or other aquatic life and aquatic plants. Do not contaminate surface waters or ditches with chemical or used container

Protective clothing/Label precautions
• A, C, H
• 6a, 18, 22a, 29, 36, 37, 53, 63, 66, 70, 78

Latest application/Harvest Interval(HI)
• Before grain milky-ripe (GS 71)

---

454   tebuconazole + triazoxide
A triazole and benzotriazine fungicide seed treatment for use in barley

Products					
1	Raxil S	Bayer	20:20 g/l	LS	06974
2	Raxil S	Zeneca	20:20 g/l	LS	07484

Uses   Leaf stripe, loose smut in BARLEY.

Notes   **Efficacy**
- Best applied through recommended seed treatment machines
- Evenness of seed cover improved by simultaneous application of equal volumes of product and water or dilution of product with an equal volume of water
- Diluted product must be used immediately
- Drill treated seed in the same season

Crop Safety/Restrictions
- Maximum number of treatments 1 per batch of seed
- Slightly delayed and reduced emergence may occur but this is normally outgrown
- Field emergence which is delayed for any reason may be accentuated by treatment
- Do not use on seed with more than 16% moisture content, or on sprouted, cracked or skinned seed

Special precautions/Environmental safety
- Harmful to fish or other aquatic life. Do not contaminate surface waters or ditches with chemical or used container
- Treated seed not to be used as food or feed, nor to be applied from the air
- Treated seed harmful to game and wildlife. Bury spillages
- Product supplied in returnable (1000 l) container. See label for guidance on handling, storage and protective clothing [2]

Protective clothing/Label precautions
- A, H [1, 2]
- 20, 69, 70, 71a, 72 [2]; 29, 53, 63, 73, 74, 75, 76, 77 [1, 2]; 67 [1]

Latest application/Harvest Interval(HI)
- Before drilling

455   tebuconazole + tridemorph
A broad spectrum fungicide mixture for cereals

Products   Allicur   Bayer   125:165 g/l   EC   06468

Uses   Botrytis, brown rust, fusarium ear blight, glume blotch, powdery mildew, septoria leaf spot, sooty moulds, yellow rust in WINTER WHEAT. Brown rust, net blotch, powdery mildew, rhynchosporium, yellow rust in BARLEY.

Notes   **Efficacy**
- For best results apply at an early stage of disease development before infection spreads to new growth
- Repeat treatments may be necessary under high disease pressure or when reinfection occurs
- To protect flag leaf and ear from Septoria diseases apply from flag leaf emergence (GS 37) but before grain is milky ripe (GS 73)

FOR FULL CONDITIONS OF USE ALWAYS READ THE PRODUCT LABEL

Crop Safety/Restrictions
* Maximum total dose 4 l/ha per crop
* Occasional slight temporary leaf speckling may occur on wheat but this has not been shown to reduce yield response or disease control
* Do not treat durum wheat

Special precautions/Environmental safety
* Harmful if swallowed
* Irritating to eyes
* Harmful to fish or other aquatic life and aquatic plants. Do not contaminate surface waters or ditches with chemical or used container

Protective clothing/Label precautions
* A, C, H
* 5c, 6a, 16, 18, 22, 29, 36, 37, 43 (14 d), 53, 63, 66, 70, 78

Withholding period
* Keep livestock out of treated areas for at least 14 d following treatment

Latest application/Harvest Interval(HI)
* Before grain is milky ripe (GS 73)

---

456   tebufenpyrad
A pyrazole mitochondrial electron transport inhibitor

Products	Masai		Cyanamid	20% w/w	WB	07452

Uses   Damson-hop aphid, two-spotted spider mite in HOPS. Red spider mites in APPLES. Two-spotted spider mite in PROTECTED ROSES.

Notes

Efficacy
* Acts on eggs (except winter eggs) and all motile stages of spider mites up to adults
* Treat spider mites from 80% egg hatch but before mites become established
* Repeat treatment may be necessary after 2-4 wk
* Treat hops before they reach tops of the wires and not later than end of burr stage. For best results good spray coverage of the whole bine is necessary
* Product can be used in a programme to give season-long control of damson/hop aphids coupled with mite control

Crop Safety/Restrictions
* Maximum total dose 1.0 kg/ha per yr for top fruit; 6.0 kg/ha per yr for protected roses, hops
* Do not treat apples between pink bud and end of flowering
* Product has no effect on fruit quality or finish
* Small-scale testing of rose varieties to establish tolerance recommended before use
* Where repeat treatment would result in exceeding maximum total dose another approved acaricide should be used
* Product not recommended for use in hand-held sprayers

Special precautions/Environmental safety
* Dangerous to fish or other aquatic life. Do not contaminate surface waters or ditches with chemical or used container

* Harmful to bees. Do not apply to crops in flower or to those in which bees are foraging, except as directed on hops. Do not apply when flowering weeds are present.
* Do not allow direct spray from ground based vehicle mounted/drawn sprayers to fall within 6 m of surface waters or ditches. Do not allow direct spray from air-assisted sprayers to fall within 18 m of surface waters or ditches. Direct spray away from water

Protective clothing/Label precautions
* A, C, H
* 14, 22, 25, 26, 29, 32, 35 (7 d), 50, 52, 54a, 63, 67

Latest application/Harvest Interval(HI)
* End of burr stage for hops
* HI apples 7d

---

457   tebutam

A pre-emergence amide herbicide used in brassica crops

Products	Comodor 600	Agrichem	600 g/l	EC	06808

Uses

Annual dicotyledons, annual grasses, volunteer cereals in BROCCOLI, BRUSSELS SPROUTS, CABBAGES, CALABRESE, CAULIFLOWERS, FODDER RAPE, KALE, SWEDES, TURNIPS, WINTER OILSEED RAPE.

Notes

Efficacy
* Acts via roots and inhibits germination of weeds. Emerged weeds are not controlled
* Best results are achieved on seedbeds which are weed free and have a good tilth without clods
* Effectiveness requires adequate soil moisture. Under very dry conditions light rain or irrigation may be needed but heavy rain on light soils may result in loss of control
* On some heavy soils there may be a reduction in weed control. Do not use on soils with more than 10% organic matter
* Recommended for use in tank-mixture with propachlor and as a tank-mix or in sequence with trifluralin. Apply trifluralin tank-mix as a pre-plant incorporated treatment
* Emerged weeds can be controlled pre-crop emergence by tank mixtures with paraquat products. See label for details

Crop Safety/Restrictions
* Maximum number of treatments 1 per crop
* Apply to brassica crops before drilling or planting in tank-mixture with trifluralin as a soil incorporated treatment. Do not use on brassica seedbeds
* Apply to oilseed rape or sown or transplanted brassicas shortly after drilling or planting but before weeds or crop germinate
* Transplants must be hardened off before treatment. Do not use on block or modular propagated brassicas
* Soil must be deep mouldboard ploughed before drilling or planting any following crop except brassicas

---

FOR FULL CONDITIONS OF USE ALWAYS READ THE PRODUCT LABEL

- In the event of crop failure only brassicas, oilseed rape, dwarf beans, maize, peas or potatoes may be grown in the same season

Special precautions/Environmental safety
- Irritating to eyes and skin
- Flammable
- Harmful to fish or other aquatic life. Do not contaminate surface waters or ditches with chemical or used container

Protective clothing/Label precautions
- A
- 6a, 6b, 12c, 18, 21, 28, 29, 36, 37, 53, 63, 66

Latest application/Harvest Interval(HI)
- Pre-crop emergence for oilseed rape and other drilled named brassicas; 3 wk after transplanting for transplanted brassicas

---

458   tecnazene

A protectant chlorobenzene fungicide and potato sprout suppressant

See also carbendazim + tecnazene

Products					
	1 Atlas Tecgran 100	Atlas	10% w/w	GR	05574
	2 Atlas Tecnazene 6% Dust	Atlas	6% w/w	DP	06351
	3 Bygran F	Wheatley	10% w/w	GR	00365
	4 Bygran S	Wheatley	10% w/w	GR	00366
	5 Fusarex Granules	Zeneca	10% w/w	GR	06668
	6 Hickstor 10 Granules	Hickson & Welch	10% w/w	GR	03121
	7 Hickstor 3	Hickson & Welch	3% w/w	DP	03105
	8 Hickstor 5 Granules	Hickson & Welch	5% w/w	GR	03180
	9 Hortag Tecnazene 10% Granules	Hortag	10% w/w	GR	03966
	10 Hortag Tecnazene Double Dust	Hortag	6% w/w	DP	01072
	11 Hortag Tecnazene Dust	Hortag	3% w/w	DP	01074
	12 Hortag Tecnazene Potato Granules	Hortag	5% w/w	GR	01075
	13 Hystore 10	Agrichem	10% w/w	GR	03581
	14 Hytec	Agrichem	3% w/w	DP	01099
	15 Hytec 6	Agrichem	6% w/w	DP	03580
	16 Nebulin	Wheatley	300 g/l	HN	01469
	17 New Hickstor 6	Hickson & Welch	6% w/w	DP	04221
	18 New Hystore	Agrichem	5% w/w	GR	01485
	19 Tripart Arena 10G	Hickson & Welch	10% w/w	GR	05603
	20 Tripart Arena 3	Hickson & Welch	3% w/w	DP	05605
	21 Tripart Arena 5G	Hickson & Welch	5% w/w	GR	05604
	22 Tripart New Arena 6	Hickson & Welch	6% w/w	DP	05813

Uses	Dry rot, sprout suppression in WARE POTATOES [1-22]. Dry rot in SEED POTATOES [1-22].

Notes

Efficacy
- Apply dust or granules evenly to potatoes with appropriate equipment (see label for details) during loading. Works by vapour phase action so clamps or bins should be covered to avoid loss of vapour
- Best results achieved on dry, dirt-free potatoes. Fungicidal activity best at 12-14°C, sprouting control best at 9-12°C. For best sprouting control maintain temperature at 10-14°C for first 14 d, then reduce to 7°C [5]
- Under suitable storage conditions sprouting prevented for 4-6 mth
- Sprouting not controlled in tubers which have already broken dormancy
- Treatment can be used in sequence with chlorpropham treatment [1, 4]
- Treatment gives some control of skin spot, gangrene and silver scurf

Crop Safety/Restrictions
- Maximum number of treatments 1 per batch of potatoes
- Air tubers for 6 wk before planting; chitting should have commenced
- Treated tubers must not be removed for sale or processing, including washing, for at least 6 wk after application
- To prevent contamination of other crops remove all traces of treated potatoes and soil after the store has been emptied

Special precautions/Environmental safety
- Dangerous to fish or other aquatic life. Do not contaminate surface waters or ditches with chemical or used container [1]
- Do not remove from store for sale or processing (including washing) for at least 6 wk after application

Protective clothing/Label precautions
- 28, 63, 67 [1-22]; 29, 35, 65, 70 [3-12, 14-22]; 30 [1, 2, 13]; 52 [1]; 54 [2-22]

Latest application/Harvest Interval(HI)
- 6 wk before removal from store for sale or processing

Approval
- One product formulated for application by thermal fogging [16]. See label for details. See notes in Section 1 on ULV application

Maximum Residue Levels (mg residue/kg food)
- potatoes 10 (ACP review - see PR 1995, Issue 7); lettuce 2

459   tecnazene + thiabendazole

A fungicide and potato sprout suppressant

Products					
1	Hytec Super	Agrichem	6:1.8% w/w	DP	01100
2	Storite SS	MSD AGVET	300:100 g/l	FS	02034

Uses   Dry rot, gangrene, silver scurf, skin spot, sprout suppression in WARE POTATOES. Dry rot, gangrene, silver scurf, skin spot in SEED POTATOES.

FOR FULL CONDITIONS OF USE ALWAYS READ THE PRODUCT LABEL

Notes   **Efficacy**
- Apply dust evenly with suitable dusting machine (not by hand) as tubers enter store and cover as soon as possible [1]
- Apply liquid with suitable low-volume mist applicator or spinning disc [2]
- Treat tubers as soon as possible after lifting, no longer than 14 d
- Under suitable storage conditions sprouting prevented for 3-6 mth. Effectiveness may be decreased with excessive ventilation or inadequate covering
- Best results achieved on dry, dirt-free tubers
- Sprouting not controlled on tubers that have broken dormancy

Crop Safety/Restrictions
- Maximum number of treatments 1 per batch
- Treated tubers must not be removed for sale or processing, including washing, for at least 6 wk after application
- Air tubers for 6 wk before planting, chitting should have commenced
- Do not mix Storite SS with thiabendazole (Storite Clear Liquid)

Special precautions/Environmental safety
- Harmful to fish or other aquatic life. Do not contaminate surface waters or ditches with chemical or used container

Protective clothing/Label precautions
- A, C, D, H, M [2]; A, C, D, H [1]
- 21, 65, 72, 76 [2]; 28, 53, 63 [1, 2]; 29, 67, 70 [1]

Latest application/Harvest Interval(HI)
- 6 wk before removal from store for sale or processing (including washing)

Maximum Residue Levels (mg residue/kg food)
- see tecnazene and thiabendazole entries

460   teflubenzuron

A benzoylurea insecticide for use on ornamentals

Products   Nemolt   Fargro   150 g/l   SC   04274

Uses   Browntail moth, caterpillars, sciarid flies, western flower thrips, whitefly in BEDDING PLANTS, CHRYSANTHEMUMS, FUCHSIAS, PELARGONIUMS, POINSETTIAS, POT PLANTS, ROSES.

Notes   **Efficacy**
- Product acts by preventing development of chitin, thus interfering with insect development
- Best results achieved when used as preventative treatment or when first signs of pest seen. Monitoring with use of sticky traps advised
- Adult insects are not killed and addition of a suitable 'knockdown' chemical recommended if desired
- Use 2 applications against caterpillars, 3 against whitefly
- For sciarid fly control add to compost and repeat after 8 wk
- May be used in conjunction with integrated pest management

Crop Safety/Restrictions
- Maximum number of treatments 3 per yr

- Test specific varieties before carrying out extensive treatments

Protective clothing/Label precautions
- 30, 54, 60, 63, 67

461   tefluthrin

A soil acting pyrethroid insecticide seed treatment for sugar and fodder beet

Products					
	Force ST	Zeneca	200 g/l	CS	06665

Uses   Carrot fly in OUTDOOR CARROTS *(Off-label)*, OUTDOOR PARSNIPS *(Off-label)*. Millipedes, pygmy beetle, springtails, symphylids in FODDER BEET, SUGAR BEET.

Notes   **Efficacy**
- Apply during process of pelleting seed (needs specialist equipment -consult firm for details)

Crop Safety/Restrictions
- Maximum number of treatments 1 per batch of seed
- Sow treated seed as soon as possible. Do not store treated seed from one drilling season to next

Special precautions/Environmental safety
- Irritant. May cause sensitization by skin contact
- Extremely dangerous to fish or other aquatic life. Do not contaminate surface waters or ditches with chemical or used container

Protective clothing/Label precautions
- B, C, D, H
- 10a, 14, 16, 18, 21, 29, 36, 37, 51, 63, 65, 72, 73, 74, 75, 76, 77

Withholding period
- Keep livestock out of areas drilled with treated seed for at least 80 d

Latest application/Harvest Interval(HI)
- Before drilling seed

Approval
- Off-label approval unlimited for use on carrots and parsnips (OLA 0537/95)[1]

462   terbacil

A residual uracil herbicide for use in asparagus

Products					
	Sinbar	DuPont	80% w/w	WP	01956

FOR FULL CONDITIONS OF USE ALWAYS READ THE PRODUCT LABEL

Uses   Annual dicotyledons, annual grasses, bent grasses, couch in ASPARAGUS, OUTDOOR HERBS *(off-label)*, PROTECTED HERBS *(off-label)*.

Notes   **Efficacy**
* Has some foliar activity but uptake mainly through roots
* Adequate rainfall needed to ensure penetration to weed root zone
* Best results achieved by application in early spring, before mid-Apr

Crop Safety/Restrictions
* Maximum number of treatments 1 per yr
* Apply to asparagus established at least 2 yr. Do not apply after first spears emerge
* Do not use on Sands, Loamy Sand or gravels or soils with less than 1% organic matter
* Do not use for at least 2 yr prior to grubbing and replanting

Protective clothing/Label precautions
* 21, 28, 29, 54, 63, 65

Latest application/Harvest Interval(HI)
* Before spears emerge for asparagus; pre-emergence of new growth in spring for herbs

Approval
* Off-label approval unlimited for use in range of herbs, see off-label approval notice for details (OLA 0082/93)

---

463   terbuthylazine

A triazine herbicide available only in mixtures

See also cyanazine + terbuthylazine
diflufenican + terbuthylazine
isoxaben + terbuthylazine

---

464   terbuthylazine + terbutryn

A pre-emergence herbicide for peas, beans, lupins and potatoes

Products   Opogard 500 SC   Ciba Agric.   150:350 g/l   SC   05850

Uses   Annual dicotyledons, annual grasses in BROAD BEANS, COMBINING PEAS, FIELD BEANS, LUPINS, POTATOES, VINING PEAS.

Notes   **Efficacy**
* Active via root uptake and with foliar activity on cotyledon stage weeds
* Best results achieved by application to fine, firm, moist seedbed, preferably at weed emergence, when rain falls after spraying
* Effectiveness may be reduced by excessive rain, drought or cold
* Residual control lasts for up to 8 wk on mineral soils. Effectiveness reduced on highly organic soils and subsequent use of post-emergence treatment recommended
* Do not cultivate after treatment
* Do not use on soils which are very cloddy or have more than 10% organic matter

Crop Safety/Restrictions
- Maximum number of treatments 1 per crop
- Apply to spring sown peas, beans and lupins as soon as possible after drilling
- Crop seed must be covered by at least 25 mm of settled soil
- Vedette and Printana peas may be damaged by treatment. Do not use on forage peas
- Heavy rain after application may cause damage to peas on light soils
- Do not treat peas, beans or lupins on soils lighter than loamy fine sand, potatoes on soils lighter than loamy sand or on very stony soils or lupins on silty clay soil
- Subsequent crops may be sown or planted after 12 wk (14 wk if prolonged drought)

Special precautions/Environmental safety
- Harmful if swallowed
- Harmful to fish or other aquatic life. Do not contaminate surface waters or ditches with chemical or used container

Protective clothing/Label precautions
- A, C
- 5c, 18, 21, 28, 29, 36, 37, 53, 63, 66, 70

Latest application/Harvest Interval(HI)
- 3 d before emergence for peas, beans and lupins; before 10% of plants emerged for potatoes

Approval
- May be applied through CDA equipment for use on peas, beans and lupins. See label for details. See notes in Section 1 on ULV spraying

---

465   terbutryn

A residual triazine herbicide for cereals and aquatic weed control

See also fomesafen + terbutryn
      prometryn + terbutryn
      terbuthylazine + terbutryn

Products					
1 Clarosan 1 FG	Ciba Agric.	1% w/w	GR	03859
2 Prebane SC	Ciba Agric.	490 g/l	SC	07364

Uses   Annual dicotyledons, annual meadow grass, blackgrass, perennial ryegrass, rough meadow grass in WINTER BARLEY, WINTER WHEAT [2]. Aquatic weeds in AREAS OF WATER [1].

Notes   **Efficacy**
- Apply pre-weed emergence in autumn. Best results given on fine, firm seedbed in which any trash or ash dispersed. Do not use on soils with more than 10% organic matter [2]
- Effectiveness reduced by long, dry period after spraying or by waterlogging [2]
- Do not harrow after treatment [2]
- For aquatic weed control apply granules to water surface with suitable equipment, when growth active but before heavy infestation develops, usually Apr-May, sometimes to Aug. Do not use if maximum water flow exceeds 1 m/3 min [1]
- Effectiveness reduced in water with peaty bottom [1]

FOR FULL CONDITIONS OF USE ALWAYS READ THE PRODUCT LABEL

Crop Safety/Restrictions
* Apply to cereal crops after drilling but before emergence and before 30 Nov [2]
* Only use lower rate on durum wheat, winter oats, triticale or winter rye. Higher rate used to control blackgrass and ryegrass should only be used on winter wheat and barley (not on barley on Sands or very light soils) [2]
* Do not use on spring barley cultivars sown in autumn [2]
* Before spraying direct-drilled crops ensure seed well covered and drill slots closed [2]
* Risk of crop damage on stony or gravelly soils especially if heavy rain falls soon after treatment [2]

Special precautions/Environmental safety
* Do not use treated water for irrigation purposes within 7 d of treatment [1]
* Concentration in water must not exceed 10 ppm [1]
* Harmful to fish or other aquatic life. Do not contaminate surface waters or ditches with chemical or used container [2]

Protective clothing/Label precautions
* 29, 63 [1, 2]; 53, 60, 66 [2]; 55, 56, 67 [1]

Approval
* Approved for aquatic weed control. See notes in Section 1 on use of herbicides in or near water [1]
* Approved for aerial application on cereals, rye, triticale [2]. See notes in Section 1
* May be applied through CDA equipment. See label for details. See notes in Section 1 on ULV application [2]

---

466   terbutryn + trietazine

A residual triazine herbicide for peas, potatoes and field beans

Products	1 Senate	AgrEvo	250:250 g/l	SC	07279
	2 Senate	Schering	250:250 g/l	SC	04275

Uses   Annual dicotyledons, annual meadow grass in FIELD BEANS, PEAS, POTATOES.

Notes

Efficacy
* Product acts mainly through roots of germinating seedlings but also has some contact effect on cotyledon stage weeds
* Weeds germinating from deeper than 2.5 cm may not be completely controlled
* Weed control is improved by light rain after spraying but reduced by excessive rain, drought or cold
* Persistence may be reduced on soils with high organic matter content

Crop Safety/Restrictions
* Maximum number of treatments 1 per crop
* Apply between drilling and 5% emergence (peas), pre-emergence (field beans) between sowing and 10% emergence (potatoes), provided weed not beyond cotyledon stage
* May be used on spring sown vining, combining or forage peas and field beans covered by not less than 3 cm soil
* Do not spray any winter sown peas or beans sown before Jan
* May be used on early and maincrop potatoes, including seed crops

- On stony or gravelly soils there is risk of crop damage, especially if heavy rain falls soon after treatment. Frost after application may also check crop
- Do not use on soils classed as Sands
- Plough or cultivate to 15 cm before sowing or planting another crop. Any crop may be sown or planted after 12 wk (14 wk with drilled brassicas)
- If sprayed pea crop fails redrill without ploughing but do not respray

Special precautions/Environmental safety
- Harmful if swallowed
- Harmful to fish or other aquatic life. Do not contaminate surface waters or ditches with chemical or used container

Protective clothing/Label precautions
- A, C
- 5c, 18, 21, 29, 36, 37, 53, 63, 66, 70, 78

Latest application/Harvest Interval(HI)
- Pre-emergence of crop for field beans; 5% emergence for peas; 10% emergence for potatoes

467   terbutryn + trifluralin
A residual, pre-emergence herbicide for winter cereals

Products	Ashlade Summit	Ashlade	150:200 g/l	SC	06214

Uses   Annual dicotyledons, annual grasses, annual meadow grass, blackgrass, chickweed, mayweeds, speedwells in WINTER BARLEY, WINTER WHEAT.

Notes   **Efficacy**
- Apply as soon as possible after drilling, before weed or crop emergence when soil moist
- Best results achieved by application to fine, firm, moist seedbed when adequate rain falls after application. Do not use on crops drilled after 30 Nov
- Do not use on uncultivated stubbles or soils with trash or more than 10% organic matter. Do not harrow after treatment
- Effectiveness reduced in mild, wet winters, especially in south-west

Crop Safety/Restrictions
- Maximum number of treatments 1 per crop
- Crops should only be treated if drilled 25-35 mm deep
- Do not use on sandy soils (Coarse Sandy Loam - Coarse Sand) or on spring-drilled crops
- Do not treat undersown crops or crops to be undersown
- Crops suffering stress due to waterlogging, deficiency, poor seedbed preparation or pest problems may be damaged
- Any crop may be sown following harvest of treated crop. In the event of crop failure only resow with wheat or barley

Special precautions/Environmental safety
- Irritating to skin, eyes and respiratory system

FOR FULL CONDITIONS OF USE ALWAYS READ THE PRODUCT LABEL

- Harmful to fish or other aquatic life. Do not contaminate surface waters or ditches with chemical or used container

Protective clothing/Label precautions
- A, C
- 6a, 6b, 6c, 18, 21, 25, 28, 29, 36, 37, 53, 63, 66

Latest application/Harvest Interval(HI)
- Pre-emergence of crop

---

468 tetradifon

A bridged-diphenyl acaricide for use in horticultural crops

See also dicofol + tetradifon

Products	Tedion V-18 EC	Hortichem	80 g/l	EC	03820

Uses Red spider mites in APPLES, APRICOTS, BLACKBERRIES, BLACKCURRANTS, CHERRIES, CUCUMBERS, FLOWERS, FRENCH BEANS, GOOSEBERRIES, GRAPEVINES, HOPS, LOGANBERRIES, NECTARINES, ORNAMENTALS, PEACHES, PEARS, PEPPERS, PLUMS, PROTECTED MELONS, RASPBERRIES, STRAWBERRIES, TOMATOES.

Notes **Efficacy**
- Chemical active against summer eggs and all stages of red spider larvae but does not kill adult mites
- Apply as soon as first mites seen and repeat as required. See label for details of timing
- May be used in conjunction with biological control
- Resistant strains of red spider mite have developed in some areas

Crop Safety/Restrictions
- Some rose cultivars are slightly susceptible. Do not treat cissus, dahlia, ficus, kalanchoe or primula

Special precautions/Environmental safety
- Harmful in contact with skin and if swallowed. Irritating to eyes and skin
- Flammable

Protective clothing/Label precautions
- A, C
- 5a, 5c, 6a, 6b, 18, 21, 29, 36, 37, 54, 63, 65, 70, 78

---

469 tetramethrin

A contact acting pyrethroid insecticide

See also phenothrin + tetramethin
phenothrin + tetramethrin
resmethrin + tetramethrin

Products	Killgerm Py-Kill W	Killgerm	20 g/l	EC	H4632

Uses Flies in AGRICULTURAL PREMISES, LIVESTOCK HOUSES.

Notes **Efficacy**
* Dilute in accordance with directions and apply as space or surface spray

Crop Safety/Restrictions
* Do not apply directly on food or livestock
* Remove exposed milk before application. Protect milk machinery and containers from contamination
* Do not use space sprays containing pyrethrins or pyrethroid more than once per wk in intensive or controlled environment animal houses in order to avoid development of resistance. If necessary, use a different control method or product

Special precautions/Environmental safety
* Harmful to fish or other aquatic life. Do not contaminate surface waters or ditches with chemical or used container

Protective clothing/Label precautions
* 12c, 21, 28, 29, 53, 63, 65

470   thiabendazole

A systemic, curative and protectant benzimidazole (MBC) fungicide

See also *carboxin + imazalil + thiabendazole*
carboxin + thiabendazole
ethirimol + flutriafol + thiabendazole
gamma-HCH + thiabendazole + thiram
imazalil + thiabendazole
tecnazene + thiabendazole

Products

1	Hykeep	Agrichem	2% ww	DP	06744
2	Hymush	Agrichem	60% w/w	WP	01092
3	Storite Clear Liquid	MSD AGVET	220 g/l	LS	02032
4	Storite Flowable	MSD AGVET	450 g/l	FS	02033
5	Tecto Flowable Turf Fungicide	Vitax	450 g/l	SC	06273

Uses Black scurf and stem canker, skin spot in SEED POTATOES *(pre-planting)*[1]. Dollar spot, fusarium patch, red thread in AMENITY GRASS, TURF [5]. Dry bubble, wet bubble in MUSHROOMS [2]. Dry rot, gangrene, silver scurf, skin spot in SEED POTATOES *(post-harvest)*, WARE POTATOES *(post-harvest)* [1, 3, 4]. Dutch elm disease in ELM TREES *(injection - off-label)* [3]. Fusarium basal rot in NARCISSI [3].

Notes **Efficacy**
* Apply post-harvest treatment to potatoes using suitable equipment within 24 h of lifting. See label for details
* Apply to turf as spray after mowing and do not mow for at least 48 h. Best results on red thread obtained in combination with fertilizer [5]
* Apply to bulbs as dip. Ensure bulbs are clean [3]

FOR FULL CONDITIONS OF USE ALWAYS READ THE PRODUCT LABEL

- Apply to mushroom casing as drench, by mechanical incorporation or by spraying between flushes [2]

Crop Safety/Restrictions
- Maximum number of treatments 1 per batch for seed treatments and as dip and module drench for asparagus; 10 per crop for drench in field grown asparagus
- Maximum total dose 680 g per 100 sq m [2]

Special precautions/Environmental safety
- Harmful to fish or other aquatic life. Do not contaminate surface waters or ditches with chemical or used container
- Treated seed potatoes must not be used for food or feed [1, 3, 4]

Protective clothing/Label precautions
- A, C, D, H, M [1, 2]
- 28, 37 [1, 3-5]; 29, 65 [3-5]; 30 [1, 2]; 53, 63, 67 [1-5]; 62 [1]

Latest application/Harvest Interval(HI)
- Before planting for seed potatoes; 21 d before removal from store for sale, processing or consumption for ware potatoes.
- HI asparagus 6 mth; mushrooms 5 d

Approval
- One product formulated for application with ULV equipment. See label for details. See notes in Section 1 [3]
- Off-label Approval unlimited for use elm trees (OLA 0990/95)[3]

Maximum Residue Levels (mg residue/kg food)
- potatoes 5

---

471 thiabendazole + thiram
A fungicide seed dressing mixture for field and vegetable crops

Products

1 Hy-TL	Agrichem	225:300 g/l	FS	06246
2 Hy-Vic	Agrichem	255:255 g/l	FS	06247

Uses

Ascochyta, damping off in PEAS [1]. Ascochyta, damping off in BROAD BEANS, FIELD BEANS [2]. Damping off, seed-borne diseases in CABBAGES, CARROTS, CAULIFLOWERS, FLAX, GRASS SEED, KALE, LEEKS, LINSEED, LUPINS, OILSEED RAPE, ONIONS, PARSNIPS, SWEDES, TURNIPS [2].

Notes

Efficacy
- Dress seed as near to sowing as possible
- Dilution may be needed with particularly absorbent types of seed. If diluted material used, seed may require drying before storage

Crop Safety/Restrictions
- Maximum number of treatments 1 per batch

Special precautions/Environmental safety
- Harmful if swallowed and in contact with skin. Irritating to eyes, skin and respiratory system
- Treated seed not to be used as food or feed. Do not re-use sack. Bury spillages

• Dangerous to fish or other aquatic life. Do not contaminate surface waters or ditches with chemical or used container

Protective clothing/Label precautions
• A, C, D, H, M [1, 2]
• 5a, 5c, 6a, 6b, 6c, 18, 21, 29, 36, 37, 52, 63, 65, 70, 78 [1, 2]; 72, 73, 75, 76, 77 [1]

Latest application/Harvest Interval(HI)
• Before drilling
• Not less than 3 mth before grazing for vetches and grass [2]

Maximum Residue Levels (mg residue/kg food)
• see thiabendazole entry

---

472   thifensulfuron-methyl
A translocated sulfonylurea herbicide
See also metsulfuron-methyl + thifensulfuron-methyl

Products					
1 Crackshot	Cyanamid	75% w/w	WB	07542
2 Prospect	DuPont	75% w/w	WB	06541

Uses   Docks in ESTABLISHED GRASSLAND.

Notes   **Efficacy**
• Best results achieved from application to young green dock foliage when growing actively
• Only broad-leaved docks (*Rumex obtusifolius*) are controlled; curled docks (*Rumex crispus*) are resistant
• Apply 7-10 d before grazing and do not graze for 7 d afterwards
• Docks with developing or mature seed head should be topped and the regrowth treated later
• Established docks with large tap roots may require follow-up treatment
• High populations or poached grassland will require further treatment in following yr
• Ensure good spray coverage and apply to dry foliage

Crop Safety/Restrictions
• Maximum number of treatments 1 per calender yr
• Do not treat new leys in year of sowing
• Do not treat where nutrient imbalances, drought, waterlogging, low temperatures, lime deficiency, pest or disease attack have reduced sward vigour
• Do not roll or harrow within 7 d of spraying
• Product may cause a check to both sward and clover which is usually outgrown

Special precautions/Environmental safety
• Extremely dangerous to aquatic higher plants. Do not contaminate surface waters or ditches with chemical or used container
• Do not allow direct spray from ground-based vehicle mounted/drawn sprayers to fall within 6 m, or from hand-held sprayers to within 2 m, of surface waters or ditches. Direct spray away from water

---

FOR FULL CONDITIONS OF USE ALWAYS READ THE PRODUCT LABEL

- Take extreme care to avoid drift onto broad-leaved plants outside the target area or onto surface waters or ditches, or land intended for cropping
- Use recommended procedure to clean out spraying equipment
- Only grass or cereals may be sown within 4 wk of treatment

Protective clothing/Label precautions
- 21, 28, 29, 43 (7 d), 51a, 54a, 63, 67

Withholding period
- Keep livestock out of treated areas for at least 7 d following treatment

Latest application/Harvest Interval(HI)
- Before 1 Aug

473   thifensulfuron-methyl + tribenuron-methyl
A mixture of two sulfonylurea herbicides for cereals

Products	Duk 110	DuPont	50:25% w/w	WG	06266

Uses   Annual dicotyledons, charlock, chickweed, mayweeds in BARLEY, WHEAT.

Notes   **Efficacy**
- Apply after 1 Feb when weeds are small and actively growing
- Ensure good spray cover of the weeds
- Ensure that weeds present are those that are susceptible. See label
- Follow label mixing instructions
- Effectiveness reduced by rain within 4 h of treatment

Crop Safety/Restrictions
- Maximum number of treatments 1 per crop
- May be sprayed on all varieties of wheat and barley from 3 leaf stage (GS 13) up to and including flag leaf fully emerged (GS 39)
- Igri winter barley may suffer damage during period of rapid growth and must not be treated before leaf sheath erect stage (GS 30)
- Do not apply to cereals undersown with grass, clover or other legumes
- Do not apply within 7 d of rolling
- Various tank mixtures recommended to broaden weed control spectrum. Other mixtures are specifically excluded. See label
- Do not apply in sequence or in tank mixture with a product containing any other sulfonyl urea
- Do not apply to any crop suffering from stress or not actively growing
- Take particular care to avoid damage by drift onto broad-leaved plants outside the target area or onto surface waters or ditches
- Only cereals, field beans or oilseed rape may be sown in the same calendar year as harvest of a treated crop. In the event of failure of a treated crop sow only a cereal crop within 3 mth of product application

Special precautions/Environmental safety
- Extremely dangerous to aquatic higher plants. Do not contaminate surface waters or ditches wirth chemical or used container

• Do not allow direct spray from ground-based vehicle mounted/drawn sprayers to fall within 6 m, or from hand-held sprayers to within 2 m, of surface waters or ditches. Direct spray away from water

Protective clothing/Label precautions
• 21, 28, 29, 51a, 54a, 63, 67

Latest application/Harvest Interval(HI)
• Before flag leaf ligule first visible (GS 39)

474   thiodicarb

A carbamamate insecticide with molluscicide uses

Products	Genesis	RP Agric.	4% w/w	RB	06168

Uses   Slugs, snails in BARLEY, DURUM WHEAT, OATS, OILSEED RAPE, TRITICALE, WHEAT.

Notes   **Efficacy**
• Apply pelleted bait as broadcast treatment

Crop Safety/Restrictions
• Maximum number of treatments 3 per crop

Special precautions/Environmental safety
• Product contains an anticholinesterase carbamate compound. Do not use if under medical advice not to work with such compounds
• Dangerous to fish or other aquatic life. Do not contaminate surface waters or ditches with chemical or used container

Protective clothing/Label precautions
• A, D, H
• 2, 5c, 6a, 10a, 18, 25, 29, 36, 37, 46, 52, 63, 67, 70, 78

Latest application/Harvest Interval(HI)
• First node detectable stage (GS 31) for cereals; stem extension stage (GS 2,5) for oilseed rape

475   thiophanate-methyl

A carbendazim precursor fungicide with protectant and curative activity

See also gamma-HCH + thiophanate-methyl
           iprodione + thiophanate methyl
           iprodione + thiophanate-methyl

Products					
1 Mildothane Liquid	RP Agric.	500 g/l	SC	06211
2 Mildothane Turf Liquid	RP Environ.	500 g/l	SC	05331

FOR FULL CONDITIONS OF USE ALWAYS READ THE PRODUCT LABEL

Uses Botrytis in DWARF BEANS [1]. Canker, powdery mildew, scab, storage rots in APPLES [1]. Chocolate spot in FIELD BEANS [1]. Dollar spot, fusarium patch, red thread in TURF [2]. Powdery mildew, scab in PEARS [1]. Powdery mildew in CUCUMBERS [1].

Notes **Efficacy**
* Number and timing of sprays varies with crop and disease, see label for details. On tree fruit sprays should be repeated at 14 d intervals
* Spray treatments can reduce wood canker on apples
* Apply a post-blossom spray or dip fruit to reduce storage rot diseases
* Apply spray or drench treatment on cucumbers. High volume sprays are not recommended where Phytoseiulus is used to control red spider mites
* Apply turf liquid during period of active growth and do not mow for 48 h
* Do not mix with MCPB herbicides or copper. See label for compatible mixtures

Crop Safety/Restrictions
* Maximum number of treatments (including applications of products containing carbendazim or benomyl) 12 per crop as pre-harvest spray on apples and pears, 1 per batch as post-harvest dip; 1 per crop for dwarf beans; 2 per crop for field beans; 6 per crop for cucumbers
* Spraying apples from blossom to fruitlet stage may increase russeting on prone cultivars
* Do not use on fruit trees at petal fall or early fruitlet stage when preceded by a protectant such as captan

Special precautions/Environmental safety
* To dispose of tip empty solution into an approved soakaway, not into drains, open ditches or soakaways

Protective clothing/Label precautions
* A, B, C, H, K, M [1]; A, C, H, M [2]
* 30, 54, 63, 67 [1, 2]

Latest application/Harvest Interval(HI)
* Before end of flowering for field beans; before pods fully formed for dwarf beans
* HI apples, pears 7 d; cucumbers 2 d

476 thiram

A protectant, dithiocarbamate fungicide and animal repellant
See also carboxin + gamma-HCH + thiram
fenpropimorph + gamma-HCH + thiram
gamma-HCH + thiabendazole + thiram
gamma-HCH + thiram
metalaxyl + thiabendazole + thiram
metalaxyl + thiram
permethrin + thiram
thiabendazole + thiram

Products

1 Agrichem Flowable Thiram	Agrichem	600 g/l	FS	06245
2 FS Thiram 15% Dust	Ford Smith	15% w/w	DP	07428
3 Unicrop Thianosan DG	Unicrop	80% w/w	WG	05454

Uses Botrytis, bottom rot, downy mildew in PROTECTED LETTUCE [2, 3]. Botrytis, cane spot, spur blight in RASPBERRIES [3]. Botrytis, rust in CHRYSANTHEMUMS, PROTECTED

CARNATIONS [3]. Botrytis in FREESIAS, OUTDOOR LETTUCE, POT PLANTS, PROTECTED FLOWERS, STRAWBERRIES, TOMATOES [3]. Damping off in BROAD BEANS *(seed treatment)*, CABBAGES *(seed treatment)*, CARROTS *(seed treatment)*, CAULIFLOWERS *(seed treatment)*, DWARF BEANS *(seed treatment)*, FIELD BEANS *(seed treatment)*, GRASS SEED *(seed treatment)*, LEEKS *(seed treatment)*, LETTUCE *(seed treatment)*, MAIZE *(seed treatment)*, OILSEED RAPE *(seed treatment)*, ONIONS *(seed treatment)*, PEAS *(seed treatment)*, RADISHES *(seed treatment)*, RUNNER BEANS *(seed treatment)*, TURNIPS *(seed treatment)* [1]. Fire in TULIPS [3]. Rust in BLACKCURRANTS [3]. Scab in PEARS [3]. Seed-borne diseases in CARROTS *(seed soak)*, CELERY *(seed soak)*, FODDER BEET *(seed soak)*, MANGELS *(seed soak)*, PARSLEY *(seed soak)*, RED BEET *(seed soak)*, SUGAR BEET *(seed soak)*[1].

Notes

Efficacy
* Spray before onset of disease and repeat every 7-14 d. Spray interval varies with crop and disease. See label for details [3]
* Apply as seed treatment for protection against damping off [1]
* Apply either as a soil treatment pre-planting or as a post-planting dust [2]
* Do not spray when rain imminent [3]
* For use on tulips, chrysanthemums and carnations add non-ionic wetter [3]

Crop Safety/Restrictions
* Maximum number of treatments 3 per crop for protected winter lettuce (thiram based products only [2, 3]; 2 per crop if sequence of thiram and other EBDC fungicides used) [2, 3]; 2 per crop for protected summer lettuce [2, 3]; 1 per crop as pre-planting soil treatment [2]; 1 per batch of seed for seed treatments [1]
* Do not apply to hydrangeas [3]
* Notify processor before dusting or spraying crops for processing [2, 3]
* Do not dip roots of forestry transplants [3]
* Do not treat seed of tomatoes, peppers or aubergines [1]

Special precautions/Environmental safety
* Irritating to eyes, skin and respiratory system
* Dangerous to fish or other aquatic life. Do not contaminate surface waters or ditches with chemical or used container [1, 3]
* Do not contaminate surface waters or ditches with chemical or used container [2]
* Do not harvest crops for human or animal consumption within 21 d of last dusting [2]

Protective clothing/Label precautions
* A, C [1]; A, D, E, H [2]; A [3]
* 5a, 5c, 78 [1]; 6a, 6b, 6c, 18, 21, 28, 29, 36, 37, 54, 63 [1-3]; 52 [3]; 65, 70 [1, 3]; 67 [2]

Latest application/Harvest Interval(HI)
* 21 d after planting out or 21 d before harvest, whichever is earlier, for protected winter lettuce; 15 d after planting out or 21 d before harvest, whichever is earlier, for protected summer lettuce.
* HI protected lettuce 21 d; outdoor lettuce 14 d; apples, pears, blackcurrants, raspberries, strawberries, tomatoes 7 d

FOR FULL CONDITIONS OF USE ALWAYS READ THE PRODUCT LABEL

477 tolclofos-methyl

A protectant organophosphorus fungicide for soil-borne diseases

See also prochloraz + tolclofos-methyl

Products					
1 Basilex	Levington	50% w/w	WP	07494
2 Fisons Basilex	Levington	50% w/w	WP	02847
3 Rizolex	AgrEvo	10% w/w	DS	07271
4 Rizolex	Schering	10% w/w	DS	03826
5 Rizolex Flowable	AgrEvo	500 g/l	FS	07273
6 Rizolex Flowable	Schering	500 g/l	FS	03719

Uses

Black scurf and stem canker in SEED POTATOES [3-6]. Bottom rot in LETTUCE, PROTECTED LETTUCE [1, 2]. Damping off and wirestem in LEAF BRASSICAS [1, 2]. Foot rot, root rot in ORNAMENTALS, SEEDLINGS OF ORNAMENTALS [1, 2]. Rhizoctonia in SEED POTATOES *(off-label)* [6].

Notes

Efficacy
* Dust seed potatoes during planting with automatic potato planter (see label for details of suitable applicator) or apply normally during hopper loading [3, 4]
* Apply flowable formulation to clean tubers with suitable misting equipment over a roller table. Spray as potatoes taken into store (first earlies) or as taken out of store (second earlies, maincrop, crops for seed) pre-chitting [5, 6]
* Do not mix flowable formulation with any other product [5, 6]
* To control rhizoctonia in vegetables and ornamentals apply as drench before sowing, pricking out or planting or incorporate into compost [2]
* On established seedlings and pot plants apply as drench and rinse off foliage [2]

Crop Safety/Restrictions
* Maximum number of treatments 1 per batch for seed potatoes; 1 per crop for pre-transplanted lettuce and brassicas; 1 at each stage of growth (ie sowing, pricking out, potting) to a maximum of 3, for ornamentals
* Only to be used with automatic planters [3, 4]
* Not recommended for use on seed potatoes where hot water treatment used or to be used [5, 6]
* Do not apply as overhead drench to vegetables or ornamentals when hot and sunny [2]
* Do not use on heathers [2]

Special precautions/Environmental safety
* This product contains an anticholinesterase organophosphorus compound. Do not use if under medical advice not to work with such compounds
* Dangerous to fish or other aquatic life. Do not contaminate surface waters or ditches with chemical or used container
* Treated tubers to be used as seed only, not for food or feed [3-6]

Protective clothing/Label precautions
* A, C, F, H, M [1, 2]; A, D, H [3-6]
* 1, 28, 29, 52, 63, 67, 70 [1-6]; 2 [1, 2]; 60, 66, 72, 73, 74, 76 [5, 6]

Latest application/Harvest Interval(HI)
* At planting of seed potatoes [3, 4]; pre-chitting of seed potatoes [5, 6]; before transplanting for lettuce and brassicas [2]

Approval
* May be applied by misting equipment mounted over roller table. See label for details. See notes in Section 1 [5, 6]
* Off-label Approval to June 1997 for use on chitted potatoes with a 16 wk harvest interval (OLA 0609/92) [6]

478 tralkoxydim

A foliar applied oxime herbicide for grass weed control in cereals.

Products	Grasp	Zeneca	250 g/l	SC	06675

Uses Blackgrass, wild oats in DURUM WHEAT, SPRING BARLEY, SPRING WHEAT, TRITICALE, WINTER BARLEY, WINTER RYE, WINTER WHEAT.

Notes **Efficacy**
* Product leaf-absorbed and translocated rapidly to growing points. Best results achieved when weeds growing actively in competitive crops under warm humid conditions with adequate soil moisture
* Activity not dependent on soil type but weeds germinating after application will not be controlled
* Best control of wild oats obtained from 2 leaf to 1st node detectable stage of weeds, and of blackgrass up to 3 tillers

Crop Safety/Restrictions
* Maximum number of treatments 1 per crop
* Apply to winter cereals in spring from 2 leaves unfolded up to and including flag leaf ligule just visible (GS 12-39)
* Apply to spring cereals from end of tillering up to and including flag leaf ligule just visible (GS 29-39)
* Only apply after 1 Mar
* Do not spray undersown crops or crops to be undersown
* Do not spray when foliage wet or covered in ice
* Do not spray if a protracted period of cold weather forecast
* Do not spray crops under stress from chemical treatment, grazing, pest attack, mineral deficiency or low fertility
* Do not roll or harrow within 1 wk of spraying
* Do not tank-mix with phenoxy hormone or sulfonylurea herbicides
* Oats or ryegrass should not be planted in autumn following spring use of Grasp

Special precautions/Environmental safety
* Irritant. May cause sensitization by skin contact
* Harmful to fish or other aquatic life. Do not contaminate surface waters or ditches with chemical or used container

Protective clothing/Label precautions
* A, C, H
* 10a, 14, 16, 18, 21, 29, 36, 37, 53, 63, 66, 70, 78

FOR FULL CONDITIONS OF USE ALWAYS READ THE PRODUCT LABEL

Latest application/Harvest Interval(HI)
* Before start of booting (GS 39)

479 tri-allate

A soil-acting thiocarbamate herbicide for grass weed control

Products

1 Avadex BW	Monsanto	400 g/l	EC	00173
2 Avadex BW Granular	Monsanto	10% w/w	GR	00174
3 Avadex Excel 15G	Monsanto	15% w/w	GR	07117

Uses

Annual meadow grass, blackgrass, meadow grasses, wild oats in DURUM WHEAT, FIELD BEANS, PEAS, TRITICALE, WINTER WHEAT [1-3]. Annual meadow grass, blackgrass, meadow grasses, wild oats in SUGAR BEET [2, 3]. Annual meadow grass, blackgrass, wild oats in MANGE-TOUT PEAS *(off-label)*, SPRING BARLEY, WINTER BARLEY [1, 2]. Annual meadow grass, blackgrass, wild oats in BEET CROPS, BRASSICAS, BROAD BEANS, CARROTS, CLOVERS, FLAX, FODDER BRASSICAS, FRENCH BEANS, LINSEED, LUCERNE, MAIZE, ONIONS, PARSLEY, PARSNIPS, RUNNER BEANS, SAINFOIN, SEED BRASSICAS, SPRING RYE, VETCHES, WINTER OILSEED RAPE [1]. Annual meadow grass, blackgrass, wild oats in SPRING WHEAT [2]. Blackgrass, meadow grasses, wild oats in BARLEY, FODDER BEET, FORAGE LEGUMES, MANGELS, RED BEET, WINTER RYE [3].

Notes

Efficacy
* Mix thoroughly into top 1.5-2.5 cm of soil immediately after spraying and drill up to 21 d later (incorporation after drilling but pre-emergence possible with some crops) [1]
* Incorporate or apply to surface pre-emergence (post-emergence application possible in winter cereals up to 2-leaf stage of wild oats) [2, 3]
* On soils with more than 10% organic matter use higher rates [1] or do not use [2]
* Wild oats only controlled before emergence [1], up to 2-leaf stage [2, 3]
* If applied to dry soil, rainfall needed for full effectiveness, especially with granules on surface. Do not use if top 5-8 cm bone dry
* Do not apply with spinning disc granule applicator; see label for suitable types [2, 3]
* Do not apply to cloddy seedbeds
* Tank mixture with various other pre-emergence herbicides recommended to increase control of dicotyledon weeds. Mixture with Betanal E recommended early post-emergence in sugar and fodder beet (see label for details) [1]
* Use sequential treatments to improve control of barren brome and annual dicotyledons (see label for details)
* Clean start technique recommended to maximize control of sterile brome and blackgrass in winter cereals involves use of Sting CT to kill emerged weeds in stubble followed by Avadex BW Granular immediately after drilling

Crop Safety/Restrictions
* Maximum number of treatments 1 per crop (2 per crop with low doses on sugar and fodder beet [1])
* Do not apply Avadex BW to spring wheat. Apply Avadex BW Granular to spring wheat post-emergence only to crops drilled in winter
* Consolidate loose, puffy seedbeds before drilling to avoid chemical contact with seed
* Drill cereals well below treated layer of soil (see label for safe drilling depths)
* Do not use on direct-drilled crops or undersow grasses into treated crops
* Do not sow oats or grasses within 1 yr of treatment

Special precautions/Environmental safety
* Harmful in contact with skin and if swallowed, irritating to skin [1]. Irritating to eyes and skin [2, 3]
* Harmful to fish or other aquatic life. Do not contaminate surface waters or ditches with chemical or used container

Protective clothing/Label precautions
* A, C, H [1]; A [2, 3]
* 5a, 5c, 16, 28, 66, 70, 78 [1]; 6a, 67 [2, 3]; 6b, 18, 36, 37, 53, 63 [1-3]; 14, 29 [1, 3]; 21 [1, 2]

Latest application/Harvest Interval(HI)
* Pre-drilling for sugar beet, fodder beet, mangels, red beet; before crop emergence for field beans, spring barley, peas, forage legumes; before first node detectable stage (GS 31) for winter wheat, winter barley, durum wheat, triticale, winter rye [3]

Approval
* Approved for aerial application on wheat, barley, winter beans, peas. See notes in Section 1 [2, 3]
* Off-label approval to Jan 1996 for use on mange-tout peas (OLA 0011/92)[1], (OLA 0012/92)[2]

480   triadimefon

A systemic conazole fungicide with curative and protectant action

Products					
1 Bayleton	Bayer	25% w/w	WP	00221
2 Standon Triadimefon	Standon	25% w/w	WP	05673

Uses Brown rust, powdery mildew, rhynchosporium, snow rot, yellow rust in BARLEY [1, 2]. Brown rust, powdery mildew, yellow rust in WHEAT [1, 2]. Crown rust, powdery mildew, rhynchosporium in GRASSLAND [1]. Powdery mildew in OATS, RYE [1, 2]. Powdery mildew in APPLES, BLACKBERRIES *(off-label)*, BLACKCURRANTS *(off-label)*, BRUSSELS SPROUTS, CABBAGES, FODDER BEET, GOOSEBERRIES *(off-label)*, GRAPEVINES *(off-label)*, HOPS, LOGANBERRIES *(off-label)*, PARSNIPS, RASPBERRIES *(off-label)*, STRAWBERRIES *(off-label)*, SUGAR BEET, SWEDES, TURNIPS [1]. Rust in LEEKS [1].

Notes **Efficacy**
* Apply at first sign of disease and repeat as necessary, applications to established infections are less effective. Spray programme varies with crop and disease - see label
* Applications to control mildew will also reduce crown rust in oats, snow rot in winter barley and rust in beet [1]

Crop Safety/Restrictions
* Maximum number of treatments 2 per crop for wheat, oats, rye, spring barley, grassland, sugar beet, fodder beet, turnips, swedes, parsnips; 3 per crop for winter barley (1 in autumn), Brussels sprouts, cabbages, cane fruit and herbs; 4 per crop for leeks; 6 per yr for grapevines; 8 per yr for hops; 12 per yr for apples
* Continued use of fungicides from the same group in cereals may lead to reduced effectiveness against mildew

FOR FULL CONDITIONS OF USE ALWAYS READ THE PRODUCT LABEL

Special precautions/Environmental safety
* Harmful to fish or other aquatic life. Do not contaminate surface waters or ditches with chemical or used container

Protective clothing/Label precautions
* 29, 35, 41, 53, 63, 67

Withholding period
* Harmful to livestock. Keep all livestock out of treated areas for 21 d [1]

Latest application/Harvest Interval(HI)
* Before grain milky ripe (GS 71) for cereals.
* HI sugar beet, fodder beet, Brussels sprouts, cabbages, turnips, swedes, parsnips, leeks, apples, hops, cane fruit, strawberries, blackcurrants, gooseberries 14 d; grassland, herbs 21 d; grapes 6 wk

Approval
* Approved for aerial application on sugar beet, brassicas, swedes, turnips [1]; cereals [1, 2]. See notes in Section 1
* Off-label approval unlimited for use on blackcurrants, gooseberries, strawberries, outdoor grapes, raspberries, loganberries, blackberries, other Rubus hybrids (OLA 0024/95)[1]

481   triadimenol

A systemic conazole fungicide for cereals, beet and brassicas
See also fuberidazole + triadimenol
      tebuconazole + triadimenol

Products	Bayfidan	Bayer	250 g/l	EC	02672

Uses

Alternaria, light leaf spot, powdery mildew, ring spot in BRUSSELS SPROUTS, CABBAGES. Fungus diseases in CARROTS *(off-label)*, HORSERADISH *(off-label)*, PARSLEY *(off-label)*, PARSNIPS *(off-label)*, SALSIFY *(off-label)*. Powdery mildew, rhynchosporium, rust, septoria, snow rot in WHEAT. Powdery mildew, rhynchosporium, rust, snow rot in BARLEY, OATS, RYE. Powdery mildew in FODDER BEET, SUGAR BEET, SWEDES, TURNIPS.

Notes

Efficacy
* Apply sprays at first signs of disease and ensure good cover
* When applied for mildew and rust control in wheat product also gives good reduction of *Septoria tritici* but for specific protection against *S. tritici* and *S. nodorum* use tank-mix with chlorothalonil
* Treatment also reduces crown rust in oats and rust in beet
* Continued use of fungicides from same group can result in reduced effectiveness against powdery mildew. See label for recommended tank mixtures

Crop Safety/Restrictions
* Maximum number of treatments 2 per crop for spring wheat, oats, spring barley, rye, sugar beet, fodder beet, turnips, swedes, carrots, parsnips; 3 per crop (only 2 in spring) for winter barley, winter wheat, cabbages, Brussels sprouts

Special precautions/Environmental safety
* Harmful if swallowed. Irritating to eyes
* Harmful to fish or other aquatic life. Do not contaminate surface waters or ditches with chemical or used container

Protective clothing/Label precautions
* A, C
* 5c, 6a, 18, 21, 23, 24, 25, 29, 36, 37, 53, 63, 66, 70, 78

Latest application/Harvest Interval(HI)
* Before grain milky ripe (GS 71) for cereals.
* HI beet crops, brassicas 14 d; carrots, parsnips 21 d

Approval
* Off-label approval unlimited for use on outdoor crops of carrots, parsnips, parsley root, salsify, horseradish (OLA0836/95)

482   triadimenol + tridemorph
A systemic fungicide mixture with protectant and curative action

Products	Dorin		Bayer	125:375 g/l	EC	03292

Uses   Crown rust, powdery mildew, rust in OATS. Powdery mildew, rhynchosporium, rust, snow rot in BARLEY. Powdery mildew, rust in SPRING WHEAT, WINTER WHEAT. Septoria leaf spot in WHEAT.

Notes

Efficacy
* Apply at first signs of disease or as 2 or 3-spray protectant programme
* When applied for mildew and rust control in wheat product also gives good reduction of *Septoria tritici* but for specific protection against *S. tritici* and *S. nodorum* use tank-mix with chlorothalonil
* Treatment for mildew control will reduce crown rust infection on oats

Crop Safety/Restrictions
* Maximum number of treatments 3 per crop
* Crop scorch may occur if sprayed in frosty weather
* Scorch may occur on wheat if applied under very warm or drought conditions

Special precautions/Environmental safety
* Harmful if swallowed. Irritating to skin
* Dangerous to fish or other aquatic life. Do not contaminate surface waters or ditches with chemical or used container

Protective clothing/Label precautions
* A, C
* 5c, 6b, 18, 23, 24, 25, 29, 36, 37, 52, 63, 66, 70, 78

Withholding period
* Grazing livestock must be kept out of treated areas for at least 14 d

Latest application/Harvest Interval(HI)
* Before grain milky ripe (GS 71)

FOR FULL CONDITIONS OF USE ALWAYS READ THE PRODUCT LABEL

483 triasulfuron

A sulfonylurea herbicide for annual broad-leaved weed control in cereals
See also bromoxynil + ioxynil + triasulfuron

Products Lo-gran 20 WG   Ciba Agric.   20% w/w   WG   05993

Uses Annual dicotyledons, charlock, chickweed, mayweeds in BARLEY, DURUM WHEAT, OATS, RYE, TRITICALE, WHEAT.

Notes **Efficacy**
• Best results achieved on small, actively growing weeds up to 6 leaf or 50 mm growth stage (up to flower bud for charlock, chickweed and mayweed)
• Some species, although not controlled remain stunted and uncompetitive with crop

Crop Safety/Restrictions
• Maximum number of treatments 1 per crop
• Product must be applied only after 1 Feb from 1 leaf stage of crop to before 3rd node detectable (GS 11-33)
• Do not apply to undersown crops or those due to be undersown
• Do not spray in windy weather. Avoid drift onto neighbouring crops
• Do not spray during period of frosty weather, when frost imminent or onto crops under stress from frost, waterlogging or drought
• Do not spray in tank mixture, or in sequence, with a product containing any other sulfonylurea
• Ensure spraying equipment is washed thoroughly according to specific instructions. Do not allow washings to drain onto land intended for cropping or growing crops
• See label for restrictions on succeeding crops

Special precautions/Environmental safety
• Extremely dangerous to aquatic higher plants. Do not contaminate surface waters or ditches with chemical or used container
• Do not allow direct spray from ground-based sprayer vehicles to fall within 6 m, or from hand-held sprayers to within 2 m, of surface waters or ditches. Direct spray away from water

Protective clothing/Label precautions
• A, H
• 29, 54a, 63, 67

Latest application/Harvest Interval(HI)
• Before 3rd node detectable (GS 33)

484 triazophos

A contact and stomach acting anti-cholinesterase organophosphorus insecticide

Products 1 Hostathion   AgrEvo   420 g/l   EC   07327
              2 Hostathion   Hoechst   420 g/l   EC   01080

Uses American serpentine leaf miner, colorado beetle in POTATOES *(off-label)* [2]. American serpentine leaf miner in ARTICHOKES *(off-label)*, ASPARAGUS *(off-label)*, BEANS *(off-label)*, CELERY *(off-label)*, CUCURBITS *(off-label)*, FENNEL *(off-label)*, GARLIC *(off-label)*, LEEKS *(off-label)*, MANGE-TOUT PEAS *(off-label)*, ONIONS *(off-label)*, ORNAMENTALS *(off-label)*, RHUBARB *(off-label)*, SHALLOTS *(off-label)*, SUGAR BEET *(off-label)* [2]. Aphids, pea and bean weevils, pea midge, pea moth in PEAS [1, 2]. Bean beetle in BROAD BEANS [1, 2]. Capsids, codling moth, red spider mites, tortrix moths in APPLES [1, 2]. Carrot fly, cutworms in CARROTS, PARSNIPS [1, 2]. Caterpillars in BRASSICAS [1, 2]. Cutworms in CELERY, LEEKS, ONIONS, POTATOES, SUGAR BEET [1, 2]. Frit fly, leatherjackets in CEREALS, GRASSLAND [1, 2]. Frit fly in MAIZE, SWEETCORN [1, 2]. Pea and bean weevils in FIELD BEANS [1, 2]. Pod midge, seed weevil in WINTER OILSEED RAPE [1, 2]. Red spider mites, tortrix moths in STRAWBERRIES [1, 2]. Wheat-blossom midges in WHEAT [1, 2].

Notes **Efficacy**
- Chemical gives rapid kill of insects and persists for 3-4 wk, less under hot, dry conditions. It is especially active against caterpillars and fly larvae
- Rate, timing and number of sprays vary with pest and crop. See label for details
- For use on brassicas addition of non-ionic wetter is recommended
- For control of second generation carrot fly direct spray toward carrot crowns. Rain soon after application may help to distribute chemical in upper layers of soil
- Do not apply to wet crops if run-off likely or heavy rain imminent

Crop Safety/Restrictions
- Maximum number of treatments 1 per crop for celery, protected onions, garlic and shallots; 3 per crop for carrots (6 per crop at half dose); 3 per crop for other crops including protected vegetables but excluding grassland; 6 per crop for protected ornamentals
- On carrots the maximum total dose applied per crop must not exceed the equivalent of 3 (on mineral soils) or 4 (on organic soils) full dose applications
- Do not spray cereals or grassland from end of Dec to end of Mar
- To minimize hazard to bees spray oilseed rape when petal-fall complete and apply in early morning or late evening
- Do not use to control pollen beetle. Do not spray spring oilseed rape
- Do not spray crops suffering from stress (especially Desirée potatoes and fruit trees) or at temperatures above 30°C
- Do not spray Laxton's Fortune or June Wealthy apples or Malus species grown as pollinators

Special precautions/Environmental safety
- Triazophos is subject to the Poisons Rules 1982 and the Poisons Act 1972. See notes in Section 1
- This product contains an anticholinesterase organophosphorus compound. Do not use if under medical advice not to work with such compounds
- Toxic if swallowed, harmful in contact with skin
- Unprotected persons must be kept out of treated structures for at least 12 h after treatment
- Keep in original container, tightly closed, in a safe place, under lock and key
- Harmful to livestock
- Harmful to game, wild birds and animals. Do not spray cereals in areas of known wild-fowl migration

FOR FULL CONDITIONS OF USE ALWAYS READ THE PRODUCT LABEL

- Extremely dangerous to bees. Do not apply to crops in flower or to those in which bees are actively foraging, except as directed on peas. Do not apply when flowering weeds are present
- Extremely dangerous to fish or other aquatic life. Do not contaminate surface waters or ditches with chemical or used container

Protective clothing/Label precautions
- A, C, H
- 1, 2, 4c, 5a, 12c, 16, 18, 23, 24, 28, 29, 35, 36, 37, 41, 46, 48a, 51, 53, 60, 64, 66, 70, 79

Withholding period
- Keep livestock out of treated cereals or grassland for at least 4 wk

Latest application/Harvest Interval(HI)
- 24 h before handling protected ornamentals.
- HI winter oilseed rape, peas 3 wk; other crops 4 wk

Approval
- Off-label approval unlimited for aerial spraying of potatoes to control Colorado beetle as required by Statutory Notice (OLA 0049/92)[2]; unlimited for use on protected ornamentals (OLA 0630/92)[2]; to June 1996 for use on protected vegetables. Specific restrictions apply (OLA 0629/92)[2]

Maximum Residue Levels (mg residue/kg food)
- bananas, carrots, parsnips 1; Brussels sprouts, head cabbages 0.1; garlic, onions, shallots, potatoes 0.05

---

485 triazoxide

A benzotriazine fungicide available only in mixtures

See also tebuconazole + triazoxide

---

486 tribenuron-methyl

A foliar acting sulfonylurea herbicide with some root activity for use in cereals

See also thifensulfuron-methyl + tribenuron-methyl

Products					
	Quantum	DuPont	50% w/w	TB	06270

Uses Annual dicotyledons, charlock, chickweed, mayweeds in DURUM WHEAT, SPRING BARLEY, TRITICALE, WINTER BARLEY, WINTER RYE, WINTER WHEAT.

Notes **Efficacy**
- Best control achieved when weeds small and actively growing
- Good spray cover must be achieved since larger weeds often become less susceptible
- Susceptible weeds cease growth almost immediately after treatment and symptoms can be seen in about 2 wk
- Product can be used on all soil types
- Weed control may be reduced when conditions very dry
- See label for details of technique to be used for dissolving tablets in spray tank
- In tank mixing ensure tablets fully dispersed before adding other products

Crop Safety/Restrictions
* Maximum number of treatments 1 per crop
* Do not spray in tank mixture, or in sequence, with a product containing any other sulfonylurea
* Apply only in spring (after 1 Feb) from 3 leaf stage of crop up to and including flag leaf fully emerged (GS 13-39)
* Do not apply to crops undersown with grass, clover or other broad-leaved crops
* Do not apply to any crop suffering stress from any cause or not actively growing
* Do not apply within 7 d of rolling
* Special care must be taken to avoid damage by drift onto nearby broad-leaved crops, surface waters or ditches
* Special care needed in spray tank cleaning. See label for details
* In the event of crop failure sow only a cereal within 3 mth of application. See label for other restrictions on subsequent cropping

Special precautions/Environmental safety
* Irritant. May cause sensitization by skin contact
* Extremely dangerous to aquatic higher plants. Do not contaminate surface waters or ditches with chemical or used container

Protective clothing/Label precautions
* A
* 10a, 18, 21, 29, 36, 37, 54, 63, 67

Latest application/Harvest Interval(HI)
* Up to and including flag leaf ligule/collar just visible (GS 39)

487   trichlorfon

A contact and ingested organophosphorus insecticide

Products	Dipterex 80	Bayer	80% w/w	SP	00711

Uses

Browntail moth, small ermine moth in HEDGES. Cabbage root fly in BRUSSELS SPROUTS, MOOLI *(off-label)*, RADISHES *(off-label)*. Caterpillars, leaf miners in BRASSICAS. Caterpillars in ROSES. Cherry bark tortrix in APPLES. Cutworms in VEGETABLES. Earwigs in BLACKCURRANTS. Flea beetles, mangold fly in FODDER BEET, MANGELS, RED BEET, SPINACH, SUGAR BEET. Flies in MANURE HEAPS, REFUSE TIPS *(off-label)*, RUBBISH DUMPS *(off-label)*. Liriomyza huidobrensis in AUBERGINES *(off-label)*, BRASSICAS *(off-label)*, LETTUCE *(off-label)*, MARROWS *(off-label)*, ORNAMENTALS *(off-label)*, PEPPERS *(off-label)*, PROTECTED CUCUMBERS *(off-label)*, PROTECTED MELONS *(off-label)*, PROTECTED ORNAMENTALS *(off-label)*, RADICCHIO *(off-label)*, SORREL *(off-label)*, SPINACH BEET *(off-label)*, TOMATOES *(off-label)*. Strawberry tortrix in STRAWBERRIES.

Notes

Efficacy
* Apply at appearance of caterpillars or larvae for most uses and repeat at 7-10 d intervals if necessary. See label for full details
* To control cabbage root fly in Brussels sprouts buttons spray 1 mth before expected harvest and repeat twice at 7 d intervals using pendant lances to obtain good cover

FOR FULL CONDITIONS OF USE ALWAYS READ THE PRODUCT LABEL

- Addition of non-ionic wetter recommended for use on brassicas
- For cutworm control apply as drenching spray
- For cherry bark tortrix control spray trunks and main branches in mid-May, taking care to avoid leaves

Crop Safety/Restrictions
- Maximum number of treatments 1 per crop for protected spinach, spinach beet and sorrel; 3 per crop for tomatoes; 8 per crop for radishes and mooli

Special precautions/Environmental safety
- This product contains an anticholinesterase organophosphorus compound. Do not use if under medical advice not to work with such compounds
- Harmful to fish or other aquatic life. Do not contaminate surface waters or ditches with chemical or used container

Protective clothing/Label precautions
- 1, 21, 28, 29, 35, 53, 63, 67

Latest application/Harvest Interval(HI)
- HI outdoor tomatoes, protected spinach, spinach beet and sorrel 14 d; glasshouse tomatoes 3 d; other crops 2 d

Approval
- Approved for aerial application on beet crops, brassicas, spinach. See notes in Section 1
- Off-label approval to Mar 1997 for use on outdoor radishes and mooli (OLA 0185/92); unlimited for use on aubergines, ornamentals, peppers, sorrel, spinach beet, tomatoes (OLA 0400/91, 0497/91), protected ornamentals (OLA 1079/92), outdoor brassicas, outdoor and protected marrows, outdoor and protected lettuce and radicchio, protected peppers, cucumbers and melons (OLA 1542/93); unlimited for use on waste tips and rubbish dumps (OLA 0690/91)

Maximum Residue Levels (mg residue/kg food)
- cereals 0.1

---

488 Trichoderma viride
A mycoparasite for the biological control of silver leaf in fruit trees

Products				
Binab T		HCS	PT	00264

Uses Silver leaf in APPLES, CHERRIES, PEARS, PLUMS.

Notes

Efficacy
- Best control is achieved by treatment after leaf-fall
- Apply by filling horizontal holes drilled in the tree trunk with pellets to within 0.7 cm of the surface
- At least one hole should penetrate the heartwood
- Cover the mouth of the hole with thick tree paint
- Sterilise the drill bit between trees with 70% denatured alcohol
- Treat before infection too far advanced

Crop Safety/Restrictions
- Maximum number of treatments 1 per tree per yr
- Maximum storage temperature 10°C

Protective clothing/Label precautions
* 30, 54, 62, 63, 67

Approval
* Approval expiry: 31 Dec 96 [00264]

489   triclopyr

A pyridyloxy herbicide for perennial and woody weed control

See also 2,4-D + dicamba + triclopyr
   clopyralid + fluroxypyr + triclopyr
   clopyralid + triclopyr
   dicamba + mecoprop + triclopyr
   fluroxypyr + triclopyr

Products					
1	Garlon 2	Zeneca	240 g/l	EC	06616
2	Garlon 4	Nomix-Chipman	480 g/l	EC	06016
3	Timbrel	DowElanco	480 g/l	EC	05815

Uses

Brambles, broom, docks, gorse, hard rush, perennial dicotyledons, stinging nettle, woody weeds in NON-CROP AREAS [1-3]. Brambles, broom, docks, gorse, hard rush, perennial dicotyledons, stinging nettle, woody weeds in ESTABLISHED GRASSLAND, ROUGH GRAZING [1]. Brambles, broom, docks, gorse, perennial dicotyledons, rhododendrons, stinging nettle, woody weeds in FORESTRY [2, 3]. Brambles, broom, docks, gorse, perennial dicotyledons, stinging nettle, woody weeds in NON-CROP GRASS, ROADSIDE GRASS [2, 3].

Notes

Efficacy
* Apply in grassland as spot treatment or overall foliage spray when weeds in active growth in spring or summer. Details of dose and timing vary with species - see label [1]
* Apply to woody weeds as summer foliage, winter shoot, basal bark, cut stump or tree injection treatment [2, 3]
* Apply foliage spray in water when leaves fully expanded but not senescent
* Apply winter shoot, basal bark or cut stump sprays in paraffin or diesel oil. Dose and timing vary with species. See label for details
* Inject undiluted or 1:1 dilution into cuts spaced every 7.5 cm round trunk [2, 3]
* Do not spray in drought, in very hot or cold conditions
* Control may be reduced if rain falls within 2 h of application

Crop Safety/Restrictions
* Maximum number of treatments 1 per yr on non-crop land (as directed spray) [1]; 2 per yr on established grassland (including land not intended for cropping)
* Clover will be killed or severely checked by application in grassland. Do not apply to grass leys less than 1 yr old [1]
* Avoid spray drift onto edible crops, ornamental plants, Douglas fir, larches or pines. Vapour drift may occur under hot conditions
* Do not drill kale, swedes, turnips, grass or mixtures containing clover within 6 wk of treatment. Allow at least 6 wk before planting trees

FOR FULL CONDITIONS OF USE ALWAYS READ THE PRODUCT LABEL

Special precautions/Environmental safety
* Irritating to skin [1, 2]; harmful if swallowed and in contact with skin [3]; may cause sensitization by skin contact [3]. Harmful if swallowed [1]
* Flammable [1, 2]
* Not to be used on food crops
* Dangerous to fish or other aquatic life. Do not contaminate surface waters or ditches with chemical or used container
* Not to be applied in or near water
* Do not apply from tractor-mounted sprayer within 250 m of ponds, lakes or watercourses

Protective clothing/Label precautions
* A, C, H, M
* 5a, 10a, 52, [3]; 5b, 12c, 78 [2]; 5c, 63 [2, 3]; 6b, 63, [1]; 14, 18, 21, 28, 29, 34, 36, 37, 43, 66, 70 [1-3]; 52 [1, 2]

Withholding period
* Keep livestock out of treated areas for at least 7 d and until foliage of any poisonous weeds such as buttercups or ragwort has died and become unpalatable

Latest application/Harvest Interval(HI)
* 6 wk before replanting; 7 d before grazing

490 tridemorph

A systemic, eradicant and protectant morpholine fungicide

See also carbendazim + maneb + tridemorph
cyproconazole + tridemorph
epoxiconazole + tridemorph
fenpropimorph + flusilazole + tridemorph
fenpropimorph + tridemorph
flusilazole + tridemorph
propiconazole + tridemorph
tebuconazole + tridemorph
triadimenol + tridemorph

Products					
1 Calixin	BASF	750 g/l	EC	00369
2 Standon Tridemorph 750	Standon	750 g/l	EC	05667

Uses Powdery mildew in BARLEY, OATS, SWEDES, TURNIPS, WINTER WHEAT.

Notes **Efficacy**
* Apply to cereals when mildew starts to build up, normally May-early Jun and repeat once if necessary. Treatment may also be made in autumn if required
* Reduced rate recommended in tank mix with triazole fungicides (cyproconazole, flutriafol, propiconazole, triadimefon, triadimenol) for control of established mildew in barley
* Tank mix with carbendazim recommended for rhynchosporium control in barley, with other products for eyespot and yellow rust, see label for details
* Apply to swedes and turnips at first signs of disease, normally Jul-Aug, and repeat at 2 wk intervals if required
* Product is rainfast after 2 h

Crop Safety/Restrictions
* Maximum number of treatments 2 per crop for oats, winter wheat, spring barley; 3 per crop (1 in autumn) for winter barley; 3 per crop for swedes and turnips
* Permitted growth stages vary with crop and level of infection. See label for details
* Do not spray winter wheat at high temperatures or in drought
* Do not apply if frost expected
* Do not mix with other chemicals on swedes or turnips

Special precautions/Environmental safety
* Harmful if swallowed [1] and in contact with skin [2].
* Irritating to eyes and skin
* Harmful to fish or other aquatic life. Do not contaminate surface waters or ditches with chemical or used container

Protective clothing/Label precautions
* A, C [1, 2]
* 5a [2]; 5c, 6a, 6b, 16, 18, 22, 29, 36, 37, 43, 53, 63, 66, 70, 78 [1, 2]

Withholding period
* Keep all livestock out of treated areas for at least 14 d

Latest application/Harvest Interval(HI)
* HI swedes, turnips 2 wk; barley, oats 4 wk; winter wheat 6 wk

Approval
* Approved for aerial application on barley, oats, winter wheat, swedes, turnips. See notes in Section 1

491   trietazine

A triazine herbicide available only in mixtures

See also simazine + trietazine
terbutryn + trietazine

FOR FULL CONDITIONS OF USE ALWAYS READ THE PRODUCT LABEL

492   trifluralin

A soil-incorporated dinitroaniline herbicide for use in various crops

See also clodinafop-propargyl + trifluralin
    diflufenican + trifluralin
    isoproturon + trifluralin
    linuron + trifluralin
    terbutryn + trifluralin

Products

1 Alpha Trifluralin 48EC	Makhteshim	480 g/l	EC	07406
2 Ashlade Trimaran	Ashlade	480 g/l	EC	06228
3 Atlas Trifluralin	Atlas	480 g/l	EC	03051
4 MSS Trifluralin 48 EC	Mirfield	480 g/l	EC	05144
5 MTM Trifluralin	MTM Agrochem.	480 g/l	EC	05313
6 Treflan	DowElanco	480 g/l	EC	05817
7 Trigard	FCC	480 g/l	EC	02178
8 Tripart Trifluralin 48 EC	Tripart	480 g/l	EC	02215
9 Tristar	PBI	480 g/l	EC	02219

Uses

Annual dicotyledons, annual grasses in CELERIAC *(off-label)*, LINSEED *(off-label)* [9]. Annual dicotyledons, annual grasses in OUTDOOR HERBS *(off-label)* [3, 5, 6, 9]. Annual dicotyledons, annual grasses in PROTECTED HERBS *(off-label)* [3, 5]. Annual dicotyledons, annual grasses in COMBINING PEAS *(off-label)*, EVENING PRIMROSE *(off-label)*, MANGE-TOUT PEAS *(off-label)*, ORNAMENTALS *(off-label)*, SOYA BEANS *(off-label)*, SUNFLOWERS *(off-label)*, VINING PEAS *(off-label)* [6, 9]. Annual dicotyledons, annual grasses in GOLD-OF-PLEASURE *(off-label)*, NURSERY FRUIT TREES AND BUSHES *(off-label)*, RADISH SEED CROPS *(off-label)*, SOFT FRUIT *(off-label)* [6]. Annual dicotyledons, annual grasses in FRENCH BEANS [2-9]. Annual dicotyledons, annual grasses in CALABRESE [2, 3, 5-9]. Annual dicotyledons, annual grasses in NAVY BEANS [2, 3, 6, 8, 9]. Annual dicotyledons, annual grasses in PARSLEY [2, 6-9]. Annual dicotyledons, annual grasses in LINSEED [2, 6]. Annual dicotyledons, annual grasses in FIELD BEANS [1, 5, 7, 8]. Annual dicotyledons, annual grasses in BROAD BEANS, BROCCOLI, BRUSSELS SPROUTS, CABBAGES, CARROTS, CAULIFLOWERS, KALE, LETTUCE, OILSEED RAPE, PARSNIPS, RASPBERRIES, RUNNER BEANS, STRAWBERRIES, SWEDES, TURNIPS [1-9]. Annual dicotyledons, annual grasses in WINTER BARLEY, WINTER WHEAT [1-6, 8, 9]. Annual dicotyledons, annual grasses in MUSTARD, SUGAR BEET [1, 2, 4-9]. Annual dicotyledons, annual grasses in DWARF BEANS [1].

Notes

Efficacy
- Acts on germinating weeds and requires soil incorporation to 5 cm (10 cm for crops to be grown on ridges) within 30 min of spraying (2 h [8], except in mixtures on cereals). See label for details of suitable application equipment
- Best results achieved by application to fine, firm seedbed, free of clods, crop residues and established weeds
- Do not use on sand, fen soil or soils with more than 10% organic matter
- In winter cereals normally applied as surface treatment without incorporation in tank-mixture with other herbicides to increase spectrum of control. See label for details
- Follow-up herbicide treatment recommended with some crops. See label for details

Crop Safety/Restrictions
- Maximum number of treatments 1 per crop
- Apply and incorporate at any time during 2 wk before sowing or planting
- Do not apply to brassica plant raising beds
- Transplants should be hardened off prior to transplanting

- Apply in sugar beet after plants 10 cm high with 4-10 leaves and harrow into soil
- Apply in cereals after drilling up to and including 3 leaf stage (GS 13)
- Minimum interval between application and drilling or planting may be up to 12 mth. See label for details

Special precautions/Environmental safety
- Irritating to eyes, skin and respiratory system [3], to eyes and skin [1, 4, 5, 8]
- Flammable
- Harmful to fish or other aquatic life. Do not contaminate surface waters or ditches with chemical or used container

Protective clothing/Label precautions
- A, C [1, 3, 4]; A [5]
- 6a, 6b [1, 3-5, 8]; 6c [3]; 12c, 21, 25, 29, 37, 53, 63, 66 [1-9]; 18, 36 [1, 3-9]; 60 [1, 4-9]; 61, 62 [1, 4]

Latest application/Harvest Interval(HI)
- Before 4 leaves unfolded (GS 13) for cereals; up to 6 leaf stage for sugar beet [4, 6], (up to 10 leaf stage [2]); pre-sowing/planting for other crops
- Pre-sowing/pre-planting for all crops [1]

Approval
- Off-label Approval to June 1996 for use on protected herbs, unlimited for use on outdoor herbs (OLA 0737, 0738/92)[3], (OLA 0759, 0760/92)[5]; unlimited for use on combining peas, sunflower, linseed, celeriac, ornamentals (OLA 1240/92)[9], evening primrose (OLA 1276/92)[9], outdoor herbs, soya bean (OLA 0077/93)[9]; to Dec 1977 for use on vining and mange-tout peas (OLA 1241/92)[9]; unlimited for use on evening primrose (OLA 1080/92)[6], gold-of-pleasure, radish seed crops, sunflowers, soya beans, combining peas, nursery fruit trees and bushes, ornamentals, outdoor herbs (OLA 0074/93)[6], to Jan 1998 for use on vining and mange-tout peas (OLA 0075/93)[6]

---

493   triforine

A locally systemic fungicide with protectant and curative activity

See also bupirimate + triforine

Products					
1 Fairy Ring Destroyer	Vitax	190 g/l	EC	05541
2 Saprol	Cyanamid	190 g/l	EC	07016

Uses   American gooseberry mildew, currant leaf spot in BLACKCURRANTS [2]. Black spot in ROSES [2]. Fairy rings in TURF [1]. Net blotch, powdery mildew in SPRING BARLEY [2]. Powdery mildew, scab in APPLES [2]. Powdery mildew in HOPS [2].

Notes   **Efficacy**
- Apply to spring barley for mildew control as soon as disease becomes apparent
- For control of net blotch, yellow and brown rust use mixture with mancozeb
- For fairy ring control apply as high volume spray or drench as soon as infection noted and repeat twice at 14 d intervals

---

FOR FULL CONDITIONS OF USE ALWAYS READ THE PRODUCT LABEL

- Apply to apples at green cluster and repeat at 10-14 d intervals. Precede by specific scab treatment where high level of scab infection present
- Number and timing of sprays varies with crop and disease. See label for details

Crop Safety/Restrictions
- Maximum number of treatments 3 per yr for turf [1]
- Do not spray apples following frost injury or when trees are under drought or other stress. Do not use on Golden Delicious or Ingrid Marie or apply before 90% petal fall on Cox's Orange Pippin
- Apply up to 4 sprays on hops in season, not recommended as full programme

Special precautions/Environmental safety
- Harmful in contact with skin, irritating to eyes

Protective clothing/Label precautions
- A, C
- 5a, 6a, 18, 21, 28, 29, 36, 37, 54, 60, 63, 66, 70, 78

Latest application/Harvest Interval(HI)
- HI apples 7 d; blackcurrants, hops 14 d

Approval
- Approved for aerial application on spring barley. See notes in Section 1. Consult firm for details [2]

---

494 Verticillium lecanii
A fungal parasite of aphids and whitefly

Products					
1 Mycotal	Koppert	16.1% w/w	WP	04782
2 Vertalec	Koppert	20% w/w	WP	04781

Uses

Aphids, whitefly in AUBERGINES, CUCUMBERS, GLASSHOUSE CUT FLOWERS, PEPPERS, PROTECTED BEANS, PROTECTED LETTUCE, PROTECTED ORNAMENTALS.

Notes

Efficacy
- Apply spore powder as spray as part of biological control programme

Crop Safety/Restrictions
- Maximum number of treatments 3 per crop [1]; 2 per crop at high dose for chrysanthemums, 1 per crop for other crops, 12 per crop at low dose, low volume [2]

Protective clothing/Label precautions
- No information

---

495 vinclozolin
A protectant dichloroanilide (dicarboximide) fungicide
See also carbendazim + vinclozolin

Products					
Ronilan FL	BASF	500 g/l	SC	02960

585

Uses	Alternaria, botrytis, sclerotinia stem rot in OILSEED RAPE. Ascochyta, botrytis, mycosphaerella in COMBINING PEAS, VINING PEAS. Blossom wilt in APPLES. Botrytis in DWARF BEANS, NAVY BEANS, RUNNER BEANS. Chocolate spot in BROAD BEANS, FIELD BEANS.

Notes

Efficacy
* Number and timing of sprays vary with crop and disease. See label for details
* May be used at reduced rate on peas and field beans in tank-mix with Bravo 500
* Where dicarboximide resistant strains have developed product may not be effective
* Do not spray if crop wet or if rain or frost expected

Crop Safety/Restrictions
* Maximum number of treatments 2 per crop for oilseed rape, beans and peas
* Do not treat mange-tout varieties of peas
* Do not treat ornamental seedlings before the 3-4 leaf stage
* Do not treat unrooted cuttings or rooted cuttings for 3-4 d after planting
* See label for ornamental species which have been treated with no adverse effects. Do not treat ornamentals during flowering
* Apply only with tractor mounted or trailed, downward placement hydraulic sprayers

Special precautions/Environmental safety
* Irritant. May cause sensitization by skin contact
* Product presents a minimal hazard to bees when used as directed but consider informing local bee-keepers if intending to spray crops in flower
* Harmful to fish and aquatic life. Do not contaminate surface waters or ditches with chemical or used container

Protective clothing/Label precautions
* A, C, H, K, M
* 10a, 18, 21, 28, 29, 36, 37, 53, 63, 66

Latest application/Harvest Interval(HI)
* Before end of petal fall for apples
* HI ornamentals 7 d; beans, peas 2 wk; oilseed rape 7 wk

Maximum Residue Levels (mg residue/kg food)
* hops 40; grapes, cane fruits, lettuces, celery 5; tomatoes, peppers, aubergines 3; apricots, peaches, nectarines, Chinese cabbage, witloof, beans (with pods), peas (with pods) 2; pome fruits, bulb vegetables, cucumbers, gherkins, courgettes, melons, squashes, watermelons, rape seed 1; cherries 0.5; tea 0.1; citrus fruits, tree nuts, bilberries, cranberries, gooseberries, wild berries, miscellaneous fruit (except kiwi fruit), beetroot, celeriac, Jerusalem artichokes, parsnips, parsley root, salsify, sweet potatoes, turnips, yams, sweetcorn, broccoli, cauliflowers, Brussels sprouts, head cabbages, kale, kohlrabi, spinach, beet leaves, watercress, herbs, stem vegetables (except celery), mushrooms, oilseed (except rape seed), potatoes, cereals, animal products 0.05

FOR FULL CONDITIONS OF USE ALWAYS READ THE PRODUCT LABEL

496   warfarin

A hydroxycoumarin rodenticide

Products

1	Grey Squirrel Liquid Concentrate	Killgerm	0.5% w/w	CB	06455
2	Sakarat Ready-to-Use (cut wheat)	Killgerm	0.025% w/w	RB	04340
3	Sakarat Ready-to-Use (whole wheat)	Killgerm	0.025% w/w	RB	01850
4	Sakarat X Ready-to-Use Warfarin Rat Bait	Killgerm	0.05% w/w	RB	01851
5	Sewarin Extra	Killgerm	0.05% w/w	RB	03426
6	Sewarin P	Killgerm	0.025% w/w	RB	01930
7	Sewercide Cut Wheat Rat Bait	Killgerm	0.05 w/w	RB	03761
8	Sewercide Whole Wheat Rat Bait	Killgerm	0.05% w/w	RB	03759
9	Sorex Warfarin 250 ppm Rat Bait	Sorex	0.025% w/w	RB	07371
10	Sorex Warfarin 500 ppm Rat Bait	Sorex	0.05% w/w	RB	07372
11	Sorex Warfarin Sewer Bait	Sorex	0.05% w/w	RB	07373
12	Warfarin 0.5% Concentrate	B H & B	0.5%	CB	02325
13	Warfarin Ready Mixed Bait	B H & B	0.025%	RB	02333

Uses

Grey squirrels in FORESTRY [1]. Mice, rats in AGRICULTURAL PREMISES [2-13].

Notes

Efficacy
* For rodent control place ready-to-use or prepared baits at many points wherever rats and mice active. Out of doors shelter bait from weather
* Inspect baits frequently and replace or top up as long as evidence of feeding. Do not underbait
* For grey squirrel control mix with whole wheat and leave to stand for 2-3 h before use
* Use bait in specially constructed hoppers and inspect every 2-3 d. Replace as necessary

Special precautions/Environmental safety
* For use only by local authorities, professional operators providing a pest control service and persons occupying industrial, agricultural or horticultural premises
* Prevent access to baits by children and animals, especially cats, dogs and pigs
* Rodent bodies must be searched for and burned or buried, not placed in refuse bins or rubbish tips. Remains of bait and containers must be removed after treatment and burned or buried
* Bait must not be used where food, feed or water could become contaminated
* The use of warfarin to control grey squirrels is illegal unless the provisions of the Grey Squirrels Order 1973 are observed. See label for list of counties in which bait may not be used

Protective clothing/Label precautions
* 25, 29, 63, 67, 81, 82, 83, 84, 85

497 zinc phosphide
A phosphine generating rodenticide

Products
1	Grovex Zinc Phosphide	Killgerm	100%	CB	06230
2	RCR Zinc Phosphide	Killgerm	100%	CB	06231

Uses Mice, rats in AGRICULTURAL PREMISES.

Notes **Efficacy**
* Use to prepare baits either by dry or wet baiting as directed

Special precautions/Environmental safety
* Zinc phosphide is subject to the Poisons Rules 1982 and the Poisons Act 1972. See notes in Section 1
* For use only by local authorities, professional operators providing a pest control serice and persons occupying industrial, agricultural or horticultural premises
* Toxic if swallowed
* Spontaneously inflammable in contact with acid
* Keep in original container, tightly closed, in a safe place, under lock and key
* Wear respirator if mixing baits in a confined space
* Prevent access to baits by children and domestic animals, especially cats, dogs and pigs
* Search for and burn or bury all rodent bodies. Do not place in refuse bins or on rubbish tips
* Do not prepare or use baits where food or water could be contaminated
* Remove all remains of baits and bait containers after treatment and burn or bury
* Wash out all mixing equipment thoroughly at the end of every operation

Protective clothing/Label precautions
* A, D, H
* 4c, 14, 16, 18, 25, 29, 36, 37, 64, 67, 70, 81, 82, 83, 84

498 zineb
A protectant dithiocarbamate fungicide for many horticultural crops
See also ferbam + maneb + zineb

Products
Unicrop Zineb	Unicrop	70 % w/w	WP	02279

Uses Blight, leaf mould, root rot in TOMATOES. Blight in POTATOES. Botrytis in ANEMONES. Celery leaf spot in CELERY. Currant leaf spot in BLACKCURRANTS. Downy mildew in HOPS, OUTDOOR LETTUCE, PROTECTED LETTUCE. Fire in TULIPS. Rust in CARNATIONS.

Notes **Efficacy**
* Best results obtained by treatment in settled weather conditions. Do not spray if rain imminent

FOR FULL CONDITIONS OF USE ALWAYS READ THE PRODUCT LABEL

- Follow local blight warnings to ensure accuracy of first treatment on potatoes
- Recommended timing and spray volume varies with crop and disease. See label for details
- For root rot control in tomatoes apply as a trench watering at about the time of set of 5th truss
- Addition of wetting agent recommended for tulips and anemones

Crop Safety/Restrictions
- Maximum number of treatments on protected lettuce (including zineb, maneb, mancozeb, other EBDC fungicides or thiram on protected lettuce) 2 per crop post-planting up to 2 wk later, and none thereafter. If thiram-based products are used post-planting on crops that will mature from Nov to Mar, 3 treatments are permitted within 3 wk of planting out
- Do not apply pre-picking to blackcurrants intended for canning

Special precautions/Environmental safety
- Irritating to eyes, skin and respiratory system

Protective clothing/Label precautions
- A
- 6a, 6b, 6c, 18, 21, 28, 29, 36, 37, 54, 63, 67

Latest application/Harvest Interval(HI)
- HI blackcurrants 4 wk; protected lettuce 3 wk; outdoor lettuce 2 wk; other edible outdoor crops 1 wk; other edible glasshouse crops 2 d

Approval
- Approved for aerial application on potatoes. See notes in Section 1

499   ziram
A dithiocarbamate bird and animal repellent

| Products | Aaprotect | Unicrop | 32% w/w | PA | 03784 |

Uses Birds, deer, hares, rabbits in FIELD CROPS, FORESTRY, ORNAMENTALS, TOP FRUIT.

Notes
Efficacy
- Apply undiluted to main stems up to knee height to protect against browsing animals at any time of yr or spray 1:1 dilution on stems and branches in dormant season
- Use dilute spray on fully dormant fruit buds to protect against bullfinches
- Only apply to dry stems, branches or buds
- Use of diluted spray can give limited protection to field crops in areas of high risk during establishment period

Crop Safety/Restrictions
- Do not spray elongating shoots or buds about to open
- Do not apply concentrated spray to foliage, fruit buds or field crops

Special precautions/Environmental safety
- Irritating to eyes, skin and respiratory system

Protective clothing/Label precautions
- A, C
- 6a, 6b, 6c, 18, 21, 28, 29, 35, 36, 37

Latest application/Harvest Interval(HI)
• HI edible crops 8 wk

SECTION 4
APPENDICES

Appendix 1
SUPPLIERS OF PESTICIDES AND ADJUVANTS

AgrEvo: **AgrEvo UK Crop Protection Ltd**
East Winch Hall
East Winch
King's Lynn
Norfolk PE32 1HN
Tel: (01553) 841581
Fax: (01553) 841090

AgrEvo Environ.: **AgrEvo Environmental**
Health Ltd
McIntyre House
High Street
Berkhamsted
Herts HP4 2DY
Tel: (01442) 863333
Fax: (01442) 872783

Agrichem: **Agrichem (International) Ltd**
Industrial Estate
Station Road
Whittlesey
Cambs PE7 2EY
Tel: (01733) 204019
Fax: (01733) 204162

Aitken: **R. Aitken Ltd**
123 Harmony Row
Govan
Glasgow G51 3NB
Tel: (0141) 440 0033
Fax: (0141) 440 2744

Allen: **Allen Power Equipment Ltd**
The Broadway
Didcot
Oxon. OX11 8ES
Tel: (01235) 813936
Fax: (01235) 811491

American Products: **American Products**
165-171 Strand Road
Londonderry BT48 7PT
Tel: (01504) 267535/6
Fax: (01504) 269313

Antec: **Antec (AH) International**
Windham Road
Chilton Industrial Estate
Sudbury
Suffolk CO10 6XD
Tel: (01787) 377305
Fax: (01787) 310846

Aquaspersions: **Aquaspersions Ltd**
Charlestown Works
Charlestown
Hebden Bridge
W. Yorks. HX7 6PL
Tel: (01422) 843715
Fax: (01422) 845067

Armillatox: **Armillatox Ltd**
121 Main Road
Morton
Alfreton
Derbyshire DE55 6HL
Tel: (01773) 590566
Fax: (01773) 590681

Ashlade: **Ashlade Formulations Ltd**
Moorend House
Moorend Lane
Dewsbury
W. Yorks. WF13 4QQ
Tel: (01924) 409782
Fax: (01924) 410792

Atlas: **Atlas Crop Protection Ltd**
P O Box 38
Low Moor
Bradford
W. Yorks. BD12 0JZ
Tel: (01274) 693707
Fax: (01274) 693708

B H & B: **Battle Haywood & Bower Ltd**
Victoria Chemical Works
Crofton Drive
Allenby Road Industrial Esate
Lincoln LN3 4NP
Tel: (01522) 529206/54124
Fax: (01522) 538960

Barclay: **Barclay Plant Protection**
28 Howard Street
Glossop
Derbyshire SK13 9DD
Tel: (01457) 853386
Fax: (01457) 853557

BASF: **BASF plc.**
Agricultural Divison
PO Box 4
Earl Road, Cheadle Hulme
Cheadle
Cheshire SK8 6OG
Tel: (0161) 485 6222
Fax: (0161) 485 2229

Batson: **Joseph Batson & Co. Ltd**
Dudley Road
Tipton
W. Midlands DY4 8EH
Tel: (0121) 557 2284
Fax: (0121) 557 8068

Bayer: **Bayer plc.**
Crop Protection Business Group
Eastern Way
Bury St Edmunds
Suffolk IP32 7AH
Tel: (01284) 763200
Fax: (01284) 702810

Brian Jones: **Brian Jones and Associates Ltd**
Fluorocarbon Building
Caxton Hill
Hertford
Herts. SG13 7NH
Tel: (01992) 550731
Fax: (01992) 584697

Bromine & Chem.: **Bromine and Chemicals Ltd**
6 Arlington Street
St James's
London SW1A 1RE
Tel: (0171) 493 9711
Fax: (0171) 493 9714

Cheminova: **Cheminova Agro (UK) Ltd**
27 Marlow Road
Maidenhead
Berks. SL6 7AE
Tel: (01628) 770030
Fax: (01628) 784215

Chemsearch: **Chemsearch (UK) Ltd**
Landchard House
Victoria Street
West Bromwich
W. Midlands B70 8ER
Tel: (0121) 525 1666
Fax: (0121) 500 5386

Chemtech: **Chemtech (Crop Protection) Ltd**
The Arable Centre
Winterbourne Monkton
Swindon
Wilts. SN4 9NW
Tel: (01672) 539591

Chiltern: **Chiltern Farm Chemicals Ltd**
11 High Street
Thornborough
Buckingham MK18 2DF
Tel: (01280) 822400
Fax: (01280) 822082

Ciba Agric.: **Ciba Agriculture**
Whittlesford
Cambridge CB2 4QT
Tel: (01223) 833621
Fax: (01223) 835211

Clifton: **Clifton Chemicals Ltd**
119 Grenville Street
Edgeley
Stockport
Cheshire SK3 9EU
Tel: (0161) 476 1128
Fax: (0161) 476 1280

CMI: **CMI Ltd (incorporating**
Collingham Marketing)
United House
113 High Street
Collingham
Newark
Notts. NG23 7NG
Tel: (01636) 892078
Fax: (01636) 893037

Coalite: **Coalite Chemicals**
PO Box 152
Buttermilk Lane
Bolsover
Chesterfield
Derbyshire S44 6AZ
Tel: (01246) 826816
Fax: (01246) 240309

Coventry Chemicals: **Coventry Chemicals Ltd**
Woodhams Road
Siskin Drive
Coventry CV3 4FX
Tel: (01203) 639739
Fax: (01203) 639717

Cyanamid: **Cyanamid of Great Britain Ltd**
Crop Protection Division
Cyanamid House, Fareham Road
PO Box 7
Gosport
Hants. PO13 0AS
Tel: (01329) 224000
Fax: (01329) 224335

Darmycel: **Darmycel (UK)**
Broadway
Yaxley
Peterborough PE7 3EJ
Tel: (01733) 244626
Fax: (01733) 245020

Dax: **Dax Products Ltd**
P O Box 119
76 Cyprus Road
Nottingham NG3 5NA
Tel: (0115) 960 9996
Fax: (0115) 960 4620

De Sangosse: **De Sangosse (UK) SA**
PO Box 135
Market Weighton
York YO4 3YY
Tel: (01430) 872525
Fax: (01430) 873123

Deosan: **Deosan Ltd**
Weston Favell Centre
Northampton NN3 4PD
Tel: (01604) 414000
Fax: (01604) 406809

Dewco-Lloyd: **Dewco-Lloyd Ltd**
Cyder House
Ixworth
Suffolk IP31 2HT
Tel: (01359) 230555
Fax: (01359) 232553

Doff Portland: **Doff Portland Ltd**
Benneworth Close
Hucknall
Nottingham NG16 6EL
Tel: (0115) 963 2842
Fax: (0115) 963 8657

DowElanco: **DowElanco Ltd**
Latchmore Court
Brand Street
Hitchin
Herts. SG5 1HZ
Tel: (01462) 457272
Fax: (01462) 426605

DuPont: **DuPont (UK) Ltd**
Agricultural Products Department
Wedgwood Way
Stevenage
Herts. SG1 4QN
Tel: (01438) 734000
Fax: (01438) 734507

Elliott: **Thomas Elliott Ltd**
143A High Street
Edenbridge
Kent TN8 5AX
Tel: (01732) 866566
Fax: (01732) 864709

English Woodland: **English
Woodlands Biocontrol**
Hoyle Depot
Graffham
Petworth
W. Sussex GU28 0LR
Tel: (01798) 867574
Fax: (01798) 867574

Fargro: **Fargro Ltd**
Toddington Lane
Littlehampton
Sussex BN17 7PP
Tel: (01903) 721591
Fax: (01903) 730737

FCC: **Farmers Crop Chemicals Ltd**
Thorn Farm
Evesham Road
Inkberrow
Worcs. WR7 4LJ
Tel: (01386) 793401
Fax: (01386) 793184

Fine: **Fine Agrochemicals Ltd**
3 The Bull Ring
Worcester WR2 5AA
Tel: (01905) 748444
Fax: (01905) 748440

Fletcher: **Sam Fletcher Agricultural Specialists**
Division of Banks Odam Dennick Ltd
Fleet Road Industrial Estate
Holbeach
Spalding
Lincs. PE12 7EG
Tel: (01406) 22207
Fax: (01406) 26525

Ford Smith: **Ford Smith & Co. Ltd**
Lyndean Industrial Estate
Felixstowe Road
Abbey Wood
London SE2 9SG
Tel: (0181) 310 8127
Fax: (0181) 310 9563

Garotta: **Garotta Products Ltd**
See Sinclair Horticulture & Leisure Ltd

Graincare: **Graincare (Colchester) Ltd**
17 Woodlands
Colchester
Essex CO4 3JA
Tel: (01206) 862436
Fax: (01206) 862436

Grampian: **Grampian Pharmaceuticals Ltd**
Marathon Place
Moss Side Industrial Estate
Leyland
Lancs. PR5 3QN
Tel: (01772) 452421
Fax: (01772) 456820

Growing Success: **Growing Success Organics Ltd**
South Newton
Salisbury
Wilts. SP2 0QW
Tel: (01722) 742500
Fax: (01722) 742571

HCS: **Horticultural Consultancy Service**
104 Stonefall Avenue
Starbeck
Harrogate
N. Yorks. HG2 7NT
Tel: (01423) 880468
Fax: (01423) 880468

Headland: **Headland Agrochemicals Ltd**
Norfolk House
Gt. Chesterford Court
Gt. Chesterford
Saffron Walden
Essex CB10 1PF
Tel: (01799) 530146
Fax: (01799) 530229

Heatherington: **J.V. Heatherington Farm & Garden Supplies**
29 Main Street
Glenary
Crumlin
Co. Antrim BT29 4LN
Tel: (018494) 22227

Helm: **Helm Great Britain Chemicals Ltd**
Wimbledon Bridge House
1 Hartfield Road
London SW19 3RU
Tel: (0181) 544 9000
Fax: (0181) 544 1011

Hickson & Welch: **Hickson & Welch Ltd**
Wheldon Road
Castleford
W. Yorks. WF10 2JT
Tel: (01977) 556565
Fax: (01977) 512809

Hoechst: **Hoechst UK Ltd**
See AgrEvo UK Crop Protection

Hortag: **Hortag Chemicals Ltd**
Salisbury Road
Downton
Wilts. SP5 3JJ
Tel: (01725) 512822
Fax: (01725) 512840

Hortichem: **Hortichem Ltd**
1 Edison Road
Churchfields Industrial Estate
Salisbury
Wilts. SP2 7NU
Tel: (01722) 320133
Fax: (01722) 326799

I T Agro: **I T Agro Ltd**
805 Salisbury House
31 Finsbury Circus
London EC2M 5SQ
Tel: (0171) 628 2040

ICI Agrochem.: **ICI Agrochemicals**
See Zeneca Crop Protection

ICI Professional: **ICI Garden &**
Professional Products
See Miracle Garden Care Ltd

Ideal: **Ideal Manufacturing Ltd**
Atlas House
Burton Road
Finedon
Wellingborough
Northants. NN9 5HX
Tel: (01933) 681616
Fax: (01933) 681042

Interagro: **Interagro (UK) Ltd**
2 Ducketts Wharf
South Street
Bishop's Stortford
Herts. CM23 3AR
Tel: (01279) 501995
Fax: (01279) 501996

Intracrop: **Intracrop**
Brian Lewis Agriculture Ltd
B7 Perimeter Road
North Culham Estate
Abingdon
Oxon. OX14 3GY
Tel: (01867) 307696
Fax: (01867) 307684

Isagro: **Isagro srl.**
See Sipcam UK Ltd

ISK Biosciences: **ISK Biosciences Ltd**
4th Floor, South Block
Central Court
Knoll Rise
Orpington
Kent BR6 0JA
Tel: (01689) 874011
Fax: (01689) 874085

Keychem: **Keychem Ltd**
9 Castlemead Gardens
Hertford SD14 7JZ
Tel: (01992) 553533
Fax: (01992) 558979

Killgerm: **Killgerm Chemicals Ltd**
115 Wakefield Road
Flushdyke
Ossett
W. Yorks. WF5 9BW
Tel: (01924) 265090
Fax: (01924) 265033

Koppert: **Koppert (UK) Ltd**
1 Wadhurst Business Park
Faircrouch Lane
Wadhurst
E. Sussex TN5 6PT
Tel: (01892) 784411
Fax: (01892) 782469

Landgold: **Landgold & Co. Ltd**
PO Box 137
St Peter's House
Le Bordage
St. Peter Port
Guernsey GY1 3HW
Tel: (01481) 723575
Fax: (01481) 710147

Lever Industrial: **Lever Industrial Ltd**
PO Box 100
Runcorn
Cheshire WA7 3JZ
Tel: (01928) 719000
Fax: (01928) 714628

Levington: **Levington Horticulture Ltd**
Paper Mill Lane
Bramford
Ipswich
Suffolk IP8 4BZ
Tel: (01473) 830492
Fax: (01473) 830386

Luxan: Luxan (UK) Ltd
Sysonby Lodge
Nottingham Road
Melton Mowbray
Leics. LE13 0NU
Tel: (01664) 66372
Fax: (01664) 480137

Makhteshim: Makhteshim-Agan (UK) Ltd
Sir William Atkins House
Ashley Avenue
Epsom
Surrey KT18 5WA
Tel: (01372) 747372

Mandops: Mandops (UK) Ltd
36 Leigh Road
Eastleigh
Hants. SO50 9DT
Tel: (01703) 641826
Fax: (01703) 629106

Maxicrop: Maxicrop International Ltd
Weldon Road
Corby
Northants. NN17 5US
Tel: (01536) 402182
Fax: (01536) 204254

McMillan: McMillan Technical Services Ltd
McMillan House
54-56 Cheam Common Road
Worcester Park
Surrey KT4 8RH
Tel: (0181) 337 0731
Fax: (0181) 335 3056

Microcide: Microcide Ltd
Shepherds Grove
Stanton
Bury St. Edmunds
Suffolk IP31 2AR
Tel: (01359) 251077
Fax: (01359) 251545

Miracle Prof.: Miracle Garden Care Ltd
Salisbury House
Weyside Park
Catteshall Lane
Godalming
Surrey GU7 1XE
Tel: (01483) 410210
Fax: (01483) 410267

Mirfield: Mirfield Sales Services Ltd
Moorend House
Moorend Lane
Dewsbury
W. Yorks. WF13 4QQ
Tel: (01924) 409782
Fax: (01924) 410792

Mitchell Cotts: Mitchell Cotts Chemicals Ltd
PO Box 6
Steanard Lane
Mirfield
W. Yorks. WF14 8QB
Tel: (01924) 493861
Fax: (01924) 490972

Monsanto: Monsanto plc.
Thames Tower
Burleys Way
Leicester LE1 3TP
Tel: (0116) 262 0864
Fax: (0116) 253 0320

MSD AGVET: MSD AGVET
Hertford Road
Hoddesdon
Herts. EN11 9BU
Tel: (01992) 467272
Fax: (01992) 451059

MTM Agrochem.: MTM Agrochemicals Ltd
See United Phosphorus Ltd

Nehra: Nehra Cookes Chemicals Ltd
16 Chiltern Close
Warren Wood
Arnold
Nottingham NG5 9PX
Tel: (0115) 920 3839
Fax: (0115) 967 1734

Newman: Newman Agrochemicals Ltd
Swaffham Bulbeck
Cambridge CB5 0LU
Tel: (01223) 811215
Fax: (01223) 812725

Nickerson Seeds: Nickerson Seeds Ltd
JNRC
Rothwell
Lincs. LN7 6DT
Tel: (01472) 371661

Nomix-Chipman: **Nomix-Chipman Ltd**
Portland Building
Portland Street
Staple Hill
Bristol BS16 4PS
Tel: (0117) 957 4574
Fax: (0117) 956 3461

PBI: **Pan Britannica Industries Ltd**
Britannica House
Waltham Cross
Herts. EN8 7DY
Tel: (01992) 623691
Fax: (01992) 626452

Promark: **Promark Ltd**
See AgrEvo UK Crop Protection Ltd

Quadrangle: **Quadrangle Agrochemicals**
Bishop Monkton
Harrogate
N. Yorks. HG3 3QQ
Tel: (01765) 677146
Fax: (01765) 677168

Rentokil: **Rentokil Environmental Services**
Felcourt
East Grinstead
W. Sussex RH19 2JY
Tel: (01342) 833022
Fax: (01342) 326229

Rigby Taylor: **Rigby Taylor Ltd**
The Riverway Estate
Portsmouth Road
Peasmarsh
Guildford
Surrey GU3 1LZ
Tel: (01483) 35657
Fax: (01483) 34058

Roebuck Eyot: **Roebuck Eyot Ltd**
PO Box 321
Welwyn Garden City
Herts. AL7 1LF
Tel: (01707) 371105
Fax: (01707) 339221

Rohm & Haas: **Rohm & Haas (UK) Ltd**
Lennig House
2 Masons Avenue
Croydon
Surrey CR9 3NB
Tel: (0181) 686 8844
Fax: (0181) 686 9447

RP Agric.: **Rhone-Poulenc Agriculture Ltd**
Fyfield Road
Ongar
Essex CM5 0HW
Tel: (01277) 301301
Fax: (01277) 362610

RP Environ.: **Rhone-Poulenc
Environmental Products**
Address as above

Sandoz: **Sandoz Agro Ltd**
16/18 Princes Street
Ipswich
Suffolk IP1 1QT
Tel: (01473) 255972
Fax: (01473) 258252

Sandoz Speciality: **Sandoz Speciality Pest
Control Ltd**
SGS House
217-221 London Road
Camberley
Surrey GU15 3EY
Tel: (01276) 25425
Fax: (01276) 25769

Sanofi: **Sanofi Animal Health Ltd**
7 Awberry Court
Hatters Lane
Watford
Herts. WD1 8YJ
Tel: (01923) 212212
Fax: (01923) 243001

Schering: **Schering Agriculture**
See AgrEvo UK Crop Protection Ltd

Service Chemicals: **Service Chemicals Ltd**
Lanchester Way
Royal Oak Industrial Estate
Daventry
Northants. NN11 5PH
Tel: (01327) 704444
Fax: (01327) 71154

Sinclair: **Sinclair Horticulture & Leisure Ltd**
Firth Road
Lincoln LN6 7AH
Tel: (01522) 537561
Fax: (01522) 513609

Sipcam: **Sipcam UK Ltd**
100 Chalk Farm Road
London NW1 8EH
Tel: (0171) 284 4218

Sorex: **Sorex Ltd**
St Michael's Industrial Estate
Hale Road
Widnes
Cheshire WA8 8TJ
Tel: (0151) 420 7151
Fax: (0151) 495 1163

Sphere: **Sphere Laboratories (London) Ltd**
The Yews
Main Street
Chilton
Oxon. OX11 0RZ
Tel: (01235) 831802
Fax: (01235) 833896

Spraydex: **Spraydex Ltd**
Brapack
Moreton Avenue
Wallingford
Oxon. OX10 9DE
Tel: (01491) 825251
Fax: (01491) 825409

Standon: **Standon Chemicals Ltd**
48 Grosvenor Square
London W1X 9LA
Tel: (0171) 493 8648
Fax: (0171) 493 4219

Steele & Brodie: **Steele & Brodie (1983) Ltd**
The Beehive Works
25 Kilmany Road
Wormit
Newport-on-Tay
Fife DD6 8PG
Tel: (01382) 541728
Fax: (01382) 543022

Stefes: **Stefes Plant Protection Ltd**
Huntingdon Business Centre
Blackstone Road
Huntingdon
Cambs. PE18 6EF
Tel: (01480) 435101
Fax: (01480) 420041

Stoller: **Stoller Chemical Ltd**
53 Bradley Hall Trading Estate
Bradley Lane
Standish
Lancs. WN6 0XQ
Tel: (01257) 427722
Fax: (01257) 477888

Sumitomo: **Sumitomo Corporation (UK) Ltd**
Vintner's Place
68 Upper Thames Street
London EC4V 3BJ
Tel: (0171) 246 3600
Fax: (0171) 246 3925

Techsol: **Techsol Manufacturing**
Stanhope Road
Swadlincote
Burton-on-Trent
Staffs. DE11 9BE
Tel: (01283) 221044
Fax: (01283) 225731

Tripart: **Tripart Farm Chemicals Ltd**
Swan House
Beulah Street
Gaywood
King's Lynn
Norfolk PE30 4DN
Tel: (01553) 674303
Fax: (01553) 674422

Truchem: **Truchem Ltd**
Brook House
30 Larwood Grove
Sherwood
Nottingham NG5 3JD
Tel: (0115) 926 0762
Fax: (0115) 967 1153

Unicrop: **Universal Crop Protection Ltd**
Park House
Cookham
Maidenhead
Berks. SL6 9DS
Tel: (01628) 526083
Fax: (01628) 810457

Uniroyal: **Uniroyal Chemical Ltd**
Kennet House
4 Langley Quay
Slough
Berks. SL3 6EH
Tel: (01753) 603000
Fax: (01753) 603077

United Phosphorus: **United Phosphorus Ltd**
18 Liverpool Road
Great Sankey
Warrington
Cheshire WA5 1QR
Tel: (01925) 633232
Fax: (01925) 652679

Vass: **L.W. Vass (Agricultural) Ltd**
Springfield Farm
Silsoe Road
Maulden
Bedford MK45 2AX
Tel: (01525) 403041
Fax: (01525) 402282

Vitax: **Vitax Ltd**
Owen Street
Coalville
Leicester LE67 3DE
Tel: (0116) 251 0060
Fax: (0116) 251 0299

Wheatley: **Wheatley Chemical Co.**
19 White House Gardens
Tadcaster Road
York YO2 2DZ
Tel: (01904) 629657
Fax: (01904) 629657

Whyte: **Whyte Chemicals Ltd**
Gateway House
322 Regents Park Road
Finchley
London N3 2UA
Tel: (0181) 346 6279
Fax: (0181) 349 4589

Yule Catto: **Yule Catto Consumer
Chemicals Ltd**
Stanhope Road
Swadlincote
Burton-on-Trent
Staffs. DE11 9BE
Tel: (01283) 221044
Fax: (01283) 225731

Zeneca: **Zeneca Crop Protection**
Fernhurst
Haslemere
Surrey GU27 3JE
Tel: (01428) 656564
Fax: (01428) 657385

Zeneca Prof.: **Zeneca Professional Products**
See Miracle Garden Care Ltd

Appendix 2
USEFUL CONTACTS

Agricultural Training Board
Head Office
Stoneleigh Park Pavilion
National Agricultural Centre
Kenilworth
Warwickshire CV8 2UG
Tel: (01203) 696996
Fax: (01203) 696732

BASIS Limited
Bank Chambers
2 St John Street
Ashbourne
Derbyshire DE6 1GL
Tel: (01335) 343945
Fax: (01335) 346488

BCPC Publications Sales
Bear Farm
Binfield
Bracknell
Berks. RG12 5QE
Tel/Fax: (01734) 341998

British Agrochemicals Association Ltd
4 Lincoln Court
Lincoln Road
Peterborough
Cambs. PE1 2RP
Tel: (01733) 349225
Fax: (01733) 62523

British Beekeepers' Association
National Agricultural Centre
Stoneleigh
Kenilworth
Warwickshire CV8 2LZ
Tel: (01203) 696679
Fax: (01203) 690682

British Pest Control Association
3 St James Court
Friar Gate
Derby DE1 1ZU
Tel: (01332) 294288
Fax: (01332) 295904

Department of Agriculture Northern Ireland
Pesticides Section
Dundonald House
Upper Newtownards Road
Belfast BT4 3SB
Tel: (01232) 524704

Forestry Commission
231 Corstophine Road
Edinburgh EH12 7AT
Tel: (0131) 334 0303
Fax: (0131) 334 3047

Health and Safety Executive
Information Centre
Broad Lane
Sheffield
Yorkshire S3 7HQ
Tel: (0114) 289 2345/6
Fax: (0114) 289 2333

Health and Safety Executive
Pesticides Registration Section
Magdalen House
Bootle
Merseyside L20 3QZ
Tel: (0151) 951 4000

Health and Safety Executive – Books
PO Box 1999
Sudbury
Suffolk CO10 6FS
Tel: (01787) 881165
Fax: (01787) 313995

HMSO
Publications Centre
PO Box 276
London SW8 5DT
Tel: (0171) 873 9090 (orders)
(0171) 873 0011 (enquiries)
Fax: (0171) 873 8200

National Farmers Union
Agriculture House
164 Shaftesbury Avenue
London WC2H 8HL
Tel: (0171) 235 5077/331 7200
Fax: (0171) 235 3526

National Poisons Information Service
and regional centres: See p. 63

National Proficiency Tests Council (NPTC)
Tenth Street
National Agricultural Centre
Stoneleigh
Kenilworth
Warwickshire CV8 2LG
Tel: (01203) 696553

National Rivers Authority
Rivers House
Waterside Drive
Aztec West
Almondsbury
Bristol BS12 4UD
Tel: (01454) 624400
Fax: (01454) 624409

Pesticides Safety Directorate
Mallard House
King's Pool
3 Peasholme Green
York YO1 2PX
Tel: (01904) 640500
Fax: (01904) 455733

Processors and Growers Research Organization (PGRO)
The Research Station
Great North Road
Thornhaugh
Peterborough
Cambs. PE8 6HJ
Tel: (01780) 782585
Fax: (01780) 783993

UK Agricultural Supply Trade Association (UKASTA)
3 Whitehall Court
London SW1A 2EQ
Tel: (0171) 930 3611
Fax: (0171) 930 3952

Appendix 3
KEYS TO CROP AND WEED GROWTH STAGES

Decimal code for the growth stages of cereals

Illustrations of these growth stages can be found in the reference indicated below and in some company product manuals.

0 Germination
00 Dry seed
03 Imbibition complete
05 Radicle emerged from caryopsis
07 Coleoptile emerged from caryopsis
09 Leaf at coleoptile tip

1 Seedling growth
10 First leaf through coleoptile
11 First leaf unfolded
12 2 leaves unfolded
13 3 leaves unfolded
14 4 leaves unfolded
15 5 leaves unfolded
16 6 leaves unfolded
17 7 leaves unfolded
18 8 leaves unfolded
19 9 or more leaves unfolded

2 Tillering
20 Main shoot only
21 Main shoot and 1 tiller
22 Main shoot and 2 tillers
23 Main shoot and 3 tillers
24 Main shoot and 4 tillers
25 Main shoot and 5 tillers
26 Main shoot and 6 tillers
27 Main shoot and 7 tillers
28 Main shoot and 8 tillers
29 Main shoot and 9 or more tillers

3 Stem elongation
30 ear at 1 cm
31 1st node detectable
32 2nd node detectable
33 3rd node detectable
34 4th node detectable
35 5th node detectable
36 6th node detectable
37 Flag leaf just visible
39 Flag leaf ligule/collar just visible

4 Booting
41 Flag leaf sheath extending
43 Boots just visibly swollen
45 Boots swollen
47 Flag leaf sheath opening
49 First awns visible

5 Inflorescence
51 First spikelet of inflorescence just visible
52 $1/4$ of inflorescence emerged
55 $1/2$ of inflorescence emerged
57 $3/4$ of inflorescence emerged
59 Emergence of inflorescence completed

6 Anthesis
60
61 } Beginning of anthesis

64
65 } Anthesis half way

68
69 } Anthesis complete

7 Milk development
71 Caryopsis watery ripe
73 Early milk
75 Medium milk
77 Late milk

8 Dough development
83 Early dough
85 Soft dough
87 Hard dough

9 Ripening
91 Caryopsis hard (difficult to divide by thumb-nail)
92 Caryopsis hard (can no longer be dented by thumb-nail)
93 Caryopsis loosening in daytime

(From Tottman, 1987. *Annals of Applied Biology*, **110**, 441–454)

Stages in development of oilseed rape

Illustrations of these growth stages can be found in the reference indicated below and in some company product manuals.

0 Germination and emergence

1 Leaf production
1,0   Both cotyledons unfolded and green
1,1   First true leaf
1,2   Second true leaf
1,3   Third true leaf
1,4   Fourth true leaf
1,5   Fifth true leaf
1,10 About tenth true leaf
1,15 About fifteenth true leaf

2 Stem extension
2,0   No internodes ('rosette')
2,5   About five internodes

3 Flower bud development
3,0   Only leaf buds present
3,1   Flower buds present but enclosed by leaves
3,3   Flower buds visible from above ('green bud')
3,5   Flower buds raised above leaves
3,6   First flower stalks extending
3,7   First flower buds yellow ('yellow bud')

4 Flowering
4,0   First flower opened
4,1   10% all buds opened

4,3   30% all buds opened
4,5   50% all buds opened

5 Pod development
5,3   30% potential pods
5,5   50% potential pods
5,7   70% potential pods
5,9   All potential pods

6 Seed development
6,1   Seeds expanding
6,2   Most seeds translucent but full size
6,3   Most seeds green
6,4   Most seeds green-brown mottled
6,5   Most seeds brown
6,6   Most seeds dark brown
6,7   Most seeds black but soft
6,8   Most seeds black and hard
6,9   All seeds black and hard

7 Leaf senescence

8 Stem senescence
8,1   Most stem green
8,5   Half stem green
8,9   Little stem green

9 Pod senescence
9,1   Most pods green
9,5   Half pods green
9.9   Few pods green

(From Sylvester-Bradley, 1985. *Aspects of Applied Biology*, **10**, 395–400)

Stages in development of peas

Illustrations of these growth stages can be found in the reference indicated below and in some company product manuals.

0 Germination and emergence
000 Dry seed
001 Imbibed seed
002 Radicle apparent
003 Plumule and radicle apparent
004 Emergence

1 Vegetative stage
101 First node (leaf with one pair leaflets, no tendril)
102 Second node (leaf with one pair leaflets, simple tendril)
103 Third node (leaf with one pair leaflets, complex tendril)
.
.
10x X nodes (leaf with more than one pair leaflets, complex tendril)
.
.
10n Last recorded node

2 Reproductive stage (main stem)
201 Enclosed buds
202 Visible buds
203 First open flower
204 Pod set (small immature pod)
205 Flat pod
206 Pod swell (seeds small, immature)
207 Pod fill
208 Pod green, wrinkled
209 Pod yellow, wrinkled (seeds rubbery)
210 Dry seed

3 Senescence stage
301 Desiccant application stage. Lower pods dry and brown, middle yellow, upper green. Overall moisture content of seed less than 45%
302 Pre-harvest stage. Lower and middle pods dry and brown, upper yellow. Overall moisture content of seed less than 30%
303 Dry harvest stage. All pods dry and brown, seed dry

(From Knott, 1987. *Annals of Applied Biology*, **111**, 233–244)

Stages in development of faba beans

Illustrations of these growth stages can be found in the reference indicated below and in some company product manuals.

0 Germination and emergence
000 Dry seed
001 Imbibed seed
002 Radicle apparent
003 Plumule and radicle apparent
004 Emergence
005 First leaf unfolding
006 First leaf unfolded

1 Vegetative stage
101 First node
102 Second node
103 Third node
.
.
10x X nodes
.
.
10n N, last recorded node

2 Reproductive stage (main stem)
201 Flower buds visible
203 First open flowers
204 First pod set
205 Pods fully formed, green
207 Pod fill, pods green
209 Seed rubbery, pods pliable, turning black
210 Seed dry and hard, pods dry and black

3 Pod senescence
301 10% pods dry and black
.
.
305 50% pods dry and black
.
.
308 80% pods dry and black, some upper pods green
309 90% pods dry and black, most seed dry. Desiccation stage.
310 All pods dry and black, seed hard. Pre-harvest (glyphosate application stage)

4 Stem senescence
401 10% stem brown/black
.
405 50% stem brown/black
.
.
409 90% stem brown/black
410 All stems brown/black. All pods dry and black, seed hard.

(From Knott, 1990. *Annals of Applied Biology*, **116**, 391–404)

Stages in development of potato

Illustrations of these growth stages can be found in the reference indicated below and in some company product manuals.

0 Seed germination and seedling emergence
000 Dry seed
001 Imbibed seed
002 Radicle apparent
003 Elongation of hypocotyl
004 Seedling emergence
005 Cotyledons unfolded

1 Tuber dormancy
100 Innate dormancy (no sprout development under favourable conditions)
150 Enforced dormancy (sprout development inhibited by environmental conditions)

2 Tuber sprouting
200 Dormancy break, sprout development visible
21x Sprout with 1 node
22x Sprout with 2 nodes
·
·
29x Sprout with 9 nodes
21x(2) Second generation sprout with 1 node
22x(2) Second generation sprout with 2 nodes
·
·
29x(2) Second generation sprout with 9 nodes

Where x = 1, sprout < 2 mm;
2, 2–5 mm; 3, 5–20 mm;
4, 20–50 mm; 5, 50–100 mm;
6, 100–150 mm long

3 Emergence and shoot expansion
300 Main stem emergence
301 Node 1
302 Node 2
·

319 Node 19
Second order branch
321 Node 1
·
·
Nth order branch
3N1 Node 1
·
·
3N9 Node 9

4 Flowering
Primary flower
400 No flowers
410 Appearance of flower bud
420 Flower unopen
430 Flower open
440 Flower closed
450 Berry swelling
460 Mature berry
Second order flowers
410(2) Appearance of flower bud
420(2) Flower unopen
430(2) Flower open
440(2) Flower closed
450(2) Berry swelling
460(2) Mature berry

5 Tuber development
500 No stolons
510 Stolon initials
520 Stolon elongation
530 Tuber initiation
540 Tuber bulking (> 10 mm diam)
550 Skin set
560 Stolon development

6 Senescence
600 Onset of yellowing
650 Half leaves yellow
670 Yellowing of stems
690 Completely dead

(From Jefferies & Lawson, 1991. *Annals of Applied Biology*, **119**, 387–389)

Stages in development of annual grass weeds

Illustrations of these growth stages can be found in the reference indicated below and in some company product manuals.

0 Germination and emergence
00 Dry seed
01 Start of imbibition
03 Imbibition complete
05 Radicle emerged from caryopsis
07 Coleoptile emerged from caryopsis
09 Leaf just at coleoptile tip

1 Seedling growth
10 First leaf through coleoptile
11 First leaf unfolded
12 2 leaves unfolded
13 3 leaves unfolded
14 4 leaves unfolded
15 5 leaves unfolded
16 6 leaves unfolded
17 7 leaves unfolded
18 8 leaves unfolded
19 9 or more leaves unfolded

2 Tillering
20 Main shoot only
21 Main shoot and 1 tiller
22 Main shoot and 2 tillers
23 Main shoot and 3 tillers
24 Main shoot and 4 tillers
25 Main shoot and 5 tillers
26 Main shoot and 6 tillers
27 Main shoot and 7 tillers
28 Main shoot and 8 tillers
29 Main shoot and 9 or more tillers

3 Stem elongation
31 First node detectable
32 2nd node detectable
33 3rd node detectable
34 4th node detectable
35 5th node detectable
36 6th node detectable
37 Flag leaf just visible
39 Flag leaf ligule just visible

4 Booting
41 Flag leaf sheath extending
43 Boots just visibly swollen
45 Boots swollen
47 Flag leaf sheath opening
49 First awns visible

5 Inflorescence emergence
51 First spikelet of inflorescence just visible
53 $1/4$ of inflorescence emerged
55 $1/2$ of inflorescence emerged
57 $3/4$ of inflorescence emerged
59 Emergence of inflorescence completed

6 Anthesis
61 Beginning of anthesis
65 Anthesis half-way
69 Anthesis complete

(From Lawson & Read, 1992. *Annals of Applied Biology*, **121**, 211–214)

Growth stages of annual broad-leaved weeds
Preferred descriptive phrases

Illustrations of these growth stages can be found in the reference indicated below and in some company product manuals.

Pre-emergence
Early cotyledons
Expanded cotyledons
One expanded true leaf
Two expanded true leaves
Four expanded true leaves
Six expanded true leaves
Plants up to 25 mm across/high
Plants up to 50 mm across/high
Plants up to 100 mm across/high
Plants up to 150 mm across/high
Plants up to 250 mm across/high
Flower buds visible
Plant flowering
Plant senescent

(From Lutman & Tucker, 1987. *Annals of Applied Biology,* **110**, 683–687)

Appendix 4
A. KEY TO LABEL REQUIREMENTS FOR PROTECTIVE CLOTHING

Engineering control of operator exposure must be used where practical in addition to the items of protective clothing specified on the product label, but may replace the protective clothing if it provides an equal or higher standard of protection. The series of letters entered in the profiles under the heading **Protective clothing/Label precautions** refers to the protective items listed below:

A   Suitable protective gloves (the product label should be consulted for any specific requirements about the material of which the gloves should be made)
B   Rubber gauntlet gloves
C   Face-shield
D   Approved respiratory protective equipment
E   Goggles
F   Dust mask
G   Full face-piece respirator
H   Coverall
J   Hood
K   Rubber apron
L   Waterproof coat
M   Rubber boots
N   Waterproof jacket and trousers
P   Suitable protective clothing

Appendix 4
B. KEY TO NUMBERED LABEL PRECAUTIONS

The series of numbers listed in the profile refer to the numbered precautions below. Where the generalised wording includes a phrase such as "... for xx days" the specific requirement for each pesticide is given in the **Special precautions/Environmental safety** section of the profile.

Medical Restrictions on Use

1   This product contains an anticholinesterase organophosphorus compound. DO NOT USE if under medical advice NOT to work with such compounds

2   This product contains an anticholinesterase carbamate compound. DO NOT USE if under medical advice NOT to work with such compounds

Hazard Classification

3   Very toxic

3a   Very toxic: In contact with skin

3b   Very toxic: By inhalation

3c   Very toxic: If swallowed

VERY TOXIC

4   Toxic

4a   Toxic: In contact with skin

4b   Toxic: By inhalation

4c   Toxic: If swallowed

TOXIC

5   Harmful

5a   Harmful: In contact with skin

5b   Harmful: By inhalation

5c   Harmful: If swallowed

HARMFUL

6   Irritant

6a   Irritating to eyes

6b   Irritating to skin

6c   Irritating to respiratory system

IRRITANT

7   Corrosive

8   Causes burns

9   Risk of serious damage to eyes

10a May cause sensitization by skin contact

10b May cause sensitization by inhalation

CORROSIVE

11   Danger of serious damage to health by prolonged exposure

12a Extremely flammable

12b Highly flammable

12c Flammable

12d Oxidising agent. Contact with combustible material may cause fire

13   Explosive when mixed with oxidizing substances

Protection of User

14   Wash all protective clothing thoroughly after use, especially the inside of gloves/Avoid excessive contamination of coveralls and launder regularly

15   Wash splashes off gloves immediately

16   Take off immediately all contaminated clothing

17   Take off immediately all contaminated clothing and wash underlying skin. Wash clothes before re-use

18   When using do not eat, drink or smoke

19   When using do not eat, drink , smoke or use naked lights

20   Handle with care and mix only in a closed container

20a Open container only as directed

21   Wash concentrate/dust from skin or eyes immediately

22   Wash any contamination from skin or eyes immediately/Wash splashes from skin or eyes immediately
22a   Wash any contamination from eyes immediately
23   After contact with skin or eyes wash immediately with plenty of water
24   In case of contact with eyes rinse immediately with plenty of water and seek medical advice
24a   In case of contact with skin rinse immediately with plenty of water and seek medical advice
25   Avoid all contact by mouth
26   Avoid all contact with skin
27   Avoid all contact with eyes
28   Do not breathe gas/fumes/vapour/spray mist/dust/Avoid working in spray mist
29   Wash hands and exposed skin before eating, drinking or smoking/before meals and after work
30   Wash hands before meals and after work
31   Before entering treated crops, cover exposed skin areas, particularly arms and legs
32   Do not touch sachet with wet hands or gloves
33   Ventilate treated areas thoroughly when smoke has cleared/Ventilate treated rooms thoroughly before occupying

Protection of Consumers
34   Not to be used on food crops
35   Do not harvest for human or animal consumption for at least xx days/weeks after last application
36   Keep away from food, drink and animal feeding-stuffs

Protection of Public, Livestock, Wildlife etc.
37   Keep out of reach of children
38   Keep unprotected persons out of treated areas for at least 24 hours/other intervals
39   Cover water storage tanks before application
40   Dangerous to livestock. Keep all livestock out of treated areas for at least xx days/weeks. Bury or remove spillages
41   Harmful to livestock. Keep all livestock out of treated areas for at least xx days/weeks. Bury or remove spillages
42   Keep poultry out of treated areas for at least xx days/weeks
43   Keep livestock out of treated areas/Keep livestock out of treated areas for at least xx weeks and until foliage of any poisonous weeds such as ragwort has died and become unpalatable
44   Do not feed treated straw or haulm to livestock within 4 days of spraying
44a   Do not use on crops if the straw is to be used as animal feed/bedding
45   Dangerous to game, wild birds and animals
46   Harmful to game, wild birds and animals
47   Harmful to animals. Paraquat may be harmful to hares, where possible spray stubbles early in the day
48   Dangerous to bees. Do not apply at flowering stage. Keep down flowering weeds/Do not apply to crops in flower or to those in which bees are actively foraging. Do not apply when flowering weeds are present
48a   Extremely dangerous to bees. Do not apply at flowering stage. Keep down flowering weeds/Do not apply to crops in flower or to those in which bees are actively foraging. Do not apply when flowering weeds are present
49   Dangerous to bees. Do not apply at flowering stage except as directed on peas etc. Keep down flowering weeds in all crops/ Do not apply to crops in flower or to those in which bees are actively foraging except as directed on peas etc. Do not apply when flowering weeds are present
50   Harmful to bees. Do not apply at flowering stage. Keep down flowering weeds/ Do not apply to crops in flower or to those in which bees are actively foraging. Do not apply when flowering weeds are present
51   Extremely dangerous to fish. Do not contaminate ponds, waterways or ditches with chemical or used container/Extremely dangerous to fish or other aquatic life. Do not contaminate surface waters or ditches with chemical or used container
51a   Extremely dangerous to aquatic higher plants. Do not contaminate surface waters or ditches with chemical or used container

52 Dangerous to fish. Do not contaminate ponds, waterways or duties with chemical or used container/Dangerous to fish or other aquatic life. Do not contaminate surface waters or ditches with chemical or used container

53 Harmful to fish. Do not contaminate ponds, waterways or ditches with chemical or used container/Harmful to fish or other aquatic life. Do not contaminate surface waters or ditches with chemical or used container

54 Do not contaminate ponds, waterways or ditches with chemical or used container/Do not contaminate surface waters or ditches with chemical or used container

54a Do not allow direct spray from vehicle mounted/drawn hydraulic sprayers to fall within 6m of surface waters or ditches/Do not allow direct spray from hand-held sprayers to fall within 2m of surface waters or ditches. Direct spray away from water

55 Do not dump surplus herbicide in water or ditch bottoms

55a Prevent any surface run-off from entering storm drains

56 Do not use treated water for irrigation purposes within xx days/weeks of treatment

57 Avoid damage by drift onto susceptible crops or water courses

58 Store away from seeds, fertilizers, fungicides and insecticides

59 Store well away from corms, bulbs, tubers and seeds

60 Store away from frost

61 Store away from heat

61a Flammable. Do not store near heat or open flame

62 Store under cool, dry conditions

63 Keep in original container, tightly closed, in a safe place

64 Keep in original container, tightly closed, in a safe place, under lock and key

65 Wash out container thoroughly and dispose of safely

66 Wash out container thoroughly, empty washings into spray tank and dispose of safely

67 Empty container completely and dispose of safely

68 Empty container completely and dispose of it in the specified manner

69 Return empty container as instructed by supplier

70 Do not re-use this container for any purpose

71 Do not burn this container

71a Do not rinse out the container

Sack Labels for Treated Seed

72 Do not handle seed unnecessarily

73 Treated seed not to be used as food or feed

74 Do not re-use sack for food or feed

75 Harmful to game and wild life. Bury spillages

76 Wash hands and exposed skin before meals and after work

77 Do not apply treated seed from the air

Medical Advice

78 If you feel unwell seek medical advice (show label where possible) - for 'harmful' pesticides

79 In case of accident or if you feel unwell, seek medical advice immediately (show label where possible) - for 'toxic' and 'very toxic' pesticides

80 If swallowed, seek medical advice immediately and show container or label

Precautions when using Rodenticides

81 Prevent access to baits/powder by children or domestic animals, particularly cats, dogs, pigs and poultry

82 Do not lay baits/powder where food, animal feed or water could become contaminated

83 Remove all remains of bait, tracking powder or bait containers after use and burn or bury

84 Search for and burn or bury all rodent bodies. Do not place in refuse bins or on rubbish tips

85 Use bait containers clearly marked POISON at all surface baiting points

Appendix 5
KEY TO ABBREVIATIONS AND ACRONYMS

The abbreviations of formulation types in the following list are used in Section 3 (Pesticide Profiles) and are derived from the Catalogue of Pesticide Formulation Types (GIFAP Technical Monograph 2, 1989).

1 Formulation Types

AE	Aerosol generator
AL	Other liquids to be applied undiluted
BB	Block bait
CB	Bait concentrate
CG	Encapsulated granule (controlled release)
CR	Crystals
CS	Capsule suspension
DP	Dustable powder
DS	Powder for dry seed treatment
EC	Emulsifiable concentrate
ES	Emulsion for seed treatment
EW	Oil in water emulsion
FG	Fine granules
FP	Smoke cartridge
FS	Flowable concentrate for seed treatment
FT	Smoke tablet
FU	Smoke generator
FW	Smoke pellets
GA	Gas
GB	Granular bait
GE	Gas-generating product
GG	Macrogranules
GP	Flo-dust
GR	Granules
GS	Grease
HN	Hot fogging concentrate
KK	Combi-pack (solid/liquid)
KL	Combi-pack (liquid/liquid)
KN	Cold-fogging concentrate
KP	Combi-pack (solid/solid)
LA	Lacquer
LI	Liquid, unspecified
LS	Solution for seed treatment
ME	Microemulsion
MG	Microgranules
OL	Oil miscible liquid
PA	Paste
PC	Gel or paste concentrate
PS	Seed coated with a pesticide
PT	Pellet

RB	Ready-to-use bait
RH	Ready-to-use spray in hand-operated sprayer
SA	Sand
SC	Suspension concentrate (= flowable)
SE	Suspo-emulsion
SG	Water soluble granules
SL	Soluble concentrate
SP	Water soluble powder
SS	Water soluble powder for seed treatment
SU	Ultra low-volume suspension
TB	Tablets
TC	Technical material
TP	Tracking powder
UL	Ultra low-volume liquid
VP	Vapour releasing product
WB	Water soluble bags
WG	Water dispersible granules
WP	Wettable powder
WS	Water dispersible powder for slurry treatment of seed
XX	Other formulations

2 Other Abbreviations and Acronyms

ACP	Advisory Committee on Pesticides
ACTS	Advisory Committee on Toxic Substances
ADAS	Agricultural Development and Advisory Service
a.i.	active ingredient
BAA	British Agrochemicals Association
CDA	controlled droplet application
cm	centimetre
COPR	Control of Pesticides Regulations 1986
COSHH	Control of Substances Hazardous to Health Regulations
d	day(s)
EBDC	ethylene-bis-dithiocarbamate fungicide
FEPA	Food and Environment Protection Act 1985
g	gram(s)
GIFAP	*Groupement International des Associations Nationales de Fabricants de Pesticides*
GS	growth stage (but also "grease" as a formulation)
h	hour(s)
ha	hectare(s)
HBN	hydroxybenzonitrile herbicide
HI	harvest interval
HMIP	HM Inspectorate of Pollution
HSE	Health and Safety Executive
ICM	integrated crop management
IPM	integrated pest management
kg	kilogram(s)
l	litre(s)
m	metre(s)
MAFF	Ministry of Agriculture, Fisheries and Food

MBC	methyl benzimidazole carbamate fungicide
MEL	Maximum Exposure Limit
min	minute(s)
mm	millimetre(s)
MRL	Maximum Residue Level
mth	month(s)
NA	notice of approval
NFU	National Farmers' Union
NRA	National Rivers Authority
OES	Occupational Exposure Standard
OLA	off-label approval
PPE	personal protective equipment
PPPR	Plant Protection Products Regulations 1995
PR	*The Pesticides Register*
PSD	The Pesticides Safety Directorate
SOLA	specific off-label approval
ULV	ultra-low volume
wk	week(s)
w/v	weight/volume
w/w	weight/weight
yr	year(s)

INDEX OF PROPRIETARY NAMES
OF PESTICIDES

The references are to entry numbers, not pages. Adjuvant names are referred to as *Adj* and are listed separately on pages 43-60.

50/50 Liquid Mosskiller . . . 196
9363 363
Aaprotect 499
Aaterra WP 248
Accendo 365
Acer . . . . *Adj*
Acquit 5
Actellic D 413
Actellic Dust 413
Actellic Smoke
Generator No 20 413
Actellifog 413
Actipron . . . . *Adj*
Activator 90 . . . . *Adj*
Actol . . . . *Adj*
Actrilawn 10 322
Acumen 32
Adder . . . . *Adj*
Addstem 67
Addwett . . . . *Adj*
Adherbe . . . . *Adj*
Adjust 99
Admire 317
Advance 54
Advizor 97
Afalon 336
Afugan 427
Agral . . . . *Adj*
Agrichem DB Plus 179
Agrichem Flowable
Thiram 476
Agrichem Mancozeb 80 . . . 340
Agrichem Maneb 80 345
Agrichem MCPA-25 348
Agricorn 500 348
Agricorn D 165
Agricultural Slug Killer . . . 363
Agropen . . . . *Adj*
Agstock Addwett . . . . *Adj*
Aitken's Lawn Sand 277
Aitken's Lawn Sand Plus . 197
Alfacron 10 WP 20
Alfadex 428
Aliette 295
Alistell 178
Alkyl 90 . . . . *Adj*
Allicur 455

Ally 378
Ally WSB 378
Alpha Bromotril P 49
Alpha Chlorotoluron 500 . 116
Alpha Isoproturon 650 325
Alpha Linuron 50 SC 336
Alpha Linuron 50 WP 336
Alpha Prometryne 50WP . 416
Alpha Propachlor 50SC . . . 418
Alpha Trifluralin 48EC . . . 492
Alphachloralose
Concentrate 91
Alphachloralose Pure 91
Alto 100 SL 161
Alto Combi 70
Alto Elite 110
Alto Major 163
Amazon TP 136
Ambush C 160
Amcide 15
Amos CMPP 353
Amos Maneb 80 345
Amos Talunex 7
Angle 567 SC 154
Antec Durakil 255
Apache 305
Apex 5E 371
Aphox 412
Aplan 161
Apollo 50 SC 138
Applaud 60
Apron Combi FS 361
Aquaflow 209
Ardent 217
Arelon 325
Arena 10G 458
Arena 3 458
Arena 5G 458
Arena Plus 81
Argylene 446
Arma . . . . *Adj*
Armillatox 152
Arresin 380
Arsenal 315
Arsenal 50 315
Arsenal 50 F 315
Ashlade 460 CCC 99

Ashlade 5C 101
Ashlade 700 5C 101
Ashlade 700 CCC 99
Ashlade Adjuvant Oil . . . . *Adj*
Ashlade Carbendazim FL . . 67
Ashlade CP 98
Ashlade Cypermethrin 10
EC 160
Ashlade Dimethoate 220
Ashlade Flotin 274
Ashlade Linuron FL 336
Ashlade Mancarb FL 75
Ashlade Mancarb Plus 69
Ashlade Mancozeb FL 340
Ashlade Maneb Flowable . 345
Ashlade Simazine 50 FL . 441
Ashlade SMC Flowable . . . 147
Ashlade Solace 158
Ashlade Sulphur FL 449
Ashlade Summit 467
Ashlade Tol-7 116
Ashlade Trimaran 492
Asset 25
Asulox 18
Atlacide Soluble Powder . 443
Atladox HI 172
Atlas 2,4-D 165
Atlas 3C:645
Chlormequat 99
Atlas 460:46 101
Atlas 5C Chlormequat 101
Atlas Adherbe . . . . *Adj*
Atlas Adjuvant Oil . . . . . . *Adj*
Atlas Atrazine 19
Atlas Bandrift . . . . *Adj*
Atlas Brown 125
Atlas Chlormequat 46 99
Atlas Chlormequat 700 . . . 99
Atlas CIPC 40 118
Atlas CMPP 353
Atlas Courier . . . . *Adj*
Atlas Dimethoate 40 220
Atlas Electrum 95
Atlas Gold 123
Atlas Indigo 126
Atlas Linuron 336
Atlas MCPA 348

The references are to entry numbers, not pages.

Atlas Pink C 121
Atlas Prometryne 50 WP . 416
Atlas Propachlor 418
Atlas Protrum K 406
Atlas Quintacel 101
Atlas Red 120
Atlas Sheriff 128
Atlas Simazine 441
Atlas Solan 40 402
Atlas Somon 445
Atlas Steward 299
Atlas Sulphur 80 FL 449
Atlas Tecgran 100 458
Atlas Tecnazene 6%
   Dust 458
Atlas Terbine 99
Atlas Thor 245
Atlas Tricol 99
Atlas Trifluralin 492
Atol 116
Atrinal 218
Auger 325
Aura 750 EC 264
Autumn Kite 329
Avadex BW 479
Avadex BW Granular 479
Avadex Excel 15G 479
Avenge 2 212
Axiom . . . . *Adj*
B-Nine 175
Bactospeine WP 21
Bandrift . . . . *Adj*
Banlene Plus 188
Barbarian 305
Barclay Actol . . . . *Adj*
Barclay Bezant 23
Barclay Canter 153
Barclay Carbosect 84
Barclay Champion 94
Barclay Clinch 127
Barclay Clinger . . . . *Adj*
Barclay Cluster 230
Barclay Corrib 500 108
Barclay Cypersect XL 160
Barclay Dart 305
Barclay Desiquat 226
Barclay Dingo 30
Barclay Dodex 237
Barclay Dryfast XL . . . . . . *Adj*
Barclay Eyetak 414
Barclay Gallup 305
Barclay Gallup Amenity . 306
Barclay Guideline 325
Barclay Hat-Trick 188

Barclay Haybob 165
Barclay Holdup 99
Barclay Hurler 284
Barclay Keeper 200 245
Barclay Manzeb 80 340
Barclay Manzeb Flow 346
Barclay Meadowman 348
Barclay Mecrop 353
Barclay Metaza 366
Barclay Pirate 430
Barclay Pirimisect 412
Barclay Piza 500 426
Barclay Shandon 161
Barclay Total 396
Barclay Winner 279
Barleyquat B 99
BAS 421F 264
BAS 438 111
BAS 46402F 270
Basagran 30
Basamid 176
BASF Dimethoate 40 220
BASF MCPA Amine 50 . 348
Basilex 477
Bastion T 285
Bavistin DF 67
Bayfidan 481
Bayleton 480
Baytan Flowable 297
Beetup 406
Bellmac Plus 349
Bellmac Straight 352
Benazalox 26
Benlate Fungicide 29
Berelex 303
Beret Gold 281
Beta 406
Beta 2 406
Betanal E 406
Betapal Concentrate 384
Betosip 406
Betosip Combi 246
Bettaquat B 99
Better DF 94
Better Flowable 94
Bezant 23
BH CMPP/2,4-D 171
BH MCPA 75 348
Biggles 436
Binab T 488
Biotrol Plus Outdoor Rat
   Killer 48
Birlane 24 93
Birlane Granules 93

Birlane Liquid Seed
   Treatment 93
BL 500 118
Blade 40
Blex 413
Blight Spray 158
Bolda FL 76
Bolero 216
Bombardier 108
Bond . . . . *Adj*
Bonzi 394
Borocil K 46
BR Destral 13
Bravo 500 108
Bravo 720 108
Bravocarb 68
Bray's Emulsion 152
Brestan 60 SP 273
Brevis 99
Briotril Plus 19/19 55
Broadshot 169
Bromotril P 49
Bronze 402
Brown 125
Burtolin 339
Butisan S 366
Bygran F 458
Bygran S 458
C-Flo 67
C/12 . . . . *Adj*
Calidan 73
Calixin 490
Campbell's Bellmac Plus . 349
Campbell's Bellmac
   Straight 352
Campbell's Carbendazim
   50% Flowable 67
Campbell's CMPP 353
Campbell's DB Straight . . . 177
Campbell's Dioweed 50 . . . 165
Campbell's Disulfoton
   FE 10 228
Campbell's DSM 182
Campbell's Field
   Marshal 188
Campbell's Grassland
   Herbicide 188
Campbell's Linuron 45%
   Flowable 336
Campbell's MC Flowable . . 75
Campbell's MCPA 50 348
Campbell's Metham
   Sodium 364

The references are to entry numbers, not pages.

Campbell's New
  Camppex 170
Campbell's Nicosoap 386
Campbell's Redipon
  Extra.................201
Campbell's Redlegor 179
Campbell's Trifluron 337
Campbell's X-Spor SC....345
Camppex 170
Canter 153
Carbate Flowable 67
Carbetamex..............83
carbon dioxide 85
Carbosect...............84
Casoron G..............194
Casoron G-SR..........194
Casoron G4.............194
Castaway Plus..........301
CAT 80 ....*Adj*
CCC....................99
CCC 640 99
CCC 700 99
CCC 720 99
CDA Spasor 306
Centex 443
Cerone 104
Certan 21
Challenge 304
Champion 94
Charge 353
Checkmate..............440
Cheetah Super..........258
Childion................207
Chipko Diuron 80.......232
Chipman Diuron
  Flowable..............232
Chloropicrin Fumigant 107
Chlorotoluron 500 116
Chrome 122
Chryzotek Beige.........319
Chryzotop Green 319
Cirrus..................406
Citadel 279
Citowett................*Adj*
Clarion 305
Clarosan 1 FG 465
Cleanrun 171
Clenecorn 353
Clifton Alkyl 90 ....*Adj*
Clifton Glyphosate
  Additive ....*Adj*
Clifton Wetter...........*Adj*
Clinch 127
Clinger.................*Adj*

Cloverkiller 353
Clovotox 353
Cluster.................230
Codacide Oil...........*Adj*
Cogito 423
Colstar.................265
Combinex 404
Commando 278
Comodor 600 457
Compass 324
Compatibility Agent 2 ....*Adj*
Complete...............305
Comulin................*Adj*
Condox 190
Contact ....*Adj*
Contact 75..............108
Contest 5
Contrast 71
Cooke's Weedclear......443
Coopex Maxi Smoke
  Generators 403
Coopex Mini Smoke
  Generators 403
Coopex WP.............403
Corbel 264
Corbel CL..............111
Corral 279
Corrib 500..............108
Cosmic FL 77
Courier ....*Adj*
Cover..................*Adj*
Crackdown 180
Crackshot 472
Cropoil.................*Adj*
Croptex Bronze.........402
Croptex Chrome 122
Croptex Fungex 145
Croptex Pewter.........119
Croptex Steel 445
Crusader S 51
Cudgel.................292
Cultar..................394
Cuprokylt 146
Cuprokylt L 146
Cuprosana H 146
Curb Crop Spray Powder ...6
Curzate M 158
Cutinol.................*Adj*
Cymag.................444
Cyper 10 160
Cyperkill 10 160
Cyperkill 5 160
Cypersect XL 160

Cypertox 160
Cytro-Lane 356
Daconil Turf 108
Dacthal W-75 132
Dagger.................314
Dairy Fly Spray.........428
Danadim Dimethoate 40 .220
Darlingtons Diazinon
  Granules 184
Darlingtons Dichlorvos ...202
Dart 305
Dash 304
Dazide 175
DB Straight............177
Decade 500 EC.........269
Decimate 133
Decis 180
Decisquick..............181
Decoy 368
Defensor FL 67
Deloxil.................55
Delsene 50 DF 67
Deltaland...............180
Deosan Chlorate
  Weedkiller (Fire
  Suppressed) 443
Deosan Fly Spray.......408
Deosan Rataway 210
Deosan Rataway Bait
  Bags 210
Dephend 406
Derosal WDG 67
Desiquat 226
Devcol Liquid Derris 439
Devcol Morgan 6%
  Metaldehyde Slug
  Killer 363
Devrinol 385
Dextrone X 396
Dexuron................234
Dicofen 255
Dicotox Extra 165
Dimilin WP.............213
Dingo..................30
Dioweed 50 165
Dipel 21
Dipterex 80.............487
Disyston P 10 228
Dithane 945 340
Dithane Dry Flowable 340
Dithianon Flowable 230
Diuron 80 WP 232
Dodex 237

The references are to entry numbers, not pages.

Doff Agricultural Slug
Killer with Animal
Repellent 363
Doff Horticultural Slug
Killer Blue Mini
Pellets...............363
Doff Metaldehyde Slug
Killer Mini Pellets......363
Doff Sodium Chlorate
Weedkiller 443
Dorado.................431
Dorin..................482
Dormone 165
Dosaflo 376
Dow Shield............139
Doxstar................286
Drat...................106
Drat Bait..............106
Draza..................368
Draza ST..............368
Dryfast XL.............*Adj*
Du Pont Adjuvant.......*Adj*
Dual...................75
Duk 110 473
Duk 51 266
DUK 747 287
DUK-880..............334
Duplosan 354
Durakil 255
Dursban 4 127
Dynamec................1
Dyrene................16
Eagle 9
Early Impact............72
EC-Kill 255
Eclipse................240
Electrum 95
Elliott's Lawn Sand 277
Elliott's Mosskiller......277
Elvaron WG...........195
Embark 355
Emerald...............*Adj*
Encore 327
Endeavour.............423
Endorats 106
Enforcer...............196
Enhance...............*Adj*
Enhance Low Foam ......*Adj*
Epic 239
Escar-Go 6 363
Estrad 283
Ethokem ....*Adj*
Ethokem C/12..........*Adj*
Ethrel C...............104

Exell ....*Adj*
Exit 368
Extratect Flowable 313
Eyetak 414
F238 236
Faber 108
Fairy Ring Destroyer 493
Falcon 420
Fargro Chlormequat 99
Fastac 5
Favour 600 SC 362
Fazor 339
Ferrax 244
Fettel 191
Field Marshal 188
Filex 419
Fisons Basilex..........477
Fisons Filex 419
Fisons Greenmaster
Autumn 277
Fisons Greenmaster Extra .350
Fisons Greenmaster
Mosskiller..............277
Fisons Helarion.........363
Fisons Octave 414
Fisons Tritox...........188
Fisons Turfclear 67
Flarepath 387
Flexidor 125 330
Flotin 274
Flotin 480 274
Flowable Thiram 476
Flowable Turf Fungicide .470
Fly Spray RTU 408
Folicur................452
Folio 575 SC 113
Fongarid 25 WP 298
Fonofos Seed Treatment .292
Force ST 461
Forester 165
formaldehyde 293
Forte 406
Fortrol 153
Freeway 232
Frigate ....*Adj*
FS Liquid Derris 439
FS Thiram 15% Dust 476
Fubol 58 WP 341
Fubol 75 WP 341
Fumat 245
Fumite Dicloran Smoke ...205
Fumite Dicofol Smoke 206
Fumite Lindane 10 299
Fumite Lindane 40 299

Fumite Lindane Pellets....299
Fumite Permethrin
Smoke 403
Fumite Pirimiphos
Methyl Smoke 413
Fumite Propoxur Smoke .425
Fumyl-O-Gas 374
Fungaflor..............311
Fungaflor Smoke 311
Fungazil 100 SL........311
Fungex.................145
Fungo 196
Fungo Exterior 196
Fusarex Granules 458
Fusilade 250 EW 279
Fusion 288
Fydulan G 174
Fytospore..............158
Fyzol 11E ....*Adj*
Galben M 23
Galion ....*Adj*
Gallery 125 330
Gallup 305
Gallup Amenity 306
Gamma-Col............299
Gamma-Col Turf 299
Gammasan 30 299
Garlon 2 489
Garlon 4 489
Garvox 3G 28
Gastratox 6G Slug
Pellets...............363
Gaucho 317
Gemini................270
Genesis 474
Genie 287
Gesagard 50 WP........416
Gesaprim 500SC.........19
Gesatop 50 WP.........441
Gesatop 500 SC 441
Gladiator 94
Gladiator DF...........94
Glint 500 EC 269
Glyphogan.............305
Glyphosate Additive .....*Adj*
Gold..................123
Golden Malrin Fly Bait ...370
Goltix WG 365
Grain Store Insecticide....256
Gramonol Five 381
Gramoxone 100 396
Grasp 478
Grassland Herbicide 188
Grazon 90 144

The references are to entry numbers, not pages.

Green-Up Weedfree Spot
Weedkiller for Lawns ...166
Greenmaster Autumn 277
Greenmaster Extra 350
Greenmaster Mosskiller ...277
Greenshield 68
Grey Squirrel Liquid
Concentrate 496
Gro-Stop 126
Grovex Zinc Phosphide ...497
Growing Success Slug
Killer 8
GS 800 ....*Adj*
Guard.................*Adj*
Guardian 109
Guardsman 6
Guardsman STP Seed
Dressing Powder 6
Guideline...............325
Hallmark 332
Halo...................112
Halophen RE 49.........196
Hardy..................363
Harlequin 500 SC.......328
Harmony M 379
Harvest 304
Hat-Trick...............188
Hawk..................137
Haybob 165
Headland Addstem 67
Headland Charge 353
Headland Dephend 406
Headland Dual 75
Headland Guard ....*Adj*
Headland Inorganic
Liquid Copper 146
Headland Intake ....*Adj*
Headland Judo 426
Headland Quilt ....*Adj*
Headland Relay 188
Headland Spear.........348
Headland Spirit........345
Headland Staff 165
Headland Sulphur 449
Headland Zebra Flo 340
Headland Zebra WP 340
Helarion................363
Helosate................305
Herbasan 406
Herrisol 188
Hickson Phenmedipham ...406
Hickstor 10 Granules 458
Hickstor 3 458
Hickstor 5 Granules 458

Hickstor 6 + MBC 81
Hilite 306
Hinge..................67
Hispor 45 WP...........80
Hoegrass 203
Holdfast D 193
Holdup.................99
Hortag Tecnacarb Dust 81
Hortag Tecnazene 10%
Granules.............458
Hortag Tecnazene
Double Dust..........458
Hortag Tecnazene Dust ...458
Hortag Tecnazene Potato
Granules.............458
Hortichem 2-
Aminobutane 10
Hortichem Spraying Oil ...405
Horticultural Slug Killer .363
Hostaquick 309
Hostathion.............484
Hotspur 141
Hurler 284
HY-D..................165
HY-MCPA 348
Hy-TL 471
Hy-Vic.................471
Hyban 190
Hydon 47
Hydraguard 302
Hygrass 190
Hykeep 470
Hymec 353
Hymush 470
Hyprone................188
Hyquat 70 99
Hysede FL 300
Hyspray...............*Adj*
Hystore 10 458
Hysward 188
Hytec..................458
Hytec 6 458
Hytec Super 459
Hyvar X 45
I T Bentazone 480 30
I T Glyphosate 305
Imber.................449
Impact Excel...........112
Indar 5EW 252
Indigo 126
Insektigun 39
Intake.................*Adj*
Intracrop BLA ....*Adj*

Intracrop Non-Ionic
Wetter................*Adj*
Intrepid 186
Invader 222
Iprofile................325
Ipso 326
ISK 375...............108
Isoproturon 500 325
Jogral.................*Adj*
Jolt...................327
Jubilee (WSB) 378
Jubilee 20 DF 378
Judo..................426
Jupital 108
Karamate Dry Flo.......340
Karate Ready-to-Use Rat
& Mouse Bait 106
Karate Ready-to-Use
Rodenticide Sachets.....106
Karathane Liquid 223
Karmex 232
Keeper 200 245
Keetak 264
Kelthane 206
Kemifam E 406
Kerb 50 W 426
Kerb Flo 426
Kerb Granules..........426
Keychem SWK 333 166
Kickdown 305
Kickdown 2 305
Killgerm Py-Kill W 469
Killgerm Rat Rods 210
Killgerm Seconal 432
Killgerm ULV 400 428
Killgerm ULV 500 408
Killgerm Wax Bait 210
Knave 229
Knot Out 330
Kombat WDG...........74
Konker................82
Kotol FS 299
Krenite................294
Kumulus DF 449
Lambda-C 332
Landgold Cyproconazole
100 161
Landgold Deltaland 180
Landgold Diquat.........226
Landgold Fenoxaprop.....257
Landgold Flutriafol.......289
Landgold Glyphosate 360 .305
Landgold Lambda-C......332
Landgold Metsulfuron ...378

The references are to entry numbers, not pages.

Laser 155
Lawn Sand 277
Lawn Sand Plus 197
Leatherjacket Pellets 299
Legend 262
Legion 75
Lentagran WP 430
Lentus . . . .*Adj*
Levington Filex 419
Levington Octave 414
Lexone 70DF 377
Leyclene 52
LI-700 . . . .*Adj*
Lightning . . . .*Adj*
Lindane Flowable 299
Lindex-Plus FS 267
Linnet 337
Linuron Flowable 336
Liquid Curb Crop Spray . . .6
Liquid Mosskiller 196
Littac 5
Lo-Dose . . . .*Adj*
Lo-gran 20 WG 483
Longlife Cleanrun 171
Longlife Renovator 167
Lorate 20 DF 378
Lorsban T 127
Ludorum 116
Ludorum 700 116
Luxan 9363 363
Luxan Chloridazon 94
Luxan CMPP 353
Luxan Cypermethrin 10 . . .160
Luxan Gro-Stop 126
Luxan Isoproturon 500
    Flowable 325
Luxan Mancozeb
    Flowable 340
Luxan Maneb 80 345
Luxan Maneb 800 WP 345
Luxan MCPA 500 348
Luxan Metaldehyde 363
Luxan Micro-Sulphur 449
Luxan Non-Ionic Wetter . *Adj*
Luxan Phenmedipham 406
Luxan Talunex 7
Lyric 287
Lyrol . . . .*Adj*
Magnum 96
Mainstay 108
Mallard 750 EC 260
Mancarb FL 75
Mancarb Plus 69

Mandops Chlormequat
    700 99
Maneb 80 345
Manex 346
Manipulator 99
Mantis 250 EC 422
Mantle 425 EC 269
Manzate 200 340
Manzeb 80 340
Manzeb Flow 346
Marshal 10G 86
Marshal/suSCon 86
Masai 456
Mascot Cloverkiller 353
Mascot Mosskiller 196
Mascot Selective
    Weedkiller 171
Mascot Super Selective-P .189
Mascot Systemic 67
Match 212
Matrikerb 143
Maxicrop Moss Killer &
    Conditioner 277
Mazide 25 339
Mazide Selective 187
Mazin 347
MC Flowable 75
Meadowman 348
Mecrop 353
Medimat 2 246
Medipham 406
Meothrin 259
Merit 401
Metarex 363
Metarex RG 363
Metaza 366
Meteor 102
Methyl Bromide 100 373
Methyl Bromide 98 375
MH 18 339
Microcarb Suspendable
    Powder 66
Microgran 2 277
Micromite 255
Microsul Flowable
    Sulphur 449
Microthiol Special 449
Midstream 226
Mifaslug 363
Mildothane Liquid 475
Mildothane Turf Liquid . . .475
Minax . . . .*Adj*
Minder . . . .*Adj*
Mini Slug Pellets 363

Mirage 40 EC 414
Mircam 192
Mircam Plus 189
Mircell 101
Miros DF 108
Mistral 264
Mitac HF 11
Mixture B . . . .*Adj*
Mocap 10G 247
Monarch 400
Monceren DS 398
Monceren Flowable 398
Monceren IM 312
Monceren IM Flowable . . .312
Monicle 263
Moot 163
Morestan 434
Mosskiller 196
MSS 2,4-D Amine 165
MSS 2,4-D Ester 165
MSS 2,4-DB + MCPA 179
MSS 2,4-DP 199
MSS 2,4-DP+MCPA 201
MSS Aminotriazole
    Technical 12
MSS Atrazine 50 FL 19
MSS Atrazine 80 WP 19
MSS Chlormequat 40 99
MSS Chlormequat 460 99
MSS Chlormequat 60 99
MSS Chlormequat 70 99
MSS CIPC 40 EC 118
MSS CIPC 5 G 118
MSS CIPC 50 LF 118
MSS CIPC 50 M 118
MSS CMPP 353
MSS CMPP Amine 60 353
MSS Diuron 50 FL 232
MSS Flotin 480 274
MSS Iprofile 325
MSS Linuron 50 336
MSS MCPA 50 348
MSS MCPB + MCPA 349
MSS MH 18 339
MSS Mircam 192
MSS Mircam Plus 189
MSS Mircell 101
MSS Optica 354
MSS Simazine 50 FL 441
MSS Sugar Beet
    Herbicide 123
MSS Sulphur 80 449
MSS Trifluralin 48 EC 492
MTM CIPC 40 118

The references are to entry numbers, not pages.

MTM Dimethoate 40 220
MTM Disulfoton P 10 228
MTM Diuron 80 232
MTM Malathion 60 338
MTM Phorate 410
MTM Sugar Beet
  Herbicide 123
MTM Sulphur Flowable . . .449
MTM Trifluralin 492
Multi-W FL 75
Multispray 428
Mycotal 494
Myriad 253
Nebulin 458
Nemolt 460
Neokil 210
Neosorexa Ratpacks 210
Neosorexa Ready to Use
  Rat and Mouse Bait 210
Neporex 2SG 164
New 5C Cycocel 101
New Arena 6 458
New Arena Plus 81
New Camppex 170
New Estermone 166
New Hickstor 6 458
New Hystore 458
New Squadron 75
Nex 84
Nico Soap 386
Nicosoap 386
Nicotine 40% Shreds 386
Nimrod 58
Nimrod-T 59
Nomix Turf Selective
  Herbicide 171
Nomix-Chipman
  Mosskiller 196
Non-Ionic Wetter . . . .*Adj*
Nortron 245
Nu Film P . . . .*Adj*
Nuvan 500 EC 202
Nuvan N 500 SC 321
Octave 414
Octolan 110
Opogard 500 SC 464
Optica 354
Optimol 363
Opus 239
Opus Plus 241
Opus Team 240
Orbis . . . .*Adj*
Osprey 58 WP 341
Oxytril CM 55

Output . . . .*Adj*
Pacer 72
Palette 72
Panacide M 196
Panacide M21 196
Panacide TS 196
Panaclean 736 196
Pancil T 388
Panoctine 307
Panoctine Plus 308
Panther 215
Parable 227
paraffin oil 395
paraformaldehyde 293
Paramos 34
Pastor 142
Pasturol 188
Patafol 342
Patrol 260
PBI Slug Pellets 363
PDQ 227
Peaweed 417
Pentac Aquaflow 209
Permasect 10 EC 403
Permasect 25 EC 403
Pewter 119
Phantom 412
Phenoxylene 50 348
Phostek 7
Phostoxin 7
Pilot 436
Pink C 121
Pirate 430
Pirimisect 412
Pirimor 412
Piza 500 426
Planet . . . .*Adj*
Plantvax 20 393
Plantvax 75 393
Plover 211
Podquat 103
Pointer 289
Pommetrol M 126
Poraz 414
PP 375 112
PP Captan 80 WG 64
Prebane SC 465
Prefix D 194
Pride 426
Profalon 124
Prospect 472
Prosper 228
Protrum K 406
Pugil 325

Pulsar 33
Punch C 71
Py-Kill W 469
Pybuthrin 33 428
Pynosect 30 Fogging
  Solution 429
Pynosect 30 Water
  Miscible 429
Pyramin DF 94
Q 900 . . . .*Adj*
Quad MCPA 50% 348
Quad Mini Slug Pellets . . .363
Quad-Ban 188
Quad-Fast . . . .*Adj*
Quadrangle Chlormequat
  700 99
Quadrangle Q 900 . . . .*Adj*
Quadrangle Quad-Fast . . . .*Adj*
Quantum 486
Quickstep 36
Quilt . . . .*Adj*
Quintacel 101
Quintozene WP 435
Racumin Bait 151
Racumin Tracking
  Powder 151
Radar 422
Radspor FL 237
Rampart 84
Ramrod Flowable 418
Ramrod Granular 418
Rapide . . . .*Adj*
Rapier 426
Rat & Mouse Bait 106
Rat Rods 210
Ratak 210
Ratak Wax Blocks 210
Rataway 210
Ratpacks 210
Raxil S 454
RCR Zinc Phosphide 497
Ready to Use Rat and
  Mouse Bait 210
Recoil 343
Red 120
Redipon Extra 201
Redlegor 179
Reflex T 291
Reglone 226
Regulex 303
Regulox K 339
Relay 188
Reldan 50 130
Remtal SC 442

The references are to entry numbers, not pages.

Renardine 43
Renovator 167
Restore 422
Rhizopon A Powder 318
Rhizopon A Tablets 318
Rhizopon AA Powder
(0.5%)................ 319
Rhizopon AA Powder
(1%) 319
Rhizopon AA Powder
(2%) 319
Rhizopon AA Tablets..... 319
Rhizopon B Powder
(0.1%)................ 383
Rhizopon B Powder
(0.2%)................ 383
Rhizopon B Tablets 383
Ridomil mbc 60 WP 78
Ridomil Plus 50 WP 148
Rimidin 251
Ringmaster 393
Ripcord 160
Ripost Pepite........... 159
Rizolex 477
Rizolex Flowable 477
Rizolex Nova 415
Ronilan FL 495
Ronstar 2G 390
Ronstar Liquid 390
Root-Out 15
Roundup 305
Roundup Biactive 305
Roundup Biactive Dry 305
Roundup Four 80 305
Roundup Pro 306
Roundup Pro Biactive 306
Rover 500 108
Rovral Flo............. 323
Rovral Green 323
Rovral Liquid FS 323
Rovral WP 323
Royal MH 180 339
Rubigan 251
Ruby Rat.............. 106
Ryda .... Adj
Sabre 325
Sakarat Ready-to-Use
(cut wheat) 496
Sakarat Ready-to-Use
(whole wheat) 496
Sakarat X Ready-to-Use
Warfarin Rat Bait 496
Salute................. 245
Salvo 280

Sambarin 312.5 SC 115
Sanction................ 287
Sapecron 240 EC 93
Saprol 493
Satisfar 249
Satisfar Dust 249
Savall................. 433
Savona................. 250
Scoot 6
Scuttle 448
Scythe LC 396
Seconal 432
Seedox SC 28
Selective Weedkiller...... 171
Semeron 25WP 183
Senate 466
Senator Flowable 147
Sencorex WG 377
Sentinel 418
Sentinel 2 418
Sequel 271
Setter 33 27
Sewarin Extra 496
Sewarin P 496
Sewercide Cut Wheat
Rat Bait 496
Sewercide Whole Wheat
Rat Bait 496
Shandon................ 161
Sheen 550 EC.......... 262
Sheriff 128
Shirlan 280
SHL Lawn Sand 277
SHL Lawn Sand Plus..... 197
SHL Turf Feed and
Weed 201
SHL Turf Feed and
Weed + Mosskiller 200
Sibutol 42
Sickle.................. 53
Siege II 160
Silvacur 453
Sinbar 462
Sipcam UK Rover 500.... 108
Sistan................. 364
Skirmish 495 SC 331
Slayer 279
Slaymor................ 48
Slaymor Bait Bags 48
Slippa .... Adj
SM99 .... Adj
SMC Flowable 147
Smite 130
Sobrom BM 100........ 373

Sobrom BM 98......... 375
Solace 158
Solan 40 402
Solar .... Adj
Solfa 449
Somon................. 445
Sorex Fly Spray RTU 408
Sorex Super Fly Spray 408
Sorex Warfarin 250 ppm
Rat Bait 496
Sorex Warfarin 500 ppm
Rat Bait 496
Sorex Warfarin Sewer
Bait................ 496
Sorex Wasp Nest
Destroyer 438
Sorexa CD Ready to Use .. 62
Source 339
Sovereign 400SC 399
Spannit 127
Spannit Granules 127
Sparkle 45 WP 80
Spasor 306
Spear 348
Spearhead 140
Spectron 96
Speedway 396
Spirit 345
Sponsor 261
Sporgon 50WP 414
Sportak 45............. 414
Sportak Alpha........... 79
Sportak Delta 460....... 162
Sprayfast .... Adj
Sprayfix............... Adj
Spraymac.............. Adj
Sprayprover............ Adj
Spread and Seal .... Adj
Spreader .... Adj
Springcorn Extra........ 188
Sprint................. 268
Stacato 305
Staff.................. 165
Stamina .... Adj
Stampede 305
Stance 104
Standon Aldicarb 10G 3
Standon Bentazone 30
Standon Carbosulfan 10G .. 86
Standon Chlorothalonil
50 108
Standon Chlorpyrifos 127
Standon Cymoxanil Extra . 158
Standon Cypermethrin 160

The references are to entry numbers, not pages.

Standon Cyproconazole ...161
Standon Deltamethrin 180
Standon Diquat 226
Standon Ethofumesate
200 245
Standon Fenpropimorph
750 264
Standon Fluazifop-P 279
Standon Fluroxypyr 284
Standon Flusilazole Plus ..71
Standon Glyphosate 360 .305
Standon Lambda-C 332
Standon Mecoprop-P 354
Standon Metazachlor 50...366
Standon Metsulfuron 378
Standon Paraquat 396
Standon Pirimicarb 50 412
Standon Pirimicarb H.....412
Standon Propiconazole 422
Standon Propyzamide 50 .426
Standon Tebuconazole 452
Standon Triadimefon 480
Standon Tridemorph 750 .490
Standup 700 99
Starane 2 284
Starter Flowable 94
Stay Off 6
Steel 445
Stefes 7G 365
Stefes Banlene Plus 188
Stefes Biggles 436
Stefes Blight Spray.......158
Stefes C-Flo 67
Stefes Carbendazim Flo 67
Stefes CAT 80 ....Adj
Stefes CCC 99
Stefes CCC 640 99
Stefes CCC 700 99
Stefes CCC 720 99
Stefes Chloridazon 94
Stefes Complete 305
Stefes Cover ....Adj
Stefes Cypermethrin 160
Stefes Cypermethrin 2 160
Stefes Derosal WDG 67
Stefes Diquat 226
Stefes Flamprop 278
Stefes Fluroxypyr 284
Stefes Forte 406
Stefes Fumat 245
Stefes Glyphosate 305
Stefes IPU 325
Stefes Kickdown.........305
Stefes Kickdown 2 305

Stefes Lenacil 333
Stefes Maneb DF 345
Stefes Mecoprop-P 354
Stefes Mecoprop-P2 354
Stefes Medimat 2 246
Stefes Medipham 406
Stefes Mepiquat 105
Stefes Metamitron 365
Stefes Paraquat 396
Stefes Phenoxylene 50 348
Stefes Pirimicarb 412
Stefes Poraz 414
Stefes Pride 426
Stefes Restore 422
Stefes Slayer 279
Stefes Spread and Seal ....Adj
Stefes Stance...........104
Stefes Toluron 116
Sterilite Hop Defoliant 17
Sterilite Tar Oil Winter
Wash 60% Stock
Emulsion 451
Sterilite Tar Oil Winter
Wash 80% Stock
Emulsion 451
Steward 299
Stik-It ....Adj
Sting CT 305
Stirrup 306
Stoller Flowable Sulphur .449
Stomp 400 SC 399
Storite Clear Liquid 470
Storite Flowable 470
Storite SS 459
Stratos 155
strychnine 447
Sugar Beet Herbicide 123
Sulfosate 305
sulfuric acid 450
Sulphur Flowable 449
Sumi-Alpha 242
Sumithrin 10 Sec 407
Summit 467
Super Fly Spray 408
Super Mosstox 196
Super Selective-P 189
Super-Tin 4L 274
Superflor 6%
Metaldehyde Slug
Killer Mini Pellets......363
Supertox 30............171
SuSCon Green Soil
Insecticide 127
Swipe P 56

Swirl ....Adj
SWK 333 166
Sydex..................171
Syford 165
Synergol 320
Systemic Insecticide 220
Systhane 6 Flo 382
Systhane 6 W 382
Systol M 158
Tachigaren 70 WP 310
Takron.................94
Talon 127
Talstar 37
Talunex 7
Tattoo 344
Tattoo C 114
Taylors Lawn Sand 277
Teal 57
Tecgran 100 458
Tecnacarb Dust 81
Tecto Flowable Turf
Fungicide 470
Tedion V-18 EC 468
Telone II 198
Temik 10G 3
Terbine 99
Tern 750 EC...........260
Terpal 105
Terraclor 20D 435
Thianosan DG..........476
Thinsec 66
Thiodan 20 238
Thiodan 35 EC 238
Thiovit.................449
Thor...................245
Thripstick 180
Thuricide HP 21
Tigress.................204
Tilt 250 EC............422
Tilt Turbo 475 EC 424
Timbrel 489
Tipoff 383
Tol-7 116
Toluron 116
Tomcat Rat and Mouse
Bait...................225
Tomcat Rat and Mouse
Blox 225
Top-Cop 150
Topas 100 EC...........397
Topas C 50 WP 65
Topas D 275 SC........231
Topik 240EC 135
Toppel 10 160

The references are to entry numbers, not pages.

Topshot 31
TopUp Surfactant . . . .*Adj*
Tordon 22K 411
Torque 254
Total 396
Totem 117
Totril 322
Touchdown 305
Touchdown LA 306
Touché 233
Tracker 185
Treflan 492
Tribunil 367
Tribute 188
Tricol 99
Trifluron 337
Trifolex-Tra 349
Trigard 492
Trik 14
Trimanzone WP 276
Trimaran 492
Tripart Accendo 365
Tripart Acer . . . .*Adj*
Tripart Arena 10G 458
Tripart Arena 3 458
Tripart Arena 5G 458
Tripart Arena Plus 81
Tripart Beta 406
Tripart Beta 2 . . . . :406
Tripart Brevis 99
Tripart Defensor FL 67
Tripart Faber 108
Tripart Gladiator 94
Tripart Imber 449
Tripart Legion 75
Tripart Lentus . . . .*Adj*
Tripart Ludorum 116
Tripart Ludorum 700 116
Tripart Minax . . . .*Adj*
Tripart Mini Slug Pellets .363
Tripart New Arena 6 458
Tripart Nex 84
Tripart Orbis . . . .*Adj*
Tripart Pugil 325
Tripart Senator Flowable .147
Tripart Sentinel 418

Tripart Sentinel 2 418
Tripart Trifluralin 48 EC .492
Tripart Ultrafaber 108
Tripart Victor 69
Tristar 492
Tritox 188
Trump 327
Trustan 159
Trustan WDG 159
Turbair Grain Store
  Insecticide 256
Turbair Permethrin 403
Turbair Resmethrin Extra .437
Turbair Rovral 323
Turbair Systemic
  Insecticide 220
Turf Feed and Weed 201
Turf Selective Herbicide .171
Turf Tonic 277
Turfclear 67
Turfclear WDG 67
Twinspan 129
Twister Flow 66
TWK Total Weedkiller . . .443
Ultrafaber 108
Unicrop 6% Mini Slug
  Pellets 363
Unicrop Fenitrothion 50 . . .255
Unicrop Flowable
  Atrazine 19
Unicrop Flowable Diuron .232
Unicrop Flowable
  Mancozeb 340
Unicrop Flowable Maneb .345
Unicrop Flowable
  Simazine 441
Unicrop Leatherjacket
  Pellets 299
Unicrop Mancozeb 340
Unicrop Maneb 80 345
Unicrop Thianosan DG . . .476
Unicrop Zineb 498
Upgrade 100
UPL Diuron 80 232
UPL Linuron 45%
  Flowable 336

Uplift 99
Vangard 406
Vassgro Cypermethrin
  Insecticide 160
Vassgro Spreader . . . .*Adj*
Venzar 333
Venzar Flowable 333
Vertalec 494
Victor 69
Vindex 50
Vitaflo 90
Vitaflo Extra 89
Vitavax RS 88
Vitax Lawn Sand 277
Vitax Microgran 2 277
Vitax Turf Tonic 277
Vitax Weed 'N' Feed
  Extra 171
Vitesse 73
Vizor 333
Vydate 10G 392
Walkover Mosskiller 277
Warefog 25 118
Warfarin 0.5%
  Concentrate 496
Warfarin Ready Mixed
  Bait 496
Wasp Nest Destroyer 438
Wax Bait 210
Wayfarer . . . .*Adj*
Weed 'N' Feed Extra 171
Weed and Brushkiller
  (New Formulation) 168
Weedazol-TL 12
Wetcol 3 44
Widgeon 750 EC 264
Winner 279
X-Spor SC 345
XL-All Insecticide 386
XL-All Nicotine 95% 386
Yaltox 84
Zebra Flo 340
Zebra WP 340
Zulu 550 EC 262

THE UK PESTICIDE GUIDE 1996
RE-ORDER FORM

Surname _____ Initials _____ Dr/Mr/Mrs/Ms

Address _____

_____

Postcode _____ Country _____

Date _____ Signed _____

Please send me _____ more copies of *The UK Pesticide Guide 1996*

☐ Payment enclosed, cheques made payable to BCPE Ltd.

☐ Please invoice (please attach your official company order)

☐ Please debit my Access/Master Card/Eurocard/Visa Card/American Express account

by the sum of _____ Name of cardholder _____

Card No _____ Expiry Date _____

Address of cardholder (if different from above) _____

_____

OR TELEPHONE YOUR ORDER NOW AND ASK FOR THE SALES DEPARTMENT

List price £17.95	
Discount for bulk sales	
No. of copies	Discount
100+	30%
50–99	25%
10–49	15%
1–9	No discount

STANDING ORDERS

☐ Yes, I wish to take out a standing order for _____ copy/copies of each new edition of *The UK Pesticide Guide*

1997 EDITION

☐ Please send me advance price details for the 1997 Edition of *The UK Pesticide Guide* as soon as they are available.

Please detach and return to:
British Crop Protection Enterprises Ltd.
BCPC Publications Sales
Bear Farm, Binfield, Bracknell, Berkshire RG42 5QE, UK
Telephone: (01734) 342727
Fax: (01734) 341998